Building the Network of the Future

Getting Smarter, Faster, and More Flexible with a Software Centric Approach

T0201056

Building the Network of the Future

Getting Smarter, Faster, and More Flexible with a Software Centric Approach

by

John Donovan and Krish Prabhu

CRC Press
Taylor & Francis Group
Boca Raton London New York

CRC Press is an imprint of the
Taylor & Francis Group, an **Informa** business

A CHAPMAN & HALL BOOK

CRC Press
Taylor & Francis Group
6000 Broken Sound Parkway NW, Suite 300
Boca Raton, FL 33487-2742

First issued in paperback 2020

© 2017 by Taylor & Francis Group, LLC
CRC Press is an imprint of Taylor & Francis Group, an Informa business

No claim to original U.S. Government works

ISBN 13: 978-0-367-57345-4 (pbk)
ISBN 13: 978-1-138-63152-6 (hbk)

Library of Congress Cataloging-in-Publication Data

Names: Donovan, John, 1960- author. | Prabhu, Krish, 1954- author.
Title: Building the network of the future : getting smarter, faster, and more
flexible with a software centric approach / by John Donovan, Krish
Prabhu.
Description: Boca Raton : Taylor & Francis, CRC Press, 2017. | Includes
bibliographical references and index.
Identifiers: LCCN 2017009175 | ISBN 9781138631526 (hardback : alk. paper)
Subjects: LCSH: Computer networks--Technological innovations. | Cloud
computing. | Virtual computer systems.
Classification: LCC TK5105.5 .B8454 2017 | DDC 004.6--dc23
LC record available at https://lccn.loc.gov/2017009175

Visit the Taylor & Francis Web site at
http://www.taylorandfrancis.com

and the CRC Press Web site at
http://www.crcpress.com

Contents

Foreword vii

Acknowledgments...ix

Authors..xi

Chapter 1   The Need for Change 1

               John Donovan and Krish Prabhu

Chapter 2   Transforming a Modern Telecom Network—From All-IP to Network Cloud 9

               Rich Bennett and Steven Nurenberg

Chapter 3   Network Functions Virtualization...25

               John Medamana and Tom Siracusa

Chapter 4   Network Functions Virtualization Infrastructure 49

               Greg Stiegler and John DeCastra

Chapter 5   Architecting the Network Cloud for High Availability...............................67

               Kathleen Meier-Hellstern, Kenichi Futamura, Carolyn Johnson, and Paul Reeser

Chapter 6   Software-Defined Networking 87

               Brian Freeman and Han Nguyen

Chapter 7   The Network Operating System: VNF Automation Platform....................103

               Chris Rice and Andre Fuetsch

Chapter 8   Network Data and Optimization 137

               Mazin Gilbert and Mark Austin

Chapter 9   Network Security...171

               Rita Marty and Brian Rexroad

Chapter 10 Enterprise Networks..201

               Michael Satterlee and John Gibbons

Chapter 11 Network Access...223

               Hank Kafka

Chapter 12 Network Edge 249

Ken Duell and Chris Chase

Chapter 13 Network Core 265

John Paggi

Chapter 14 Service Platforms 293

Paul Greendyk, Anisa Parikh, and Satyendra Tripathi

Chapter 15 Network Operations 331

Irene Shannon and Jennifer Yates

Chapter 16 Network Measurements 353

Raj Savoor and Kathleen Meier-Hellstern

Chapter 17 The Shift to Software 389

Toby Ford

Chapter 18 What's Next? 401

Jenifer Robertson and Chris Parsons

Index 407

Foreword

It's easy to forget just how thoroughly mobile connectivity has changed our relationship with technology. Any new piece of hardware is judged not just for the processor, memory, or camera it contains but for how well it connects. Wi-Fi, fiber, 4G—and soon 5G—have become the new benchmarks by which we measure technological progress.

In many ways, this era of high-speed connectivity is a lot like the dawn of the railroads or the superhighways in the nineteenth and twentieth centuries, respectively. Like those massive infrastructure projects, connectivity is redefining society. And it is bringing capabilities and knowledge to all facets of society, enabling a new era of collaboration, innovation, and entertainment.

Yet most people take networks for granted because so much of what happens on them is hidden from view. When we use our phones or tablets or stream video to our connected cars, we expect the experience to be simple and seamless. The last thing we should care about is how they are connecting or onto what networks. And all the while we continue to place ever-heavier loads on these networks, which now include 4K video and virtual reality.

Keeping up with these demand curves is a monumental task. In fact, knowing how to manage, secure, transport, and analyze these oceans of data might be the single greatest logistical challenge of the twenty-first century.

Needless to say, this is a relatively new phenomenon. It used to be that the biggest test for any telephone network was the ability to handle the explosion of calls on Mother's Day. Now, any day can be the busiest for a network—when an entire season of a popular show is released or when the latest smartphone OS upgrade goes live. A network today must always be prepared.

This paradigm shift requires a new kind of network, one that that can respond and adapt in near-real time to constant and unpredictable changes in demand while also being able to detect and deflect all manner of cyber threats.

The old network model completely breaks down in the face of this. That's why software-defined networking (SDN) and network function virtualization (NFV) are moving from concept to implementation so rapidly. This is the network of the future: open-sourced, future-proofed, highly secure, and flexible enough to scale up to meet any demand.

This book lays out much of what we have learned at AT&T about SDN and NFV. Some of the smartest network experts in the industry have drawn a map to help you navigate this journey. Their goal is not to predict the future but to help you design and build a network that will be ready for whatever that future holds. Because if there's one thing the last decade has taught us, it's that network demand will always exceed expectations. This book will help you get ready.

Randall Stephenson
Chairman and Chief Executive Officer, AT&T

Acknowledgments

The authors would like to acknowledge the support from and the following contributions by Mark Austin, Rich Bennett, Chris Chase, John DeCastra, Ken Duell, Toby Ford, Brian Freeman, Andre Fuetsch, Kenichi Futamura, John Gibbons, Mazin Gilbert, Paul Greendyk, Carolyn Johnson, Hank Kafka, Rita Marty, John Medamana, Kathleen Meier-Hellstern, Han Nguyen, Steven Nurenberg, John Paggi, Anisa Parikh, Chris Parsons, Paul Reeser, Brian Rexroad, Chris Rice, Jenifer Robertson, Michael Satterlee, Raj Savoor, Irene Shannon, Tom Siracusa, Greg Stiegler, Satyendra Tripathi, and Jennifer Yates.

The authors would also like to acknowledge the contributions of Tom Restaino and Gretchen Venditto in the editing of the book.

All royalties will be donated to the AT&T Foundation and will be used to support STEM education.

Authors

John Donovan is chief strategy officer and group president AT&T Technology and Operations (ATO). Donovan was executive vice president of product, sales, marketing, and operations at VeriSign Inc., a technology company that provides Internet infrastructure services. Prior to that, he was chairman and CEO of inCode Telecom Group Inc. Mr. Donovan was also a partner with Deloitte Consulting, where he was the Americas Industry Practice director for telecom.

Krish Prabhu retired as president of AT&T Labs and chief technology officer for AT&T, where he was responsible for the company's global technology direction, network architecture and evolution, and service and product design. Krish has an extensive background in technology innovation from his time at AT&T Bell Laboratories, Alcatel, Morgenthaler Ventures, and Tellabs. He also served as COO of Alcatel, and CEO of Tellabs.

1 The Need for Change

John Donovan and Krish Prabhu

CONTENTS

References..8

Although the early history of electrical telecommunication systems dates back to the 1830s, the dawn of modern telecommunications is generally viewed as Alexander Graham Bell's invention of the telephone in 1876. For the first 100 years, the telecom network was primarily "fixed"—the users were restricted to being stationary (other than some limited mobility offered by cordless phones) in locations that were connected to the telecom infrastructure through fixed lines. The advent of mobile telephony in the 1980s, made it possible to move around during a call, eventually ushering in the concept of "personal communication services," allowing each user to conduct his or her own call on his or her own personal handset, when and where he or she wants to. To facilitate this, the fixed network was augmented with special mobility equipment (towers with radio antennas, databases that tracked the movement of users, etc.) that interoperated with the existing infrastructure.

The invention of the browser launched the Internet in the mid-1990s. Like mobile networks, the Internet also builds on top of the fixed telecom network infrastructure, by augmenting it with special equipment (modems, routers, servers, etc.). However, the development of the Internet has not only made it possible to bring communication services to billions of people, but has also dramatically transformed the business and the social world. Today, the widespread use of the Internet by people all over the world and from all walks of life has made it possible for Internet Protocol (IP) to be not only the data network protocol of choice, but also the technology over which almost all forms of future networks are built, including the traditional public telephone network.

At a high level, a telecom network can be thought of as comprising two main components: a core network with geographically distributed shared facilities interconnected through information transport systems, and an access network comprising dedicated links (copper, fiber, or radio) connecting individual users (or their premises) to the core network. The utility provided by the infrastructure is largely dependent on availability of transport capacity. The Communications Act of 1934 called for "rapid, efficient, Nation-wide, and world-wide wire and radio communication service with adequate facilities at reasonable charges" to "all the people of the United States." This principle of universal service helped make telephone service ubiquitous. Large capital investments were made to deploy transport facilities to connect customers who needed telephone service, across wide geographic areas. The deployed transport capacity was much above what was needed for a basic voice service, and as new technology solutions developed (digital subscriber, optical fiber, etc.), it became clear that the network was capable of doing much more than providing voice services. The infrastructure that supported ubiquitous voice service, evolved to support data services and provide broadband access to the Internet. Today, according to the Federal Communications Commission (FCC), high-speed (broadband) Internet is an essential communication technology, and ought to be as ubiquitous as voice. The quality of political, economic, and social life is so dependent on the telecom network that it is rightly deemed as a nation's critical infrastructure.

The history of mobile telephony and Internet has been one of continuous evolution and ever increasing adoption. The pace of innovation has been exponential—matching the explosion of service offerings with those of new applications. Starting with the 1G mobility system that were first introduced in the 1970s and reached a peak subscriber base of 20 million voice customers, today's 4G systems serve a global base of nearly 2 billion smartphone customers.[1]

"The transformative power of smartphones comes from their size and connectivity."[2] Size makes it easy to carry them around; connectivity means smartphones can not only connect people to one another, but also can deliver the full power of online capabilities and experiences, spawning new business models as with Uber and Airbnb. In addition, the fact that there are 2 billion smartphones spread out all over the world, and each one of them can "serve as a digital census-taker," makes it possible to get real-time views of what people like and dislike, at a very granular level. But the power of this framework is limited by the ability to interconnect the devices and web sites—at the necessary interconnection speeds and manage the latency requirements for near-instantaneous services. This is made possible by a transport layer, with adequate bandwidth and throughput between any two end points.

Transport technology has continually evolved from the early local access and long-distance transport networks. Fundamental advances in the copper, radio, and fiber technologies have provided cheaper, more durable, and simpler to manage transport. The traffic being transported has evolved from analog to digital (time division multiplexed) to IP. Layered protocols have enabled many different services to share a common transport infrastructure. The hub and spoke, fixed bandwidth, fixed circuits, and relatively long duration switched sessions (an architecture that grew with the growth of voice services) are no longer the dominant patterns but one of many different traffic patterns driven by new applications such as video distribution, Internet access, and machine-to-machine communication, with mobility being an essential part of the offering. This evolution has now made it possible for the global telecom network to be the foundation for not just voice, data, and video communication among people, but also for a rich variety of social and economic activity transacted by billions of people all over the world. The next thing on the horizon is the Internet of Things (IoT)—driven by the ability to connect and control tens of billions of remote devices, facilitating the next round of productivity gains as operations across multiple industries get streamlined.

Traditionally, the users of carrier networks were fairly static and predictable, as was the traffic associated with them. As a result, the underlying architecture that carriers relied on at the time was, more often than not, based on fixed, special purpose, networking hardware. Since the operating environment within a Telecom operator followed an organizational approach that separated the IT and Network groups, the growth in traffic and the variety of services that needed to be supported, created an environment where the ability to respond to business needs was increasingly restricted. The networks of today support billions of application-heavy, photo-sharing, video-streaming, mobile devices, and emerging IoT applications, all of which require on-demand scaling, high resiliency, and the capability to do on-the-fly service modifications.

Network connectivity is at the core of every innovation these days, from cars to phones to video and more. Connectivity and network usage are exploding—data traffic on AT&T's mobile network alone grew by nearly 150,000% between 2007 and 2014. This trend will continue, with some estimates showing wireless data traffic growing by a factor of 10 by 2020. Video comprises about 60% of AT&T's total network traffic—roughly 114 petabytes per day as of this writing. To put that into perspective, 114 petabytes equals about 130 million hours of HD video. As it stands right now, sending a minute of video takes about 4 megabytes of data. In comparison, sending a minute of virtual reality (VR) video takes hundreds of megabytes. With VR becoming increasingly popular, having an easy and convenient way to access this data-hungry content is essential. At the same time, IoT is also generating mountains of data that have to be stored and analyzed. By 2020, there will be an estimated 20–50 billion connected devices.

Public networks and especially those that cover large areas have always been expensive to deploy. To enable the growth and the use of networks (the value of a network increases as more users get on it), operators have relied on technology advances to continue to improve performance and drive costs down. Compared to a public network in the twentieth century, today's all-IP network with advanced services radically changes many of the requirements. The number of different services or applications available at any one time has grown by several orders of magnitude and services are constantly evolving, driven by mobile device capabilities, social media, and value created by available content and applications.

To keep pace with these new applications, and to be able to provide reliable networking services in the face of growing traffic and diminishing revenue, we need a new approach to networking. Traditional telecom network design and implementation has followed a tried and tested approach for several decades. When new network requirements are identified, a request for proposals (RFPs) is issued by the carrier. The vendors respond with proprietary solutions that meet interoperability standards, and each network operator then picks the solution they like. The networks are built by interconnecting physical implementation of functions. Today's telecom networks have over 250 distinct network functions deployed—switches, routers, access nodes, multiplexors, gateways, servers, etc. Most of these network functions are implemented as stand-alone appliances—a physical "box" with unique hardware and software that implements the function and conforms to standard interfaces to facilitate interoperability with other boxes. For operational ease, network operators prefer to use one or two vendors, typically, for a given class of appliances (routers, for example). The appliance vendor often uses custom hardware to optimize the cost/performance; the software is mated with hardware, and the product can be thought of as "closed." This creates vendor lock in, and since most deployed appliances are seldom replaced, the result is a platform lock in, with limited options for upgrading with technology advances. Figure 1.1 qualitatively depicts the evolution of cost over a 10-year period. For purposes of illustration, the unit cost signifies the cost for performing one unit of the network function, (e.g., packet processing cost for 1 GB of IP traffic). Due to platform lock in, any cost improvement comes from either hardware redesigns at the plug-in level with cheaper components, or price concessions made by the vendor to maintain market position. On the other hand, technology advances characterized by advances in component technology (Moore's Law) and competitive market dynamics, facilitate a much faster decline in unit cost, more in-line with what is needed to keep up with the

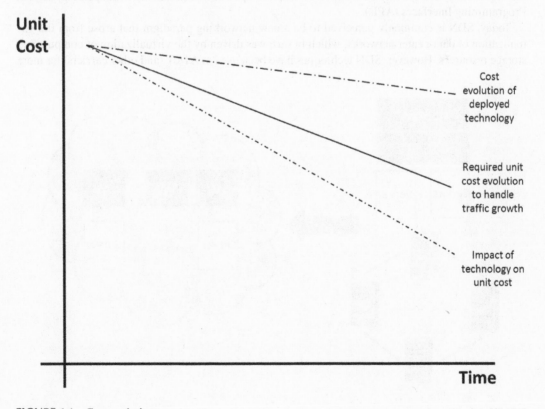

FIGURE 1.1 Cost evolution.

growth in traffic. A second problem with the current way of network design and implementation is that in the face of rampant traffic growth, network planners deploy more capacity than what is needed (since it could take several months to enhance deployed capacity, beyond simple plug add-ons), thus creating a low utilization rate.

We need to transition from a hardware-centric network design methodology to one that is software centric. Wherever possible, the hardware deployed ought to be one that is standardized and commoditized (cloud hardware, for example) and can be independently upgraded so as to benefit from technology advances as and when they occur. The network function capability ought to be largely implemented in software running on the commodity hardware. The hardware is shared among several network functions, so that we get maximum utilization. Network capacity upgrades occur in a fluid manner, with continuous deployment of new hardware resources and network function software as and when needed. In addition, the implementation of software-defined networking (SDN) would provide three key benefits: (1) separation of the control plane from the data plane allowing greater operational flexibility; (2) software control of the physical layer enabling real-time capacity provisioning; and (3) global centralized SDN control with advanced and efficient algorithms coupled with real-time network data allowing much better multilayer network resource optimization on routing, traffic engineering, service provisioning, failure restoration, etc. This new network design methodology (depicted in Figure 1.2) has a much lower cost structure in both capital outlays and ongoing operating expenses, greater flexibility to react to traffic surges and failures, and better abilities to create new services.

In order to become a software-centric network, and tap into the full power of SDN, several steps need to be taken—rethinking the IT/Network separation, disaggregation of hardware and software, implementing network functions predominantly in software and capable of executing on a commodity cloud hardware platform, and a high degree of operational automation. Add to this the requirement that the implementation be done on open source platforms with open Application Programming Interfaces (APIs).

Today, SDN is commonly perceived to be a new networking paradigm that arose from the virtualization of data-center networks, which in turn was driven by the virtualization of compute and storage resources. However, SDN techniques have been used at AT&T (and other carriers) for more

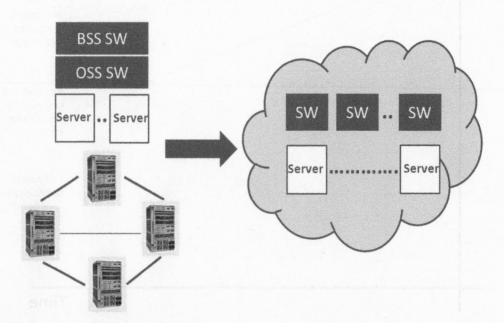

FIGURE 1.2   Transforming the network.

than a decade. These techniques have been embedded in the home grown networking architecture used to create an overlay of network-level intelligence provided by a comprehensive and global view of not only the network traffic, resources, and policies, but also the customer application's resources and policies. The motivation was to enable creation of customer application-aware, value-added service features that are not inherent in the vendor's switching equipment. This SDN-like approach has been repeatedly applied successfully to multiple generations of carrier networking services, from traditional circuit-switching services through the current IP/Multi-Protocol Label Switching (MPLS) services.

As computer technology evolved, the telecommunication industry became a large user of computers to automate tasks such as setting up connections. This also launched the first application of software in telecom—software that allowed the control of the telecommunications specific hardware in the switches. As the number and complexity of the routing decisions grew, there came a need to increase the flexibility of the network—to make global decisions about routing a call that could not be done in a distributed fashion. The first "software-defined network controller" was invented, namely the Network Control Point (NCP). Through a signaling "API," the controller could make decisions in real time about whether a request for a call should be accepted, which end point to route the call to, and who should be billed for the call. These functions could not be implemented at the individual switch level but needed to be done from a centralized location since the subscriber, network, and service data were needed to decide how to process the call.

Initially, the NCP was a relatively static service-specific program. It had an application to handle collect calls, an application for toll-free 800 services, an application for blocking calls from delinquent accounts, etc. It soon became clear that many of the service application shared a common set of functions. The "service logic" was slightly different but there were a set of primitive functions that the network supported, and a small set of logical steps that tied those functions together. The data varied and the specific parameters used for a service might vary but there was a reusable set. The programmable network and, in particular, the programmable network controller was born.[3]

The Direct Services Dialing Capability (DSDC) network was a new way to do networking. Switches, known as action points, had an API that could be called over a remote transport network (CCIS6/SS7) and controllers, known as control points, could invoke the programs in a service execution environment. The service execution environment was a logical view of the network. Switches queried controllers, controllers executed the program and called APIs into the switches to bill and route calls. The controller had a function to set billing parameters and to route calls for basic functions. Advanced functions used APIs to play announcements and collect digits, to look up digits for validation and conversion, and to temporarily route to a destination and then reroute the call. These functions allowed different services to be created.

The Toll Free Service (i.e., 800-service) that translated one dialed number into another destination number and set billing parameters to charge the called party for the call was another example of using SDN-like techniques in a pre-SDN world. The decisions on which destination number was controlled by the customer for features such as area code routing, time of day routing, etc. were implemented using software that controlled the switches. An entire multibillion dollar industry grew out of the simple concept of reversing the charges and the power of letting the customer decide which data center to route the call to, for reducing their agent costs or providing better service to their end customers.

For enterprise customers, AT&T provided a trademarked service called AT&T SDN ONENET, which provided private dialing plans, authentication codes and departmental charging capability as a virtual Time-Division Multiplexing (TDM) network for enterprise customers. With or without a Private Branch Exchange (PBX), the service exposed a programmable virtual network to enterprises. The DSDC network also provided the ability to add or remove resources to the call for advanced functions and mechanisms for dealing with federated controllers. Service Assist was a mechanism to service-chain an announcement and digit collection node into a call; this grew into Interactive Voice Response (IVR) technology with speech recognition but it was an early form of service chaining for virtual networks. When more than one controller was needed to handle the load (or to perform the entire set of desired features) NCP transfer was used, which today we would see as controller-to-controller federation. Many variations of these types of service were created, across the globe, as both the ability for software control and the availability of computing to do more processing, allowed the service providers to let their customers manage their business.

These SDN-like networks in the circuit-switched world were followed by a similar capability of providing network-level intelligent control in the packet-routed world, where the routers replace the switches. The development and implementation of SDN for AT&T's IP/MPLS network—generally known as the Intelligent Routing Service Control Platform (IRSCP)—was launched in 2004.[4] It was driven with the goal of providing AT&T with "competitive service differentiation" in a fast-maturing and commoditized general IP networking services world. The IRSCP design uses a software-based network controller to control a set of specially enhanced multiprotocol Border Gateway Protocol (BGP) (MP-BGP) route reflectors and dynamically distribute selective, fine-grained routing and forwarding controls to appropriate subsets of IP/MPLS routers. The essential architectural construct of IRSCP enables on-demand execution of customer-application specific routing and forwarding treatment of designated traffic flows, with potential control triggers based on any combination of the network's and the customer-application's policies, traffic and resource status, etc. This yields a wide range of value-added features for many key networking applications such as Content Distribution Network (CDN), networked-based security (Distributed Denial of Service [DDoS] mitigation), Virtual Private Cloud, etc.

Apart from SDN, the other key aspect of the transformation is network function virtualization with transitioning to an architecture that is cloud-centric. AT&T started its own cloud journey in late 2010 with multiple efforts, each focused on different objectives—a VMware-based cloud for traditional IT business applications, a Compute-as-a-Service (CaaS) offering for external customers, and an OpenStack-based effort for internal and external developers. These disparate clouds were subsequently merged into a common cloud platform, which evolved into the AT&T Integrated Cloud (AIC). Today, this is the cloud environment in which AT&T's business applications run. It is also the platform for the cloud-centric network function virtualization effort. By 2020, it is anticipated that 75% of the network would be virtualized and running in the cloud.

The methods and mechanisms for deploying applications on a dedicated server infrastructure are well known, but a virtualized infrastructure has properties, such as scalability and active reassignment of idle capacity, which are not well understood. If applications are not structured to make use of these capabilities, they will be more costly and less efficient than the same application running on a dedicated infrastructure. Building services that are designed around a dedicated infrastructure concept and deploying them in a virtualized infrastructure fails to exploit the capabilities of the virtualized network. Furthermore, building a virtualized service that makes no use of SDN to provide flexible routing of messages between service components adds significantly to the complexity of the solution when compared to a dedicated application. Virtualization and SDN are well defined, but the key is to enable using virtualization and SDN together to simplify the design of virtualized services. Integration of virtualization and SDN technologies boils down to understanding how decomposition, orchestration, virtualization, and SDN can be used together to create, manage, and provide a finished service to a user. It enables services to be created from modular components described in recipes where automated creation, scaling, and management are provided by Open Network Automation Platform (ONAP), an open source software platform that powers AT&Ts SDN. ONAP enables

- Independent management of applications, networking, and physical infrastructure
- A service creation environment that is not limited by a fixed underlying network or compute infrastructure
- The automatic instantiation and scaling of components based on real-time usage
- The efficient reuse of modular application logic
- Automatic configuration of network connectivity via SDN
- User definable services

ONAP will benefit service providers by driving down operations costs. It will also give providers and businesses greater control of their network services becoming much more "on demand." At the end of the day, customers are the ones who will benefit the most. The idea of creating a truly personalized secure set of services on demand will change the way we enable consumer applications and business services.

Today's telecom networks are rich treasure troves of data—especially pertaining to the world of mobility. On an average day, AT&T measures 1.9 billion network-quality check points from where our wireless customers are actually using their service. This data go into network analytics to better understand the customers' true network experience. The program is internally referred to as Service Quality Management; sophisticated analytics is used to make sense of this massive volume of network data and discern what customers are experiencing. These technologies have transformed how we manage our network—driving smarter decisions and resolving issues quicker than ever before.[5]

For instance, if two areas have disrupted cell towers—restoring service to all customers quickly is a key operational objective. Based on the real-time data and historical measurements that are available, AT&T's home developed algorithm called TONA—the Tower Outage and Network Analyzer—factors in current users, typical usage for various times of the day, population and positioning of nearby towers, to assess which towers can respond by off-loading the traffic from the disrupted towers. TONA has created a 59% improvement in identifying customer impact, and it shortens the duration of network events for the greatest number of customers. This is but one example of the power of using real-time data to improve network performance.

This transformation is not limited to the network and its associated technologies. The transformed network also needs a software-centric workforce. Reskilling the employees in a variety of software and data specialties is as important as transforming the architecture of the network. Besides, the traditional RFP approach, used today to procure new products or introduce new vendors, gets replaced by a more iterative and continuous approach for specifying and procuring software. It also calls for active involvement with the open source community.[6] Open source speeds up innovation, lower costs, and helps to converge quickly on a universal solution, embraced by all. Initially conceived to be something people can modify and share, because the design is publicly accessible, "open source" has come to represent open exchange, collaborative participation, rapid prototyping, transparency, meritocracy, and community-oriented development.[7] The phenomenal success of the Internet can be attributed to open source technologies (Linux, Apache Web server application, etc.) Consequently, anyone who is using the Internet today benefits from open source software. One of the tenets of the open source community is that you do not just take code. You contribute to it, as well.

AT&T strongly believes that the global telecom industry needs to actively cultivate and nurture open source. To spur this, we have decided to contribute the ONAP platform to open source. Developed by AT&T and its collaboration partners, the decision to release ONAP into open source was driven by the desire to have a global standard on which to deliver the next generation of applications and services. At the time of this writing, there are at least four active open source communities pertaining to the Network Functions Virtualization (NFV)/SDN effort:

- OpenStack
- ON.Lab
- OpenDaylight
- OPNFV

AT&T has been active in all four, and will continue to do so. Many other telecom companies, as well as equipment vendors, have also been equally active.

Any significant effort to re-architect the telecom network to be software-centric cannot ignore the needs of wireless (mobile) systems. The fourth generation wireless system, Long-Term Evolution (LTE), which was launched in late 2009, made significant advances in the air interface, as well as the architecture of the core network handling mobile traffic. The driver behind this was the desire to build a network that was optimized for smartphones and the multitude of applications that runs on them. As of this writing, fifth generation mobile networks are being architected and designed and their rollout is expected in 2020. Compared to the existing 4G systems, 5G networks offer

- Average data rate to each device in the range of 50 Mb/s, compared to the current 5 Mb/s
- Peak speeds of up to 1 Gb/s simultaneously to many users in the same local area
- Several hundreds of thousands of simultaneous connections for massive wireless sensor network
- Enhanced spectral efficiency for existing bands
- New frequency bands with wider bandwidths
- Improved coverage
- Efficient signaling for low latency applications

New use cases such as the IoT, broadcast-like services, low latency applications, and lifeline communication in times of natural disaster will call for new networking architectures based on network slicing (reference). NFV and SDN will be an important aspect of 5G systems—for both the radio access nodes as well as the network core.

The chapters that follow explore the concepts discussed in this chapter, in much more detail. The transformation that is upon us will achieve hyper scaling, not unlike what the web players have done for their data centers. We think this is one of the biggest shifts in networking since the creation of the Internet.

REFERENCES

1. https://www.statista.com/statistics/330695/number-of-smartphone-users-worldwide/
2. http://www.economist.com/news/leaders/21645180-smartphone-ubiquitous-addictive-and-transformative-planet-phones
3. 800 service using SPC network capability—special issue on stored program controlled network, *Bell System Technical Journal*, Volume 61, Number 7, September 1982.
4. J. Van der Merwe et al. Dynamic connectivity management with an intelligent route service control point, *Proceedings of the 2006 SIGCOMM Workshop on Internet Network Management*. Pisa, Italy, pp. 147–161.
5. http://about.att.com/innovationblog/121014datapoints
6. http://about.att.com/innovationblog/061714hittingtheopen
7. https://opensource.com/resources/what-open-source

2 Transforming a Modern Telecom Network—From All-IP to Network Cloud

Rich Bennett and Steven Nurenberg

CONTENTS

2.1   Introduction 9
2.2   Rapid Transition to All-IP 10
2.3   The Network Cloud 10
2.4   The Modern IP Network 11
      2.4.1   The Open Systems for Interconnection Model 11
      2.4.2   Regulation and Standards 11
2.5   Transforming an All-IP Network to a Network Cloud 12
      2.5.1   Network Functions Virtualization (NFV) 13
      2.5.2   NFV Infrastructure 13
      2.5.3   Software-Defined Networking 14
      2.5.4   Open Network Automation Platform 15
      2.5.5   Network Security 15
      2.5.6   Enterprise Customer Premises Equipment 16
      2.5.7   Network Access 17
      2.5.8   Network Edge 17
      2.5.9   Network Core 18
      2.5.10 Service Platforms 19
      2.5.11 Network Data and Measurements 22
      2.5.12 Network Operations 23
References 24

This chapter provides context for and introduces topics covered in subsequent chapters. We summarize major factors that are driving the transformation of telecom networks, characteristics of a modern all-IP network, and briefly describe how the characteristics change as an all-IP network transforms to a Network Cloud. The brief descriptions of change include references to succeeding chapters that explore each topic in more detail.

2.1   INTRODUCTION

The dawn of the Internet began with industry standards groups such as the International Telecommunications Union (ITU) and government research projects, such as ARPANET, in the United States proposing designs based on loosely coupled network layers. The layered design combined with the ARPANET open, pragmatic approach to enhancing or extending the design proved to have many benefits including the following: integration of distributed networks without central control; reuse of layers for multiple applications; incremental evolution or substitution of technology without disruption to other layers; and creation of a significant sized network in parallel with the development of the standards.

Tremendous growth and commercialization of the network was triggered by the ubiquitous availability of the Transmission Control Protocol/Internet Protocol (TCP/IP) network stack on Windows and UNIX servers and the development of web browsers and servers using the Hypertext Transfer Protocol (HTTP). This convergence of events created the World Wide Web where islands of information could be linked together by distributed authors and seamlessly accessed by anyone connected to the Internet. The Public Telecom network became the transport network for Internet traffic. Analog telephony access circuits were increasingly used with modems for Internet data, driving redesign or replacement of access to support broadband data. Traffic on core network links associated with Internet and IP applications grew rapidly.

The IP application protocols were initially defined to support many application types such as email, instant messaging, file transfer, remote access, file sharing, audio and video streams, etc. and to consolidate multiple proprietary protocol implementations on common network layers. As the HTTP protocol became widely used across public and private networks, usage grew beyond the original use to retrieve HTML Hypertext. HTTP use today includes transport of many IP application protocols such as a file transfer, the segments of a video stream, application-to-application programming interfaces, and tunneling or relaying lower level datagram packets between private networks. A similar evolution happened with mobile networks that started providing mobile voice and today with 4G Long-Term Evolution (LTE) wireless networks, IP data are the standard transport and are used for all applications including use for voice telephony. IP has become the universal network traffic standard and modern telecom networks are moving to all-IP.

2.2   RAPID TRANSITION TO ALL-IP

The convergence of technologies to all-IP is driving major changes in telecom networks. The price and performance of mobile computing devices and the emergence of a vibrant open software ecosystem is creating an exponentially increasing demand for mobile applications and data access. Telecom service providers have made large investments to meet mobile applications demand with wireless network technologies such as 4G LTE having embraced the use of IP as a foundational architectural element. The Third Generation Partnership Project (3 GPP) standards provided a blue print for integrating the existing telecom services and future IP service with an IP multimedia subsystem (IMS) and alternatives for IP services platforms have emerged from adjacent industries. Public compute, storage, content delivery clouds accessed via IP protocols have changed the transport usage patterns for both business and consumer users. Business applications leverage shared resources in the cloud rather than private, distributed data centers and consumers increasingly interact more with the content in the public clouds. Today, a new telecom network, deployed in a "greenfield" fashion, will exclusively designed, built, and operated in an all-IP manner end to end.

2.3   THE NETWORK CLOUD

The declining significance of transport and emergence of an all-IP network platform with diverse IP services demand creates an opportunity, or in some cases an imperative, for telecommunications providers, to remain relevant by embracing a new framework for the delivery of services. The essential elements of the new framework include refactoring hardware elements into software functions running on commodity cloud computing infrastructure; aligning access, core, and edge networks with the traffic patterns created by IP-based services; integrating the network and cloud technologies on a software platform that enables rapid, highly automated, deployment and management of services, and software-defined control so that both infrastructure and functions can be optimized across change in service demand and infrastructure availability; and increasing competencies in software integration and a DevOps operations model. We call this the *"Network Cloud."*

The benefits of a Network Cloud framework include lower unit cost of providing services, faster delivery and flexibility to support new services compared to the traditional approach (where a specific service requires dedicated infrastructure with low utilization to ensure high availability), and the opportunity to automate network operations.

2.4 THE MODERN IP NETWORK

The modern IP network as illustrated in Figure 2.1, is formed using high-capacity fiber optic transmission, optical switching and multiplexing, and packet switching technology. Each of these is combined to create local, regional, national, and international networks. In the past, virtually all telecommunications companies tended to specialize in both service type and geographic reach. With newer technologies, it has become possible to use the same network to offer a range of voice, video, data, and mobile services leading to significant competition.

2.4.1 THE OPEN SYSTEMS FOR INTERCONNECTION MODEL

The primary model for networking used today is the Open Systems for Interconnection (OSI) model, which breaks network functionality down into layers (see Figure 2.2). This allows technology enhancements and change to occur without disrupting the entire stack (combination of layers). It also allows interoperation between networks (or network segments) that use dissimilar networking technology as long as they connect using a common interface at one of the layers. This is the method that allowed the Internet and its lower layer data networks to evolve so quickly. Each of the different local area networks, or LANs, could evolve as long as they all connected using the common Internet Protocol.

2.4.2 REGULATION AND STANDARDS

No matter how organized, telecommunications networks need to be built against a set of standards in order to provide value. Internationally, the United Nations has historically had an arm called the ITU that ensured the creation of standards-based telecommunications networks. With the creation of newer technologies, additional technology groups or consortia were created to facilitate the advancement of specific approaches. For the Internet, this is the Internet Engineering Task Force (IETF); for mobile networks, this is the Third Generation Partnership Project or 3GPP.

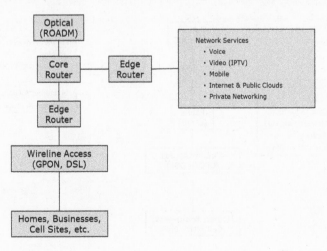

FIGURE 2.1 Illustration of a modern IP network.

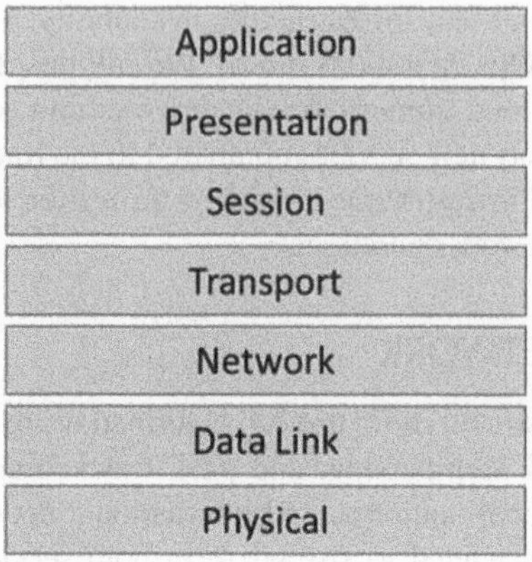

| Application |
| Presentation |
| Session |
| Transport |
| Network |
| Data Link |
| Physical |

FIGURE 2.2 OSI layers.

To help put the transformation of public networking into context, this chapter describes two examples—a traditional "appliance"-based network and a contemporary software-defined and virtualized network (software-defined networking [SDN]/NFV) at a high level. This is to convey that the new SDN/NFV network is more than just faster networking and certainly different than previous efforts to add intelligence and programmed control.

2.5 TRANSFORMING AN ALL-IP NETWORK TO A NETWORK CLOUD

Since every part of a modern telecom network can be architected to be all-IP, it can also be transformed to a Network Cloud architecture. This transformation needs to be addressed individually for each of the components of the network—customer premise, access, edge, core, and service platforms. This is because, the benefits of NFV and the degree of SDN control varies from part to part. But there is a great deal of commonality—the NFV infrastructure (NFVI), the SDN control framework, cloud orchestration, network management, security considerations, data collection and analysis, and a new operations paradigm for the software-controlled network. This is illustrated in Figure 2.3, where traditional routers are replaced by a network fabric and where the various dedicated elements are replaced by software running on common off the shelf (COTS) servers to implement the data, control, and management planes.

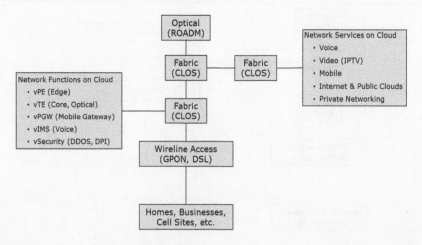

FIGURE 2.3 Transformation to network cloud.

2.5.1 Network Functions Virtualization (NFV)

Network functions virtualization (NFV) draws equally from information technology (IT) and network. Over the past decade, IT environments have gone through multiple stages of evolution, with the most recent being the use of virtualization that allows multiple software environments to share the same compute and storage hardware. The network aspect comes about when this same technique is applied to functionality used in networking.

In the past, networking functionality was creating using purpose-built equipment that contained the hardware and software bundled together to meet the customer's specifications. Examples include switches/routers (OSI layers 2 and 3), firewalls, session border controllers (used for voice and media networking), and mobility management entities (MME, used for signaling in mobile networks). Depending on the equipment, the hardware ranged from being a basic server, a server with special cards, or a custom-designed element using proprietary application-specific integration circuits (ASICs). In most cases, the underlying hardware can be separated from the operating software through an abstraction layer and the hardware platform deployed could be COTS hardware, which has very low unit cost because of widespread use across many industries. However, networking does have a requirement for special purpose hardware, either for specific physical interfaces such as optical or radio, or for throughput reasons (exceeding 10 Gb/s). In these, functional separation can be used to minimize the use of specialized hardware.

NFV leverages the IT environment, specifically the cloud paradigm, separating network software from hardware and leveraging the fact that today's modern servers are sufficiently scalable for most network functions where logic is employed. NFV-related work accelerated when a group of network operators collaborated with each other on requirements establishing a study group under the auspices of the European Standards Institute (ETSI).[2] This group was established in November 2012 and quickly gathered industry-wide participation from carriers, vendors, chip designers, academic institutions, and other standards bodies. It laid out the logical framework for NFV and the critical functional blocks along with their interaction and integration points. Transforming physical network functions (PNF) into virtualized network functions (VNF) introduces a host of challenges, which are discussed in more detail in Chapter 3.

2.5.2 NFV Infrastructure

Using the NFV approach, traditional network elements are decomposed into their constituent functional blocks. Traditionally, in a telecom network built with appliances, software is used in two major ways—network element providers acquire and/or develop the software components to support the operation and management of the network element, and service providers acquire and/or develop the software components needed to integrate the network elements into business and operations support systems (OSS). However, in an NFV model, the potential for greater efficiency is expected from more standardization and reuse of components that go into the network functions and as a result, the effort to integrate functions and create and operate services is lower because they can be supported in a common NFVI platform.

Software plays a very important role in the transformation of a traditional telecom network to a Network Cloud. The software discipline provides extensible languages and massive collaboration processes that enable solutions to complex problems; reusing solution components to automate tasks and find information and meaning in large amounts of raw data; and insulating solutions from current hardware design in a way that allows solutions to evolve independently of hardware and to benefit from a declining cost curve of commodity computing infrastructure.

In the new architecture, PNF are supplanted by VNF that run on NFVI, which is the heart of the Network Cloud. Key drivers for the platform are the ability to utilize open source software, create sharable resources, and provide for the use of modern implementation practices such as continuous integration/continuous deployment (CI/CD). The Open Stack environment is an obvious choice for NFVI. Open Stack provides the functionality to schedule compute tasks on servers and administrate the environment. In addition, it provides a fairly extensive set of features; this is discussed in more detail in Chapter 4.

2.5.3 SOFTWARE-DEFINED NETWORKING

While the general perception is that SDN for networks is a relatively new phenomenon with roots in data centers, the reality is that telecom networks have been using these techniques for several decades. An example is the use of techniques to control the data plane in circuit-switched networks, via the control plane from a controller, to provide 800 number services (for long-distance automatic routing of calls and reverse billing).

In the Network Cloud, SDN is further advanced by applying these learnings and those from cloud data centers, to create a layered distributed control framework. A "master" (or global) controller that has a network-wide view of connectivity, controls subsidiary controllers associated with different functional parts of the network. Each controller may also be replicated for scale. For the IP/Multiple Protocol Label Switching (MPLS) network fabric, the Network Cloud uses a hybrid approach retaining distributed control planes and protocols for responsiveness and using the centralized control plane for optimization and complex control.

One very unique aspect of Network Cloud SDN is the shift to model-driven software and network configuration. In the past, these tasks were done by creating detailed requirements documents that network engineers then had to translate into the specific vendor configuration language of the associated network element. Any introduction of new network capabilities took extra time waiting for documentation, testing configuration, and implementing the configuration in OSS. The new process involves defining capabilities using the Yet Another Next Generation (YANG) modeling language.[1] YANG templates, which are vendor neutral, create a portable and more easily maintained definitive representation of desired network functionality. Another benefit is that simulation can be used to verify correctness before using the templates with actual network functionality.

In AT&T's approach to SDN, all SDN controllable elements are treated similarly, whether they are implemented as classic PNF or the more modern virtualized network functions (VNF). The Network Cloud SDN controller has a number of subsystems and is designed to operate in conjunction with the Open Network Automation Platform (ONAP). It combines a service logic interpreter (SLI) with a real-time controller based on Open Daylight (ODL) components. The SLI executes scripts that define actions taken on lifecycle events such as service requests and closed-loop control events. SLI scripts that are frequently used or complex can be encapsulated as a java class that runs in the ODL framework then reused as a single script operation. Another function is network resource autonomous control, which is used to associate (assign) network resources with service instances.

At the bottom of the controller are adapters that can interact with a wide variety of network element control interfaces. These can range from older element management system (EMS) style "provisioning" to real-time network transactions using Border Gateway Protocol (BGP).

No control framework would be complete without a policy function. Policy rules can be defined based on events collected from NFVI and services running on it. Events can trigger changes in algorithms at enforcement points in and/or control actions on the NFVI/SDN Network Cloud. For example, a high utilization of a network link might trigger a traffic engineering (TE) routing change or an unresponsive service function might trigger a restart of the service function.

Using the advanced SDN controller capabilities, it is possible to create not only conventional network services such as Internet, virtual private networks, real-time media services, etc., but also more complex on-demand services. Examples described in more detail in Chapter 6 are VPN service ordering, bandwidth calendaring, and flow redirection.

2.5.4 OPEN NETWORK AUTOMATION PLATFORM

The movement to network functions virtualization changes many aspects of the way network infrastructure and services are managed over the life cycle of operation. The initial design of a service is no longer a vertically integrated optimized infrastructure, but must assume use of a distributed cloud hardware infrastructure and reuse network functions from any source that best meets the needs in the service design. The initial installation, configuration, turn up, and then changes in response to lifecycle events must be formally described in ways that can be automated such as defining work flows and actions that reference run-time sources of information and adjust use of infrastructure resources. The infrastructure as well as the service functions must expose software interfaces that allow monitoring, external control, and specification of policies that change run-time behavior. Both short-term TE and long-term capacity planning decisions must consider a broad range of services and scenarios making use of the common infrastructure.

Traditional operations support systems (OSS) and business support systems (BSS) were designed to integrate monolithic network element components and to add capabilities necessary to deliver and support the operation of customer services. This approach has a number of limitations. Network element components lack standard interfaces for some lifecycle operations and tend to be optimized for a particular service or not easily shared across different services. This increases cost and time to perform lifecycle operations such as initial delivery, upgrades, routine maintenance, repair of faults, etc. and requires dedicated infrastructure and skills. The time to design, integrate, and deploy infrastructure with staff trained to operate it at scale limits the flexibility of a service provider to deliver new services. New or emerging service volumes and uncertainty of growth make it hard to justify the investment and the long lead time to deliver increases risk of missing market opportunities.

The ETSI NFV[2] effort described above produced a specification for VNF management and orchestration (MANO). ONAP expands on ETSI MANO by adding a comprehensive service design studio to onboard resources, create services, and define lifecycle operations; closed-loop control based on telemetry, analytics, and policy-driven lifecycle management actions; a model-driven platform to accelerate and reduce costs of onboarding functions and eliminate VNF-specific management systems; and support for PNF.

A consistent plan that includes both how ONAP capabilities replace traditional OSS/BSS systems such as fault correlation, performance analysis, and element management and how these traditional systems are phased out is critical for the transition period where there is a mix of traditional physical network (PNF) and new virtualized (VNF) infrastructure. Without a clear plan for both, there is a risk that traditional systems and design assumptions get integrated with service designs on the ONAP platform, thus decreasing the benefit of highly automated operation and increasing maintenance costs.

Chapter 6 describes the ONAP software platform that supports the design and operation of services in a Network Cloud. This platform is visible as APIs to software designers and developers creating services and virtual functions; operations to view data exposed on the run-time state, events, and performance; to business and OSS that interact with real-time events; and customers in using, higher level interfaces where they need to configure services and integrate with their private infrastructure.

2.5.5 NETWORK SECURITY

Security for networks is a multidisciplinary problem—starting with the classic security problems such as confidentiality and integrity, coupled with the problem of availability, that is, ensuring the service and network work, and can withstand a hostile attempt to bring it down. With Network Cloud, there is the need to understand security through the combined lens of software and network, in an operating cloud environment. For example, using the security technique of "defense in depth" and "separations of concerns," it is a pragmatic approach to categorize VNF keeping each category type from running on the same server simultaneously. For these and other reasons, security is a key architecture and design factor for the Network Cloud.

Security manifests itself from two different perspectives—the protection of the infrastructure itself and the ability to deliver security functionality in a service. The basic structure for infrastructure security is to ensure each component is individually assessed and designed with security in mind. For the Network Cloud, the fabric component contains functionality such as access control lists (ACLs) and virtual private networks to limit and separate traffic flows. The servers run operating systems with hypervisors that provide separate execution environments with memory isolation. Around this, operational networks are used to provide access to managements ports and use firewalls to provide a point of control and inspection.

Other aspects of security architecture also come into play. While the desire is to prevent a security problem, a good security approach also provides mechanisms for mitigation, recovery, and forensics. Mitigation approaches for infrastructure include overload controls and diversion capabilities. Overload controls (which also help in the case of network problems) prioritize functionality so that resources are best applied to control plane and high priority traffic. Diversion is the ability, typically using specific portions of the IP header to identify candidate network traffic, to redirect packets for subsequent processing, rate limiting, and when necessary, elimination.

Forensics is fundamental to security. The ability to log activity provides the ability to analyze past incidents to determine root cause and to develop prevention and mitigation solutions. It also plays a role in active security enforcement by acting as an additional point of behavior inspection, which may indicate either the failure or a weakness in the security design or implementation.

Within security processes, two key components are automation and identity management. Automation allows for the administration of complex sequences and eliminates the human element from configuration that is a typical source of security problems. All it takes is entering the wrong IP address in an ACL to create a vulnerability. Identity management ensures that both people and software are authorized to view, create, delete, or change records or settings. The Network Cloud uses a centralized approach for identity verification. This prevents another weakness of local passwords, which are more easily compromised. These topics are discussed in more detail in Chapter 9.

2.5.6 Enterprise Customer Premises Equipment

Enterprise networks are the environment that businesses use to tie together their people, IT, and operating infrastructures. From offices, warehouses, factories, and on-the-go people in trucks, cars, and while visiting their customers, enterprises need comprehensive solutions for communications. Typically, they use a range of voice, video, data, and mobile services. All of their operating locations need some form of on-premises equipment, called customer premises equipment (CPE). In the past, CPE followed the same approach as network equipment, that is, implemented as appliances. However, CPE is undergoing the same transformation using NFV. This allows a single device to provide multiple functions when needed, under software control.

When redesigning CPE using NFV, the approach taken was to allow functions to operate either on the premise inside the new virtualized CPE or in the network, on the Network Cloud. For the on-premises network functions within the CPE, innovation was required to create a suitable execution environment. Unlike the Network Cloud where multiple servers are available to scale up or down VNF, the CPE environment is limited to typically a single CPU chipset. To allow for software portability and to leverage the open source ecosystem, again Kernel-based Virtual Machine (KVM) was selected as the hypervisor to allow multiple VNF to share the CPU in individual virtual machines. (There are less security concerns here since CPE is dedicated to a single customer.)

One of the most challenging aspects for CPE is management. Since CPE is located on-premises at the point between the end of a customer's wide area network (WAN) service and the local network, it is important to provide an operational network connection capability. This is done by sharing the WAN service segregating the traffic with special virtual LAN (VLAN) or IP addresses. However, before the service is initiated or during service after a failure, the WAN connection may not be available. The solution was to leverage mobile access to the Network Cloud and provide a "call home" capability. This second connection allows for remote "test and turn-up" procedures and in the case of WAN failure, diagnosis and fault location determination. This is discussed in more detail in Chapter 10.

2.5.7 NETWORK ACCESS

Access is used to describe the portion of the network that goes from homes and businesses to network infrastructure buildings called central offices or points of presence. This "last mile" of the network is also the most expensive investment a carrier must make since it provides connectivity across the geography of the service area involving the installation of media and equipment termed "outside plant." Historically, most access was based on copper twisted-pair cabling that followed the roadways outwards. Modern networks use fiber as the preferred media for "fixed" or "wired" access with the latest approach called Gigabit Passive Optical Networking (or GPON). (The older metallic copper or for cable networks, coax, is still used into the home or business except where a rebuild has happened that delivers fiber all of the way as a "fiber to the home" or FTTH). With the emergence of mobile (cellular) communications as a method for virtually anywhere communications, many have decided to forgo traditional wired service and go completely wireless. Like fixed networks, these networks have also undergone major changes, the most recent being the LTE or 4G (for fourth generation), which is based on an all-IP structure.

The goal of the access network is to provide each customer with reliable and high-performing connectivity to the network in as economical a manner as possible. This is done by maximizing the amount of the infrastructure that can be shared. For GPON, sharing is done by combining up to 64 service connections onto the same single strand of fiber. For LTE, it is done by sophisticated radio control protocols that shares radio spectrum as efficiently as possible.

With the Network Cloud, access technology is being fundamentally transformed by leveraging the distributed network fabric, compute and storage capabilities to host access-related network control and management functions. The data plane of access, due to its unique aspects, will continue to the use of specialized hardware that needs to be very close to the access media, but even here it is envisioned that programmability within that hardware can be brought to bear to provide for flexible evolution. Other access functions can run in the Network Cloud providing management and control functionality, but must typically be located nearby (e.g., under 10 miles) in order to be able to operate in the necessary timescales.

Chapter 11 gives a detailed overview of modern access technology and how it is being transformed using SDN and NFV technologies and the Network Cloud.

2.5.8 NETWORK EDGE

In packet networks, the edge platform is where network connectivity services are delivered. Before the advent of SDN and NFV, this was done using specialized network elements that converted and multiplexed the bit streams of customer access onto interoffice trunks while applying service functions. With IP networks, this is done using a router, which implements the provider edge (or PE) functionality. A router was designed in a chassis with processor cards to handle control and management functions, fabric cards that interconnected cards across the chassis, and line cards that could be used to connect to customers or to other portions of the network (typically PE connects to P core routers). The PE is configured for the type of service purchased such as Internet or private network.

With the Network Cloud, three transformations occur. First, physical connectivity is shifted onto a fabric built using merchant silicon that offers higher density and lower cost than the proprietary and specialized PE routers. Second, routing functions are disaggregated with control and management shifted to software on the Network Cloud. Third, the customer data plane is split between the fabric and software switching that runs as software. This last part allows a variety of options for providing PE services. For example, instead of deploying a proprietary PE dedicated to Internet or private networking, multiple PE software elements can run within the same server and provide the same services. If logical resource limits or specialized configurations are required, additional PE software can be executed and dedicated.

The network edge is described in more detail in Chapter 12.

2.5.9 NETWORK CORE

Chapter 13 describes in more detail, core networking for optical (layer 1) and packet (layers 2 and 3) in the OSI layer model. The optical layer provides fiber optic interconnect between remote central offices; this entails electrical to optical conversion, optical amplification, transport controls, and multiplexing/demultiplexing of multiple signals. The packet core acts as a "backbone" that ties together the periphery allowing packet traffic to enter the network virtually anywhere and be economically forwarded to its intended destination.

The electro-optical equipment used for modern fiber communications falls into two basic types: the reconfigurable optical add/drop multiplexor (ROADM) and the optical amplifier (or OA). The former is placed typically in operator central offices to allow traffic to be added and removed (similar to on and off ramps of the highways). The latter is responsible for increasing the strength of the light so that it can continue toward its destination.

Modern high-capacity optical systems work by combining a number of separate high-rate (currently 100 or 200Gbps, but soon to be 400Gbps) optical signals onto the same fiber, using a technique known as dense wavelength division multiplexing (DWDM) where each signal is assigned a unique wavelength. The ROADM adds onto this by providing the ability to add or drop specific wavelengths at each node. This forms an optical network allowing a large number of locations to send optical wavelengths between each other without requiring separate fiber optic cables.

With the advent of SDN, the optical layer is being further advanced in two key ways—disaggregation of functionality and global optimization of transport resources in conjunction with the packet layer. Global optimization is done by using SDN. The centralized controller keeps track of all of the wavelength entry and exit points and uses routing optimization algorithms and any per wavelength constraints to place them across all of the various sections that make up the network. Examples of constraints include maximum round trip delay and multiwavelength diversity (when two or more wavelengths need to have separate paths that do not coincide so that they are unlikely to fail simultaneously.)

The packet core is an integral part of the network core. The ingress packets are aggregated on to common links at different points in the network and the role of the network core is to take the packet traffic from the edge, and send it to the designated destination, utilizing the shared optical transmission layer. Since service providers also offer a number of IP services such as consumer Internet, business Internet, connections to other Internet companies (known as Peering), private business networks (known as virtual private networks), connections to cloud data centers, etc., the core network also needs to perform the task of bulk packet transport, provided in as service agnostic a manner as possible. Given the different types of IP services that need to be supported, MPLS is used as a simplified method for packet transport. MPLS uses a small header in front of every packet, which allows the core routers on either side to process and forward the packets in a fast and efficient manner. The label portion is an address known to just these two routers (i.e., it is locally significant) allowing reuse of the MPLS label address space on every link. The rest of the header contains information for traffic prioritization (also known as quality of service [QoS]) used when links become congested.

To manage traffic, a technique called Traffic Engineering (TE) creates one or more tunnels from every core router to every other core router. Multiple tunnels are used to allow different end-to-end paths to more effectively use core capacity. Inbound packets are mapped on to the tunnel that is appropriate for its destination. Intermediate core routers forward MPLS packets solely based on the tunnel mapping. In case of a link failure, two mechanisms come into play. First, every link has a predefined backup reroute path to its neighboring core routers. Second, as the reachability information is propagated across the network, all of the tunnels redetermine their paths so that they no longer expect to use the now failed link. This global repair allows for better global optimization of core capacity. Using SDN, all global traffic information and link state can be accumulated in the SDN controller for even better global optimization. This allows further tuning of the network. This hybrid approach of using both distributed and centralized control was selected to best balance speed of initial restoration with maximum efficiency. In the near term, a new approach called segment routing (or SR), is being introduced. SR allows simplification of the network since multiple control plane protocols can be consolidated. Essentially SR works by allowing each packet to carry path routing information. Thus, at the origination point, some or all of the intermediate hops can be predetermined. Thus, SR supplants the need for separate protocols such as Resource Reservation Protocol for Traffic Engineering (RSVP-TE) to execute fast reroute (FRR) local repair and label distribution protocol (LDP) for label distribution. This is discussed in more detail in Chapter 13.

The final component of the core is the route reflector (RR). All of the edges that connect to the core need the ability to communicate reachability of service with each other. This is done using BGP. But how do you let hundreds of edge routers communicate control plane information with each other? By using an aggregation and distribution node known as a RR. Every edge router in the network connects to a pair of RRs. For scale, multiple pairs of RR can be deployed for different services or geographic regions. Each of these take in the BGP messages from the edge routers they serve and replicate and distribute the messages to the other edge routers. This offloads the work from the edge and core routers allowing for scale and separation of control and data plane processing.

2.5.10 SERVICE PLATFORMS

For most of the twentieth century, the basic capabilities of a network were the customer service, a fixed point-to-point circuit between two customer locations or the ability to request a temporary narrow band voice connection between two network end points through a switch. Investment in a service platform beyond the core network was largely focused on the OSS, BSS systems, and processes to support the delivery and monetization of the network. Business functions included customer ordering, billing, and directory assistance based on a subscription business model. Operational functions included inventory, planning, installation, maintenance, and diagnostics. The customer services of transport and switching were an integral part of the network design, expanding the availability of this basic universal service was more attractive than enhancing services that would require investment to upgrade many end points.

As fundamental improvements in transport and computing technology reduced unit cost and increased capabilities of transport and end point devices, enhanced services became feasible. Circuits were replaced by packets, switching replaced by session establishment, and service platforms based on IP protocols emerged to support: simpler and faster connection setup; a range of end point devices that were programmable; increased security of data in transit; interoperability between different services and networks; policies to control network behaviors based on service or customer; multiple types of standard data and media streams; and delivery of rich context to enhance communication.

As technology advanced and architectures converged on IP, customers themselves demanded more than circuits and media sessions, and the need for a common platform became evident from the isolated islands of infrastructure dedicated to a service. The following paragraphs describe some common platform component capabilities used in an all-IP network, two examples platforms that integrate common components, and some examples of customer services. The component and capabilities include directories, public key encryption, signaling protocols, gateways, communication optimizations, and policy rules and enforcement. The platform examples include the 3GPP IMS and a content delivery network (CDN).

While the network was the service, the method for finding an end point was to assign unique numbers to physical end points, provide some level of assurance that the end-to-end physical connection was secure, have the end user or device find and remember unique numbers they wanted to connect with, and involve the end user to verify the identity of the other side of the connection. As technologies such as IP protocols and mobile computing devices became available, a better method was required given the diverse range of services expanding number of end points and identities.

Two major classes of technologies support finding and trusting an end point—directories and public key encryption. In modern IP network services, directories are used at multiple levels such as to translate invariant, global unique identifiers to a current IP address, to find the next hop to move a packet destined for a particular IP address toward the destination, and to represent the current devices that are registered and available for a particular type of real-time communications session. Directories may also include rich information about the end point of a connection that assists in the selection and insures the connection is secure. Public key encryption relies on a key pair, one publicly advertised (in directories or other sharing mechanisms) and one held securely by a connection end point. One side of a communicating link between end points, can encrypt information using the public key and the other can decrypt it using the private key. This technique along with a trust hierarchy of directories and hardware methods for securing private keys in devices are used to insure the identity of end points and protect private data in transit across public or untrusted networks. For example, the Universal Integrated Circuit Card (UICC) in a mobile phone contains a subscriber identification module (SIM) with keys used to identify and authenticate subscribers.

Another capability in services platform is a standard protocol for connection establishment, referred to as a signaling or control protocol. The signaling protocols enable rapid connection setup, optimization of network resources by deferring allocation of resources until there is agreement between end points or users that communications can begin, setting expectations with end users or devices about how a session will be initiated, negotiating to characteristics that both end points can support, and adjusting communication service algorithms to work acceptably over varying network conditions. Examples of signaling protocols include Session Initiation Protocol (SIP) that is semantically similar to the traditional caller/called party interaction over a telephone network augmented with rich session descriptor options to support many types of real-time media; extensible messaging and presence protocol (XMPP) a streaming message protocol that has been used for short messaging, presence indications, and to coordinate real-time media connection establishment; and real-time media control protocol (RTCP) that enables monitoring and feedback on the performance of a real-time connection where packet loss, available bandwidth, and jitter may change and the service adapts use of the network to handle these variations and with minimal impact to user experience.

Another common capability in a service platform are gateways. These are used to maintain compatibility with legacy services as new services are introduced and to limit, enhance, and/or capture information exchanged across network boundaries. Some classic examples of gateways include a circuit switched telephony gateway that translates control protocols and transcodes media between a circuit switched connection and an IP packet-oriented connection; a session border gateway that performs similar functions as the circuit switch gateway and limits the exposed internal end points; and a proxy server that terminates internal or external sessions, forwards information between the two sessions, and restricts, captures, or enhances the information being forwarded.

In an all-IP network, optimizing higher level communications patterns is another platform capability. Some common optimizations include mixing or bridging many media streams into one combined stream that can be sent back to all sources (e.g., mix all real-time audio session in a conference, forward the last N active video session frames to all sources in a video conference), caching frequently accessed files near the edge of a network to reduce capacity required over the WAN with a content distribution network, and accelerating perceived application performance by constructing a full web page near the sources of data when the page is composed of many small, separate objects, that would take much longer to assemble from a distant location where each session establishment is delayed by end-to-end protocol exchanges.

To support different services in the Network Cloud and react in a coordinated way to real-time events requires an ability to adjust algorithms distributed through the network based on a service, specific customer, network condition, etc. These capabilities are referred to generally as policy and includes both the capability to define policy rules and to execute these rules at distributed policy enforcement points.

The context for communications established in advance of an exchange of information can greatly improve the efficiency and effectiveness of communication. An early example of this was the capability to provide caller identification for incoming voice calls to a business and support controls to transfer a connection from a public network user to different location, agent, and/or automated application based on the caller identity. Communication services and platforms continue to expand in their ability to provide a rich context for communication enabled by mobile access networks; sensors present in mobile hand-held and embedded devices; willingness of users to share information on their location, activities, etc.; data collection and analytics used to infer context from prior activity; service provider privacy policies and information controls that allow a user to opt-in to different levels of context data handling.

The 3GPP-specified IMS[3] is one example of the above components being used together in a platform. IMS architecture refers to components and layers as service applications, session control and management, access networks, and end point devices or user equipment. This layering integrates traditional telecom services as well as provide the platform framework for future multimedia services over IP on mobile phones. IMS uses SIP to negotiate and establish connections, a home subscriber server (HSS) directory, public key encryption with private keys stored in the HSS and UICC/SIM of a connected device, gateways between networks that adapt media, separate media and control, and maintain the security and integrity of the platform (Figure 2.4).

Significant services and platforms have emerged that do not rely on a complete IMS framework. The session control and management layer is not always needed and/or alternatives for components in this layer may be less costly and complex to deliver a wide variety of services. Capabilities such as signaling and identity are frequently tailored to and tightly integrated with a web application, community, and usage scenarios.

A content distribution network (CDN) is another example of a service platform used to optimize the delivery of content that many customers want at the same time in the same general location. A CDN uses components and capabilities such as directories that map a request for content to the closest edge location, streaming media protocols that adapt to network availability, and optimize performance under varying conditions (Figure 2.5).

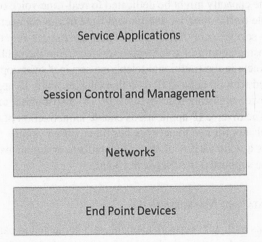

FIGURE 2.4   IMS layered architecture.

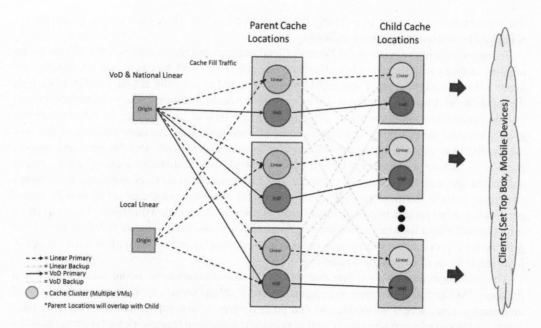

FIGURE 2.5 CDN supporting IP video distribution: VoD/linear IP television (IPTV). (AT&T example design.)

Common examples of business customer services include Internet access; virtual private networks; public and private voice/video networks; and customer-dedicated edge configurations that allow connections to a virtual private network at their premise, a public cloud data center, and/or from a wireless device.

Internet access ranges from best effort access over copper wire, fiber, or wireless access point to various levels of managed Internet access that guarantee a higher level of service. This may include guarantees on bandwidth, latency, jitter, loss, availability, as well as monitoring to detect and protect against security threats. For example, when a distributed denial of service attack is detected adjust routing rules at the ingress points of the attack to drop packets before they are routed to a customer end point.

MPLS virtual private network service allows a customer to design and control a multilocation secure network and establish quotas for how much capacity is devoted to different types of traffic. For example, 40% of the capacity might be dedicated to real-time voice or video to insure an occasional burst of deferrable traffic does not disrupt real-time media connections.

Custom voice/video services are offered to business customers as part of their virtual private network and/or by exposing control of voice/video services from the public telecom network.

Consumer services include the traditional telephone services over copper, fiber, or radio and in the last decade enhanced with broadband IP data that enables many IP applications including those from a large mobile device application ecosystem, IP TV distribution, and monitoring systems at home and in vehicles. The diverse IP applications enable monetization of network services through multiple business models not limited to network customer subscription.

Chapter 14 describes a wide variety of services, platform components that enable multiple services, and how these are supported in a Network Cloud.

2.5.11   Network Data and Measurements

In traditional IP networks built with network elements and dedicated to particular services, data and measurements are performed in a relatively static configuration by collecting data from network elements and/or inserting probes at the points necessary to support the requirements of service.

For example, one might collect data on real-time media sessions, number of sessions established per unit time, attempts to establish sessions when all capacity is being used, packet loss, jitter, and latency through a session border gateway. By continuously collecting this data, it can be used to measure and improve the QoS and predict when additional transport or switching capacity is needed. Analysis of data in real-time is not critical since the actions one could take based on the data collected involved administrative, planning, and engineering tasks that occur over days or months.

Moving to a Network Cloud using SDN and NFV creates both challenges to achieving the same level of data collection and measurements as would exist in a traditional IP network and new opportunities to benefit from real-time analytics performed on measurements. The challenges stem from the implementation of a service on flexible infrastructure that is simultaneous supporting multiple services where the mix may change over short periods of time. Real-time analytics across shared, dynamically changing components create opportunities to optimize across multiple services and dimensions. These dimensions can include tradeoffs such as performance of a service, cost to deliver a service, timing of an infrastructure investment, excess capacity to eliminate emergency maintenance, etc. An additional benefit is that tightly coupled to network element service, network, and component measurements are now decoupled and can be done once in a general way within the ONAP platform, reducing the cost, complexity, and time to create services.

Real-time control to optimize the network relies on SDN in the most general sense that all resources must be controllable via software and information is more centralized in near real-time for making decisions. For example, this includes controlling and configuring tunnels, packet flows, load balancers, directories, optical wavelengths mapped to packet layers, etc. and extends all the way to customer services and applications that support interfaces to allow a customer to defer or forecast demands in exchange for better price or performance.

To facilitate appropriate real-time optimization, the capture of network data and executing the measurements is an integral part of the Network Cloud. The infrastructure to capture, retain, and retire network data is described in Chapter 8, while the efficacy of techniques to do the required measurements is discussed in Chapter 16.

2.5.12   Network Operations

The transition of network operations from supporting traditional and IP networks to a Network Cloud is perhaps the largest challenge and key to realize the benefits from the transformation.

The network operational challenges include growing new skills and managing through an interval where there is a mixture of legacy and cloud infrastructure. Significant new software engineering and quality assurance skills are needed to support continuous delivery and integration of new or rapidly changing capabilities, a role commonly referred to as DevOps. Investment to replace current technology infrastructure and refactoring network functions to operate in an automated, closed-loop control platform will require operating in a mixed infrastructure for multiple years.

The benefits derived from a successful transformation of operations include reduced costs; reduced cycle time and increased flexibility to use the Network Cloud infrastructure supporting new or rapidly changing demand for services; and an increased utilization of infrastructure that is not dedicated but shared with software service when needed.

Operations scenarios that change to realize these benefits include the following: manual administration, configuration, and monitoring of infrastructure or services is highly automated, emergency maintenance is replaced by treating faults as a reduction in capacity that can be addressed during normal, periodic upgrades; planning, testing, and turn-up of new capabilities is routine and highly automated through extensive regression tests delivered with software functions; and organizationally the need for staff aligned to particular services is reduced or eliminated and common technical skills can be shared across all services. Chapter 15 describes the challenges, changes, and approach for operating a Network Cloud.

Thinking more broadly than network operations and automating or optimizing a task in the next year to other business functions and creating a culture that embraces change, innovates, and continues to create new value, it is important to consider other types of change. As software has transformed other industries, there are examples where major impacts are attributed to a software mentality or discipline that is different than what might exist in a telecom provider culture today. Two examples of what might sustain leadership as a Network Cloud platform provider are empowering individuals who have the capability to contribute to open source software platform components; and committing to the use and continuous improvement of an open source platform in collaboration with other community members. Chapter 17 covers many examples of successful operations in a software world and contrasts them with where a telecom provider may be coming from.

REFERENCES

1. https://tools.ietf.org/html/rfc6020
2. http://www.etsi.org/technologies-clusters/technologies/nfv
3. http://www.etsi.org/deliver/etsi_es/282000_282099/282007/02.00.00_50/es_282007v020000m.pdf.

3 Network Functions Virtualization

John Medamana and Tom Siracusa

CONTENTS

3.1 Virtualization...26
    3.1.1 Network Virtualization..26
    3.1.2 Compute Virtualization 28
    3.1.3 Virtualization of Network Functions..28
    3.1.4 Benefits of NFV...29
3.2 NFV and Software-Defined Networking...31
    3.2.1 Relationship between NFV and SDN 31
3.3 Decomposition of VNFs..32
    3.3.1 Decoupling Virtual Functions 32
        3.3.1.1 Data Plane 33
        3.3.1.2 Control Plane 35
        3.3.1.3 Management Plane...35
    3.3.2 Service Chaining 37
    3.3.3 Overlay, Underlay, and vSs/vRs...38
    3.3.4 Reusability...39
    3.3.5 Multi and Single Tenancy 39
3.4 NFV Resiliency and Scaling 40
    3.4.1 Multipath and Distributed VNF Designs 41
    3.4.2 vPE Example for VNF-Specific Resiliency Design 42
3.5 NFV Economics 44
    3.5.1 Hardware Costs..45
    3.5.2 Software Costs 45
    3.5.3 Operational Costs 45
3.6 NFV Best Practices 46
Acknowledgment 47
References..47

Network functions virtualization (NFV) is a set of software capabilities created by developments in two separate disciplines: networking and operating systems (OSs). In the early days, both networks and computers were very expensive. Hence, sharing a networked computer became a necessity—while the cost per user reduced considerably, every time a new user was added to the network, the "value" of the network increased in proportion to the number of users (Metcalfe's Law). As the number of computers increased, the design of the network became more complicated, with switches and routers being used to direct traffic from source to destination, and networking technologies that facilitated traffic flow in local and wide-area networks.

The networking technologies referred to here are Ethernet (Spurgeon and Zimmerman, 2000) and IP (Comer, 2014), which enable the creation of networks that are shared by multiple users. Ethernet switches and IP routers are built using hardware and software to implement the underlying networking protocols. However, neither Ethernet nor IP was designed to provide adequate isolation among users (i.e., hosts) or user groups. Isolation requires privacy and protection of one user from another unless the two users are directly communicating with each other. Ethernet relies on a broadcast discovery protocol. The addressing architecture of IP routing results in every user being "visible" to, or "reachable" by every other user on the same network. NFV enables safe and efficient sharing of networks among large number of users while providing a high degree of isolation among groups of users.

In addition to sharing and isolation, NFV requires a third property which we refer to as functional decoupling. The switches and routers used to implement Ethernet and IP are monolithic elements that are built to deliver very high performance—high packet throughput and very fast packet switching time (sum of packet processing, queueing, and transmission delays). As the Internet grew, the switches and routers became increasingly complex. They use complex, custom-built ASICs for packet forwarding and purpose-built OSs to implement various networking protocols and functions. NFV requires the decoupling of the networking functions (forwarding, control, and management plane) from the underlying hardware. Also, NFV implements packet forwarding in software. The hardware becomes generic (e.g., x86 or merchant silicon) and the software is decoupled from hardware. We will survey these technologies before we dive deeper into NFV.

3.1 VIRTUALIZATION

In this section, we discuss virtualization as it relates to networks and compute environments and then focus on the virtualization of network functions, which applies to both network and compute virtualization.

3.1.1 NETWORK VIRTUALIZATION

Let us begin by understanding the concept of closed user groups (CUGs). A CUG allows a set of users to communicate among each other in isolation from other users of the network. A network should allow the creation of multiple CUGs. Each CUG behaves as though the entire network belongs to the particular CUG with no interference from other CUGs that may exist on the same network. The technology that enables such CUGs is called network virtualization. Each CUG operates on its own virtual network, coexisting with other virtual networks on the same physical network.

Ethernet and IP networks were designed to allow multiple users to communicate and share a common physical network. However, the original definition of Ethernet and IP did not have the notion of CUGs.

Ethernet defines the notion of a broadcast domain over a shared medium. Virtual LAN (VLAN) is a broadcast domain that is partitioned and isolated from other VLANs and associated broadcast domains. To partition a network into VLANs, the network switch or router needs to be configured. VLANs allow network administrators to group hosts together into a broadcast domain or virtual network. VLAN membership can be configured through software. Without VLANs, partitioning users (i.e., hosts) necessitates physically relocating network switches and cables that connect stations to switches. VLANs greatly simplify network design and deployment by virtualizing the LAN (see Figure 3.1).

Virtual private wire service (VPWS) and virtual private LAN service (VPLS) (Rekhter, 2007) are widely used layer-2 virtual network services for large campuses or multilocation wide area networks (Anderson, 2006). VPWS defines point-to-point services and VPLS defines multipoint services. Ethernet VPN (EVPN) (Sajassi, 2015) is the newest standard for building and delivering CUG services for Ethernet end points.

IP routing defines a set of protocols that compute the path that a data packet follows in order to travel across multiple networks from the packet's source to its destination. Data are routed from its source to its destination through a series of routers, and often across multiple networks. The IP-routing protocols enable routers to build a forwarding table that correlates destinations with next hop addresses. In basic IP routing, all sources and destinations are visible to each other by means of routing protocols.

In order to create CUGs over IP networks, we need to use an approach that is commonly referred to as overlay networks. An overlay network is a virtual network implemented over an underlying IP

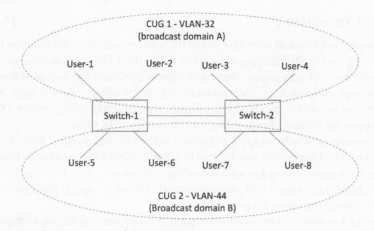

FIGURE 3.1 Simple Ethernet virtual networks using VLANs.

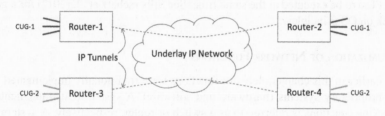

FIGURE 3.2 Virtual network connections using IP tunneling.

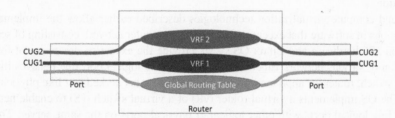

FIGURE 3.3 Virtualization of IP networks using VRFs.

network. There are two widely used technologies for creating overlay IP networks: IP tunnels and virtual route forwarding (VRF).

IP tunnels (Simpson, 1995) are used for connecting two virtual IP end points that do not have a routing path to each other (see Figure 3.2). The tunnel uses a routing protocol across a common intermediate IP transport network. The IP tunnel is a virtual overlay network connection between the end points and the common intermediate transport network, which is sometimes referred to as the underlay network. A set of IP tunnels can be used to create a CUG among a group of IP end points. This is called a virtual private network (VPN).

VRF is a technology that enables the creation of multiple instances of routing tables in an IP network (Rosen, BGP/MPLS IP Virtual Private Networks (VPNs), RFCs 4364 and 2547, 2006). Routing tables are learned and maintained by each router in an IP network. Routing table lookup is performed at each hop along the path of a packet to determine the next hop of the packet. In the original IP network architecture, routing logic is defined using a single-routing table, sometimes referred to as the global routing table. VRF defines multiple instances of routing table where each instance may define a CUG. This is a very effective way of creating VPNs (see Figure 3.3).

3.1.2 COMPUTE VIRTUALIZATION

Numerous innovations in OSs software have paved the way for compute virtualization (Rosenblum, 2004). Multitasking and multithreading are two of the earliest OS innovations that significantly improved the efficiency of computers. The basic idea of multitasking is that software applications contain multiple tasks that are independently scheduled entities. As the logic within a single-software process became complex OS developers created multitasking within OS processes. Multitasking creates the illusion of concurrent execution of parallel tasks.

Another OS innovation was to segment a system into multiple independent OSs. A new supervisory software layer called hypervisor was created to allow multiple OSs to run as independent tasks. These OSs in turn use multitasking and multithreading to efficiently share the machine among multiple users. This enables the creation of multiple virtual systems or virtual machines (VMs). Each VM has its own isolated execution environment.

Multiprocessing was another important innovation. Multiprocessing allows multiple processors (CPUs) to use multitasking and multithreading capabilities to enable real concurrency of the underlying processes. Multiprocessing allows two or more tasks and threads to be allocated to different processors (CPUs) to be executed at the same time. See Silberschatz et al. (2012) for a good general reference book on OS principles.

3.1.3 VIRTUALIZATION OF NETWORK FUNCTIONS

As described earlier in this chapter, network functions were traditionally implemented in purpose-built, vendor-proprietary systems (hardware and software). A system that implements layer-2 or layer-3 networking functions is referred to as a switch or router, respectively. A system that implements other networking functions is traditionally referred to as a network appliance. A typical enterprise network has many switches, routers, and appliances, sometimes several such systems in a single location.

Network and compute virtualization technologies described earlier allow the implementation of network functions in software that executes on a generic computer network consisting of servers typically based on x86 hardware and Linux OS. VMs provide the execution environment for software implementation of various network functions. Each VM has a logical port that behaves like a physical port on a switch, router, or appliance. Logical ports can be networked just like physical ports are networked. The OS implements a virtual router (vR) or a virtual switch (vS) to enable network connections that link logical ports with other logical or physical ports on the same server. The network connections that terminate at a logical port are virtual overlay connections described in Section 3.1.1. The network functions that are implemented in this manner are referred to as virtualized network functions (VNFs).

Figure 3.4 shows the use of VNFs running on VMs to implement a network connection (dotted line) that is used to carry packet flows for a given source-destination pair. VNF-1 and VNF-2 denote network application software running on two different VMs. The data center fabric (for Ethernet/IP forwarding) and the servers provide the infrastructure, built using commercial off-the-shelf (COTS) hardware. For example, VNF-1 may be an edge router and VNF-2 may be a network analyzer used to troubleshoot the traffic flow.

Most network functions can be implemented in software that runs on COTS platform such as an x86 server. Since general purpose computers are relatively inefficient in implementing packet-forwarding functions, NFV may become impractical for VNFs that require very high throughput. Network functions that require aggregate rates below 10 Gbps are generally not a problem and can be easily virtualized. VNFs that require aggregate rates above 50 Gbps are not suitable for virtualization at the time of writing this book. However, server technology and packet-processing technology are improving rapidly. We expect the 50 Gbps limit to rise over time. All VNFs except core routers (aka P routers) in large service provider networks can be virtualized.

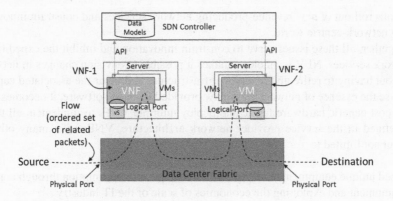

FIGURE 3.4   VNFs that implement a network connection.

VNFs can be classified into two general categories: host VNFs and middle box VNFs. Host VNFs are sources and sinks of packet traffic. Host VNFs originate or terminate end-user network connections. Examples of host VNFs are as follows:

- Domain Name Service (DNS)
- RADIUS
- Any HTTP client

Middle box VNFs do not terminate end-user network connections as implied by the name. They sit in the middle of a network connection. Packets go in and come out. These VNFs most likely will modify the packet header. Layer-2 switching or layer-3 routing are the two most common functions that are implemented by using middle box VNFs. Examples of middle box VNFs are

- Customer edge (CE) router
- Provider edge (PE) router
- Firewall
- Secure gateway
- Proxy
- NAT
- Load balancer
- WAN acceleration
- Service architecture evolution (SAE) gateway
- Proxy
- Cache (content distribution network [CDN])

3.1.4   BENEFITS OF NFV

A typical service provider's network consists of a large number of proprietary hardware switches, routers, and appliances. Creating and launching a new network service often requires adding yet another variety of proprietary systems. In addition to the capital expense required, the space and power these hardware appliances consume represents a significant operating expense. These challenges are further compounded by the increasing costs of energy and the scarcity of the specialized skills needed to design, integrate, and operate complex hardware-based appliances. Moreover, appliances rapidly reach end of life, which leads to a procure–design–integrate–deploy cycle that has to be continuously repeated with little or no revenue benefit. In addition, hardware lifecycle is becoming shorter as technology changes and service innovation accelerates, and this can inhibit the expeditious roll out of new revenue-producing network services and constrain innovation in an increasingly network-centric world.

Taken together, all these issues serve to constrain innovation and inhibit the expeditious rollout of new network services. NFV technology makes it possible to envision changes in network architecture without having to retire and redeploy infrastructure and incur the associated capital investment. Because the essence of networking is now provided through software, it becomes possible to build a low-cost generic hardware fabric and deploy multiple VNFs to implement all the network functions defined in the service provider network architecture. VNFs offer many other benefits including, but not limited to

- Reduced unique equipment diversity and reduced power consumption through consolidating equipment and exploiting the economies of scale of the IT industry.
- Drive independent scaling and innovation by architecturally decoupling the network function (based in software) from the support infrastructure (based in hardware).
- Faster time to market by minimizing the typical cycle of innovation for the network operator. Economies of scale required to cover investments in hardware-based functions are strongly mitigated through software-based deployment, which more or less follows a variable cost model and is far more fluid.
- Reduced costs by sharing resources across many services and different customer bases. Ability to run multiple versions of a network function is simple since the functions are implemented in software. Virtualization enables multitenancy, which allows the use of a single platform for different applications, users, and tenants.
- Targeted service introduction based on geography or customer sets is possible. Services can be rapidly scaled up or down as required.
- Ability to deploy systems that elastically support various network functional demands and which allow directing the capacity of a common resource pool against a current mix of demands in a flexible manner.
- Enable a wide variety of ecosystems and encourage openness. NFV opens the virtual appliance market to pure software entrants, small players, and academia—thus encouraging innovation in bringing more new services and new revenue streams quickly, at a much lower risk.
- Enable new types of network services. NFV can readily be applied to the control and management plane in addition to the data plane. This allows virtual networks to be created and managed by end users and third parties using the tools and capabilities heretofore reserved only for native network operators.
- Faster testing of new services. Testing and certification of complex services is a time-consuming endeavor. Often, it is difficult to test the field of use in a laboratory environment. The multitenancy aspect of NFV enables the service provider to test new services and updates in a production setting without risking real customer traffic.
- Higher reliability; software-based systems can be made resilient more easily and inexpensively. Hardware-based reliability tends to be more expensive due to the need to deploy special-purpose redundant hardware and proprietary logic.

To achieve these benefits, there are several technical challenges that need to be addressed:

- Achieving high-performance virtualized network functions, which are portable between different hardware vendors, and with different hypervisors.
- Achieving coexistence with custom hardware-based network platforms while enabling an efficient migration path to fully virtualized network platforms. Similarly, transitioning from existing business support system (BSS) and operations support system (OSS) to nimbler DevOps and orchestration approaches.

- Managing and orchestrating many virtual network appliances while ensuring security from attack and misconfiguration.
- Ensuring an appropriate level of resilience to hardware and software failures.
- Integrating multiple virtual appliances from different vendors. Service providers prefer the use of "mix and match" hardware from different vendors, hypervisors from different vendors, and VNFs from different vendors without incurring significant integration costs.

3.2 NFV AND SOFTWARE-DEFINED NETWORKING

Software-defined networking (SDN) is an architectural framework for creating intelligent networks that are programmable, application aware, and open. A key aspect of the architectural framework is the separation of data forwarding from the control plane, and establishment of standard protocols and abstractions. However, the term SDN is also applied to a number of other approaches that espouse more open, software-centric methods of developing new abstractions for both the control plane as well as the data plane of networks and the ability to expose network capabilities via APIs. SDN will be discussed in more detail in Chapter 6. Here, we look at the relationship between SDN and NFV.

3.2.1 RELATIONSHIP BETWEEN NFV AND SDN

SDN can act as an enabler for NFV, since the separation of control and data planes enables the virtualization of the separated control plane software. NFV can also act as an enabler for SDN since the separation between data plane and control plane implementations is simplified when one or both of them are implemented in software running on top of standard hardware. The term SDN has come to mean different things based on one's perspective as illustrated in Figure 3.5.

The term SDN was coined in 2009 based on work being done in academia. The focus was on separating the control plane from the forwarding plane for Ethernet switches to centralize forwarding decisions in a controller. The goal was to allow for more agility and flexibility on forwarding rules, while actually simplifying the forwarding devices without the need for complicated distributed routing protocols. OpenFlow (Foundation) emerged as a new protocol to support the communication between a central controller and forwarding devices.

The first production use case for these concepts was large data centers. Sophisticated orchestration emerged to support the deployment and mobility of VMs, primarily supporting IT applications. The data center network, however, lacked the agility to support this environment. If one moved a workload from one server to another, the network VLANs/IP addressing had to be manually

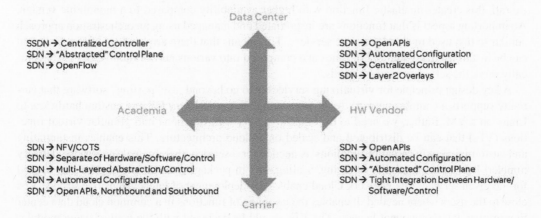

FIGURE 3.5 SDN views.

configured to support the workload in its new locations. This typically meant reconfiguring all the switches in the path. To overcome this on an existing network, new overlay technologies emerged to support virtual networks over the top of the existing network fabric. These technologies have included LISP (Cisco), VM-Ware NSX (VMWare), OpenStack Neutron (OpenStack), TRILL, and even Multi-Protocol Label Switching (MPLS).

This opened the door for hardware vendors to create new fabric switches that could natively support these overlay protocols. The new switches, while they do create more network agility, tend to be vendor proprietary and also tightly couple the hardware, software, and control. While this might satisfy the requirements for a single-enterprise data center, it is not robust for a carrier infrastructure that needs to support network functions from many different suppliers. It is the joining of SDN with NFV on a common COTS hardware platform that is the key for implementing an end-to-end network for service providers. The decision choice on hardware, software functions, and control needs to be independent and based on open-source architectures with well-defined APIs.

3.3   DECOMPOSITION OF VNFs

Software that implements a network function is either a one-to-one mapping of an existing appliance function or alternately some combination of network functions designed for cloud computing. For example, functions designed for cloud computing might be combined or leverage distributed data services to eliminate a set of hardware-based network appliances. In addition, they use software logic and characteristics of cloud computing to eliminate fault-tolerant or one-for-one failover with spare appliance hardware requirements. Many of the initial opportunities in developing cloud networking are centered around using a network functions virtualization infrastructure (NFVI)—which closely resembles cloud computing infrastructure—and developing new applications that support many of the existing monolithic control plane elements, such as route reflectors (RRs), DNS servers, and Dynamic Host Control Protocol (DHCP) servers. NFVI will be covered in Chapter 4.

In this section, we will discuss the virtualization of functions and how to interconnect them over a virtual network.

3.3.1   DECOUPLING VIRTUAL FUNCTIONS

Over time, more network edge functions and middle box functions are expected to migrate to this infrastructure. These functions include SAE gateways, Broadband Network Gateways, IP edge routers for services like IP-VPN and Ethernet, security functions, and load balancers and distributers. Because these elements do not typically need to forward large aggregates of traffic, their workload can be distributed across a number of servers—each of which adds a portion of the capability; overall, this creates an elastic function with higher availability compared to a monolithic version. An important aspect is that functions are instantiated and managed using an orchestration approach similar to that used in cloud compute services. This means that there are catalogs of functions that can be instantiated any number of times and composed into various network topologies to dynamically serve the service provider's needs.

A key design principle for virtualizing services is to go beyond just "porting" software that currently supports a complex network function running on a proprietary OS and custom hardware, to Linux on a VM. Rather, we need to decompose the network functions into granular virtual functions (VFs) that can be distributed and scaled on a cloud architecture. This enables instantiating and customizing only essential functions as needed for a service, thereby making service delivery nimbler. It provides flexibility of sizing, scaling, and in packaging and deploying VFs as needed for the given service. The Network Cloud enables the deployment of latency-sensitive service VFs close to the users where needed. It enables the grouping of functions in a common cloud data center to minimize intercomponent latency. The VFs should be designed with the goal of being modular and reusable, facilitating the use of best-in-breed vendors. Decoupling subscriber data and state improves the reliability of virtualized services. It also improves scalability as the data tier can scale independently of the network function.

The decomposition of network functions should address the following guidelines:

- Decompose if the functions have significantly different scaling characteristics (e.g., signaling vs. media functions, control vs. data plane functions, etc.)
- Decomposition should enable customizing a specific aspect of the network function on instantiations (e.g., the interworking function may need to be customized specific to each carrier interconnect instantiation)
- Decomposition should enable instantiating only the functionality that is needed for the service (e.g., if transcoding is not needed, it should not be instantiated)

Deploying NFV in a cloud service platform has provided operators new opportunities to minimize their costs and deploy new advanced value-added services. The NFV implementation ought to be cloud-native, and not just an adaptation of software from dedicated vendor hardware to a x86-based COTS platform. A native cloud implementation ensures efficient sharing of storage and computing resource. To fully utilize the resource, NFV needs to be both dynamic and flexible, as well as

- Scalable
- Interoperable
- Reusable
- Distributed
- Maintain state in the event of a VF failure

Enabled by SDN and virtualization, NFV function decomposition (disaggregation) is a key pillar of the new network design. One of the benefits of NFV disaggregation (i.e., decompose a NFV into small components) is to enable operators to achieve efficient capacity scaling and reusability. Especially, it allows operators to scale different components based on different load profiles. Another key defining characteristic and benefit of decomposed NFV systems is that it allows function sharing across different VNFs.

The idea of breaking large applications into small components can be traced back to service-oriented architecture (SOA). We have also seen it help scale big data applications across thousands of servers with technologies such as Hadoop and Storm. The key benefit is to optimally utilize resources such as individual central processing unit (CPU) cores or memory on a cloud host. Host resources can be more effectively utilized by small VM instances. Finding capacity for a small VM is much easier than for a large VM. Further, large instances can result in more inefficient allocation of unused resources at hosts, lowering the engineering economics of cloud implementation.

Finally, decomposition enables higher resiliency and scalability. Smaller VFs are natively able to scale independently. If a component fails or is out of capacity, another one can easily be spun up without impact to other components. Service orchestrators can scale up just the single VF that needs to handle a specific workload, instead of scaling up the entire VNF.

Network functions fall into three general categories: data plane, control plane, and management plane.

3.3.1.1 Data Plane

The forwarding plane (sometimes called the data plane) is defined as functions implemented in hardware, software, or both, which relate to packet handling of end-to-end user communications. The role of the forwarding plane in a router is to decide what to do with packets arriving on an inbound interface. Most commonly, the router maintains a table in which to look up the destination addresses of the incoming packets and retrieve the information necessary to determine the right outgoing interface to send the packet to.

Today's service provider networks need to support more than just basic packet forwarding. A large service provider maintains many separate routing domains for the different services and different customer networks it supports. Many service providers utilize MPLS technology to maintain separate forwarding planes for different services that they offer—Internet services, Ethernet services, and IP VPN services. To accomplish this, separate routing instances called VRF instances are implemented in a network device by maintaining distinct routing tables known as forwarding information bases (FIBs).

A MPLS-based VPN service will have the following basic security characteristics—derived directly from the strengths of MPLS and MPLS VPN-related standards (e.g., RFC 4364):

- *Containment*: Traffic (and routing information) sent between CE routers on the same VPN always stay within that specific VPN—no spillover or "leakage" can occur.
- *Isolation*: No customer's VPN can in any way materially affect or influence the content or privacy of another customer's VPN.
- *Availability*: Aside from the basic security-related attributes of MPLS and MPLS VPNs, service providers carefully engineer shared resources to meet the highest levels of availability and mitigate potential denial of service activities through additional methods such as access control lists (ACLs), route filters, turning off unnecessary services, and other infrastructure-hardening techniques.
- *Simplicity*: MPLS networks allow for simplified provisioning in both the customer and service provider domains (and hence can help to avoid security-related configuration mistakes). First, MPLS VPNs are much simpler for customers to configure than legacy layer 1 (e.g., private line), layer 2 (e.g., frame relay or ATM) point-to-point solutions, or layer 3 (e.g., Internet Protocol Security [IPSec] VPNs). Second, MPLS VPNs allow for much more scalable service provider architectures, unlike some other VPN solutions (e.g., L2TP) based on ACLs and separating customer address space. A service provider network using ACLs or separate IP spaces as the primary method to create VPN separation has a very difficult task to manage. In this scenario, every new site or route that is added can potentially require a change on every other router in the network to ensure security. This is not a scalable solution and can lead to errors in configuration and potential security breaches.

Lastly, additional requirements are imposed on the data plane to support traffic prioritization so that service-level agreements (SLAs) can be achieved for different voice, video, and data applications. This requires deep packet inspection (DPI) so that forwarding decisions and priorities can be determined by the application, to achieve quality of service (QoS) requirements. Similarly, rate limiting could be employed on interfaces to control the allowable traffic by class or application. These put greater stress on the CPU because packets need to be analyzed beyond the destination address field.

Basic packet forwarding makes heavy utilization of network I/O and memory read/write operations. When moving this to a server-based architecture, the I/O capabilities of the server can be the limiting factor in forwarding throughput. Moving these workloads to a server-based architecture can prove challenging if we continue to build routers as monolithic functions that need to support all these features. Many network workloads will require high data throughput that is hard to achieve on top of hypervisors and overlay networks where the network function does not have direct access to the network interface. Overlay routing or switching can impede the throughput a VF can achieve via virtual platform. Techniques such as single root I/O virtualization (SRIOV) (Intel) can be employed to bypass these overlays and instead directly pass data to the network interface card (NIC).

SRIOV is a network interface that allows a single physical peripheral component interconnect (PCI) express hardware interface to be shared by different VMs. Allocating a VF to VM instance enables network traffic to bypass the hypervisor and flow directly between the VF and the VM. This allows near line rate performance without dedicating a single physical NIC to each VM.

The downside is that this VNF is now bound to that specific server so the ability to move that function between servers, becomes more challenging. Over time, higher throughput will be achieved through overlay routers and switches when they leverage emerging APIs that provide higher data plane throughput. An example is the Data Plane Development Kit (DPDK),* which provides a set of libraries and drivers for fast packet processing.

3.3.1.2   Control Plane

The control plane is characterized by logic, which generally relates to handling of communications between network functions, and not directly related to end-to-end user communications. This includes signaling protocol handling, session management or authorization, authentication, and accounting (AAA) functions, and routing protocols. These functions are expected to have lower transaction rates with a commensurate increase in the complexity of processing required for each transaction, so network I/O is generally not expected to be as heavily utilized as for a data plane workload. Instead, CPU resources and/or memory resources are more likely to be bottlenecks.

The most common control plane workload is routing protocols such as open shortest path first (OSPF), BGP, LDP, etc. Routing protocols are employed to serve the data plane. Routing in IP networks generally uses distributed control planes, where each local router is responsible for creating its own forwarding tables based on the control plane data it receives from other routers or from centralized control plane functions such as RRs. AT&T's core network is based on MPLS technology. At a high level, MPLS networks use a combination of multiprotocol BGP and label switching for differentiating and segregating routing information and traffic flows between VPNs. RRs are used to distribute routes to the edges. Additional control plane elements exist, namely, DNS, DHCP, Signaling for VoIP, etc.

As we migrate control plane functions to the cloud, existing standalone control plane elements such as RR and DNS servers are the ideal first candidates since they are already disaggregated and perform a unique core function. PE routers for VPN services are more complex because their primary function is packet processing, yet they support many control plane protocols and workloads.

Currently, the routing and forwarding engines can be disaggregated and supported on different cores. But control plane protocols like BGP are still integrated into the PE instance. This will evolve over time so that the control and data planes can be completely disaggregated.

3.3.1.3   Management Plane

The management plane involves functions, such as orchestration, configuration management, fault and event reporting, and monitoring. These workloads require high resiliency and tend to be centralized. Generally, they support the operations, administration, and maintenance (OA&M) functions in the network. Historically, management plane workloads have been vendor proprietary for configuration, fault, and performance management. Even Simple Network Management Protocol (SNMP) (International), which is an open standard, is not interoperable across vendors. SNMP is an Internet-standard protocol for collecting and organizing information about managed IP devices.

Going forward, the goal is to move away from proprietary management protocols in favor of data model-driven architectures with open API interfaces. This will dramatically change the way service delivery and service assurance are accomplished across many disparate VFs. A significant shift is to move away from command line interface (CLI) code to support the configuration of network functions (e.g., routers) in the network. The industry is being driven to a common data model for key services such as Ethernet, utilizing a modeling language called Yet Another Next Generation (YANG) (IETF).

YANG is a standardized modeling language to define a service construct. For example, Ethernet can be defined by ports and VLANs. Ports are characterized by interface types, speed, tagging or not, etc. VLANs include speed, VLAN tag, QoS parameters, etc. This standard model can then be

---

* DPDK is a set of data plane libraries and network interface controller drivers for fast packet processing.

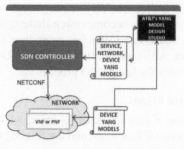

- YANG is a standardized modeling language
  - YANG = Yet Another Network Generator

- AT&T's YANG Model Design Studio helps build YANG Models efficiently

FIGURE 3.6 YANG and NETCONF.

pushed to a forwarding device for implementation. Figure 3.6 illustrates how YANG models can be designed and stored in a catalog. (AT&T has created a YANG model design studio to help a service designer create the YANG models for their service.) The model can then be pushed to the VF via a protocol called NETCONF (IETF). NETCONF is defined to install, manipulate, and delete the configuration of network devices via remote procedure calls (RPC).

YANG models are extensible to many software models for more abstracted layers of network functions. As Figure 3.7 depicts, we can use YANG to define the data model for a service at the device layer (e.g., single-forwarding device), network layer (e.g., VLANs across a metropolitan area network [MAN] or wide area network [WAN]), service layer (end-to-end service definition like Ethernet WAN), and finally the API layer, which defines the software interface we can expose to request the end-to-end service for a customer.

Two additional workload types can be found in VNF implementations, but to a lesser degree:

Signal processing workloads—defined as workloads related to digital signal processing (such as Fast Fourier Transform [FFT] decoding/encoding in a Cloud Radio Access Network [C-RAN] base band unit [BBU] [Jun Wu]). These are typically very CPU intensive and highly delay sensitive. In previous discussions, we have tended to characterize media transcoding as a signal processing workload, although given the definitions in this specification, it would appear that transcoding has more in common with data plane workloads involving encryption. It is expected to have increased CPU utilization compared to other data plane workloads, but is no more delay sensitive than general packet forwarding.

Storage workloads—defined as workloads involving disk storage, subdivided in the specification into intensive and nonintensive.

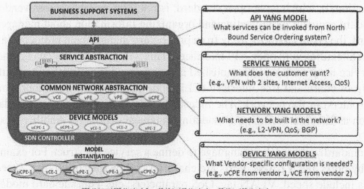

FIGURE 3.7 YANG abstraction layers.

The process of disaggregating a VNF is complex and a detailed discussion of all the engineering considerations and possible tradeoffs is beyond the scope of this chapter. At a high level, the process involves

- Identifying and classifying the external interfaces and protocols by workload.
- Separating the functions of the VNF around those interfaces and protocols.
- Identifying and separating any storage or management functions that are candidates for disaggregation.
- Identifying candidate functions for disaggregation for reasons other than separation of workloads, (e.g., separation of front-end processing from data storage for better fault handling, component reuse, reducing individual VNF complexity, …).
- Identifying suitable interfaces between candidate disaggregated functions.
- Evaluating whether the benefits of disaggregating each function outweigh the potential costs, considering factors such as
  - Efficiency tradeoff
  - Fault domains and fault handling
  - Management and orchestration complexity

3.3.2 SERVICE CHAINING

A network service might be composed of a source, a destination, and a set of intermediate interconnected VNFs that process the traffic from the source to the destination. This is commonly referred to as *service chaining* (Central). This chaining of VNFs might occur in a number of scenarios, for example,

- Chaining based on a network service designed according to a VNF forwarding graph (VNF-FG)
- Chaining based on customer policy/service

Several VNFs can be defined in a chain including host and middle box VNFs as defined in Section 3.1.3.

Middle boxes can alter the flow of traffic. For example, a firewall can drop packets or an application gateway can change the path based on the type of traffic. A NAT function can modify the packet. So a service chain replaces the traditional cabling that was used to interconnect functions, it does not necessarily dictate the end-to-end flow across all VFs for every packet flow.

Service chains are not a new concept; we have been chaining functions together in service provider networks for many years. The challenge in the existing architecture is that those chains were static and dictated by the physical cabling installed between the devices. The key difference now is that these chains can be defined in software and implemented using SDN capabilities within the NFVI. This allows for on-demand deployment of services where interconnection between multiple functions is needed to provide an end-to-end service. For example, consider the introduction of a security service into an enterprise VPN or extending a VPN into a cloud service provider. These are capabilities supported today with AT&T Flexware and Netbond services, respectively; this is covered in more detail in Chapter 10.

Figure 3.8 depicts an example of a service chain that can be defined through a series of physical functions.

When these functions become VFs on an NFVI, the connection between them can be defined in software and controlled by the SDN infrastructure. This is illustrated in Figure 3.9 and will be covered in more detail in Chapter 4.

Service chains require connectivity between functions that may be distributed on different network segments, different servers, and maybe even located at different physical locations. In order

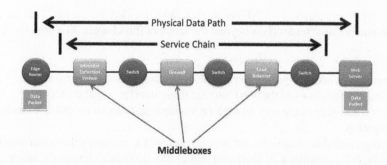

FIGURE 3.8 Physical service chain.

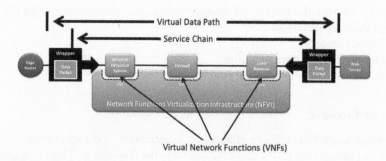

FIGURE 3.9 Service chain on NFVI.

to provide next-hop connectivity between functions, overlay networks are required. Traditional Ethernet, MPLS, and IP networks provide the foundational underlay of a service provider's network and it can be based on static configurations and static IP addressing. In order to connect two or more VNFs together, we need to use virtual addresses for the VNFs that can be defined in real time and a way to tunnel or encapsulate the traffic over the top of the underlay. This facilitates the decoupling of the network service from the underlying infrastructure. Traditionally, the overlay could be done with generic routing encapsulation (GRE) or encrypted IPSEC tunnels, but these require explicit configurations, typically via CLI code. VPN technologies like MPLS, provide a way to support the tunneling with routing protocols like BGP, but even these, traditionally, need to be explicitly configured. In an SDN environment, these need to be instantiated in real time through a centralized SDN controller. MPLS continues to be an appealing overlay technology because it does rely on dynamic routing protocols. This allows for the creation of resilient architectures. VNFs can be deployed in pairs with active–active or active–backup configurations where fast restorations and segment routing (IETF, Segment Routing) techniques can be employed.

3.3.3 Overlay, Underlay, and vSs/vRs

Figure 3.10 reintroduces a diagram from Section 3.1.3 to highlight the use of overlay networks (dashed line) on top of the data center fabric that supports the underlay. To support this on demand virtual network, a vS or vR is used on the server and it sits on the hypervisor to provide connectivity to specific VNFs running on VMs. OpenStack environments provide some basic overlay functionality, specifically single tenant VLAN support. This is defined in the OpenStack Neutron (OpenStack) specification. At the time of this writing, Neutron lacks advanced networking capabilities such as multitenant layer 2 and layer 3 IP service chains. It also lacks support for QoS that is needed to prioritize different traffic streams. In a service provider network, network functions, while virtual, can still be multitenant in that the function can support several virtual customer networks. Therefore, VLAN tagged Ethernet and layer 3 VPNs must be supported on the overlay to maintain

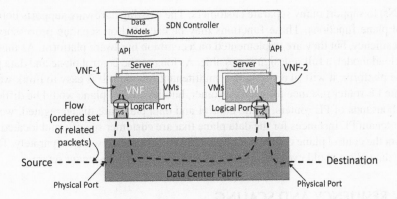

FIGURE 3.10 Overlay and underlay.

network separation even when attaching to a common and shared network function. OpenContrail (Apache) can be used as a complement to Neutron to support these more advanced network constructs. OpenContrail is an Apache project that is built using standard protocols like MPLS to provide the necessary components for advanced network virtualization. In a VMWare cloud, VMWare NSX (VMWare, VMWare NSX), can provide the overlay functionality and it is integrated with the hypervisor. NSX would be the common choice for a service provider or enterprise data center where VMWare is already being used to manage the compute and storage resources.

3.3.4 REUSABILITY

VNFs need to be reusable and location independent. Figure 3.11 illustrates network solutions whereby a VNF could be deployed at a customer premises (Model A), in the service provider's network on the access circuit supporting a single site (Model B), or centrally located in the service provider's network supporting multiple sites (Model C).

Consider the example of a firewall for an enterprise customer, who has an Internet connection to support Guest Wi-Fi, or employee Internet access. If this is done at a branch office, then a firewall is likely needed at the site or in the network supporting that Internet gateway and support all their sites via a larger network-based firewall. In all cases, a firewall needs to be orchestrated and managed on the service provider's network. This function needs to be stored in a catalog where it can be instantiated in the appropriate location to meet the customer requirements. This is covered in more detail in Chapter 9.

3.3.5 MULTI AND SINGLE TENANCY

Many existing network functions are multitenant by design. For example, a PE router supporting MPLS VPN services is multitenant; it supports many enterprise customers on a common platform. The existing PE router has many physical interfaces, typically Ethernet, that are virtualized

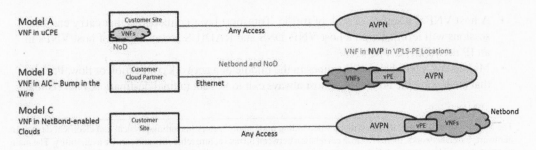

FIGURE 3.11 VNF deployment models.

(e.g., VLANs) to support many separate customers. The existing hardware supports both data plane and control plane functions. These functions may be executed on separate processors to support scale and resiliency, but they are implemented on a common hardware platform. As these functions move to a cloud model, a full redesign is possible. As long as the control plane and data plane reside on the same platform, it will likely remain a multitenant function. It is easy to think we could turn up a separate PE router instance for each customer, but the control plane would be difficult to scale if we had thousands of PE routers. If the control and data plane are disaggregated, we could consider single-tenant PE instances for the data plane that are customer specific and located close to the customer, yet the control plane could be multitenant and centralized scaled separately. Taking a step further, the data plane could even be distributed across many single-tenant servers.

3.4   NFV RESILIENCY AND SCALING

VNFs enable on-demand deployment and elastic growth of network services. The shift from a physical hardware-appliance-based architecture to a software-based, multivendor, and potentially open-source-based architecture creates many benefits (as outlined in Section 3.1) while also introducing new challenges. Delivering service continuity and predictable performance, on parity with today's carrier-grade network environment, will require a combination of better software engineering and an architecture that leverages a dynamic on-demand infrastructure. It will also require VNF-specific logic in some cases. As we will see in this section, both the Network Cloud platform and the specific VNF will need to be architected to achieve high performance and reliability.

Resiliency (Smith, 2011) is the property of a VNF that enables very high availability of the services. VNFs are complex software functions implemented on servers running Linux and an appropriate hypervisor. In the simplest form of implementation, VNF availability will be less than or equal to server availability (this is discussed in more detail in Chapter 5). When a server fails, all software tasks that run on that server will fail and need to be restarted. Servers can fail due to a number of reasons. The most common reasons are hardware failures, OS failures, hypervisor failures, and planned events for taking the server out of service for maintenance and upgrades. Server failures are infrequent events and can be modeled using an exponential distribution. Average server availability in a service provider data center is in the range of 99.9%–99.99% (50–500 unavailable minutes in a year). This range is derived from server mean time between failure (MTBF) of approximately 50,000 h and assumptions regarding restoration time in the event of failures.[*]

Service providers typically deliver 99.99%–99.999% availability for most network services. The fundamental challenge is to deliver very high availability of network functions that are running on data center infrastructure, which has significantly lower availability. This means that the VNF will need to continue operating in the event of a server failure. In theory, continuous operation can be achieved by running VFs on redundant servers organized as a high-availability cluster. Both VMWare and KVM have capabilities to organize a group of servers as a high-availability cluster.

For the purpose of understanding resiliency, we refer to the VNF classification from Section 3.1.3—host VNFs and middle box VNFs.

- A host VNF is a source or sink of traffic. Transport layer connections that carry end-user sessions will terminate on a host VNF. DNS and RADIUS are examples of host VNFs in an IP network.
- Middle box VNFs behave like pipes in the middle of a network connection or flow. Packets that enter a middle box VNF almost always exit to be sent to final destination. Packets that

---

[*] The MTBF and availability figures will vary depending upon many factors, including physical and electrical design. In hardware-based networks, there is a close correlation between infrastructure reliability and service availability. The main point of this section is to explain how software-centric networks incorporate the use of COTS hardware and physical design limitations and deliver high availability for services.

transit through a middle box VNF may undergo changes such as header transformation. Proxies often change HTTP header fields to ensure a desirable effect such as making a web page display on a mobile device more readable. Modification of time to live (TTL) or adding/removing an encapsulation header field by an edge router, are other examples of header modification by a VNF. Middle box VNFs have at least two interfaces. Connecting a trusted network (e.g., a private network) to an untrusted network (e.g., the Internet) is an example of the use of two interfaces. Connecting a customer network to the core network is another example. A middle box VNF always resides in between two (or more) networks that are reachable via the VNF. Therefore, routing packets through the middle box VNF is a fundamental attribute of such VNFs. See Section 3.1.3 for examples of middle box VNFs.

Resilient design of host VNFs is straightforward. The designer can generally rely on the under-lying cloud platform (server clusters and cloud software) to deliver high availability by relocat-ing the VNF to another server in the event of a server failure. VMWare's Distributed Resource Scheduler (VMWare, Introduction to VMware DRS and VMware HA Clusters, n.d.) and KVM's Live Migration (KVM, n.d.) are examples of cloud platform capabilities that are used by host VNFs in order to achieve high availability.

Resilient design of a middle-box VNF is a harder problem. Middle-box VNFs typically imple-ment stateful functions. VNFs need to maintain states needed for one or more of the following features:

- Routing
  - Static routes, default routes, BGP routes, OSPF routes
  - DHCP reservations
- QoS
  - Traffic shapers (e.g., Leaky buckets and peak rate shapers for enforcing SLAs)
- Statistics
  - Counters for tracking usage-based billing or OAM functions
- Connection state
  - Flow state (User datagram protocol [UDP] or TCP flows) for firewall or network address translation (NAT)
  - IPSEC sessions (credentials, session keys)
  - Compression state (block-level compression) for WAN acceleration

3.4.1 Multipath and Distributed VNF Designs

Middle box VNFs may have to support forwarding of packets over multiple network paths that exist between a source and destination. Such parallel paths are called equal-cost paths. The VNF needs to manage control information used for routing packets over multiple equal-cost paths.

Clouds enable resiliency and scale out by allowing VFs to be distributed across a cluster of servers so that the failure of a single instance of a VF does not interrupt service to end users. Managing certain networking states requires a single instantiation of the VNF and therefore may not easily support splitting and distributing the VNF function across multiple instantiations of the VNF. Peak-rate shaping is a good example of this constraint. Consider an end-to-end flow that requires peak-rate shaping feature. The flow is split into two or more load-balanced subflows that are processed by two more instantiations of the VNF over two separate paths (see Figure 3.12). The rate computation logic (e.g., leaky bucket algorithm) and packet processing operations (buffering, forwarding, discarding, etc.) will have to be split across two or more VNFs in real time in order to deliver end-to-end rate control for the aggregate flow. The two VNFs have to communicate and coordinate rate-control statistics and operations. The latency associated with such coordination will lead to inaccurate aggregate rate control.

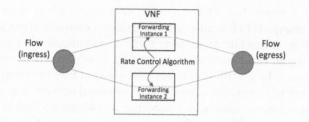

FIGURE 3.12 Problem of performing accurate rate control coordination across two forwarding VNF instances.

There are other less egregious examples of the difficulty of splitting VNFs as multiple instantiations. As stated earlier in this section, a network flow often uses multiple equal-cost paths. Counters and statistics at multiple points along the path of flow may need to be combined in order to meet OAM requirements. Accurate computations that are needed for OAM may be difficult due to synchronization and discontinuity issues. For example, interval statistics for two paths may be added together to compute the total number of packets transmitted for the flow during a given hour. Time-synchronization across the multiple counting tasks will not be precise. Another example is managing states of TCP sessions (NAT, firewall [FW], proxy) and point-to-point entities. If session management is split across multiple VNFs, state has to be copied across instances. If one instance fails, the end-to-end session will fail. This may actually result in lower resiliency.

Despite the complexities described above, operating multiple, parallel instances is the only way to achieve very high availability for middle-box VNFs. In most cases, an individual flow will be preserved over a single path with a single-forwarding instance of the VNF. Middle box VNFs will require application-specific logic in the VNF to resolve the complexities associated with distributing flows across multiple forwarding instances. The VNF design cannot simply rely on VM movement by means of cluster management implemented by cloud software. This is illustrated in the example of virtual provider edge (vPE) router. The vPE is a complex middle box VNF. It is widely used to provide CUG capability for many different services, implementing the VRF capability, described in Section 3.1. Enterprise VPN service, routing functions needed for mobile core and VoIP services rely heavily on vPE.

3.4.2 vPE Example for VNF-Specific Resiliency Design

vPE is a compound VNF that consists of multiple instances of at least two types of VNFs that are integrated into a coordinated system. vPE forwarding and control plane functions are separated as two distinct VNFs. A single vPE instance will consist of multiple forwarding plane VNFs and at least two control plane VNFs. Forwarding plane VNF instances enable scale by distributing customer interfaces and traffic flows across multiple VMs. Such a design enables scale out by allowing small subsets of flows belonging to the entire traffic flow to be implemented on separate forwarding VNF instances. As more customers are added to the vPE, new forwarding VNF instances can be added to accommodate growth. This also provides better resiliency by reducing the "blast radius"* in the event of a server failure. In order to accomplish this, the various forwarding VNF instances need to be distributed across multiple servers in the metro cluster. Customers may use single-homed or multihomed access links to connect to the network. Majority of the customers use single-homed access since access is usually a major element of the total cost of a customer's network. Buying a second access link for the purpose of high availability is expensive; customers choose dual-homed access only for larger locations with high uptime requirement.

---

* Blast radius is an informal term used to describe the extent of impact of a failure event. Network design attempts to keep blast radius small so that a failure event impacts the smallest possible number of users.

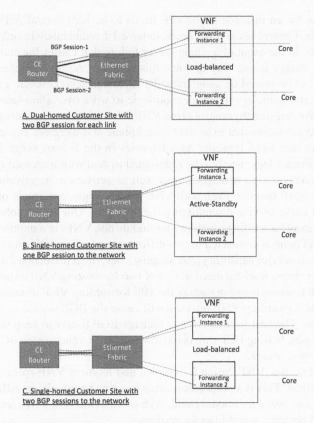

FIGURE 3.13 (A, B, and C): Distributed forwarding VNF instances.

Figure 3.13 shows three different design options for distributing forwarding plane functions over multiple instances of forwarding VNFs.

Using multiple instances of forwarding VNF is straightforward for a customer who has dual-homed access links. Each access link will connect to a separate instance of the forwarding VNF. BGP sessions will be established between customer router and each of the VNF instances, enabling a load-balanced design. If forwarding VNF instance 1 fails, traffic will migrate in a hitless manner, to forwarding VNF instance 2. This is depicted in Figure 3.13A.

Using multiple forwarding instances is more difficult in the case of a customer who has only a single-access link to the network. There are two design choices as depicted in Figure 3.13B and C. The first option is an active-standby design for the two forwarding VNF instances. The network cloud data center fabric will terminate the access line and manage two connections, one each to each of the two instances of forwarding VNFs. There will be only one BGP session between the CE router and forwarding VNF instance 1. If the forwarding VNF instance 1 fails, traffic flow will stop. A new BGP session will be established to forwarding VNF instance 2 and traffic flow will resume after a short duration outage. The second option is a load-balanced design where the CE router will use VLANs over the single-access link and maintain two separate BGP sessions to the two forwarding VNF instances, as depicted in Figure 3.13C. This allows a load-balanced design, allowing hitless failover of traffic in the event of a failure of either of the two forwarding VNF instances. This option will require the customer to implement a new design that is capable of taking advantage of load-balanced traffic forwarding in the cloud. Active-standby forwarder VNF instance design has disadvantages. First, the service behavior is not exactly the same as that in a single-homed design. Customers who use static routing will not be able to take advantage of the active-standby design. Finally, active-standby design is more expensive since vPE capacity will need to be reserved for the standby forwarding VNF instance.

The control plane for an instance of the vPE needs to be kept centralized due to the need to manage routing scale. Control plane VNF needs to have 1:1 redundancy in order to avoid the failure of the entire vPE, in the event of a single-server failure that causes the entire control plane to fail. Active-standby design is one way to solve this problem. Active- and standby-control plane VNF instances need to be placed on geographically separated servers inside a metro cluster. This would require the active-standby synchronization logic to work over a low-latency metro connection between the active and standby control plane VNFs. Traditional router software design assumes active and standby components that exist over a backplane and therefore expect very low latency (<100 µs). Metro-distance links typically have latencies in the 1–5 ms range. Hence, the control plane redundancy software logic needs to be redesigned to deal with increased latency.

Automation for configuration management as well as activation, deactivation, and movement of vPE will become VNF specific. Even with VNF-specific logic, movement of forwarding plane VNF instances will cause service discontinuity (i.e., outages). One of the solutions is to provide continuous (i.e., nonstop) forwarding capability for middle box VNFs in a multihomed design where a single customer end point is connected to two different forwarding VNF instances. The problem of managing quality-of-service capability (rate shaping), described in Section 3.4.1 forces an active-standby design rather than a load-balanced design of two forwarding VNF instances. This is due to the existence of BGP sessions between each of the vPE forwarding VNF instances and end customers. Movement of the forwarding VNF instance will cause the BGP session to reset, resulting in a short-duration outage. The best we can do is to manage BGP timers to keep the duration of such outages to 10 of seconds. Setting BGP timers too low can cause unnecessary BGP resets if a circuit has a very short duration outage.[*]

In summary, middle box VNF design is complex and requires VNF-specific logic in order to make the design resilient. This is a point that is often lost when network virtualization is compared to running Web services over a traditional cloud. Web services use host VNFs and therefore can rely entirely on the cloud platform capabilities for resiliency.

3.5   NFV ECONOMICS

Traditional networks follow a near-fixed-cost model. Network design utilizes dedicated hardware for each function, independently scaled at each location. Capacity is added incrementally, by adding plugs in deployed shelves, and since the time needed to enhance capacity by deploying new shelves can be substantial, there is a tendency to deploy more shelf capacity than needed. On the other hand, NFV allows a variable-cost model that has been widely supported in cloud architectures for IT workloads.

Service providers have been on an aggressive multiyear journey to migrate IT applications to the cloud. The size and complexity of this effort is one of the largest in the world. Migrating IT applications to a cloud infrastructure includes refactoring these applications to run natively on the cloud. There are significant savings in retiring legacy infrastructure and moving to an automated cloud. In addition, the agility provided by the cloud to scale in and scale out, chain services and use robust APIs facilitates flexibility. The migration to a cloud architecture meets the following high-level objectives:

- Increase utilization of IT assets by optimizing cloud footprint
- Reduce total cost of ownership (TCO) by reducing compute cores by greater than 50% through the virtualization of assets, and utilizing open source technologies and shared infrastructure and being able to oversubscribe CPU capacity
- Increase flexibility and improve the IT group's ability to respond to business demand by creating hardware and software independence

---

[*] It is possible to have a truly hitless operation by mirroring TCP/BGP states across the active and standby forwarding VNF instances. However, this will make the VNF design even more complex. Also TCP state mirroring in an expensive process.

- Software and tool standardization for those migrated applications driving lower costs and consistency among application teams

IT applications that provide the best return on investment are identified as the top candidates for migration to the cloud. Each application is analyzed to understand its needs based on the current location, system requirements, and changes required to migrate to standard operating environments like Linux.

Some applications cannot be moved to the cloud, at least initially, based on three main reasons:

- Database requirements for high I/O, Oracle RAC, or Veritas Clustering (requires capability enablement)
- Software requirements from vendors that do not provide virtualized options (requires vendors to change)
- Networking requirements for low latency performance (requires capability enablement)

In each case listed above, dedicated hardware configurations continue to be used for these applications. As there are improvements made to hardware and software solutions, more applications will be migrated to the cloud. It is easy to see that such an evolution can be adopted for network functions. Like IT applications, there will be an evolution of the types of functions that can migrate to the cloud. It will be a continuous journey as the platforms mature, higher data planes are achieved, and vendors evolve their software methodology to align with a distributed cloud model.

Automation of infrastructure management is critical to the success of open source cloud solution deployments because the cloud solution is expected to reduce operational costs, while delivering critical business services. To assess the viability of moving network functions to the Network Cloud, one needs to look independently at the hardware, software, and operational costs.

3.5.1 HARDWARE COSTS

The move to commodity hardware should greatly reduce the overall cost of equipment. The rationale for the cost savings is twofold. First, you remove the costs associated with custom application specific integrated circuits (ASICs) used in the physical implementation of a network function. Readily available merchant silicon chip sets can be used on commodity server hardware and support many different VNFs (this is covered in more detail in Chapter 12). Second, and more importantly, the capacity management for hardware is done once across the many VNFs or workloads. This allows one to oversubscribe the hardware and gain economies of scale efficiencies that the sharing of a common platform can provide.

3.5.2 SOFTWARE COSTS

The costs for software needs to include the cost of the actual VNF software plus the recurring licensing fees for each function. Most VNF vendors will look to recoup their revenue loses (attributed to commodity hardware replacing custom hardware) with higher costs for software and licenses. The move to software also drives more competition among vendors and allows new entrants to emerge. This competition will drive down software costs. Lastly, this paradigm shift to software has caused the creation of new open-source communities focused on different functional areas (covered in more detail in Chapter 17).

3.5.3 OPERATIONAL COSTS

The transition to a cloud architecture probably has the highest cost benefit when it comes to network operations. Service delivery and assurance can now be done through centralized software. This creates more agility in managing operations procedures, and also reduces the time to market for

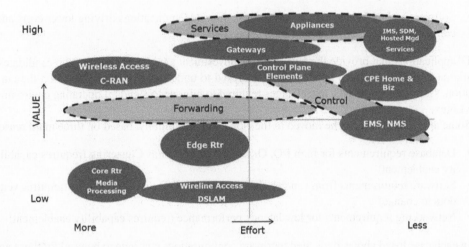

FIGURE 3.14   VNF zones of advantage.

introduction of new services. It also enables services to be delivered globally or targeted at specific locations, without the need to ship and deploy new hardware. This lowers the risk for deploying new services at scale, because the cost to shut down service, once it is deployed, is *de minimis* since the deployed hardware can be repurposed for other services.

Figure 3.14 answers the question as to which VNFs to deploy on the cloud. At a high level, core network elements that require very high data plane support and see low churn in the network, are likely to initially remain on dedicated hardware platforms. There is little value in virtualizing these elements and meeting the high data plane requirements is difficult on an x86 server platform. There is great value in virtualizing elements that are closer to the edge and provide specific service features to customers. These are typically appliances. For example, firewalls that provide a security service or WAN acceleration devices that provide caching and prepositioning of common data elements. Control plane elements like RRs, DNS are also great candidates since they can be centralized and do not require a very high data plane requirement. Edge routing functions, such as PE routers are also likely to be virtualized.

3.6   NFV BEST PRACTICES

The Network Cloud will provide a base set of capabilities that can be used by VFs running on a service provider's network. The network cloud will provide and support the host, host OS, OpenStack, KVM hypervisor, and vSwitch/vRouter. It will provide an environment that allows each VM to run on its own partition, isolated from other VMs, on one physical server. The AT&T Open Network Automation Platform (ONAP) framework, described in Chapter 7, provides the service orchestration and VF instantiation and configuration capabilities and it provides configuration management at instantiation as well as when in service. It collects fault, performance, application KPIs, and probe data from VFs and makes decisions to perform auto-recovery or auto-scaling of VFs based on collected data and vendor-provided engineering rules. It will trigger dynamic instantiation of new VFs for auto-recovery or auto-scaling as needed in conjunction with the service orchestrator. The network is primarily IPv6 based but is able to interact with IPv4-based components.

The network cloud supports a variety of VFs with varying performance needs. Response time characteristics of supported VFs, can be categorized as follows:

- *Real time* is measured in milliseconds/microseconds and is the responsiveness required to support, for example, the low timing threshold for session initiation protocol (SIP) queries and responses

- *Near-real time* is a response measured in seconds
- *Nonreal time* is a response measured in minutes, hours, or days

To ensure efficient VNF interoperation, and since VNFs can be sourced from multiple vendors, AT&T has compiled the following best practices for VNF design:

- VFs must be agnostic to the details of the cloud platform (e.g., hardware, host OS, hypervisor) and must run on a shared standard cloud with acknowledgment to the paradigm that the cloud platform will continue to rapidly evolve and the underlying components of the platform will change regularly.
- The VF design must use cloud-based paradigms to enable standardization of technology, scalability, and reliability.
- Decomposition of network functions must be supported.
- The ability to reuse VFs must be supported so that services can be created rapidly by chaining the VFs based on service needs.
- Long-lived state and end-user (subscriber/customer) data should be decoupled from processing logic.
- Geo-resiliency and the ability to deploy with local and geo-redundancy should be supported for VFs.
- Common platform solutions (e.g., cloud-based load-balancers, databases, resiliency solutions, etc.) instead of vendor-proprietary solutions, should be supported where possible.
- Standardized mechanisms for fault configuration, accounting, performance, and security (FCAPS) functions must be supported by VFs.
- The VF design should meet the resiliency, availability, and performance (e.g., real-time response) requirements of the service.
- Transition to a cloud-based design should be transparent to the end user.
- VFs must be able to be instantiated and controlled via ONAP functions. A VF or its component should not interact directly with the OpenStack infrastructure.
- Open and standard APIs must be supported. Idempotent interfaces should be implemented in all possible cases.
- VFs should run without modifications on standard guest OS images. Vendor-provided guest OS images should be supported with restrictions.

ACKNOWLEDGMENT

The authors would like to acknowledge Chris Chase for his contribution of ideas to this chapter.

REFERENCES

Anderson, L. 2006. Framework for Layer 2 Virtual Private Networks (L2VPNs), RFC 4664. Retrieved from IETF: https://www.ietf.org/rfc/rfc4664.txt
Apache. n.d. OpenContrail. www.opencontrail.org
Central, S. n.d. Service Chaining. https://www.sdxcentral.com/sdn/network-virtualization/definitions/what-is-network-service-chaining/
Cisco. n.d. LISP Overview. http://lisp.cisco.com/lisp_over.html
Comer, D. E. 2014. *Computer Networks and Internets*. Upper Saddle River, New Jersey: Pearson Education, Inc. ISBN: 978-0133587937.
Foundation, O. N. n.d. https://www.opennetworking.org/sdn-resources/openflow
IETF. n.d. NETCONF. https://tools.ietf.org/html/rfc6241
IETF. n.d. Segment Routing. https://tools.ietf.org/html/draft-ietf-spring-segment-routing-04
IETF. n.d. YANG. https://tools.ietf.org/html/rfc6020

Intel. n.d. SR-IOV. http://www.intel.com/content/www/us/en/pci-express/pci-sig-sr-iov-primer-sr-iov-technology-paper.html

International, S. R. n.d. SNMP. http://www.snmp.com/

Jun Wu, Z. Z. n.d. Cloud Radio Access Network (C-RAN): A Primer. http://www.ntu.edu.sg/home/ygwen/Paper/WZHW-WCM-15.pdf

KVM. n.d. *Live Migration*. Retrieved from http://www.linux-kvm.org/page/Migration

Openstack. n.d. Neutron. http://docs.openstack.org/developer/neutron/

Rekhter, K. K. 2007. *Virtual Private LAN Service (VPLS) Using BGP for Auto-Discovery and Signaling (RFC 4761)*. Retrieved from IETF: https://tools.ietf.org/html/rfc4761

Rosen, E. 2006. *BGP/MPLS IP Virtual Private Networks (VPNs), RFCs 4364 and 2547*. Retrieved from https://tools.ietf.org/html/rfc4364

Rosenblum, M. 2004. The reincarnation of virtual machines. *ACM Queue*. July–August. 2(5): 34–40.

Sajassi, A. 2015. *BGP MPLS-Based Ethernet VPN*. Retrieved from IETF: https://tools.ietf.org/html/rfc7432

Silberschatz, A., Galvin, P. B., and Gange, G. 2012. *Operating System Concepts*. Hoboken, New Jersey: John Wiley and Sons. ISBN: 978-0470128725.

Simpson, W. 1995. *IP in IP Tunneling, RFC 1853*. Retrieved from https://tools.ietf.org/html/rfc1853

Smith, P. 2011. Network resilience: A systematic approach. *IEEE Communications Magazine*. 49(7): 88–97.

Spurgeon, C. E. and Zimmerman, J. 2000. *Ethernet: The Definitive Guide: Designing and Managing Local Area Networks*, 2nd Edition. Sebastopol, California: O'Reily Media. ISBN: 978-1449361846.

VMWare. n.d. *Introduction to VMware DRS and VMware HA Clusters*. Retrieved from VMWare vSphere 5 Documentation Center: https://pubs.vmware.com/vsphere-55/index.jsp

VMWare. n.d. VMWare NSX. http://www.vmware.com/products/nsx.html

4 Network Functions Virtualization Infrastructure

Greg Stiegler and John DeCastra

CONTENTS

4.1   Network Functions Virtualization Infrastructure 50
4.2   Components of NFVI 51
    4.2.1   Physical Components 51
        4.2.1.1   Compute: Server, Computer Processing Unit, Graphical Processing
                  Unit, Network Interface Controller, I/O 52
        4.2.1.2   Network 52
        4.2.1.3   Storage 52
    4.2.2   Virtual Infrastructure Manager 53
    4.2.3   VIM Solutions 53
        4.2.3.1   Commercial 53
        4.2.3.2   Commercial Open Source 53
        4.2.3.3   Pure Open Source 54
    4.2.4   VIM Components 55
        4.2.4.1   Virtualization 56
        4.2.4.2   Virtual Machine 56
        4.2.4.3   Containers 56
        4.2.4.4   Hypervisor 57
        4.2.4.5   Storage Virtualization 57
    4.2.5   Orchestrator 58
4.3   Building an NFVI Solution 58
    4.3.1   Operational Changes 58
        4.3.1.1   Agile 58
        4.3.1.2   DevOps 59
        4.3.1.3   Continuous Integration/Continuous Delivery 59
    4.3.2   Innovation and Integration 59
        4.3.2.1   Design 59
        4.3.2.2   Build 60
        4.3.2.3   Manage 60
        4.3.2.4   Integrate 60
        4.3.2.5   Operate 60
4.4   NFVI Deployment 61
    4.4.1   Meeting Zone Demand 61
    4.4.2   Fault Tolerance 61
    4.4.3   Infrastructure Resiliency 62
    4.4.4   Application Resiliency 62
4.5   Leveraging NFVI for VNFs 63
    4.5.1   VNF Performance Profiles 63
    4.5.2   Scalability 64
    4.5.3   VNF Management 65

4.6   Summary 65
References...66

4.1   NETWORK FUNCTIONS VIRTUALIZATION INFRASTRUCTURE

The use of network functions virtualization (NFV) for the design and implementation of a telecom network requires a distributed cloud infrastructure on which the various virtualized network functions (VNFs) run. Chapter 3 described the design and operating constraints of VNFs. As a refresher, a VNF is a virtualized task formerly carried out by proprietary, dedicated hardware. VNFs move network functions out of dedicated hardware devices and into software. This allows specific functions that required hardware devices in the past to operate on standard servers (1, n.d.).

This chapter explores the distributed cloud infrastructure needed to operate the VNFs, and implement an end-to-end network function, specifically an infrastructure capable of running network workloads. The infrastructure is generically referred to as network functions virtualization infrastructure (NFVI). NFV seeks to provide economies of scale, thus NVFI is grounded in widely available and low-unit-cost, standardized computing components. This is critical for service providers to remain competitive by keeping infrastructure overhead costs to a minimum.

Integrating NFVI with existing capabilities allows businesses to bring new products and services to market faster than ever before. This additional agility and flexibility is an enormous competitive advantage over traditional, siloed, enterprises. An example of this new paradigm is a software-defined offering like AT&T's Network on Demand (NOD) solution. Where a customer previously had to phone in and request changes, with NOD they are given the flexibility to change their services on demand.

Early cloud implementations were disparate, implemented primarily to accommodate differing client needs. Building, managing, and maintaining multiple cloud platforms and solutions is costly and cumbersome, thus, a strong industry trend has begun to consolidate cloud platforms into a single, common cloud platform that can provide all the compute needs of both internal and external customers. This then begs the question—can the same platform be adapted to perform the functions needed by NFVI?

This was the approach that AT&T took with the creation of the AT&T Integrated Cloud (AIC), a cloud capable of performing network workloads and having a unified management plane. AIC is AT&T's globally distributed cloud platform that comprises a single, integrated codebase, used to support both enterprise and network workloads. AIC is deployed in dozens of geographically distributed locations, each known as an "AIC site." Within each site, typically located in a city, resides one or more "AIC zones." An AIC zone is the term used to describe the collective set of AIC hardware that is managed as a single unit within an AIC site.

At the heart of each zone is a collective set of hardware that creates the shared infrastructure platform for the hosting tenant of application virtual machines (VMs), virtual networks, and virtual storage used to provide carrier-grade enterprise and network workloads. The zones may have different reference models or cloud offerings. For example, datacenter, mobility, and compact.

- *Datacenter*—used to automate typical manual data center operations
- *Mobility*—similar to a "datacenter," but the support is focused on mobility-related network functions
- *Compact*—provides network functions to customers in various market areas and is generally a central office or another location near the customer location

AIC zones and reference models are built using a local control plane (LCP) and a distributed control plane (DCP). Each LCP is local to the AIC site. Within the LCP is a set of control servers, VMs, and control applications, including OpenStack (virtual infrastructure manager [VIM]), ONAP (orchestrator), etc., as well as the managed components, such as, servers, switches, and storage. There are no proximity requirements between LCPs and the managed components. For instance, two compute servers that are in the same rack and connected to the same leaf switches, can be managed by different LCPs. LCPs provide a variety of functions:

- API management
- Orchestration
- Image management
- Automation
- Authentication

The DCP is a collection of orchestration management servers that control or provide services to multiple AIC locations. Redundant DCPs host centralized components of AIC that are integrated with back office systems to provide information technology (IT) provisioning. The DCP provides a variety of functions, including, but not limited to

- Monitoring
- Automation
- Service orchestration
- LCP resource management

With the aforementioned AIC as a reference throughout, this chapter explains how NFVI can be utilized to form a common, distributed cloud platform that is responsible for enterprise workloads, carrier grade network workloads, VNFs, and VNF performance profiles.

4.2 COMPONENTS OF NFVI

NFVI is a distributed data center with many local sites. At a high level, NFVI is a set of physical and virtual resources that are used to host and connect virtual functions (2, n.d.). Beyond the physical components of the infrastructure, there are two key components of the NFVI—NFV orchestrator and VIMs (3, n.d.). The framework for NFVI is depicted in Figure 4.1.

4.2.1 Physical Components

As technology continues to evolve and new virtual technologies emerge, the foundation of any network infrastructure is still reliant on its physical resources. To that point, NFVI is no different in

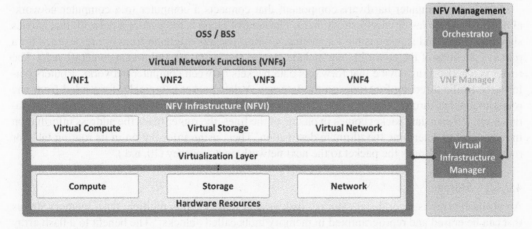

FIGURE 4.1 NFVI framework. (33, n.d. Retrieved from https://www.sdxcentral.com/wp-content/uploads/2014/09/opnfv-nfv-open-source-network-functions-virtualization-diagram.jpg)

that it depends on physical components to provide the basis. For the sake of simplicity, we will break the components into three main categories—compute, network, and storage.

4.2.1.1 Compute: Server, Computer Processing Unit, Graphical Processing Unit, Network Interface Controller, I/O

Computing components provide the functionality of the other programs or devices within the network. The most commonly known compute resource is a server. Servers can provide various functionalities, often called "services," such as sharing data or resources among multiple clients, or performing a computation for a client. A single server can serve multiple clients, and a single client can use multiple servers. A client process may run on the same device or may connect over a network to a server on a different device. Typical servers are database servers, file servers, mail servers, print servers, web servers, game servers, and application servers (4, n.d.). Computer processing units (CPUs) and graphical processing units (GPUs) are two of the main components that allow servers to process data and move it across the network. A CPU is the electronic circuitry within a computer that carries out the instructions of a computer program by performing the basic arithmetic, logical, and control operations specified by the instructions (5, n.d.). A GPU is similar to a CPU but is a specialized electronic circuit designed to rapidly manipulate and alter memory to accelerate the creation of images in a frame buffer intended for output to a display. GPUs are used in embedded systems, mobile phones, personal computers, workstations, and game consoles. Modern GPUs are very efficient at manipulating computer graphics and image processing, and their highly parallel structure makes them more efficient than general-purpose CPUs for algorithms where the processing of large blocks of data is done in parallel (6, n.d.). Together, the CPU and GPU work to provide the powerhouse behind the server that communicates with another component or a person in a process known as input/output, or I/O. Common I/O devices used by humans to communicate with computers or servers are keyboards, a mouse, monitors, or printers. Common I/O devices used for communication between computers are modems and network interface controllers (NICs).

4.2.1.2 Network

Networking hardware, also known as network equipment or computer networking devices, are physical devices which are required for communication and interaction between devices on a computer network. Specifically, they mediate data in a computer network (7, n.d.). There are many variations and types of networking hardware and configurations; here we will focus on NICs, control planes, and routers.

A NIC is a computer hardware component that connects a computer to a computer network, either by using cables or wirelessly (8, n.d.). The control plane is the part of a network that carries signaling traffic and is responsible for routing. Control packets originate from or are destined for a router. Functions of the control plane include system configuration and management (9, n.d.). The router is a networking device that forwards data packets between computer networks. Routers perform the traffic directing functions on the Internet. A data packet is typically forwarded from one router to another through the networks that constitute the Internet work until it reaches its destination node. When a data packet comes in on one of the lines, the router reads the address information in the packet to determine the ultimate destination. Then, using information in its routing table or routing policy, it directs the packet to the next network on its journey (10, n.d.).

4.2.1.3 Storage

An all-flash array is a solid-state storage disk system that contains multiple flash memory drives that can be erased and reprogrammed in memory units called "blocks." The benefit to a flash array is that it can transfer data to and from solid-state drives at a higher rate than traditional disk drives because they do not have to spin up additional hard disk drives at each time of need.

Block storage is a type of data storage typically used in storage-area network (SAN) environments where data are stored in volumes, also referred to as blocks. Each block acts as an individual hard drive and is configured by the storage administrator. Because the volumes are treated as individual hard disks, block storage works well for storing a variety of applications such as file systems and databases. While block storage devices tend to be more complex and expensive than file storage, they also tend to be more flexible and provide better performance (11, n.d.).

Object storage, also called object-based storage, is a generic term that describes an approach to addressing and manipulating discrete units of storage called objects. Like files, objects contain data—but unlike files, objects are not organized in a hierarchy. Every object exists at the same level in a flat address space called a storage pool and one object cannot be placed inside another object (12, n.d.). Both files and objects have metadata associated with the data they contain, but objects are characterized by their extended metadata. Each object is assigned a unique identifier, which allows a server or end user to retrieve the object without needing to know the physical location of the data.

4.2.2 VIRTUAL INFRASTRUCTURE MANAGER

One of the most important and influential pieces of the NFVI is the VIM. The VIM is responsible for managing the virtualized infrastructure of an NFV-based solution. VIM operations include the following:

1. It keeps an inventory of the allocation of virtual resources to physical resources. This allows the VIM to orchestrate the allocation, upgrade, release, and reclamation of NFVI resources and optimize their use (13, n.d.).
2. It supports the management of VNF forwarding graphs by organizing virtual links, networks, subnets, and ports. The VIM also manages security group policies to ensure access control (13, n.d.).
3. It manages a repository of NFVI hardware resources (compute, storage, and networking) and software resources (hypervisors), along with the discovery of the capabilities and features to optimize the use of such resources (13, n.d.).

The VIM performs other functions as well—such as collecting performance and fault information via notifications; managing software images (add, delete, update, query, and copy). In summary, the VIM is the management glue between hardware and software in the NFV world (13, n.d.).

4.2.3 VIM SOLUTIONS

When choosing a VIM, the market provides three categorical solutions —commercial, commercial open source, and pure open source. Each solution has its own benefits and shortfalls and should be vetted based on the desired outcomes of the enterprise or the service provider.

4.2.3.1 Commercial

Commercial vendors provide proprietary solutions to VIM demand. These proprietary solutions help customers transform existing IT infrastructures into private clouds, but at the cost of losing the ability to customize the environment to suit their needs. The most notable current vendor in the commercial VIM space is VMware with their vSphere product.

4.2.3.2 Commercial Open Source

Commercial open source is a platform that is based on an open-source platform such as Apache CloudStack or OpenStack, but relies on a third party to provide an enterprise/proprietary edition or extension for the right to use software, at a cost. The proprietary nature of the software often means that the vendor is in control. As the user, you can be placed in a position where you may not have access to public documentation such as transparency to bugs, or to source code, and when source code is provided, it may not provide the test frameworks to validate production readiness. Notable current vendors in the commercial open-source space are Red Hat, VMware, and Oracle.

4.2.3.3 Pure Open Source

Pure open source as a VIM solution meets three defining characteristics—no license cost, a vibrant open foundation community, and an accessible source code that is readily available for download from the sponsoring organization. Rather than charging for the software itself, vendors can provide quality software with no license cost and then secure support arrangements and professional service assistance to generate profit. With a large, pure open-source community, you create a forum where individuals in different companies can share knowledge, bounce ideas off of one another, and spread support around more widely across a larger base. This results in faster introduction of features, maintenance of code, and quicker turnaround time on bug fixes. Also, depending on the scale of implementation, with pure open source, everyone has the ability to introduce upgrades, changes, or new ideas all together that your company can benefit directly. A current OpenStack example is shown in Figure 4.2.

Making a decision of this magnitude is not straightforward and requires many internal discussions to decide which solution is needed. Many chose to go the pure open-source route to prevent vendor lock-in while also providing a resilient cloud platform across the globe. The identification of this target state platform would enable a pivot from longstanding use of proprietary solutions to an open-source software-based environment. For the core of the NFVI platform, AT&T choose OpenStack as the de facto standard for building and automating AIC. OpenStack is one of the best-known, pure open-source solutions in the world, with support from more than 500 companies and 30,000+ individual members located in 170 counties, and with over 20 million lines of code written to date (14, n.d.) (15, n.d.).

The use of open-source products to reduce cost, while allowing for customer internal tool creation, means that new partnerships are needed with experienced third parties. Credible third parties are required to serve the industry as integrators of the open-source code, and to execute the required changes.

For instance, some organizations have contributed deployment automation capabilities to the open-source community that allow placement of OpenStack nodes on multiple racks that span multiple network segments. This capability provides greater flexibility and scalability required for larger OpenStack deployments. Operational features such as these are key to maturing any new technology. Additional functionality, such as Neutron, the project that provides networking-as-a-service

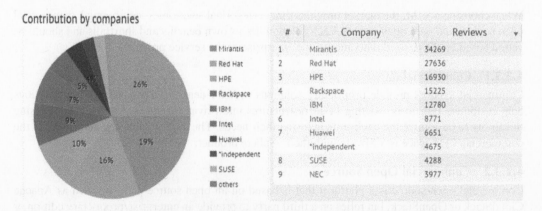

FIGURE 4.2 Example open-source contributions. (34, n.d. Retrieved from http://stackalytics.com/)

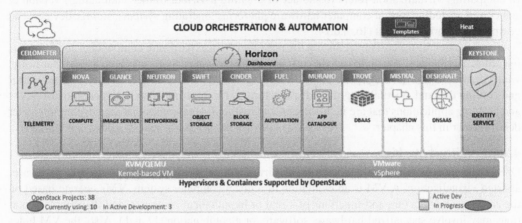

FIGURE 4.3 OpenStack components currently used in AIC.

between interface devices (virtual network interface cards or vNICs) that are managed by other OpenStack services, are being contributed to operationally mature OpenStack.

The OpenStack "projects" typically used in an AIC production platform are as follows:

- *Heat*—provides orchestration services for multiple composite cloud applications
- *Horizon*—provides a modular web-based user interface (UI) for OpenStack services
- *Ceilometer*—provides a single point of contact for analytic and billing systems
- *Nova*—provides VMs upon demand
- *Glance*—provides a catalog and repository for virtual disk images
- *Neutron*—provides network connectivity-as-a-service between interface devices managed by OpenStack services
- *Swift*—provides a scalable storage system that supports object storage
- *Cinder*—provides persistent block storage to guest VMs
- *Fuel*—to streamline and accelerate the process of deploying, testing, and maintaining various configuration of OpenStack at scale
- *Murano*—a standard set of software stacks to support application and infrastructure requirements
- *Keystone*—provides authentication and authorization for all the OpenStack services (Figure 4.3)

4.2.4 VIM COMPONENTS

The VIM is responsible for controlling and managing the NFVI compute, storage, and network resources, usually within one operator's infrastructure domain. These functional blocks help standardize the functions of virtual networking to increase interoperability of software-defined networking elements (13, n.d.).

Enterprise utilization of a VIM allows organizations to have an open, flexible, and modular architecture that serves the business purposes of scaling to meet the explosive demand at lower cost, while increasing the speed of feature delivery and providing much greater agility. With a VIM, providers can rapidly make changes to the platform and extend real-time management of the service to the customer through digital portals, such as dedicated Internet on-demand bandwidth adjustments. This is made possible by the practice of virtualization.

4.2.4.1   Virtualization

In computing, virtualization refers to the act of creating a virtual (rather than actual) version of something, including virtual computer hardware platforms, operating systems, storage devices, and computer network resources (16, n.d.).

Virtualization limits costs by reducing the need for physical hardware components and improves operational efficiency. Virtualizing the infrastructure provides more efficient use of hardware, which in turn, lowers the magnitude of hardware, the related maintenance costs, and reduces power and cooling demand. Virtualizing the infrastructure is also a pivotal step in the creation of VNFs and building applications that are fault-tolerant and resilient—two topics that are covered in greater detail further in this chapter.

4.2.4.2   Virtual Machine

One of the most well-known applications of virtualization is when a VM is used in hardware-assisted virtualization. A VM is an emulation of a given computer system. VMs operate based on the computer architecture and functions of a real or hypothetical computer, and their implementations may involve specialized hardware, software, or a combination (17, n.d.). Also, the VM hides the server resources from the end users, including the number and identity of individual physical servers, processors, and operating systems. The server administrator can divide one physical server into multiple isolated virtual environments using a software application, which results in an increase in utilization of assets through the reduction of compute cores.

Utilizing VMs in hardware-assisted virtualization allows a transition from purpose-built network appliances to open, "white box" commodity hardware that is virtualized and controlled within AIC. It also allows the liberation of network functions from the same purpose-built "black box" appliances into stand-alone software components and manages the full lifecycle of these virtualized network functions within AIC, working in conjunction with local and DCP software controllers.

This is the primary approach to virtualization because it enables efficient and full virtualization with help from hardware components such as host processors to simulate the entire hardware environment. Each VM guest runs on a virtual imitation of its surrogate hardware. This allows the guest operating system to operate without any needed changes or modifications. It also allows the administrator to create guests that use different operating systems. The guest has no knowledge of the host's operating system. The untouched guest operating system executes in complete isolation, using an identical instruction set used by the host server. It does, however, require real computing resources from the host—so it uses the hypervisor to manage and direct instructions to the underlying CPU.

4.2.4.3   Containers

Operating system-level virtualization is a server virtualization technique where the kernel of an operating system allows the existence of multiple isolated user-space instances, instead of just one. Such instances, are typically referred to as containers or virtualization engines. In this approach, the operating system's kernel runs on the hardware node with several isolated guest VMs installed on top of it. The isolated guests are called containers. From the end users' point-of-view, these containers give the appearance and feel of a real server.

With container-based virtualization, there is not the overhead associated with having each guest run a completely installed operating system. This approach can also improve performance because there is just one operating system taking care of hardware calls. A disadvantage of container-based virtualization, however, is that each guest must use the same operating system the host uses.

Typically, corporate environments avoid container-based virtualization, preferring hypervisors and the option of having many operating systems. A container-based virtual environment, however, is an ideal choice for hosting providers who need an efficient and secure way to offer operating systems for customers to run services on (18, n.d.).

The vision being that containers could be used as a replacement to a full stack that is defined with a HEAT template in OpenStack. This makes it possible for developers to drop their applications into the container—test, run, and deploy with the same configurations. Furthermore, deploying containers on bare metal and controlling the environments with OpenStack would allow a service provider to treat the containers just as they treat cloud assets.

4.2.4.4   Hypervisor

A hypervisor is a program that allows multiple operating systems to share a single hardware host. Each operating system appears to have the host's processor, memory, and other resources all to itself. However, the hypervisor is actually controlling the host processor and resources, allocating what is needed to each operating system in turn and making sure that the guest operating systems (VMs) cannot disrupt each other (19, n.d.).

For the open AIC, the hypervisor of choice is KVM running on VMs because it is viewed as the industry's leading open-source hypervisor, particularly among users of OpenStack.

4.2.4.5   Storage Virtualization

A SAN is a dedicated high-speed network (or subnetwork) that interconnects and presents shared pools of storage devices to multiple servers (20, n.d.). SANs are primarily used to enhance storage devices, such as disk arrays, tape libraries, and optical jukeboxes, accessible to servers so that the devices appear to the operating system as locally attached devices. A SAN typically has its own network of storage devices that are generally not accessible through the local area network (LAN) by other devices (21, n.d.).

Storage virtualization is the pooling of physical storage from multiple network storage devices into what appears to be a single storage device that is managed from a central console. Storage virtualization helps the storage administrator perform the tasks of backup, archiving, and recovery more easily—and in less time—by disguising the actual complexity of a SAN (22, n.d.).

Many service providers will also deploy Ceph, an open-source software solution to provide distributed object store and file-based storage designed to provide excellent performance, reliability, and scalability. Ceph provides object storage and block storage under a unified solution.

Listed below are some benefits of NFVI storage virtualization for service providers and end users alike.

4.2.4.5.1   Data Migration

One of the major benefits of abstracting the host or server from the actual storage is the ability to migrate data while maintaining concurrent I/O access. The host only knows about the logical disk (the mapped logical unit number [LUN]) and so any changes to the metadata mapping is transparent to the host. This means the actual data can be moved or replicated to another physical location without affecting the operation of any client. When the data have been copied or moved, the metadata can simply be updated to point to the new location, therefore, freeing up the physical storage at the old location.

4.2.4.5.2   Improved Utilization

Utilization can be increased by virtue of the pooling, migration, and thin provisioning services. This allows users to avoid over-buying and over-provisioning storage solutions. In other words, this kind of utilization through a shared pool of storage can be easily and quickly allocated as it is needed to avoid constraints on storage capacity that often hinder application performance.

4.2.4.5.3   Fewer Points of Management

With storage virtualization, multiple independent storage devices, even if scattered across a network, appear to be a single monolithic storage device and can be managed centrally (23, n.d.). With the VIM solution OpenStack, the Swift and Cinder projects assist in the virtualization of storage needs. The OpenStack Object Store project, known as Swift, offers cloud storage software so that you can store and retrieve lots of data with a simple API. It is built for scale and optimized for durability, availability, and concurrency across the entire data set. Swift is ideal for storing unstructured data that can grow without bound (24, n.d.). Cinder is a block storage service for OpenStack. The short description of Cinder is that it virtualizes the management of block storage devices and provides end users with a self-service API to request and consume those resources without requiring any knowledge of where their storage is actually deployed or on what type of device (25, n.d.).

4.2.5 Orchestrator

Network functions virtualization orchestration is used to coordinate the resources and networks needed to set up cloud-based services and applications. This process uses a variety of virtualization software and industry standard hardware (26, n.d.). Cloud service providers often use NFV orchestration to rapidly deploy services, or VNFs, using cloud software rather than specialized hardware networks.

Resource orchestration is important to ensure there are adequate compute, storage, and network resources available to provide a network service. To meet that objective, the NFV orchestrator can work either with the VIM or directly with NFVI resources, depending on the requirements. It has the ability to coordinate, authorize, release, and engage NFVI resources independently of any specific VIM. It also provides governance of VNF instances sharing resources of the NFVI (27, n.d.).

Some of the functions that are typically performed by an NFV orchestrator are

Service coordination and instantiation: The orchestration software must communicate with the underlying NFV platform to instantiate a service, which means it creates the virtual instance of a service on the platform.

Service chaining: Enables a service to be cloned and multiplied to scale for either a single customer or many customers.

Scaling services: When more services are added, finding and managing sufficient resources to deliver the service.

Service monitoring: Tracks the performance of the platform and resources to make sure they are adequate to provide for good service (26, n.d.).

In many service provider networks, Open Network Automation Platform (ONAP) serves as OpenStack's orchestration solution for network cloud services. ONAP provides high utilization of network resources by combining dynamic, policy-enforced functions for component and workload shaping, placement, execution, and administration. These functions are built into the ONAP platform and utilize the network cloud. When combined, these provide a unique level of operational and administrative capabilities for workloads that run natively within the ecosystem.

4.3 BUILDING AN NFVI SOLUTION

In order to fully realize the lower cost and increased flexibility of NFVI and internal cloud networking, fundamental operational changes are necessary and a set of new strategic philosophies should be put in place to successfully build and deploy an NFVI solution.

4.3.1 Operational Changes

4.3.1.1 Agile

Agile software development describes a set of principles for software development under which requirements and solutions evolve through the collaborative effort of self-organizing cross-functional teams. It promotes adaptive planning, evolutionary development, early delivery, and continuous improvement, and it encourages rapid and flexible response to change. These principles support the definition and continuing evolution of many software development methods (28, n.d.). The collaborative nature of Agile can pay large dividends in a relatively short amount of time. It has been stated publicly that the closing of OpenStack operational gaps with custom extensions and automation was due in large part to the use of an Agile structure.

4.3.1.2 DevOps

"A set of guiding principles which promotes software excellence throughout its life. Barriers between software delivery disciplines are broken down to provide seamless transitions accelerating time to market while improving quality."

DevOps (a clipped compound of development and operations) is a culture, movement, or practice that emphasizes the collaboration and communication of both software developers and other IT professionals while automating the process of software delivery and infrastructure changes (29, n.d.). The usage of DevOps allows for cross-functional collaboration between operational teams, developers, quality assurance, and management teams while prioritizing operational needs earlier in the scoping process. It also establishes an environment where building, testing, and releasing software happens rapidly, frequently, and more reliably, with operations providing new features' requirements approval to ensure that ample operational support is provided at the least cost.

4.3.1.3 Continuous Integration/Continuous Delivery

Continuous integration/continuous delivery (CI/CD) is an amalgamation of two software engineering practices that aim to automate the building, testing, and releasing of software. Together, they go a long way in creating a more agile service provider.

The goal of CI is to provide rapid feedback so that if a defect is introduced into the code base, it can be identified and corrected as soon as possible (30, n.d.). With CI, iterative software changes are immediately tested and added to the larger code base. If a change creates issues, such as runaway storage use, those problem can be isolated and rolled back quickly (31, n.d.). When used properly, CI provides various benefits, such as constant feedback on the status of the software. Because CI detects deficiencies early on in development, defects are typically smaller, less complex, and easier to resolve (30, n.d.).

CD is an extension of the concept of CI. Whereas CI deals with the build/test part of the development cycle for each version, CD focuses on what happens with a committed change after that point. With CD, commits that passes the automated tests can be considered a valid candidate for release (32, n.d.).

CI/CD allows service providers to manage the delivery and meet timelines of their massive network cloud platform. CI/CD allows hundreds of developers to pull code down, to make changes, and to check the code back in. While doing this, additional controls around human reviews and approval are automated, while automated integration, deployment, and testing takes place simultaneously. This allows service providers to move very fast to meet the needs of a constantly evolving space, and continuously improve the quality of their deliverables.

4.3.2 INNOVATION AND INTEGRATION

When choosing an NFVI solution, particularly one that is open source, it is important to know that the service that you purchase will not be ready to go, "out-of-the-box." Every organization or service provider has unique needs, and simply put, no existing solution can solve everything. That is why it is important to emphasize innovation and integration when building NFVI. Utilizing Agile, DevOps, and CI/CD principles, each respective organization can create in-house extensions that tie the existing enterprise infrastructure into the VIM and NFVI.

4.3.2.1 Design

The network cloud infrastructure design tools can automate the design of zones, cutting the time for this phase by more than half, when compared to early manual designs. Some tools allow for template-based hardware allocation and rack arrangement, automated naming, automated IP address allocation, and automated cabling instructions, among others things.

4.3.2.2 Build

Once automating the design is complete, a network cloud service provider needs to effectively and efficiently virtualize OpenStack, enable the control plane to be dynamic, and reduce its physical footprint. This leads to the creation of virtual LCP (vLCP). With vLCP, one can build and automate under the cloud that is entirely virtualized, so that the OpenStack control plane runs inside VMs. This allows the operator to be able to easily move around, snapshot, rollback, and redeploy or upgrade AIC zones when needed.

4.3.2.3 Manage

Once built, AIC needs to be managed at scale. By engineering "day 2" operations into fuel, and the automation framework, capabilities ranging from minor setting changes, to deploying new plugins and enabling new features across 100 s of AIC zones, are rapidly enabled.

4.3.2.4 Integrate

In order to fully and effectively transform OpenStack environments into a Network Cloud, a few key gaps need to be filled. A VNF automation system is required to build carrier grade network functions and integrate with OpenStack Neutron APIs. Capabilities to enable and manage high-end communication services across the wide area network and integrate them with existing systems while fully automating the lifecycle of the VNF in a network cloud ecosystem, are also required. This VNF automation system is created via ONAP (see Chapter 7). ONAP provides a design time and execution time framework that tightly integrates all of the systems within the Network Cloud. ONAP is intentionally separated from AIC to create a layered software approach. ONAP provides the full lifecycle of VNF automation to an AIC, as well as enables and manages the wide area network.

4.3.2.5 Operate

The massive scale of a service provider's zone deployments calls for a management framework. This is provided by the OpenStack resource manager (ORM). The ORM framework

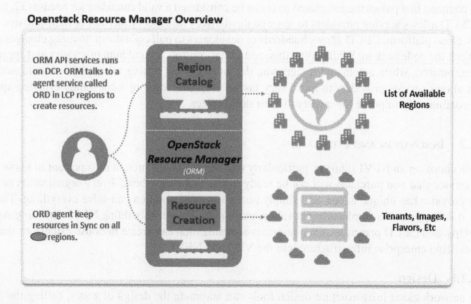

FIGURE 4.4 OpenStack resource manager.

is a collection of services on a DCP, acting as a single point of entry that provides two key functions:

- A resource creation gateway that provides APIs for the definition of new images, performance profiles, and user accounts, and then distributes those to the requisite AIC zones
- A region discovery service that returns zones that are best suited to support the tenant's needs at the time of request

A great analogy for ORM is an online hotel reservation system. By inputting the selection criteria, the service analyzes live information on thousands of hotels and ultimately provides a room recommendation. ORM essentially provides the same value for an AIC. ORM is also important because it can be used to quickly push out updates to all the AIC zones and to manage the distribution of workloads that run on the AICs. ORM is depicted in Figure 4.4.

4.4 NFVI DEPLOYMENT

This section explores the strategy and the methods for deploying cloud zones. From a business perspective, each zone location is fundamentally determined by two factors—customer demand and internal use. From a technology perspective, each zone is located based on proximity to those customers so that latency is minimized. In addition, telecom operators require carrier grade solutions that provide resiliency and fault tolerance.

4.4.1 Meeting Zone Demand

An interzonal architecture facilitates the flow of traffic from site to site. The interzonal architecture refers to functions such as network signaling/controlling, transport connections, and traffic forwarding that takes place between cloud sites, generally over some distance. This makes it possible to scale out across sites and regions instead of scaling up in a single location. Scaling out results in significantly lower costs and higher availability.

To meet the reliability demands of a carrier grade network, the AIC infrastructure management (IM) domain manages all interzonal networking. IM ties all nodes, LCP systems and applications, hypervisors, and software-defined networking (SDN)-local agents together, irrespective of their location or connectivity, into a single management plane with uniform security and access to other networks.

Chapter 5 discusses various architectures for achieving availability targets.

4.4.2 Fault Tolerance

Many organizations and enterprises want "cloud-native" VNFs, operations support systems (OSS), and business support systems (BSS). Cloud-native is a term that describes the design of systems and applications specifically for the cloud. These applications can realize numerous benefits that are enabled by the cloud environment, including improved reliability of service, simplified IT management, and strategic advantages related to service delivery. In fact, cloud-native applications are the driving force behind an organization's ability to serve customers globally, operate at an industry leading cost structure, and deliver an effortless customer experience. What is essential to understand about cloud-native applications is that they are built to expect failure. This concept introduces the need for fault tolerance.

Fault tolerance is what enables a system to continue its intended processes when some part of the system fails. At the point that an application or the infrastructure fails, the fault tolerance strategy and network design minimizes the impact of the failure by sharing the workloads across sites to achieve improved reliability (see Figure 4.5).

The fault tolerance strategy revolves around two key components—"infrastructure resiliency" and "application resiliency." These two components are the defining features of any AIC fault-tolerant building blocks (FTBBs).

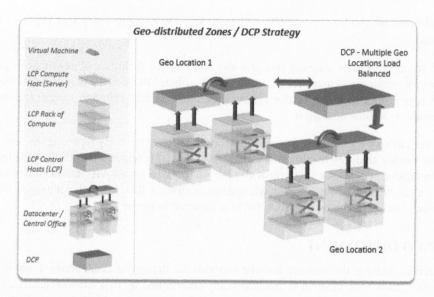

FIGURE 4.5   Fault-tolerant design of AIC zones.

4.4.3 INFRASTRUCTURE RESILIENCY

As mentioned previously, the geographic dispersal of cloud zones is essential to provide the resiliency needed to guarantee a constantly accessible network. This is done in large part to have the physical infrastructure deployed in close proximity to customers, to ensure best network experience and lowest latency.

Aside from the geographical locations of the NFVI zones, the network design and physical infrastructure itself plays a large part in creating a resilient infrastructure. This starts with the use of fault-tolerant hardware, that is, redundant network cabling from the servers to the switches, dual power supply for the servers' multiple NIC cards, and Redundant array of independent disks (RAID) internal drive configuration in the servers. RAID is a method of pairing multiple drives together virtually, to improve performance, redundancy, or both.

It is critical to have an "active–active" VNF configuration across zones. It is even better to have an active–active–active VNF configuration across zones. Active–active configuration uses data replication built into the infrastructure to increase resiliency. The data are written to a data store and automatically replicated to two other geographic locations so that it is simultaneously live in both zones and minimal latency would exist with the failover of single zone.

A service provider can also deploy the core infrastructure software in a fault-tolerant manner. For example, OpenStack components, such as Nova, Cinder, Swift, Keystone, and Neutron, are deployed on different servers in the event that either the server or software component fails. In addition, other software such as network load balancers and continuous monitoring-high available (CM-HA) systems monitor critical VMs and re-instantiate them on other hardware, if the hardware fails.

4.4.4 APPLICATION RESILIENCY

NFVI can be utilized by multiple applications and VNFs. These applications and VNFs do not need to acquire their own dedicated infrastructure. Once the infrastructure has been built in a resilient manner, it is paramount that applications also be created in a manner that allows for unobstructed performance and resiliency. Below are some application resiliency practices to consider when creating fault-tolerant NFVI applications:

1. *Service decomposition.* Structure the service as a collection of loosely coupled virtual functions, each of which realizes a well-defined portion of the final service. It should be possible to deploy and manage them independently of other virtual functions.
2. *State management.* Explicitly consider how to manage state, and in particular, how to externalize state for those virtual functions that are not stateless.
3. *Monitoring and control.* Cloud-based applications can use scale-out/scale-in as a fundamental technique for handling failures, as well as to deal with both long- and short-term load variations and platform upgrade impact mitigation. Having the ability to monitor and control service virtual functions is key to realizing this vision.
4. *Fault-tolerant features.* Fault-tolerant features such as live migration, clone, and snapshot are used by tenants to increase resiliency in their specific applications.

4.5 LEVERAGING NFVI FOR VNFS

As stated at the beginning of this chapter, NFVI represents the assets that form platforms for supporting the NFV and hosting VNFs. VNFs provide varying types of services to the enterprise; as a result, the configuration of one VNF may not meet the requirements of another VNF. These configurations are commonly referred to as "performance profiles" when speaking in generalities. The specifics of performance profile configurations and naming conventions are unique to the service provider or organization that they belong to.

4.5.1 VNF Performance Profiles

VNF performance profiles are introduced to provide carrier grade capabilities, for example, memory intensive, high I/O, graphics intensive for different throughput profiles based upon network functions. This is done by utilizing "Accelerated Data Path" methods such as single root input/output virtualization (SR-IOV) and Data Plane Development Kit (DPDK) that enhance networking, and processing efficiency technologies such as CPU pinning.

SR-IOV enables multiple VMs to share the same peripheral component interconnect express (PCIe) device, and presents the network interface card (NIC) virtually to the VM. This performance profile enables network traffic to bypass hypervisors and virtual switches to achieve low latency and improvements in packet processing time.

DPDK is a framework for fast packet processing that is performed by abstracting the network interface access. This is called environment abstraction layer (EAL) and provides a standard access to the network interface card, maximizing the network card features such as hardware acceleration. Another way that DPDK increases packet processing performance is the change from interrupt processing to polling processing, which allows the processor to operate more streamlined by not being interrupted every time a packet comes off the wire. Without the use of polling, the networking interface card sends an interrupt request (IRQ) to the CPU to request processing time. The CPU stops what it is currently doing, and allows the interrupt handler to process the interrupts based on the IRQ priority.

CPU pinning allows the packet processing to be handled by a specific CPU and is also known as processor affinity. Overcoming the hurdles of using general purpose hardware can be done by optimizing the hardware and software to focus more on processing packets. This involves assigning dedicated CPU cores to process packets through the use of CPU pinning and implementing more optimized memory management solutions. In normal processing, CPU pinning is not used, as the processing moves between the CPU cores. The processing of moving a workload from one CPU to another is costly, as CPU cache will have to be populated with data when processing moves from one CPU to another, which leads to reduced efficiency. It also mitigates the default settings for interrupt processing—in this case, CPU0 handles all interrupts in Linux before to kernel 2.6. After 2.6, the interrupts are randomly assigned, which can cause contention between shared resources, as

Flavor Series		Description
General purpose	GV	General Purpose Host supporting Kernel vRouter
Network Optimized	NV	Network-optimized hosts supporting Kernel vRouter
	NS	Network-optimized supporting SRIOV Data traffic for Guest VMs
	ND	Network-optimized host supporting DPDK vRouter

FIGURE 4.6   Example flavor series.

well as drive up the cache miss rate. The primary benefit of CPU pinning is increased processing efficiency since the traffic is being handled by the same CPU, and resolves these issues.

Service providers can categorize general purpose VNFs along with the optimization methods mentioned above, in the VNF performance profiles (named, *Flavor Series*). This is depicted in Figure 4.6. The flavor series classification is based on the specific parameter (ram, disk, ephemeral disk, swap disk) and optional parameter (bandwidth, processor) requirements of the VNF.

4.5.2   SCALABILITY

For scaling VNFs, there are two main scenarios—horizontal or vertical (also referred to as out or up). Scaling out generally refers to adding more zones to the network to distribute the processing and thus being able to handle a larger load. Scaling up refers to each zone having more power to process a larger load.

Both scaling out and scaling up are difficult to do efficiently and effectively, but typically scaling up is the more cost-effective approach. Arriving at a balance between horizontal and vertical scaling is a key component of cost efficiency, as neither horizontal nor vertical scaling is linear. There is a sweet spot range for hardware cost, where the cost of adding additional general purpose hardware is preferable to acquiring more powerful hardware. When scaling horizontally, the more nodes in the network, the more control nodes that are required, so finding the balance between power and quantity is a key cost-efficiency driver.

Scaling up involves using more powerful hardware, additional hardware, and optimizing the hardware.

Using more powerful hardware is fairly simple, adding more and faster memory, CPU cores, network interface cards, and storage. There are limitations on how far you can push adding additional hardware—physical limitations of space in the machine, power requirements, heat handling, and other factors that limit how much one system can do. The cost benefit of using general purpose hardware also decreases if each systems is loaded with maximum CPU, memory, network interface cards, and storage.

Scaling the hardware vertically can be done by simply adding more resources, or optimizing the resources already there. One way of optimizing the resources that are there is using "Accelerated Data Paths" for packet processing.

Adding additional processing zones to the virtualized carrier network, to scale out, promises to have an almost unlimited scale, as adding additional compute nodes is technically all that is required. While that sounds simple, there are limitations to this, including the management of the nodes.

Scaling out VNFs and providing resiliency can be done in a stateless or stateful manner. Stateful VNFs track and document the current traffic flows that they are processing. It then communicates these flows in the case of an additional VNF instantiation, either because of a failure, or because additional processing is required. On the other hand, the stateless VNF assumes that it is acceptable to initialize the service at the time that the additional processing or resiliency is required.

Another important consideration when scaling is the use of microservices. Microservices deconstruct an application into a series of smaller, separately deployable units. When discrete parts of a solution are separately deployed, they can also be separately scaled. This allows for a solution to run multiple instances of one service while only running a single instance of another, which is critical to flexibility and elasticity to meet surges in demand.

VNF management needs to support continuous tuning to optimize utilization and changing traffic flows. VNFs can be instantiated on demand, or leveraged out of a pool, depending on the requirements of resiliency and recovery targets. Pool-based VNFs can provide faster recovery and higher resiliency as the VNFs are already running. However, the VNFs consume resources even when not used. The on-demand VNFs are instantiated at the time of need, and when not needed, they are disposed of, which means that the VNFs only consume resources when they are needed.

4.5.3 VNF MANAGEMENT

It is important to be efficient and cost effective when deploying clouds locally or across the globe; a homogenous approach is needed.

Limiting the variances in configuration is important for adopting a homogenous approach; however, when existing performance profiles (Flavor Series) do not meet the needs of the customer, concessions and exceptions must be made. This is another important function of having a strategic governance mechanism. For example, exceptions are managed and allow for updating existing Flavor Series as appropriate and allow for temporary Flavor Series exceptions. When the desired feature becomes part of a standard Flavor Series, the exception is revoked and the standard Flavor Series is enforced.

4.6 SUMMARY

The migration to a Network Cloud has, at its core, the deployment of a NFVI. For telecom operators, leveraging any experience with the virtualization of IT infrastructure for internal and external business applications may be helpful to avoid the creation of bifurcated cloud offerings that are built for specific business purposes. NFVI depends on physical components to provide a platform to operate VNFs. The three main physical components of NFVI are compute, network, and storage. Beyond the physical components of NFVI, a complete NFVI solution also needs an NFV orchestrator and a set of VIMs. Service providers interested in open source options may use ONAP as the NFV orchestration component of NFVI; many network clouds use OpenStack, an open-source cloud solution, as the underlying technology for the VIM.

Fundamental operational changes and new strategic philosophies need to be established to successfully build and deploy an NFVI solution that lowers cost and increases flexibility. Implementing Agile development principles and a DevOps methodology to emphasize collaboration between software developers and other IT professionals to automate software delivery processes and infrastructure changes helps in keeping up with rapid technology changes.

NFVI solution deployment strategies should provide the resiliency needed to guarantee a constantly accessible network. NFVI deployment requires the physical infrastructure to be deployed in close proximity to customers to ensure the best network experience with the lowest latency. The VNFs are also deployed with an active–active–active configuration across zones to provide enhanced resiliency.

Near real-time network performance expectations require NFVI solutions to include scaling options that deliver carrier grade performance. To facilitate this, some service providers have defined standard NFVI configurations called "Flavor Series" to provide enhanced performance for

VNFs. Example Flavor Series include SR-IOV and DPDK and CPU pinning techniques to deliver the throughput needed for carrier grade networks.

REFERENCES

1. n.d. Retrieved from http://searchsdn.techtarget.com/definition/virtual-network-functions
2. n.d. Retrieved from http://searchsdn.techtarget.com/tip/NFV-Infrastructure-What-really-lies-beneath
3. n.d. Retrieved from http://searchsdn.techtarget.com/tip/NFV-Infrastructure-What-really-lies-beneath
4. n.d. Retrieved from https://en.wikipedia.org/wiki/Server_(computing)
5. n.d. Retrieved from https://en.wikipedia.org/wiki/Central_processing_unit
6. n.d. Retrieved from https://en.wikipedia.org/wiki/Graphics_processing_unit
7. n.d. Retrieved from https://en.wikipedia.org/wiki/Networking_hardware
8. n.d. Retrieved from https://en.wikipedia.org/wiki/Network_interface_controller
9. n.d. Retrieved from http://searchsdn.techtarget.com/definition/control-plane-CP
10. n.d. Retrieved from https://en.wikipedia.org/wiki/Router_(computing)
11. n.d. Retrieved from http://searchstorage.techtarget.com/definition/block-storage
12. n.d. Retrieved from http://searchstorage.techtarget.com/definition/object-storage
13. n.d. Retrieved from https://www.sdxcentral.com/nfv/definitions/virtualized-infrastructure-manager-vim-definition/
14. n.d. Retrieved from https://en.wikipedia.org/wiki/OpenStack
15. n.d. Retrieved from https://www.openstack.org/foundation/
16. n.d. Retrieved from https://en.wikipedia.org/wiki/Virtualization
17. n.d. Retrieved from https://en.wikipedia.org/wiki/Virtual_machine
18. n.d. Retrieved from http://searchservervirtualization.techtarget.com/definition/container-based-virtualization-operating-system-level-virtualization
19. n.d. Retrieved from http://searchservervirtualization.techtarget.com/definition/hypervisor
20. n.d. Retrieved from http://searchstorage.techtarget.com/definition/storage-area-network-SAN
21. n.d. Retrieved from https://en.wikipedia.org/wiki/Storage_area_network
22. n.d. Retrieved from http://searchstorage.techtarget.com/definition/storage-virtualization
23. n.d. Retrieved from https://en.wikipedia.org/wiki/Storage_virtualization
24. n.d. Retrieved from https://wiki.openstack.org/wiki/Swift
25. n.d. Retrieved from https://wiki.openstack.org/wiki/Cinder
26. n.d. Retrieved from https://www.sdxcentral.com/nfv/definitions/what-is-nfv-orchestration/
27. n.d. Retrieved from https://www.sdxcentral.com/nfv/definitions/nfv-orchestrator-nfvo-definition/
28. n.d. Retrieved from https://en.wikipedia.org/wiki/Agile_software_development
29. n.d. Retrieved from https://en.wikipedia.org/wiki/DevOps
30. n.d. Retrieved from http://searchsoftwarequality.techtarget.com/definition/continuous-integration
31. n.d. Retrieved from http://searchdatacenter.techtarget.com/photostory/4500255104/Data-center-terminology-that-will-get-you-hired/3/Define-continuous-integration-and-delivery-as-Dev-meets-Ops
32. n.d. Retrieved from http://searchitoperations.techtarget.com/definition/continuous-delivery-CD
33. n.d. Retrieved from https://www.sdxcentral.com/wp-content/uploads/2014/09/opnfv-nfv-open-source-network-functions-virtualization-diagram.jpg
34. n.d. Retrieved from http://stackalytics.com/

5 Architecting the Network Cloud for High Availability

Kathleen Meier-Hellstern, Kenichi Futamura,
Carolyn Johnson, and Paul Reeser

CONTENTS

5.1   Network Cloud Infrastructure Availability 69
    5.1.1   Single-Site Availability..69
    5.1.2   Availability to Cost Tradeoff (Bottleneck Analysis) 71
5.2   The Impact of Planned Downtime and Georedundancy 72
    5.2.1   Illustrative Example of the Impacts of Planned Downtime 72
    5.2.2   Best Design Practices for Minimizing the Impacts of Planned Downtime 74
5.3   VF Software Design 75
    5.3.1   Fault Tolerant VM Designs..76
        5.3.1.1   Hot VM Replication...76
        5.3.1.2   Warm VM Replication...77
        5.3.1.3   Passive VM Replication...77
    5.3.2   Low Software Failure Rates and Accurate Fault Detection 77
        5.3.2.1   VF Protection from External Services 77
        5.3.2.2   VF Error Handling...77
        5.3.2.3   VF Software Fault Detection 77
        5.3.2.4   VF Failure Detection and Alerting...78
        5.3.2.5   VF Fast Recovery...78
        5.3.2.6   VF Software Stability 78
    5.3.3   Software Resiliency Engineering 78
5.4   Putting It All Together: VF Classification and Examples 79
    5.4.1   Example: Stateful Network Access Services...79
    5.4.2   Example: Layer 4 Stateful Control Functions 81
    5.4.3   Example: Stateless Network Function with Multisite Design 82
5.5   Areas for Further Study...83
Acknowledgments..84
References...84

Driven by the need to maintain the reliability of the nation's critical communications infrastructure, the telecommunications industry has a long tradition of designing for high availability, sometimes referred to as "carrier grade availability" or loosely quantified as "five-nines availability" (Lancaster, 1986). Significant service disruptions must be reported to the FCC (Healy, 2016) by wireline, wireless, paging, cable, satellite, Voice over Internet Protocol (VoIP), and Signaling System 7 service providers. Communications providers must also report information regarding communications disruptions affecting Enhanced 9-1-1 facilities and airports that meet the thresholds in the FCC's rules. Detailed best practices for maintaining network continuity have been developed,[*] and switching, transport, and

---

[*] http://www.corp.att.com/ndr/pdf/cpi_5181.pdf

network infrastructures have been designed to meet these very stringent requirements. High availability has been attained using custom hardware, fault-tolerant software design, and extensive testing to ensure that the production software has few defects (Giloth, 1987). Excellent examples of highly reliable designs date back to digital switching systems such as the 4ESS (K. E. Martersteck, 1981) and 5ESS (Martersteck and Spencer, 1985), and networks have been designed with survivability as a core objective (Wu, 1992; Krishnan, 1994; Choudhury, 2004; Klincewicz, 2013; Ramesh Govindan, 2016).

As the industry evolved to IP and Voice over IP technologies, switches and routers continued to be designed with specialized hardware and software that allowed them to continue to meet "carrier grade availability" requirements (Johnson, 2004). The introduction of network functions virtualization (NFV), combined with the deployment of network functions on cloud-based platforms, has had a disruptive impact on traditional high availability hardware-based solutions. With NFV, failure of an individual hardware element may have little or no impact on service availability since virtualized network functions can be quickly (even automatically) restarted on spare cloud hardware, often within seconds or minutes. NFV is leading to significant changes in the way traditional vendor resiliency requirements are defined, measured, monitored, and managed (Rackspace, 2015). Some requirements that made sense in the traditional monolithic environment no longer apply, while others need to be redefined or renormalized. Once the appropriate NFV-based resiliency requirements are defined, the measurement and monitoring architecture must change to accommodate the new cloud-based delivery platform.

The design challenge for NFV is to create a five-nines carrier grade application using commercial cloud technology that is designed to operate in the 3–9s (99.9%) availability range (Amazon Web Services, 2013; Hoff, 2015; Butler, 2016; Taylor, 2016). In this chapter, a comprehensive, multilayer view demonstrating how carrier grade availability can be achieved in this environment is developed. Both the Network Cloud infrastructure as well as the application architecture of the virtual functions (VFs) are considered. As shown in Figure 5.1, the Network Cloud infrastructure includes the physical infrastructure (servers, network, storage, and physical plant) and the virtualization infrastructure (hypervisor and virtual machine [VM]). Availability tradeoffs among the different infrastructure components are addressed, as are the impacts of planned downtime and georedundancy. Next, the critical attributes of high availability VF designs are defined. Achieving

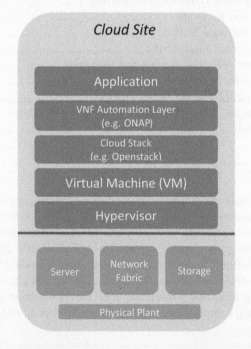

FIGURE 5.1   Layers of a network cloud.

5–9s carrier grade availability requires that VF designers use resilient design patterns that include spatial redundancy with redundant hardware, redundant information with redundant data structures, and temporal redundancy with redundant computation. Finally, the principles of high availability designs are demonstrated for different types of VFs, showing how high availability can be achieved.

For the purposes of this discussion, it is important to distinguish between availability, reliability, and resiliency. *Availability* is defined as the time-based probability that the "entity" is operational, where the entity is a hardware or software element, cloud site, functional capability, application VF, etc. Both control plane (provisioning) and service data plane (VF reachability) availability need to be considered. Downtime in minutes/year is calculated as 525,600*(1—availability). Availability and downtime can be either unplanned (due to hardware failures, software faults, etc.) or planned (due to maintenance, upgrades, etc.). *Reliability* is defined as the conditional time-dependent probability that the entity is still operational at time *t* given that it was operational at time 0. Mean-time-between-failures (MTBF) is commonly used as a simple surrogate for reliability. *Resiliency,* as used in this discussion, is the ability of application software to recover from certain types of failures and yet remain functional from the customer perspective to achieve the goals of availability. Both reliability and resiliency items addressed in this chapter are intended to focus on areas that will benefit and support application availability.

In addition to these *time-based* metrics, *transaction-based* metrics are critical in providing the full reliability picture. *Service reliability* (Tortorella, 2005a,b) provides a framework to characterize user-perceived system behavior by defining and measuring transactional defects (e.g., dropped transactions, lost sessions in progress, excessive delays, timeouts, etc.). Transactional defects fall into three categories: accessibility (*can I initiate my transaction?*), continuity (*can I complete my transaction?*), and fulfillment (*does my transaction complete in a timely manner?*). Transaction defects are typically tracked and reported in terms of defects per million (DPMs) transactions. While critical to understanding the complete picture, transaction-based service reliability metrics (as well as performance metrics) are not addressed in this chapter. This chapter also does not address *performability*, the joint analysis of performance and reliability (Wirth, 1988), or concepts such as "blast radius" (i.e., the magnitude of a given failure) (Miller, 2015) that further emphasize the need for resilient VF designs.

There are many other considerations in resiliency engineering that are not covered in depth in this chapter (Lyu, 1996; Musa, 1999). Those fall under the broader topic of software reliability engineering and include topics such as software stability testing, fault insertion testing, measurement, and fault detection.

Some key principles that are illustrated in this chapter are

- Single-site availability can be expected to be in the 3–9s to 4–9s range
- VF availability in a single site cannot exceed the site availability
- High availability can be achieved through multisite VF design
- Planned downtime further impacts availability and reinforces the need for multisite design
- High availability requires mature, fault-tolerant software
- Accurate fault detection and fast failover are critical to achieving high availability

5.1 NETWORK CLOUD INFRASTRUCTURE AVAILABILITY

5.1.1 Single-Site Availability

Within a single Network Cloud site (Figure 5.1), there are layers of infrastructure that must be balanced to achieve a target availability with an effective cost profile. Figure 5.2 shows an example of some of the key contributors to single-site VF availability. Each layer, like a "dial" on A/V equipment, can be tuned to achieve an overall availability profile. Some typical target levels of availability and the corresponding downtimes are represented by the vertical lines and are also shown in Table 5.1. As a first-order approximation, availability can be estimated by adding all of the

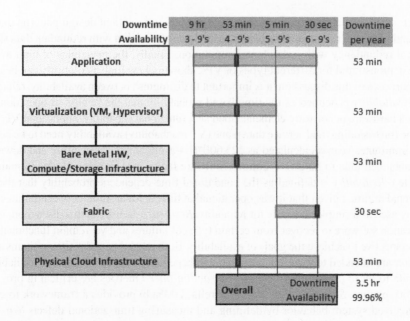

FIGURE 5.2   Key availability contributors within a single site.

TABLE 5.1

Availability to Downtime/Year Conversion

Availability[a]	Approximate Downtime/Year
3–9s	9 h
4–9s	53 min
5–9s	5 min
6–9s	30 s

[a] Based on 525,600 min/year.

component downtimes, and then converting the sum to time-based availability. The ranges shown in the figure are typical for the different layers of infrastructure.

The physical cloud infrastructure capabilities include physical properties of the plant such as generators, battery, heating, and cooling. Traditional Tier 4 data centers (DCs) supporting critical functions are designed with complete redundancy in the physical plant in order to achieve very high availability (ADC, 2006). In contrast, commercial clouds achieve a lower cost profile by using less redundancy. This is illustrated in the wide availability range shown in the figure. As carriers virtualize their networks, decisions need to be made about how reliably this layer should be engineered. On the one hand, improving the physical infrastructure layer to 5–9s or 6–9s could have little overall benefit and, in fact, may be quite costly. On the other hand, low physical layer availability may dominate the availability profile at a site. Though engineering to low availability may be less costly, the need to engineer greater capacity at neighboring sites to absorb frequent failures may offset these savings. Furthermore, low availability also has negative consequences to transaction reliability and latency. Given the availability of the commercial hardware and software that resides in the site, a middle-ground approach to physical infrastructure availability may provide a good cost/performance tradeoff.

Next, in the fabric layer, WAN access will generally have high availability due to transport layer diversity, a principle of carrier networks (Wu, 1992; Krishnan, 1994; Klincewicz, 2013). Similarly, it is standard and cost effective to design fully redundant LAN fabrics within a site, leading to high availability. Therefore, the fabric layer is generally not a significant contributor to downtime.

The compute/storage layer of the infrastructure, depending on design choices and ability of VFs to utilize resiliency in the infrastructure, can be a significant contributor to single site downtime. Because cloud infrastructure generally uses commercial compute and storage, the availability tends to be lower than for purpose-built network hardware. Cloud features that improve within-site availability include server sparing; multiple availability zones with anti-affinity rules to distribute VFs across hosts and racks to guard against shared hardware failures (Amazon, 2016); multiple control planes; and dual storage or high availability distributed storage such as Ceph (Ceph, 2016). Assuming that VFs are designed to be able to take advantage of the infrastructure resiliency features, the compute/storage layer availability can range widely at a single site.

Finally, the top two layers of the figure illustrate the impact of virtualization, which is an interaction between the application and virtualization layers. Many VFs cannot withstand short failures or "hiccups" in the hypervisor. So, even if the hypervisor appears to be highly available (e.g., the hiccups are of short duration), short disruptions could cause the VF to experience failures. When the failure detection and recovery time of the VF are factored in, the virtualization layer typically has 3– to 4–9s availability, though future improvements may move this higher. The application layer is similarly shown with a wide range of availability to indicate this.

5.1.2 Availability to Cost Tradeoff (Bottleneck Analysis)

Figure 5.3 illustrates the concept of bottleneck analysis. In this example, while the other layers achieve 4–9s (or higher) availability, the physical cloud infrastructure layer achieves only 3–9s availability. By improving this (bottleneck) layer to 4–9s, site availability can be improved from 99.87% to 99.96% (11.5–3.5 h of downtime per year)—a significant improvement. On the other hand, if the physical cloud infrastructure layer was already at 5–9s, improving this to 6–9s would have a very small impact to the site availability.

A more interesting question might be whether it is beneficial to improve the physical cloud infrastructure layer from 4– to 5–9s. In this intermediate case, the answer would depend on the specific costs of achieving this improvement. Achieving high site availability at low cost comes down to understanding the tradeoffs between cost and availability for each layer and adjusting the "dials"

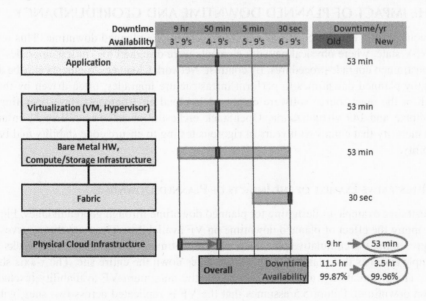

FIGURE 5.3 Bottleneck analysis.

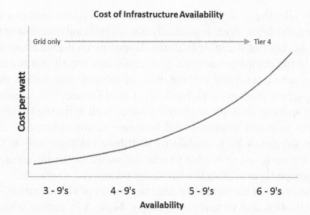

FIGURE 5.4 Infrastructure availability versus cost per watt.

accordingly. For example, the power configuration at a site could greatly affect the availability and cost. Figure 5.4 shows an illustrative cost curve for providing power to a site, ranging from using Grid only to Tier 4 or better (ADC, 2006). Depending on the power configuration used (e.g., UPS, redundant Gensets, battery backup, etc.), the costs will change. Comparing the availability to cost ratio for each power configuration against the similar availability to cost ratios for other layers will help identify where the largest impact can be achieved at the lowest cost.

By balancing these different layers, it is possible to minimize the cost of achieving a certain site availability (e.g., 4–9s). But it may be that certain VFs require higher availability (e.g., 5–9s), which may be cost prohibitive to achieve in a single site. In particular, with the introduction of commercial compute and storage, along with virtualization software, the availability of a single site generally would be no better than 3– to 4–9s. This is consistent with results reported elsewhere (Butler, 2016; Amazon Web Services, 2013; Hoff, 2015; Taylor, 2016). Factoring in VF-attributable failures, VF availability at a single site would be lower. In this case, these functions would need to make use of georedundancy across multiple sites to achieve high availability.

5.2 THE IMPACT OF PLANNED DOWNTIME AND GEOREDUNDANCY

The previous section considered VF availability assuming no planned downtime. This represents the pre-NFV state, where physical network elements were designed for "hitless upgrades" or with tightly constrained upgrade procedures. By contrast, Network Cloud environments may be designed with lengthy planned downtimes to perform infrastructure upgrades, often driven by the reality that much of the open-source software employed in cloud environments (host operating system [OS], compute and I/O virtualization, OpenStack orchestration, etc.) is still in its infancy and lacks the maturity that comes with years of rigorous testing to ensure code stability and backward compatibility.

5.2.1 Illustrative Example of the Impacts of Planned Downtime

As an illustrative example of designing for planned downtime through georedundancy, Figures 5.5 and 5.6 capture the effect of planned downtime on VF availability. There are three curves in each figure representing the cumulative planned annual downtime of 0 (hitless), 1, or 2 weeks per site. The example assumes that planned maintenance takes down the entire site. The x-axis shows the single-site cloud availability, and the y-axis shows the maximum VF availability (excluding VF application downtime). Figure 5.5 assumes that the VF is replicated across two sites, with 100% of the total required VF capacity provisioned at each site (200% total). Thus, at least one of the

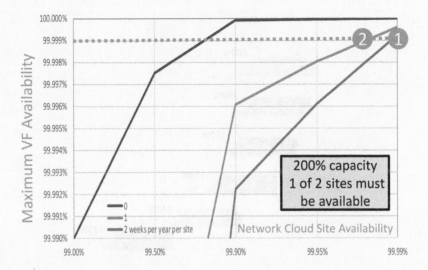

FIGURE 5.5   Maximum VF availability versus Cloud site availability assuming 2 sites.

two sites must be available for the VF to be available (i.e., have adequate capacity to serve the VF demand without performance degradation). In contrast, Figure 5.6 assumes that the VF is replicated across four sites, with 50% of the total required VF capacity provisioned at each site (again, 200% total). Thus, at least two of the four sites must now be available for the VF to be available.

Assume the VF requires 5–9s availability as represented by the dotted line in each figure. From Figure 5.5, if the planned downtime is 2 weeks per site per year (bottom curve), then the individual site availability must be at least 99.99% (labeled 1) in order for the VF to potentially achieve 5–9s availability. With 1 week per site per year of planned downtime (middle curve), 5–9s VF availability can be achieved if single-site availability is 99.97% (labeled 2), and with no planned downtime (top curve), 5–9s VF availability can be achieved if single-site availability exceeds only 99.7%.

In Figure 5.6 (four-site scenario), 5–9s VF availability can be achieved with 2 weeks per site per year of planned downtime when the single-site availability exceeds only 99.6% (labeled 3). Required single-site availability is further reduced to 99.4% with 1 week per site per year of planned downtime (green curve), and to 98.7% with 0 planned downtime "hitless" upgrades (blue curve).

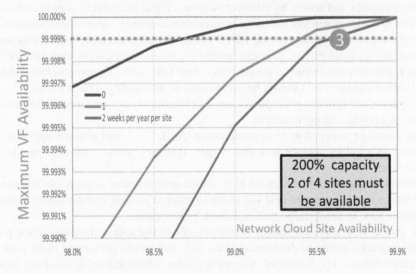

FIGURE 5.6   Maximum VF availability versus Cloud site availability assuming 4 sites.

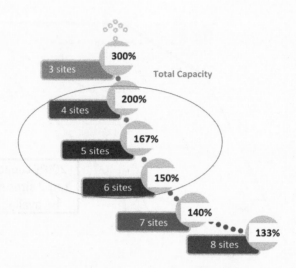

FIGURE 5.7   Capacity over-provisioning versus georedundancy tradeoff.

Thus, spreading the same capacity across four sites greatly increases VF resiliency to planned upgrades in a 3–9s cloud environment.

As mentioned above, additional capacity must be (over)provisioned in order to ensure sufficient VF capacity to tolerate simultaneous planned and unplanned site outages. In fact, as the number of redundant sites increases in the N+2 deployment scenario, the total preconfigured capacity required decreases by the ratio N/(N–2) for N ≥ 3. That is, in a 3-site scenario, the total required capacity = 3/(3–2) = 300%; in a 4-site scenario, 200%; in a 5-site scenario, 167%, etc. As shown in Figure 5.7, the "sweet spot" in terms of (over)provisioned capacity is in the 4–6 site range (subject to latency constraints).

5.2.2   Best Design Practices for Minimizing the Impacts of Planned Downtime

As demonstrated above, planned downtime can have a significant impact on VF availability. With this in mind, several common-sense "best practice" guidelines should be considered:

- Each site should support multiple compute node clusters, often called availability zones, in distinct racks and served by different switches. These availability zones can be used by tenants to achieve local site redundancy through anti-affinity capabilities (Amazon, 2016).
- The infrastructure should provide L2/L3 connectivity between the availability zones so that tenant applications can exchange heartbeats, transaction/session state data, etc.
- Minor upgrades that impact compute nodes (host OS, guest OS, compute/networking virtualization software, etc.) should be performed one availability zone at a time.
- All upgrade and rollback procedures should be thoroughly system tested before being executed in a production environment.
- Sites should be assigned to an upgrade group (A, B, C,...), and planned upgrades should be done in a rolling manner such that A sites, B sites, C sites,... are not upgraded at the same time.
- A minimum gap should be allowed between the completion of upgrades in one site group and the start of upgrades in next site group in order to provide stateful tenant applications adequate time to move traffic back and forth between sites.
- The goal is that upgrades should be transparent to the tenant (hitless). When possible, major upgrades should be decomposed into tasks that can be performed in off-peak maintenance windows. The frequency of major upgrades (where stateful applications must redirect traffic away from the site being upgraded) should be very infrequent.

- The network cloud should use DNS-based routing to facilitate traffic redirection (e.g., use of fully qualified domain names rather than static IP addresses for routing).

This section examined the interrelation between planned downtime and georedundancy, highlighting design considerations for achieving 5–9s "carrier grade" availability. Several key conclusions became evident:

- VF availability is extremely sensitive to both planned and unplanned site downtime
- At least N+1 redundancy is required to achieve 5–9s availability with a 3–9s site availability, even with no planned downtime
- At least N+2 redundancy may be required to additionally accommodate lengthy planned downtime
- Additional capacity must be (over)provisioned in order to ensure sufficient VF capacity to tolerate simultaneous planned and unplanned site outages
- Spreading the same amount of capacity across more sites greatly increases VF resiliency to lengthy planned upgrades in a 3–9s cloud environment
- An upgrade cycle framework is needed to ensure that no two sites used by the same VF are upgraded at the same time

To summarize, VF applications should be designed for a Network Cloud environment. They should support local redundancy using multiple availability zones, L2/L3 connectivity between VMs in different availability zones, and automatic local failover. Some applications should support at least an N+2 site georedundancy with automatic geo-failover and orchestrated movement of traffic between georedundant sites.

5.3 VF SOFTWARE DESIGN

Before any software design is initiated, there should be a clear understanding of the key performance indicators (KPIs), or requirements. Typical availability ranges from 3–9s for single-site VFs to 5–9s or higher for VFs with both local and geographic site redundancy. This section focuses on high availability software designs that result in no more than a few minutes per year downtime. Both the failure frequency and the downtime of the software must be carefully managed to achieve high availability.

Figure 5.8 shows a software-based services maturity curve for high availability VFs.

The previous section considered VF architectures for the Network Cloud. Implicit in those calculations was the assumption that failover for both local and georedundancy has a high probability of executing successfully and can be completed very quickly (e.g., in seconds). Also implicit was the assumption that faults are detected with very high probability (i.e., silent failures are very rare

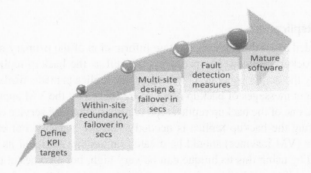

FIGURE 5.8 Software-based services maturity curve for high availability VFs.

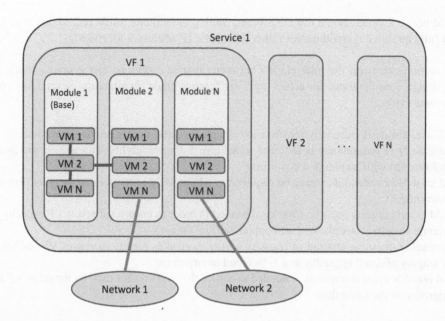

FIGURE 5.9   Virtual service architecture.

events) and that the software must be sufficiently mature so that faults are rare (e.g., less than once a year). These design considerations are covered as part of software maturity. Mature software is characterized by four traits: a low failure rate, accurate fault detection, fast failover to a redundant component, and fast recovery. In addition to these characteristics, there are a number of software techniques and fault tolerant design patterns that can be used to achieve resilient software. Most of the replication designs are based on variations of well-known replication techniques combined with solid software reliability engineering (Lyu, 1996). These techniques apply to virtual as well as physical applications.

In order to frame this section, some definitions are illustrated in Figure 5.9. A virtualized service (say a Voice over IP application) may comprise many VFs, such as signaling gateways, call processing applications, and location databases. Each of these VFs may comprise multiple modules in turn comprising one or more VMs.

5.3.1   Fault Tolerant VM Designs

Using the principles of fault-tolerant design, there are three types of replication that can be used for the VMs within a VF.

5.3.1.1   Hot VM Replication

Load is either provided to all the replicas, or state information of the primary replica is frequently transmitted to the backup replicas. The primaries as well as the backup replicas execute all the instructions, but only the output generated by the primary replica is made available to the user VM application. The output messages of backup replicas are logged by the VM application. In case the primary replica fails, one of the backup replicas can readily resume the service execution. A control scheme for designating the backup replica is needed to ensure stability. For each replica failure, an equivalent replica (VM instance) should be created on another host and its state updated. The availability obtained by using this technique can be very high, but it comes at the expense of high resource consumption. This replication design is typically reserved for applications designed for low tolerance to downtime and transaction defects.

5.3.1.2 Warm VM Replication

The state information is obtained by frequently check pointing the primary replica and buffering the input parameters between each checkpoint; replication is performed by transferring the state information to the backup replicas. The backup replicas do not execute the instructions, but save the latest state obtained from the primary replica. In case the primary replica fails, a backup replica is initiated and updated to the current state, incurring some loss in the present execution cycle and limited downtime. The availability obtained from this technique is less than that from hot replication, but the resource consumption costs are reduced since the backup replicas do not execute instructions.

5.3.1.3 Passive VM Replication

The state information of a VM instance is regularly stored on a backup. In case of a failure, another VM instance is commissioned, and it restores the last saved state. A backup can be configured to share the state of several VM instances (or it can be dedicated to a particular application), and the VM restart process can be performed based on a priority value assigned to each VM instance. This approach consumes the least amount of resources, but provides lower availability than the first two methods.

5.3.2 Low Software Failure Rates and Accurate Fault Detection

Some techniques for achieving low failure rate and accurate fault detection are described in this section. Many of these techniques can be realized using the capabilities of ONAP described in Chapter 7.

5.3.2.1 VF Protection from External Services

VFs need to protect themselves from both incoming north- and south-bound services utilized by the VF. This means that VFs need to ensure incoming requests can be distributed between multiple instances of the VFs (i.e., load balancing) if needed to maintain both service availability and performance. VFs also need to implement throttling mechanisms to limit the negative impact to the VF from either malicious or errant requestors so that the VF service is not compromised. South-bound VF interactions need to ensure isolation utilizing such mechanisms as multiple sets of thread pools. VFs also need to build these same capabilities into modules within the VF. The VF software should ensure any persistent state data (such as state tables, Java messaging service [JMS] queues, etc.) is replicated to redundant data stores to ensure service continuity in case of failure of the primary store/queue. The VF software should use redundant connection pooling to connect to any data source that can be switched between pools in an automated or scripted fashion to ensure high availability of the connection to the data source.

5.3.2.2 VF Error Handling

As a VF is migrated into the cloud environment from a traditional tightly coupled hardware or software configuration, a new potential for error situations will likely present itself. This includes, but is not limited to, network transport errors between instances of the VF, instance unavailability, local area network (LAN) latency, and software errors. VFs need to identify these classes of error conditions and potentially retry requests as appropriate to complete the requested service successfully instead of returning the client an error condition.

5.3.2.3 VF Software Fault Detection

If a VF component instance fails, that failure should be detected in seconds (or even milliseconds) and a mechanism to initiate an automated recovery of that instance should be provided. In the cloud world, that typically involves the recreation of the VM and/or the recreation of the failed instance. If automated recovery is not possible, notification mechanisms should be in place to create an alert for the failure. ONAP can be leveraged to utilize capabilities to monitor and react to failures and potentially provide recovery capabilities. In either case, care should be taken to ensure other instances can absorb and process traffic to the failed instance until a new instance is available.

5.3.2.4   VF Failure Detection and Alerting

VFs should have the ability to detect failures and/or unhealthy states of instances and generate alerts that result in corrective action. These failures should include both functional failures as well as exhaustion of resources (such as threads, queue limits, or memory structures), which may lead to catastrophic failure. In order to minimize silent failures, external probes are also an essential element for achieving accurate failure detection.

5.3.2.5   VF Fast Recovery

VF instances should support fast recovery from all failure types. Creation of recovered instances should be in the minutes or seconds range and should not impact or have dependencies upon other VFs. For example, a failed instance as part of a cluster configuration should not impact the cluster either during failure or restoration. Automation of recovery should be the desired behavior for facilitating fast VF recovery and can be achieved, for example, using ONAP closed-loop analytics and control.

5.3.2.6   VF Software Stability

VF software should be capable of supporting configuration parameter changes in such a way as to minimize the need to restart a service. Given that software can fail, just like the underlying infrastructure, the stability of the VF software should be tested with particular focus on resiliency and failover. Stability can also be achieved during upgrades by having the ability to quickly and easily revert to an earlier version if issues are found in a particular version of the VF software. In addition, stability can be achieved by providing the ability for multiple versions of VF software to coexist in order to eliminate the need to migrate all instances of the VF at one time, as opposed to rolling upgrades as needed by the business.

5.3.3   SOFTWARE RESILIENCY ENGINEERING

Software resiliency engineering is a longstanding discipline with well-known methods of evaluating software readiness for deployment (Lyu, 1996; Musa, 1999). Techniques to harden software to increase overall resiliency include software development best practices, software architecture reviews, testing methodologies, and stability prediction methods and tools. Testing methodologies are often restricted to feature functionality, during which faults may be uncovered and serve as a data source for residual fault prediction (Hoeflin, 2005; Zhang, 2009). Fault classification has been well defined in standards bodies (IEEE Computer Society, 2010), although in practice, classification is often provided by application testers as defects are discovered and vetted. In addition to fault analytics based on testing, introducing faults into software is another technique that is employed in cloud technologies. Netflix introduced the Chaos Monkey (Bennett and Tseitlin, 2012) fault insertion tool for cloud applications, and made it available in open source for other users. Tools of this type are useful for ensuring that faults can be detected and the system can recover quickly.

Maintaining high availability, reliability, and resiliency relies on a continuous cycle of measurement and improvement. ONAP is ideally positioned to assist in this regard using ONAP's data collection, analytics and events (DCAE) and associated microservices as described in Chapter 7. DCAE affords the opportunity to gather data from many sources that can be correlated and analyzed using advanced analytics and visualization techniques. DCAE metrics are processed by microservices to proactively detect anomalies, failures, and capacity exhaust. Microservices may be part of a closed loop that is used to automatically mitigate issues, such as by rebooting a VM, re-initiating failed virtualized network functions (VNFs) or spinning up new VNFs. In cases where an issue cannot be automatically resolved, the control loop should notify a network operator via ticketing or alarms so that the operator can intervene to resolve the issue. The centralized repository in DCAE also offers rich opportunities for data mining and analysis using advanced visualization capabilities such as RCloud (RCloud, 2016) that are made available to users as microservices.

5.4 PUTTING IT ALL TOGETHER: VF CLASSIFICATION AND EXAMPLES

This section ties together all the previous material by categorizing different types of VFs and the availability that can be expected. VFs can be classified into the following categories:

No site failover, no local redundancy. These VFs are typically associated with transport and network functions, where end users physically terminate their access at a single site. Examples include access routers and line-termination devices. Applications in this category must maintain state and have very demanding synchronization requirements. In the previrtualized network, high availability for these applications was achieved using purpose-built highly available hardware. In the early phases of virtualization, many of these VFs do not yet have local redundancy.

Local redundancy with no site failover. These VFs are also typically associated with transport and network access functions, where end users physically terminate their access at a single site. However, applications in this category have the ability to replicate and synchronize data within a site, enabling local redundancy.

Stateful function with site failover. Most of the critical layer 4–7 service VFs fall into this category. Examples are VFs associated with Voice over IP control plane or mobility control plane. While these VFs service end users, the users do not physically terminate access at these sites. The VFs are stateful, maintaining information about call state and customer profiles. The synchronization requirements are not quite as demanding, and service providers for these VFs have a tradition of architecting for georedundancy and local redundancy. These applications achieve high availability using established techniques at the VF layer to ensure fast failover. As the Network Cloud matures in its ability to handle network applications, there will be opportunities to use cloud orchestration and ONAP capabilities instead of custom VF capabilities for some of the resiliency needs.

Stateless function with site failover. These are the "ideal" VFs for a cloud. Because these applications are stateless, it is relatively straightforward to deploy them in many sites and dynamically distribute load across the sites. This also allows for seamless failure recovery since it can be done by taking one of the failed instances out of service and redistributing the load. An example of this type of VF is distributed Domain Name System (DNS).

This categorization is summarized in Table 5.2. The categories represent the types of VFs that exist today. As both the Network Cloud as well as the virtualized applications mature, VF architectures will migrate toward cloud-optimized implementations. This will result in more cost-optimized VFs with higher availability.

5.4.1 EXAMPLE: STATEFUL NETWORK ACCESS SERVICES

Consider the example of a stateful function that might be typical of a layer 2/3 access or transport service, such as switched Ethernet or managed Internet services. Such services could consist of customer-managed premises equipment (CPE) connected through a metro mesh network to a

TABLE 5.2
VF Categorization

VF Category	Characteristics of High Availability Solution	Expected Typical Availability Range
No site failover, no local redundancy	• Limited by availability of a single availability zone within a site; availability bound by worst-case physical infrastructure, compute, and store availability • Improvements to infrastructure availability may be required to achieve target availabilities	<3–9s
Local redundancy with no site failover	• Takes advantage of local redundancy • Availability bounded by physical infrastructure and the degree of redundancy and replication for compute/store within a site • May be able to take advantage of dual control planes and multiple availability zones within a site	3– to 4–9s
Stateful function with site failover	• VFs designed to operate using georedundancy • Data replication solutions to maintain state across multiple instances within and across sites • Failure detection and failover often addressed by the VF (do not rely on Cloud or ONAP capabilities)	4– to 5–9s
Stateless function with site failover	• VFs can be easily distributed across multiple sites • Load balancers and protocol-level retries can be used to mitigate the impact of site failures • Can achieve extremely high availability even when infrastructure availability is low	>5–9s

provider-managed edge router (PER). A common way to implement these L2/3 services is to home the CPE to a particular PER port. These services fall into the category of "no site failover with local redundancy" (3– to 4–9s). However, higher levels of resiliency (4– to 5–9s) can often be achieved by locating the redundant PER equipment in hardened (5–9s) network DCs.

Now suppose that the same services are implemented in a virtualized cloud environment, where a single Network Cloud site offers say 3– to 4–9s of availability. As shown in Figure 5.10, achieving levels of resiliency comparable to the precloud environment for such services requires automated orchestration capabilities to achieve seamless application failover following commodity cloud equipment failures. However, even with fast intrasite failover, service availability is still limited by the availability of a single cloud site itself. Thus, in order to achieve resiliency levels comparable to the precloud services for these stateful access services, a multisite design with dynamic rehoming is typically employed (shown in Figure 5.11).

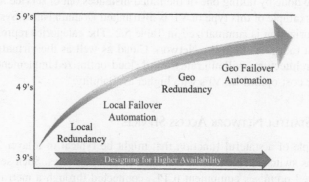

FIGURE 5.10 Stateful service design for high availability.

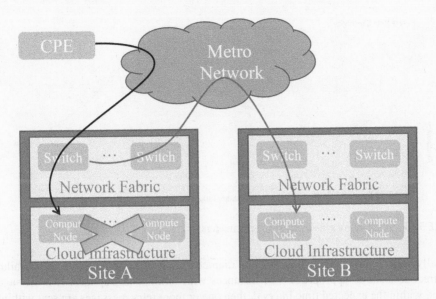

FIGURE 5.11 Multisite design with dynamic rehoming.

5.4.2 EXAMPLE: LAYER 4 STATEFUL CONTROL FUNCTIONS

Consider the case of layer 4 virtualized functions that provide control functions for real-time applications. In today's highly mobile communication networks, real-time services such as VoLTE, IP messaging, and video conferencing are provided by SIP-based solutions. The Mobility 3GPP standards organization specifies IP Multimedia System (IMS), standards including functions, interfaces, and protocols (3GPP, 2016). One of the key IMS functions is the call session control function (CSCF), which provides call control for the entire duration of sessions it handles. The CSCF monitors all signaling messages for locally registered users. The CSCF decides how to process SIP messages, including routing to application servers and network routing servers. In addition, the CSCF enforces policies specified by the service provider.

In IMS-based services, it is critical to provide high availability as well as maintain call state in order to minimize service disruptions during failures. These same needs apply equally to virtual IMS solutions. For the case of a virtual CSCF, local redundancy and georedundancy are needed to ensure high reliability. Solutions may include local redundancy implemented with load sharing across all active VMs (shown in Figure 5.12). In addition, failure detection and recovery times should be less than a second.

Accurate call state is critical to proper session processing; hence, the redundancy solution must ensure call state is maintained across local VM instances as well as across sites. Various implementations can be used to achieve call state replication with minimal lag and a control scheme to determine master call state.

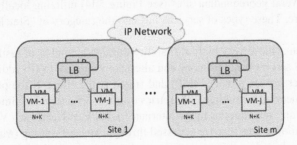

FIGURE 5.12 Example of multisite design with local redundancy.

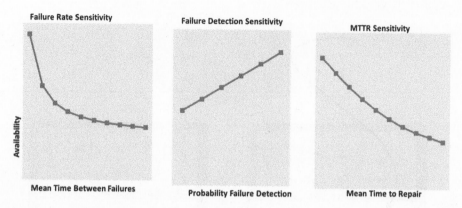

FIGURE 5.13 Availability sensitivity to key parameters.

Finally, layer 4 protocols such as SIP and diameter are used to monitor and detect failures and to redirect sessions to appropriate VM instances with minimal disruptions. If a message is not received within the expected time interval, then one or more retry messages are sent within a short time. Typically, failures can be determined within a second and failover to alternate instances within seconds.

Depending on the architecture and implementation, it is possible to achieve availability of 99.999% or higher. Conversely, if the failure rates are too high, the ability to successfully detect and recover from failures is too low, or the time to recover following a failure is too long, then the availability may be much lower and may not meet the desired overall service availability. Figure 5.13 illustrates sensitivity of availability to these key factors.

Design decisions for the level of redundancy required within sites and across sites are highly dependent on the combination of failure parameters. Reliability analysis is an important part of the design process to drive optimal solutions. In the virtual Call Session Control Function (vCSCF) example shown in Figure 5.12, there can be multiple subcomponents that make up the whole solution, and the vCSCF must interoperate with other IMS components. Redundancy combined with low failure rates, high failure detection rates, and fast recovery times are essential to achieving high availability. The complexity requires careful reliability design and test validation to ensure the desired targets are met in a cost-effective solution.

5.4.3 Example: Stateless Network Function with Multisite Design

Finally, consider a stateless service with georedundancy and fast failover, such as vDNS-R (Domain Name System Resolver). DNS is a stateless application that translates user-friendly domain names or URLs to the numerical IP addresses needed for locating/identifying computer services and devices. This is an example of a service that is perfectly suited for virtualization. This service can be hosted across several georedundant sites (see Figure 5.14) utilizing locally redundant VMs on commodity hardware. These types of services fall into the category of "Stateless function with site failover" (5–9s +).

vDNS service can achieve very high availability through a number of resiliency features. First, clients access vDNS servers using primary and alternate virtual IP (VIP) addresses, normally provided by the Internet service provider (ISP) and stored at the client. If the primary VIP does not respond to a DNS query after multiple (~5) retries over a short interval of time (<1 min), the client attempts to connect and send queries to the alternate vDNS VIP. Next, each VIP is associated with both primary and failover sites that are accessed through Anycast routing (Metz, 2002). This redirection is very fast, taking less than a second to redirect queries to the failover site. In addition to these georedundancy features, each site provides local redundancy through load balancing across

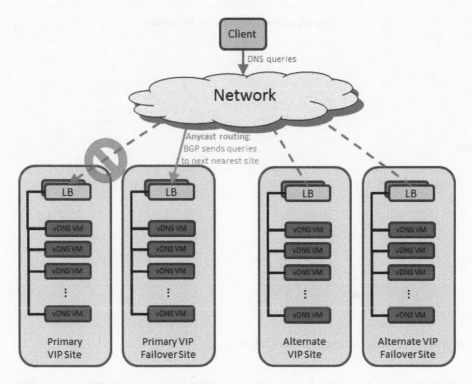

FIGURE 5.14   vDNS architecture and redundancy.

many VMs spread across multiple servers within the site. Therefore, if a VM or server fails, queries are quickly diverted to others that are still operational.

As a stateless network function in multiple sites, vDNS-R is able to take advantage of a combination of these resiliency features. Figure 5.15 illustrates the impact of each of these features on overall availability (time-based) or service reliability (transaction-based) expressed as DPMs. By employing these resiliency techniques, the vDNS-R service is able to achieve extremely high reliability.

5.5   AREAS FOR FURTHER STUDY

As seen in the examples cited in this chapter, the basic principles and mathematics of reliability analytics have not changed with the introduction of virtualized solutions. However, there are new challenges that must be addressed to ensure that the desired reliability for virtual solutions can be achieved. In this section, some topics for further exploration are identified:

1. *Software stability for Agile developments.* The Agile software development process uses short development cycles and continuous integration and deployment. This makes it very difficult to assess software stability using traditional software reliability techniques that involve rigorous and lengthy testing. Netflix developed a tool called Chaos Monkey, a software tool that is used in the operational cloud to test the resiliency and recoverability of their Amazon Web Services (Tseitlin, 2012). Telecommunications applications can be highly complex with many failure modes. Approaches that combine the benefits of Chaos Monkey style testing and more traditional stability testing would be very valuable.

2. *Mitigating the effects of planned outages.* VFs need to be able to accommodate long planned outages. It can be costly to add additional levels of replication as a guard against

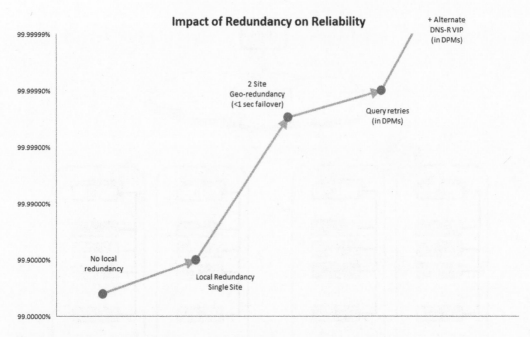

FIGURE 5.15   Resiliency features and impact on resiliency.

planned outages. With ONAP, there is the possibility to use automation to migrate VFs to locations that are not subject to the planned outage. Designing reliable and efficient techniques for migration is an area that deserves further investigation.

3. *Accurate fault identification.* The multilayer distributed nature of NFV architectures means that accurate and timely fault detection can be a challenge. Without accurate and timely fault detection, achieving high availability is impossible. Understanding what types of metrics need to be collected and how to process and correlate the metrics to detect and isolate faults is essential. There are many opportunities to apply advanced analytics techniques such as machine learning to improve fault detection.

4. *Automated failure recovery.* Failures are more frequent when using commodity hardware and software. Therefore, it is essential to handle them more efficiently by using automated failure detection and recovery. Procedures for automated detection and recovery represent a major opportunity to improve VF availability.

5. *Software robustness.* Applications designs that are insensitive to short failures and fluctuations in the Network Cloud infrastructure can greatly improve service availability.

ACKNOWLEDGMENTS

The authors would like to acknowledge the Cloud and Service realization teams who appreciate the importance of designing for reliability and ensure that the reliability recommendations are implemented. This technical work would not be possible without their partnership.

REFERENCES

3GPP. 2016. 3GPP. Retrieved from 3GPP: www.3GPP.org

ADC. 2006. TIA-942 Data Center Standards Overview—102264AE—Accu-Tech, ADC Telecommunications Whitepaper. www.accu-tech.com/hs-fs/hub/54495/file-15894024-pdf/docs/102264ae.pdf

Amazon. 2016. AWS Global Infrastructure. Retrieved from Amazon: https://aws.amazon.com/about-aws/global-infrastructure/

Amazon Web Services. 2013. Amazon EC2 Service Level Agreement. https://aws.amazon.com/ec2/sla/

Bennett, C. and Tseitlin, A. 2012. Chaos Monkey Released into the Wild, Netflix Tech Blog, July 30, 2012. http://techblog.netflix.com/2012/07/chaos-monkey-released-into-wild.html

Butler, B. 2016. And the cloud provider with the best uptime in 2015 is…. *Network World.* January 7. http://www.networkworld.com/article/3020235/cloud-computing/and-the-cloud-provider-with-the-best-uptime-in-2015-is.html

Ceph. 2016. Ceph. Retrieved from www.ceph.com

Choudhury, G. 2004. Models for IP/MPLS routing performance: Convergence, fast reroute, and QoS impact. In *Performance, Quality of Service and Control of Next-Generation Communications Networks II (Proceedings of SPIE).* Philadelphia, PA: SPIE—The International Society for Optical Engineering, pp. 1–10.

Giloth, G. F. 1987. The evolution of fault tolerant switching systems in AT&T. In H. K. A. Avizienis (ed.), *The Evolution of Fault Tolerant Computing* (pp. 37–54). Baden, Austria: Springer-Verlag/Wien.

Healy, J. 2016. Network Outage Reporting System (NORS). https://www.fcc.gov/network-outage-reporting-system-nors

Hoeflin, D. A. 2005. An integrated defect tracking model for product deployment in telecom services. In *Proceedings of the 10th IEEE Symposium on Computers and Communication (ISCC 2005)* (pp. 927–932). IEEE. June 27–30, 2005, Cartagena, Murcua, Spain.

Hoff, T. 2015. The Stunning Scale of AWS and What It Means for the Future of the Cloud. http://highscalability.com/blog/2015/1/12/the-stunning-scale-of-aws-and-what-it-means-for-the-future-o.html

Martersteck, K. E. and Spencer, A. E. Jr. 1985. The 5ESS switching system—Introduction. *AT&T Technical Journal,* 64(6), part 2, 1305–1314.

IEEE Computer Society. 2010. *IEEE Std. 1044-2099 (Revisions of IEEE Std. 1044-1993) IEEE Standard Classification for Software Anomalies.* New York: IEEE Computer Society.

Johnson, C. R. et al. 2004. VoIP reliability: A service provider's perspective. IEEE Communications, July, pp. 48–54.

Klincewicz, J. G. 2013. Designing an IP link topology for a metro area backbone network. *International Journal of Interdisciplinary Telecommunications and Networking,* 5(1), 26–42.

Krishnan, K. D. 1994. Unified models of survivability for multi-technology networks. In *Proceedings of ITC 14.* New York: Elsevier Science B.V, Volume 1A, pp. 655–665.

Lancaster, W. G. 1986. Carrier grade: Five nines, the myth and the reality. *Pipeline* (www.pipelinepub.com), 3(11), 6.

Lyu, M. R. 1996. *Handbook of Software Reliability Engineering.* Los Alamitos, CA: IEEE Computer Society Press and McGraw-Hill.

Martersteck, K. E. 1981. The 4ESS system evolution. *Bell System Technical Journal,* 60(6), part 2, 1041–1228.

Metz, C. 2002. IP anycast: Point-to-(any) point communication. *IEEE Internet Computing,* 6(2), 94–98.

Miller, R. 2015. Inside Amazon's Cloud Computing Infrastructure. http://datacenterfrontier.com/inside-amazon-cloud-computing-infrastructure/

Musa, J. 1999. *Software Reliability Engineering.* New York: McGraw-Hill.

Rackspace. 2015. *Creating Resilient Architectures on Microsoft Azure.* Windcrest, TX: Rackspace.

Ramesh Govindan (Google, USC); Ina Minei, Mahesh Kallahalla, Bikash Koley, Amin Vahdat (Google). 2016. Evolve or die: High availability design principles drawn from Google's network infrastructure. *Proceedings of the 2016 ACM Conference on Special Interest Group on Data Communication,* pp. 58–72. August 22–26, Florianopolis, Brazil.

RCloud. 2016. RCloud. Retrieved from www.rcloud.social/index.html

Taylor, M. 2016. Telco-grade service availability from an it-grade cloud. *IEEE CQR,* May 9–12 (p. 21). Stevenson, WA: IEEE.

Tortorella, M. 2005a. Service reliability theory and engineering, I: Foundations. *Quality Technology and Quantitiative Management,* 2(1), 1–16.

Tortorella, M. 2005b. Service reliability theory and engineeting, II: Models and examples. *Quality Technology and Quantitative Management,* 2(1), 17–37.

Tseitlin, C. B. 2012. Chaos Monkey Released into the Wild. http://techblog.netflix.com/2012/07/chaos-monkey-released-into-wild.html

Wirth, P. A. 1988. A unifying approach to performance and reliability objectives. ITC 12, June 1998. Torino, Italy: International Advisory Council of the International Teletraffic Congress, (Vol 4, pp. 4.2B2.1–4.2B2.7).

Wu, T.-H. 1992. *Fiber Network Survivability.* Boston, MA: Artech House, Inc.

Zhang, X. 2009. Software risk management. In Sherif, M. H. (ed.), *Handbook of Enterprise Integration.* Boca Raton, FL: CRC Press. Chapter 10, pp. 225–270.

6 Software-Defined Networking

Brian Freeman and Han Nguyen

CONTENTS

6.1 Functional Overview 87
    6.1.1 Network Control 88
6.2 Implementing Network Control...88
6.3 A New Paradigm for Network Feature Delivery 89
6.4 Network Controller Architecture...90
    6.4.1 Network Controller Software Components 91
        6.4.1.1 Compiler Function 91
        6.4.1.2 Service Logic Interpreter Function......................................92
        6.4.1.3 Network Resource Autonomous Control Function..............92
        6.4.1.4 Adapters..93
        6.4.1.5 API Handler 93
        6.4.1.6 Policy 93
        6.4.1.7 Data Collection Analytic and Events Function....................93
    6.4.2 Software Validation 94
    6.4.3 High Availability and Geo-Diversity...94
    6.4.4 Relationship to Application Service Controllers 94
    6.4.5 Federation between Network Controllers 94
    6.4.6 Abstraction Modeling 94
    6.4.7 AT&T Network Domain-Specific Language 95
6.5 YANG Service Model Example 96
6.6 YANG Network Model Example...97
6.7 Network Controller and Orchestration Use-Case Example: Customer Requesting
    VPN Service 99
6.8 Some Use-Case Examples of SDN Control..100
    6.8.1 Bandwidth Calendaring..100
    6.8.2 Flow Redirection 100
6.9 Choice of Open Source SDN Controllers 101
6.10 Topics for Further Study 101
Acknowledgment 101
References...102

The SDN architecture approach has been repeatedly applied successfully to multiple generations of AT&T networking services from traditional circuit-switching through current IP/MPLS services (see Chapter 1). AT&T's design and implementation of SDN for network functions virtualization builds on this prior development experience with the overarching goal of designing and implementing a programmable network-control platform for near real-time service management and network management—which is the subject of this chapter.

6.1 FUNCTIONAL OVERVIEW

The AT&T global network supports all of AT&T's network-connectivity services and application services for business and consumer customers worldwide. Due to the network's very large-scale

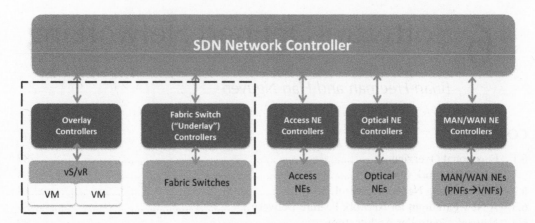

FIGURE 6.1   SDN network control architecture.

(in terms of the number of network elements) and global geographic footprint, and very high availability/reliability performance requirements, the SDN control architecture is based on a hybrid design of centralized and distributed controls, which is shown in Figure 6.1. For smaller networks, a simpler framework may suffice.

The distributed controls are provided by a set of distributed local controllers, which are responsible for the traffic forwarding (including failure detection and recovery) operations of individual network elements in their respective local domain. The centralized controls are provided by the global SDN network controller for the purpose of end-to-end service management and network resource optimization.

6.1.1 NETWORK CONTROL

Network control is responsible for the management, deployment, operation, and coordination of network connectivity services implemented on underlying network resources, which comprises virtualized network functions (VNFs) and physical network functions (PNFs). Network control includes the following:

- The establishment and management of network connectivity
- Maintaining network-wide resource inventory
- Running network-wide data collection and analysis processes
- Running a policy engine using network-wide set of policies
- Handling network-wide exceptions

In the remainder of this chapter, the following short-hand terminology is used, unless explicitly noted:

- Network services—refers to network connectivity services, which are services providing layer 0–3 communications or closely related functions
- Network controller—refers to the platform that implements network control

6.2 IMPLEMENTING NETWORK CONTROL

This chapter focuses on the platform that implements network control and allows the separation of control plane from the data plane as well as the creation of abstraction layers, which separates network services and network design constructs from physical network element constructs. The design is for a controller that provides a programmable platform for network tools, network control of service instantiation and management, and for resource and traffic management. The platform also provides a consolidated network management interface to permit the combination of real-time data from the network services and underlying network elements with real-time control of the forwarding plane. The key attributes for this platform are near real-time configuration, real-time flow setup, programmability through service and network script-like logic, extensibility for competitive differentiation, standard interfaces, and multivendor support.

In today's environment, automated provisioning is focused on non-real-time configuration management. Real-time redirection of flows is performed by each vendor's control plane, which is normally integrated with the data plane. The future direction is separating this control plane from the data plane and creating a real-time controller, which has a centralized real-time network-wide view, which may optionally complement a local controller that has a nodal view. The initial goal is not to take all intelligent control out of a specific network element, but to pull out the information needed for functions that need to be performed from a real time and/or a multilayer network perspective—for example, redirecting flows, application of security policies, and multilayer traffic engineering.

Also in today's environment, the network service and network configurations are intertwined such that if one changes the network element, the network service configuration gets impacted. A fundamental controller design principle is creating modular functions that become independent building blocks enabling the creation of specific new network services based on a specific set of building blocks. The basic building blocks consist of network services features, and network resources. The separation of network service control from network resource control enables the abstraction of the network service definition (e.g., Ethernet services) from the particular types of network resource (e.g., Ethernet over MPLS) used to implement the network service. Network service and network design are now decoupled from the actual vendor-specific implementation (e.g., Juniper MX960 for Ethernet over MPLS).

Finally, in today's environment, network access service features are intertwined with network flow service features, so that every flow service needs to specify the access services thus making provisioning very specific. The new model enables separation of access service features (e.g., broadband, cellular, T1 services) from flow service features (e.g., Ethernet, L3 VPN, internet), allowing access services to connect to different flow services very quickly. Access services and flow services can be mixed and matched dynamically. The main benefits of this implementation are as follows:

1. Faster time to market for instantiation of network elements as well as new network services
2. Ability to match the network features, performance, and capabilities to customer needs on demand
3. Ability to slice the network for an individual customer while maintaining network and operational efficiencies

Since virtualization is a key component to on-demand instantiation of network services, this controller works intimately with the infrastructure control, which instantiates a virtualized environment (compute, storage, and data center networking) for virtual applications.

6.3 A NEW PARADIGM FOR NETWORK FEATURE DELIVERY

Network control will use technology to translate formal information models into programmable logic modules that will increase feature development velocity. Formal information models will allow the separation of network service definition from the network design from a vendor-specific implementation. This will allow vendor devices to change without impacting the network design abstraction layer or the network service abstraction layer. The network abstraction layer can change from a switch to a router to an optical multiplexor, without impacting the network service layer. Control of NFV will permit new processes for software validation by using virtual routers/switches in the network.

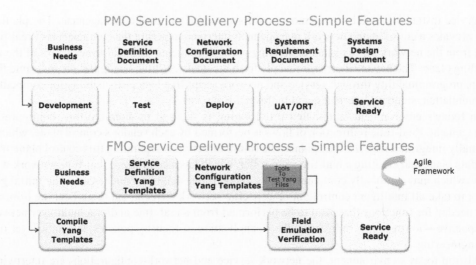

FIGURE 6.2 Service delivery.

The present mode of operation (PMO) shown in Figure 6.2 follows the waterfall development process where each function is handled by a different organization. Design specification documents are created in the first five steps from the individual organization's perspective, basically translating the preceding organization's requirements into the current organization's terminology and requirements. This method entails many pair-wise and then group-wide reviews of the different documents, and can create a lot of misinterpretation. The execution steps of development, test, deployment, operational readiness testing phases are also serially executed to get to a service-ready status. Along each of these "execution" steps, interpretation of the documents and clarification or changes to the developed service are iterated until all agree that the service is ready.

The future mode of operations (FMO) introduces a new paradigm, which drastically simplifies and reduces the process through a programmable definition of the service, thereby reducing the number of documents created for interpretation. Basically, the business needs are captured in requirements documents and the design organization uses a programming language to define the network service and/or network templates. These templates are compiled by the controller, which then can be validated through a field emulated test environment immediately. Iteration on this method can use the agile development process to get to service ready much more quickly. The designers have to pivot their skills from document writing to service programming based on business requirements.

6.4 NETWORK CONTROLLER ARCHITECTURE

This section describes the platform, environment, and the constructs for building an SDN network controller. This is based on adapting the information or data model-driven approach to network service and resource configuration management, drastically increasing the flexibility for service or feature creation and the agility in network development. Modeling of network services, network feature implementation designs, and network element device semantics allows code generation for feature delivery rather than custom coding of any new feature or function. In addition, design principles originally applied to programmable service logic for call processing networks (Network Control Points [NCPs]) are applied to make the determination of the flow logic to be specified in a script or template that can be flexibly changed without coding. The combination of programmable service or network flow logic and a model-driven approach to parameter specification should result in feature development times measured in minutes or hours instead of weeks or months. There are seven major components of this controller:

1. Compiler function, compiles models to create service logic in configuring the network resources
2. Service control interpreter function, interprets customer or API requests for instantiation of services
3. Network resource autonomous controller, interprets the real-time network map and provides the network resources necessary in creating the services requested by the service control interpreter
4. Adapters, allow multi-vendor and multi-interface configuration and data collection functions
5. API handler, manages the applications that are programming the controller through the API interface
6. Data collection, analytics, and events (DCAE), collects the data and events from the network resources, analyzes them creating a real-time network map
7. Policy, is the principle that guides the decisions in instantiating a service or network element

6.4.1 NETWORK CONTROLLER SOFTWARE COMPONENTS

Figure 6.3 provides a graphical view of the main software components for the network controller (the functional block ONAP is discussed in Chapter 7).

6.4.1.1 Compiler Function

The network controller accepts as input a set of information or data abstraction models—YANG modeling language [1], which define the service abstraction layer (e.g., Ethernet services), network abstraction layer (e.g., Ethernet over optical), the vendor-specific device definition (e.g., Ethernet over Ciena Optical), and the directed graph (DG) (the sequence of steps in configuring components) that the network controller will manage. These will be defined by the vendor or the system engineering team. The compiler takes the abstraction models and the network logic (defined in an XML file) and creates the run-time Service Logic DG that is used during event processing. Basically, the actual order and timing of the commands to configure the network elements through a network service order or through a network event will be defined by the collection of these YANG models.

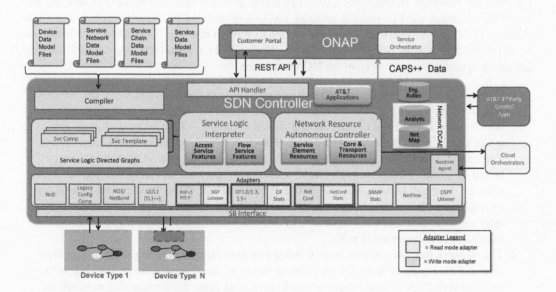

FIGURE 6.3 Network controller components.

6.4.1.2   Service Logic Interpreter Function

The service logic interpreter (SLI) is the event-driven component that deals with requests from external systems for new network services, connections, or features. SLI interprets the service requests into specific network resource requests, which are optimized based on the network resource autonomous controller results. Service requests will come from MSO and other service control functions like closed loop at run time. The system engineering definitions of a network service, network, or vendor-specific devices will be provided at design time through the YANG modules.

1. The SLI implements a key aspect of the platform flexibility. The SLI reads DGs (e.g., DGs define the order of execution of software nodes/objects); this is explained in more detail in Section 6.4.7. This is network implementation logic software that satisfies the incoming network service request. By implementing the network service to network mapping in a flexible DG or script, changes to the network implementation can be quickly created and tested without requiring changes in the engine. New functions may be needed in the engine over time as new adapters or fundamental new protocols are added to the platform, but generally only the DGs need to change as capabilities are used in new combinations. The DG operates on the network and service models to gather all the data items, selects resources based on interacting with the resource control function, and passes the data set to the adapters for changing the state of the network. The combination of the model and the DG creates a set of network service creation tools that drastically shorten the time to market for new network services or features.
2. The interpreter works in conjunction with a compiler. The compiler takes the model (e.g., YANG models) and the network logic (defined in an XML file) and creates the run time Service Logic DG that is used during event processing.
3. Formal information models will allow separation of network service definition from network abstraction definition from vendor-specific implementation. In particular, an abstraction layer with three distinct components allows for specification and flexible mapping among service, network, and device definitions. Network service and network layer abstractions are defined later in this chapter.

6.4.1.3   Network Resource Autonomous Control Function

The Network Resource Autonomous Control Function allocates (and deallocates) as well as optimizes the resources used to provide network services or features in the networks. It assigns or reassigns (PNF or VNF) network resources needed to support customer network service instances, based on real-time service requests from Service Design and Creation (SDC) or network-resource utilization optimization actions from the DCAE function.

1. Today, this is a network service order-driven process with the assumption that the network resource is in a fixed location. This new solution integrates virtualized network resources, which can be instantiated on demand and moved as the state of the network changes due to traffic or failures or because of business policies (e.g., cheaper power rates by time of day). The resource controller continually monitors and analyzes the network state telemetry collected from the adapters through DCAE and determines whether changes in the network state are required. If changes are required, it interacts with the service control interpreter to implement the intended state change. The Network Resource Control Function also interacts with resource-specific policy and inventory functions for assignment of resources in adherence to engineering rules.
2. The Network Resource Autonomous Control Function separates the functions and applications for managing core packet and optical transport (wide area network [WAN], metro area network [MAN], and access network) resources from those managing network service resources. Core transport resources are largely "hard" resources such as router and

link bandwidths, whereas the edge service resources also need to include "soft" resources like IP addresses, VLAN IDs, etc.

6.4.1.4 Adapters

Adapters interact with the network devices (e.g., network elements) and VNFs through well-defined protocols. The adapter layer contains the complexity of vendor-specific and protocol-specific implementation issues for both changing the state of the network device and reading the state of the network device.

A vendor may have their own proprietary controller (represented as the green box inside the device type N in Figure 6.3) or EMS, which interfaces with their network device and therefore the appropriate vendor adapter to interface with their controller or EMS would be needed.

1. Changing the state of the network is handled by "write mode" adapters. It can be done using traditional configuration through template or scripted CLI changes, transactional NETCONF with YANG changes or flow changes through protocols like Border Gateway Protocol (BGP), BGP-Flow-Spec or OpenFlow.
2. Reading the state of the network is handled by "read mode" adapters. This allows for the collection of data through protocols like SNMP, NETCONF, OpenFlow, BGP-Link-State, OSPF link state, etc. The data collected through those means are used by the controller for application processing in autonomous control loops, which would result in changes in the network state through the "write" mode adapters. The DCAE function collects the data periodically by polling or auto-streaming, or on-demand based on trigger events.

6.4.1.5 API Handler

The network controller exposes a programmable API used by service providers and customers to allow manipulation of network access and flow services, and the use of service, core, and transport resources. The service design and creation (SD&C) function will be the interface for the service providers and customers to create and request a service. The API exposed by the network controller is an abstraction of the network service and network so that the customer or higher level application does not need to know the details of the network implementation. Depending on the state of the resources in the network and the request, a different network implementation could be implemented in real time. Again there are two separate abstraction layers, network service, and then network. Each is defined independently to allow changes at one layer to not impact the definition of the other layer.

6.4.1.6 Policy

Policy is a principle or a set of rules that guides decisions, and is defined by the customer, the network designer, or engineer. Policy governs the resource allocation and placement for the service or network element, which is getting instantiated, so as to meet the service-level agreement (SLA). Policy also prescribes actions to be taken in the event of a resource overload or failure condition. This is discussed in more detail in Chapter 7.

6.4.1.7 Data Collection Analytic and Events Function

The network controller is a self-optimizing platform. It uses the telemetry collected from the adapters and determines whether the state of the network should change to optimize service performance and network resource utilization. The ability of the network controller to self-optimize is enabled by the presence of a local DCAE function in the network controller that encapsulates the Big Data storage and analysis functions needed for the analytics. This function provides the reference information for the Active and Available Inventory (A&AI) function. The combination of DCAE and policy feeds into the orchestration of environments, services, or network flows. This control loop, among DCAE, policy, orchestration, and SDN control continuously runs to maintain the network services and resources in optimal conditions. This is explained in more detail in Chapter 7. The Big Data infrastructure design and implementation is the topic of discussion for Chapter 8.

6.4.2 Software Validation

A benefit of the type of network control that the network controller provides is the ability to partition traffic flows using virtual routers and switches in the network into "staging" environments or safe zones where new capabilities or software, for example, new NFVs, can be safely tested without impact to the current production or test environment. This allows new capabilities or software to be introduced and evaluated and validated on the target environment in a controlled manner, without the need for dedicated staging infrastructure.

6.4.3 High Availability and Geo-Diversity

The physical location of network controller instances will be deployed into high-availability clusters of colocated servers, with each cluster having a designated "master" node. In support of geo-diversity, network controller instances will be deployed in different data centers. Traffic will be steered toward particular network controller clusters by means of data sharding [2]. Shards will replicate state to two or more peers so that upon failure of the master for the shard a new master will be elected. Members of a shard cluster will have a protocol to detect failure of the master node and elect a new master node.

6.4.4 Relationship to Application Service Controllers

The same protocols that allow the master service orchestrator to call the network controller will allow application service controllers to make requests as well. For instance, the VOIP application service controller could call the network controller to request changes in the L3 configuration supporting a VOIP service.

6.4.5 Federation between Network Controllers

Peer-to-peer interactions between similar types of network controllers may be needed. A transport flow between two customers, regions, etc. may involve two separate network controller clusters depending on the outcome of the data sharding. The network controller on one side of the flow may communicate requests to its peer network controller on the other side to dynamically change the flow between the two domains.

For example, a customer could run their own network controller with visibility of both the campus and WAN network and send requests to the AT&T network controller to make changes to the flows on the WAN under the direction of the customer controller.

6.4.6 Abstraction Modeling

AT&T's service portfolio includes network connectivity services, which functionally encompasses L0 through L3 and L4 service solutions and geographically can span over Data-Center LANs, MANs and WAN, and complex L4+ services such as VoIP. To meet the service and network management needs of this portfolio, the information modeling approach in the SDN architecture is based on a novel three-tier abstraction design in lieu of the traditional single-tier abstraction design commonly used for multivendor network management.

- The network service abstraction model describes the network service as presented to the customer application—independent of the particular network protocol architecture and technology solution chosen to implement the network service for a given customer service instance. Individual service models can be concatenated ("chained together") to create a richer end-to-end chain of respective individual services. Valid service chains are prescribed by means of a service chain graph.

- The network service implementation abstraction model describes a network protocol architecture and technology design solution used to implement the given network service—independent of vendor-specific platform used in the implementation.
- The network device abstraction model describes the set of functional networking building block capabilities supported by a given network device—independent of the device vendor's hardware and software implementation.

6.4.7 AT&T Network Domain-Specific Language

The AT&T Network Domain-Specific Language (ANDSL) is used to specify the execution logic in processing the service and network models. ANDSL defines an executable path for a set of functions or nodes also referred to as a "DG." Nodes along the path execute functions to get, update, or release data and execute functions in the network or operations support systems (OSS) based upon the available data. The data are the leaves defined in the service and network models. The functions that can exist on a path will grow over time but initially would consist of the following list:

- *Allocate*—to allocate a resource from either local or remote inventory. The input values that identify the resource being requested and the parameters that influence the decision on which resource or how much resources to be granted are part of the allocate node (and would come from the context memory structured per the network and service data model). The data returned by the allocate function would be saved in context memory and persistent storage using the name defined in the network data model.
- *Set*—to specify or calculate a value that is not an inventory resource but either determined from a calculation or algorithm on other variables or applicable engineering rules.
- *Block*—to indicate that a set of nodes should all succeed or all fail, and if one node fails then roll back the network state to that which existed before the block statement was executed.
- *Configure*—to indicate that a state change on a device should occur. The configure function will indicate which adapter family is to be used to perform the configuration, and the operation for requested action. The adapter will pull the appropriate data from the context memory per the network data model and change the state of the network. The adapter will use the device data model to map the network configuration data to the vendor-specific device model.
- *Switch*—a case statement that allows decisions in the DG to be handled by different logic depending on either service or network data model variable.
- *SendMessage*—to reply to the event using the service model defined output (or error).
- *Test*—to execute a command and/or test against the network. This is similar to the configure function but is not a permanent state change. A test node could be used after the call to a configure node to test that a configuration succeeded.
- *UserDefined Node*—an important attribute is to allow service and network designers to create new nodes. This node allows the designer to define a new node as a Java class and to pass in the attributes in advance of the engine having optimized implementation of the node. Over time, UserDefined Nodes will be pulled into the SLI Engine for efficiency

```
<service-logic module="vpn" rpc="attach-site">
  <block atomic="false">
    <allocate type="route-target-any" saveAs="$resource-route-target-any" />
    <allocate type="route-target-hub" saveAs="$resource-route-target-hub" />
    <allocate type="route-target-spoke" saveAs="$resource-route-target-spoke" />
    <allocate type="vrf-id" saveAs="$resource-vrf-id" />
```

```
      <allocate type="route-distinguisher" saveAs="$resource-route-distinguisher" />
      <allocate type="import-route-target" site-type="$site-type"
               saveAs="$resource-import-route-target" />
      <allocate type="export-route-target" site-type="$site-type"
               saveAs="$resource-export-route-target" />
   </block>
   <block atomic="true">
      <configure adaptor="PE" deviceid="$reserve-equipment-clli" />;
   </block>
</service-logic module="vpn" rpc="attach-site">
```

Creating DGs directly in xml is tedious and error prone. A graphical tool called the DGBuilder, based on Node-RED [3], has been built so that the DGs can be created by a drag drop graphical environment that permits fast creation and editing of the DGs. Actions to validate, upload, and download the DGs in either the graphical or xml format have been developed to permit service creation of the DGs.

6.5   YANG SERVICE MODEL EXAMPLE

The service model generally describes the high-level data items for a service along with the business functions that support the instantiation and lifecycle management of the service. Consequently, the main data items in a service model are the remote procedure calls (RPCs) that make up the northbound interface (NBI), along with the input and output data items for those RPCs. A typical service will have a create, update, and delete RPC method with the corresponding "customer orderable" data as input. While the service model simply captures the abstract view of the requested service, other models will assign, calculate, or engineer resources at the network and device layers to implement the requested service.

The service model RPCs also handle the lifecycle management interface for modify and delete, however, the network service model does not specifically assign or release resources. The network model provides the mapping that addresses the assignment or release of resources, and when network state change should occur. This is important because a particular RPC message may only affect the database inventory and not affect state in the network or it may affect both.

In the example service model below *for a simple VPN service definition*, there is an RPC for "create-vpn" that assigns a vpn identifier. It is not shown in the example, but there would be a separate RPC to actually "attach" a preexisting access connection to that VPN. Multiple calls to the "attach" RPC would be called to create the VPN due to two access connections. For the "create-vpn" request, the input is a custID, vpnName, and topology and the output is simply a vpnId that will be used in subsequent messages.

The service model is complete when all the inputs needed for the network model to apply engineering rules, assign resources, and drive the set of business process RPC (create, update, delete) are satisfied.

The service model is generally both network and device agnostic but will include some minimum data set of network information from the customer because all customer ordered parameters/ options must come in at the service model layer.

One of the most important aspects of the service data model is that it will be used to auto-generate the REST API northbound to the OSS/ONAP systems that initiate the RPC requests into the network controller. Adding a new parameter to the YANG model and adding it to the system through the compiler causes it to be instantly available on the NBI, available to logic in processing events and added to persistent storage

```
module vpn-network {
  namespace "urn:opendaylight:vpn";
  prefix vpn;
  import ietf-inet-types {prefix inet;}
```

```
revision "2014-06-03" {
  description "Example VPN Service Module";
}

rpc create-vpn {
  description "Create VPN request";
  input {
    leaf custId{
      type uint32;
    }
    leaf vpnName {
      type string;
    }
    leaf topology {
      type string; // hub&spoke; any-to-any -> could do as ENUM
    }
  }
  output {
    leaf vpnId {
      type uint32;
    }
  }
}
```

6.6   YANG NETWORK MODEL EXAMPLE

The network model augments the data from the service request with the resources and attributes needed for a vendor agnostic network implementation. It includes the identification of resources needed for the implementation, the relationship of the resources, and their grouping where it impacts the network resource assignments. For example, in the VPN case, the assignment of VRFs is only needed once per PE because all access connections on that PE can share the same VRF definition. Similarly, route targets for any-to-any VPNs do not need to be unique, however, for hub-and-spoke VPNs the route targets have to be unique for hubs and spokes.

Network models do not need to have RPC definitions, but there needs to be a mapping between the service model and the network model that applies to the RPC. The YANG standards do not define this mapping, and so various mechanisms are used. In the AT&T Network Controller either a YANG extension to map a container (grouping of elements in a YANG model) to a service model RPC, or a separate configuration file that defines that mapping, is used.

In the example below, we have a container for the "vpn" that is used for the create-vpn RPC from the service model. In the complete network model, there is another container for "vpn-pe" that maps to the RPC from "attach-to-vpn" RPC message in the service model.

The intent of a network model and in particular, a container in a network model (e.g., the virtual compute and storage environment in which the network function runs), is that once all the "leafs" have been assigned (implicitly completing all engineering rules for each resource to be assigned and for the resulting container) then the network state can be changed. In the network controller, the process of filling out the leaves in the network data model is performed by the DG logic. That logic uses the data items in the network data model as the "context" data storage blob that needs to be filled out for the RPC event message being processed.

One of the advantages or uses of the network YANG model is the definition of the data items that will be stored in the database repository for the customer service. Class libraries and helper functions can be autogenerated from the YANG model, but the most important aspect from a development cycle time standpoint is that the YANG model will be used to autogenerate the context structure and persistent storage structures.

```
module vpn-network {
  namespace "urn:opendaylight:vpn";
  prefix vpn;
  import ietf-inet-types {prefix inet;}
  revision "2014-06-03" {
    description "Example VPN Network Module";
  }
  container vpn {
      leaf resource-vpn-id {
        type uint32; // <allocate type="vpn-id" saveAs="$vpn-id" />
      }
      leaf cust-id{
        type uint32;
      }
      leaf vpn-name {
        type string;
      }
      leaf topology {
        type string; // hub&spoke; any-to-any -> could do as ENUM
      }
  }

  container vpn-pe {
      leaf ref-resource-vpn-id {
    type uint32;
      }
      list pe {
    key "equipment-clli";
        leaf equipment-clli {
          type string;
        }
        leaf network-type{
          type string;
        }
        leaf resource-route-target-any {
      type uint32;
          //when topology is any-to-any;
      }
        leaf resource-route-target-hub {
      type uint32;
          //when topology is hub-and-spoke;
      }
        leaf resource-route-target-spoke {
          type uint32;
          //when topology is hub-and-spoke;
      }
          list vrf {
        key "resource-vrf-id";
            leaf resource-vrf-id {
                type uint32;

            leaf resource-route-distinguisher {
                type uint32;
            }
            leaf resource-import-route-target {
              type uint32;
```

```
            }
            leaf resource-export-route-target {
                type uint32;
            }
        list site {
                key "site-id";
                leaf site-id {
                    type uint32;
                }
                leaf site-interface {
                    type string;
                }
                leaf site-l3-routing {
                    type string;
                }
            } //list-site
        } //list-vrf
    } // list-pe
  } // vpn-pe container
} // module vpn-network
```

6.7 NETWORK CONTROLLER AND ORCHESTRATION USE-CASE EXAMPLE: CUSTOMER REQUESTING VPN SERVICE

Figure 6.4 (and the associated description) shows an end-to-end use case of what happens when a customer requests a five-node VPN service.

1. Business support system (BSS) submits request to master orchestration and management function (OMF) service orchestration for 5 VPN sites for customer X.
2. OMF requests 5 VPN sites for customer X to service controller (SC), which supports VPN.

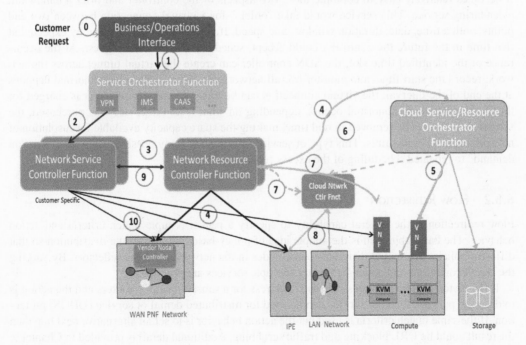

FIGURE 6.4 Customer request for VPN service.

3. VPN service controller (VPN-SC) requests from network controller (NC) if network resources are ready.
4. NC configures the set of PNFs/VNFs to support service
   a. For PNF—goes directly to the WAN PNF to configure
   b. For VNF—goes to infrastructure orchestration and control (IO&C) to create the container (compute, storage) for the VNF application
   c. NC requests a VNF from the IO&C with the specific location, policy
   d. IO&C spins up compute, storage based on VNF container policies, and loads/starts up the VNF
   e. IO&C tell NC that VNF is installed
   f. NC configures the VNFs and notifies the IO&C to configure the IPE/LAN/OVS to the VNF
   g. IO&C configures the IPE/LAN/OVS
5. NC tells SC network configured graph is complete.
6. SC sets up VPN service (VRFs) to WAN PNF, IPE, VNF.

6.8 SOME USE-CASE EXAMPLES OF SDN CONTROL

6.8.1 BANDWIDTH CALENDARING

Bandwidth calendaring is the idea that high bandwidth circuits between end points in the network (like large data centers) could be offered at lower prices during off-peak demand with perhaps special conditions in the event of a network need for the otherwise idle capacity. Implementing this type of feature with its needs for historical traffic analysis and real-time control actions is problematic in a pre-SDN environment. SDN controllers can "listen" to telemetry from the network and build up a history of utilization across the nodes and links. The SDN controller can also carve out or virtualize the network to allocate paths and bandwidth for special function via MPLS TE Tunnels using either NetConf or Path Computation Element Protocol (PCEP) signaling [4]. It becomes relatively easy to combine these two aspects of the controller and offer a bandwidth calendaring service. This service would take "orders" for a virtual connection between two end points, with a time, date, duration window, and speed. If there was available capacity expected at that time in the future the controller could accept, schedule, and save the request. At the occurrence of the identified time slot, the SDN controller can create the virtual tunnel across the network, record the start time, and monitor overall network conditions. If nothing abnormal happens at the end of the interval, the virtual connection can be removed and the customer is charged for the use. If something abnormal occurs, depending on what level of service was purchased, the virtual circuit could be removed in real time, making the spare capacity available for resolution of network capacity constraints. This type of service offering has many possible variations from "on demand" to repeated scheduling of the excess capacity.

6.8.2 FLOW REDIRECTION

Flow redirection is the general capability to specify a packet N-tuple match criteria and action behavior. The basic approach is the simple idea of policy-based targeted route distribution so that different policies are sent to different source nodes in the network via route reflectors. By varying the match conditions and action behaviors, multiple services are possible.

If the match criteria is on destination IP address for a subset of source routers and the action is to drop the packets, then the service can be used for distributed denial of service (DDOS) protection. If the same match criteria is used but the action behavior is to set an alternative next hop then the result could be URL blocking and traffic scrubbing. Additional detail is provided in Chapter 9. SDN controllers can create and distribute policies for the matching action through multiple types of adapters: NetConf or CLI, BGP, or even OpenFlow. The particular adapter is less important that the fact that the SDN controller can manage the data, translate the requested service into a network and device configuration or signaling change, and monitor the service during operation.

6.9   CHOICE OF OPEN SOURCE SDN CONTROLLERS

There are multiple open source SDN controllers available in the industry that range from research experiments to hardened and scalable production controllers. Three controllers merit closer inspection when it comes to choosing an SDN controller for a service provider. As is to be expected they have some overlapping capabilities and any one of them is likely to be an acceptable choice. However, based on the specific needs of global control, local control or embedded device control, a specific open source SDN controller could be the proper choice.

- OpenDaylight [5] is used in AT&T as a global controller. It has most of the attributes for a flexible controller needed for a mature service provider. It also supports a large variety of adapters so that it can configure and control a large number of physical and virtual devices through established and emerging protocols (BGP, PCEP [4], NetConf, OpenFlow). OpenDaylight's modeling capability is the strongest of any of the open source SDN controllers. The dynamic binding of YANG model-based java objects, makes it easy to create new services and a great fit to the configuration phase of NetConf devices that use YANG models.
- OpenContrail [6] is used in AT&T as a local controller. Primarily because of its unique approach to managing the virtual router (vRouter) on the compute nodes, the controller provides a large data center scaled solution for dealing with the overlay network, at scale. It does not have as good a modeling capability as OpenDaylight but it is a good fit for service chaining in the overlay networks.
- ONOS [7] is not directly used by AT&T but it is positioned as the embedded controller that would best fit into the control of merchant silicon switches. In this mode, the SDN controller needs to be focused on switch-control interfaces and in efficiently managing the OpenFlow-based protocols through high-level abstractions like network intents.

6.10   TOPICS FOR FURTHER STUDY

In this section, we identify and describe some topics for further exploration.

Network closed-loop control with telemetry lag: There is measurable lag in collection of telemetry from large-scale networks. There is also a massive amount of data to be analyzed to determine both the current state and whether a new state is more optimal. Between collection lag and analytic lag, centralized closed-loop control can be challenging and usually puts the centralized algorithm into the space of optimizing after distributed algorithms have already reacted. It would be interesting to look at this problem and determine whether there were better algorithms to manage the two types of lag and come up with more optimal centralized control-loop paradigms.

Model to execution software techniques: Today, the YANG models used to define service and network data structures are compiled directly to java code to generate the REST APIs. It would be interesting to look at an API engine that could directly operate off the YANG models and bypass the steps for code generation. Dynamic binding of the REST RPCs and RESTCONF GET/PUT/POST/DELETE operations against the http server based on the model would be another opportunity to take time and resources out of the feature development cycle.

ACKNOWLEDGMENT

The authors would like to acknowledge the contribution of Margaret Chiosi.

REFERENCES

1. YANG—A Data Modeling Language for the Network Configuration Protocol (NETCONF), IETF RFC6020, https://tools.ietf.org/html/rfc6020.
2. S. Agrawal, V. Narasayya, B. Yang, Integrating vertical and horizontal partitioning into automated physical database design, *Proceedings of the 2004 ACM SIGMOD International Conference on Management of Data*, June 13–18, 2004, Paris, France.
3. Node-RED—http://www.nodered.org/
4. PCEP—Path Computation Element (PCE) Communication Protocol (PCEP) RFC5440, https://tools.ietf.org/html/rfc5440.
5. ODL—http://www.opendaylight.org/
6. OpenContrail—http://www.opencontrail.org/
7. ONOS—http://onosproject.org/

7 The Network Operating System
VNF Automation Platform

Chris Rice and Andre Fuetsch

CONTENTS

7.1 ONAP: Logical Technical Architecture 105
    7.1.1 Open Network Automation Platform 106
    7.1.2 ONAP Component Roles 107
    7.1.3 European Telecommunications Standards Institute: NFV Management
        and Orchestration and ONAP Alignment 109
7.2 Master Services Orchestrator 111
7.3 SDC Environment 112
    7.3.1 Metadata-Driven Design Time and Runtime Execution 112
    7.3.2 Data Repositories 115
    7.3.3 Certification Studio 116
    7.3.4 Distribution Studio 116
7.4 Software-Defined Controllers 116
    7.4.1 Application, Network, and Infrastructure Controller Orchestration 116
    7.4.2 Infrastructure Controller Orchestration 116
    7.4.3 Network Controller Orchestration 117
    7.4.4 Application Controller Orchestration 117
7.5 Portal, Reporting, GUI, and Dashboard Functions 117
7.6 Data Collection Analytics and Events 119
    7.6.1 DCAE's Four Major Components 120
    7.6.2 Platform Approach to DCAE 121
    7.6.3 DCAE Platform Components 122
7.7 Policy Engine 123
    7.7.1 Policy Creation 125
    7.7.2 Policy Distribution 126
    7.7.3 Policy Decision and Enforcement 126
    7.7.4 Policy Unification and Organization 127
    7.7.5 Policy Technologies 127
    7.7.6 Policy Use 128
7.8 Active and Available Inventory (A&AI) System 128
    7.8.1 Key A&AI Requirements 129
    7.8.2 A&AI Functionality 130
7.9 Control Loop Systems: Working Together 130
    7.9.1 Design Framework 131
    7.9.2 Orchestration and Control Framework 131
    7.9.3 Analytic Framework 132
7.10 Legacy BSSs Interactions with ONAP 133
    7.10.1 BSS Scope 134

As discussed in Chapter 2, the network cloud leverages cloud technologies, software control, and network virtualization in the new network. The goal is to deliver this new network cloud, while reducing capital, and operations expenditures by achieving significant levels of operational automation.

Originally, AT&T created an internal network operating system, called Enhanced Control, Orchestration, Management, and Policy (ECOMP). That ECOMP software platform delivered product/service independent capabilities for the design, creation, and lifecycle management of the network cloud environment for carrier-scale, real-time workloads. It consisted of eight software subsystems covering two major architectural frameworks: a design time environment to design, define, and program the platform, and an execution time environment to execute the logic programmed in the design phase utilizing closed-loop, policy-driven automation.

In 2016, AT&T decided to open source ECOMP through the Linux Foundation with the belief that benefits to the entire ecosystem outweighed any time advantage AT&T would have by implementing a proprietary ECOMP platform. In addition, to scale the network cloud vision, something like ECOMP in open source would be required to jumpstart other service providers as they moved down the network cloud path. The Linux Foundation had a similar effort, called Open-O with members such as China Mobile, Huawei, China Telecom, Intel, and others. The main principals in ECOMP and Open-O worked cooperatively to join forces, harmonizing the open source network automation space, under the auspices of the Linux Foundation. Positively for the industry, these two projects joined forces to create the Open Network Automation Platform (ONAP) within the Linux Foundation. Throughout this chapter, and in the rest of this book, the abbreviation ONAP will be used to refer to this software-defined, open source, networking automation platform.

A virtualized network function (VNF) automation software platform like ONAP is critical for service providers in achieving the network cloud vision, delivering on the imperatives to increase the value of the network to customers by rapidly onboarding new services, by enabling the creation of a new ecosystem of cloud consumer and enterprise services, by reducing capital and operational expenditures, and by providing operational efficiencies. ONAP also delivers enhanced customer experience by allowing the customer in near real-time to reconfigure their network, services, and capacity. While ONAP does not directly support legacy physical elements, it works with traditional operations support systems (OSSs) to provide a seamless customer experience across both virtual and physical elements.

ONAP enables network agility, elasticity, and improves time to market, revenue, and scale via the Service Design and Creation (SDC) module, allowing for visual modeling and design. ONAP provides a policy-driven Operational Management Framework (OMF) for security, performance, and reliability/resiliency utilizing a metadata-driven repeating design patterns at each layer in the architecture. It reduces capital expenditure (CapEx) through a closed-loop automation approach that provides dynamic capacity and consistent failure management when and where it is needed. It facilitates operational efficiency though the real-time automation of service, network, cloud delivery, and lifecycle management provided by the OMF and application components.

ONAP provides high utilization of network resources by combining dynamic, policy-enforced functions for component and workload shaping, placement, execution, and administration. These functions are built into ONAP, and utilize the network cloud; when combined, these provide a unique level of operational and administrative capabilities for workloads that run natively within the ecosystem. They will also extend many of these capabilities into third-party cloud ecosystems as interoperability standards for network clouds evolve.

Diverse workloads are at the heart of the capabilities enabled by ONAP. The service design and creation capabilities and policy recipes eliminate many of the manual and long-running processes performed via traditional OSSs (e.g., break-fix largely moves to a plan and build function). ONAP provides external applications (OSS/business support system [BSS], customer apps, and third-party integration) with a secured, RESTful application programming interfaces (API) access control to ONAP services, events, and data.

7.1 ONAP: LOGICAL TECHNICAL ARCHITECTURE

In order to completely understand what drove the definition of the ONAP logical technical architecture and the functions ascribed to it, it is helpful to consider the initial drivers of the software-defined network (SDN) cloud and network functions virtualization (NFV) efforts.

When the network cloud effort began, cloud technology was being adopted, focusing primarily on information technology (IT) or corporate applications. The cloud technology provided the ability to manage diverse workloads of applications dynamically. Included among the cloud capabilities were

1. Real-time instantiation of virtual machines (VMs) on commercial hardware
2. Dynamic assignment of applications and workloads to VMs
3. Dynamic movement of applications and dependent functions to different VMs on servers within and across data centers in different geographies
4. Dynamic control of resources made available to applications (CPU, memory, storage)

At the same time, efforts were underway to virtualize network functions that had been realized as purpose-built appliances: specialized hardware and software (e.g., routers, firewalls, switches, etc.). NFV was focused on transforming network appliances into software applications.

The broad network cloud strategy, led largely by AT&T, is grounded in the confluence of NFV, SDN, and cloud technology. With VNFs running as cloud applications, the network cloud takes advantage of the dynamic capabilities of cloud cited above in defining, instantiating, and managing network infrastructures and services. This strategy shaped the definition of ONAP, its architecture and the capabilities and functions it provides. This strategy also shaped the cloud infrastructure that converges multiple clouds into a single enterprise cloud capable of interoperating dynamically with ONAP-controlled virtual functions (VFs) and third-party clouds.

In this new network cloud, the dynamic cloud capabilities are applied to applications—that is, VNFs—thus applying the benefits of cloud to virtual network elements. For example, VNFs, such as routers, switches, firewalls, can be "spun up" on commodity hardware, moved from one data center to another center dynamically (within the limits of physical access tie-down constraints) and resources such as CPU, memory, and storage can be dynamically controlled.

The ONAP architecture and OMF was defined to address the business/strategic objectives of the network cloud as well as new technical challenges of managing a highly dynamic environment of VNFs and services, that is, a software ecosystem where functions such as network and service provisioning, service instantiation, and network element deployment all occur dynamically in real time.

ONAP enables the rapid onboarding of new network cloud services. The reduction of operating expenditure (OpEx) and CapEx is enabled through its metadata-driven service design and creation platform and its real-time OMF—a framework that provides real-time, policy-driven automation of management functions. The metadata-driven service design and creation capabilities enable services to be defined with minimal IT development required; thus contributing to reductions in CapEx. The real-time OMF provides significant automation of network management functions enabling the detection and correction of problems in an automated fashion contributing to reductions in OpEx.

One of the challenges for service providers in a traditional telecommunications environment is that unique and proprietary interfaces are required for many network management systems (NMS) and element management systems (EMSs); these lead to significant integration, startup, and operational costs, and standards evolve slowly in this space.

As the service provider industry transitions to this new network cloud environment, there are plans to continue contributing to and leveraging code within the open source community; this approach facilitates agile and iterative standards that incorporate incremental improvements. In the network cloud ecosystem controlled by ONAP, the goal is to rapidly onboard vendor VNFs with standard processes, and then, operate these resources via vendor-independent controllers and standard management, security, and application interfaces. Configuration and management is model-driven and will utilize new standards as they become available. To this end, the network cloud should support open-cloud standards (e.g., OpenStack, TOSCA, etc.) and support cloud and network virtualization industry initiatives and protocols (e.g., NetConf, YANG, OpenConfig, OPNFV, etc.). As further standardization occurs, ONAP will incorporate these, as appropriate.

7.1.1 OPEN NETWORK AUTOMATION PLATFORM

ONAP enables product- and service-independent capabilities for design, creation, and lifecycle management. There are many requirements that must be met by ONAP to support this new network cloud vision. Of those many requirements, some are key in supporting the following foundational principles:

1. The architecture will be metadata-driven and policy-driven to ensure flexible ways in which capabilities are used and delivered
2. The architecture shall enable sourcing best-in-class components
3. Common capabilities are "developed" once and "used" many times
4. Core capabilities shall support many services
5. The architecture shall support elastic scaling as needs grow or shrink

These capabilities are provided using two major architectural frameworks: (1) a design time framework to design, define, and program the platform (uniform onboarding) and (2) a runtime execution framework to execute the logic programmed in the design time framework (uniform delivery and lifecycle management). Figure 7.1 shows ONAP frameworks.

The design time framework component is an integrated development environment with tools, techniques, and repositories for defining/describing network and service assets. The design time framework facilitates reuse; resources and services are described as models, thus improving efficiency, especially as more and more models are available for reuse. Assets include models of network cloud resources, services, and products. The models include various process specifications and policies (e.g., rule sets) for controlling behavior and process execution. Process specifications are used by ONAP to automatically sequence the instantiation, delivery, and lifecycle management aspects of network cloud-based resources, services, products, and ONAP components themselves. The design time framework supports the development of new capabilities, augmentation of existing capabilities, and operational improvements throughout the lifecycle of a service. SDC, Policy, and Data Collection, Analytics, and Events (DCAE) software development kits (SDKs) allow operations, security, third parties (e.g., vendors), and other experts to continually define/refine new

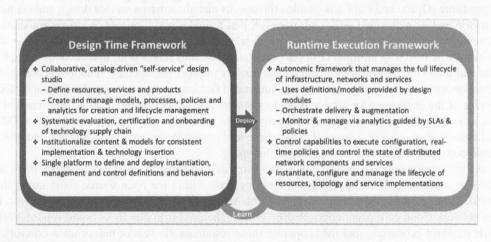

FIGURE 7.1 ONAP frameworks.

collection, analytics, and policies (including recipes for corrective/remedial action); these can be accessed using the ONAP design framework portal. Certain process specifications (aka "recipes") and policies are geographically distributed to many points of use to optimize performance and maximize autonomous behavior in the network cloud environment.

Figure 7.2 provides a high-level view of the ONAP software components—eight unique and distinct subsystems. These subsystems use micro-services to perform their roles. The platform also provides the common functions (e.g., data collection, control loops, metadata recipe creation, policy/recipe distribution, etc.) necessary to construct specific behaviors. To create a service or operational capability, it is necessary to develop service/operations-specific collection, analytics, and policies (including recipes for corrective/remedial action) using the ONAP design framework portal.

The three primary components of the design time framework are SDC, the policy creation components, and the analytic application design. SDC is an integrated development environment with tools, techniques, and repositories to define/simulate/certify the network cloud assets as well as their associated processes and policies. Each asset is categorized into one of four asset groups: resource, services, products, or offers. The policy creation component deals with polices; these are conditions, requirements, constraints, attributes, or needs that must be provided, maintained, and/or enforced. At a lower level, policy involves machine-readable rules enabling actions to be taken based on triggers or requests. Policies often consider specific conditions in effect (both in terms of triggering specific policies when conditions are met, and in selecting specific outcomes of the evaluated policies appropriate to the conditions). Analytics that define the data collection, Key Performance Indicators (KPIs), events, and alarming of the service are also defined as part of the design time framework.

The design and creation environment supports a multitude of diverse users via common services and utilities. Using the design studio, product and service designers onboard/extend/retire resources, services, products, and offers. Operations, engineers, customer experience managers, and security experts create workflows, policies, and methods to implement closed-loop automation and manage elastic scalability.

The runtime execution framework executes the rules and policies distributed by the design and creation environment. This allows us to distribute policy enforcement and templates among various ONAP modules such as the Master Service Orchestrator (MSO); Network and Application Controllers; DCAE; Active and Available Inventory (A&AI); and a security framework. These components advantageously use common services that support logging, access control, and data management.

7.1.2 ONAP COMPONENT ROLES

Orchestration is the function defined via a process specification that is executed by a work flow component. MSO automates a sequence or a set of sequences of activities, tasks, rules, and policies needed for on-demand creation, modification, or removal of network, application, or infrastructure services and resources. The MSO provides workflow orchestration at a very high level, with an end-to-end view of the infrastructure, network, and application scopes. Controllers are applications that are closely associated with cloud and network services and execute the configuration, real-time policies, and control the state of distributed components and services. Rather than using a single monolithic control layer, ONAP has chosen to use three distinct controller types that manage resources in the execution environment corresponding to their assigned controlled domain such as cloud computing resources (infrastructure controller, typically within the cloud layer), network configuration (network controller), and application (application controller).

DCAE and other ONAP components provide Fault Configuration Accounting Performance Security (FCAPS) functionality. "Finished goods" for specific services are created from the platform capabilities for specific services and products. DCAE supports closed-loop control and higher-level correlation for business and operations activities. It is the ecosystem component supporting analytics and events: it collects performance, usage, and configuration data; provides computation

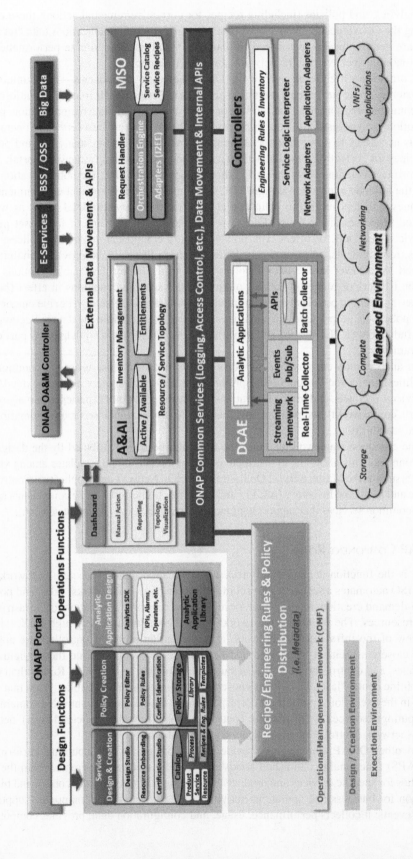

FIGURE 7.2 ONAP software components.

of analytics; aids in troubleshooting; and publishes events, data, and analytics (e.g., to policy, orchestration, and the data store).

A&AI is the ONAP component that provides real-time views of the network cloud resources, services, products, and offers, as well as their relationships to one another. The views provided by A&AI relate data managed by ONAP, BSSs, OSSs, and network applications to form a "top to bottom" view ranging from the offers in the market to the products customers buy to services that compose those products to the resources that form the raw material for creating the products. A&AI not only forms a registry of products, services, and resources, but also maintains up-to-date views of the relationships between these inventory items. To deliver the dynamic vision of the network cloud, A&AI will manage these multidimensional relationships in real time.

A&AI is updated in real time by controllers as they make changes in the network cloud environment. A&AI is metadata-driven, allowing new inventory item types to be added dynamically and quickly via SDC catalog definitions, eliminating the need for lengthy development cycles.

The platform includes a real-time dashboard, controller, and administration tools to monitor and manage all the ONAP components via an Operations, Administration, and Management (OA&M) instance of ONAP. It allows the design studio to onboard ONAP components and create recipes, and it allows the policy framework to define ONAP automation.

ONAP delivers a single, consistent user experience (UX) based on the user's role and allows role changes to be configured within the single ecosystem. This UX is managed by the ONAP Portal. The ONAP Portal provides access to design, analytics, and operational control/administration functions via a common role-based menu or dashboard. The portal architecture provides web-based capabilities including application onboarding and management, centralized access management, dashboards, as well as hosted application widgets. The portal provides an SDK to drive multiple development teams to adhere to consistent user interface (UI) development requirements by taking advantage of built-in capabilities (services/API/UI controls), tools, and technologies.

ONAP portal provides common operational services for all ONAP components including activity logging, reporting, common data layer, access control, resiliency, and software lifecycle management. These services provide access management and security enforcement, data backup, restoration, and recovery. They support standardized VNF interfaces and guidelines.

The virtual operating environment of the network cloud introduces new security challenges and opportunities. ONAP provides increased security by embedding access controls in each ONAP component, augmented by analytics and policy components specifically designed for the detection and mitigation of security violations.

7.1.3 European Telecommunications Standards Institute: NFV Management and Orchestration and ONAP Alignment

The European Telecommunications Standards Institute (ETSI) developed a reference architecture framework and specification in support of NFV management and orchestration (MANO). The main components of the ETSI-NFV architecture are orchestrator, VNF manager, and virtualized infrastructure (VI) manager. ONAP expands the scope of ETSI MANO coverage by including controller, data collection and analytics, portal, and policy components. Policy plays an important role to control and manage behavior of various VNFs and the management framework. ONAP also significantly increases the scope of ETSI MANO's resource description to include complete metadata for lifecycle management of the virtual environment (infrastructure as well as VNFs). The ONAP design framework is used to create resource/service/product/offer definitions (consistent with MANO), as well as engineering rules, recipes for various actions, policies, and processes. The metadata-driven generic VNF manager (i.e., ONAP) allows quick onboarding of new VNF types; doing this without going through long development and integration cycles and efficiently managing cross dependencies between various VNFs is key. Once a VNF is onboarded, the design time framework facilitates rapid incorporation into future services.

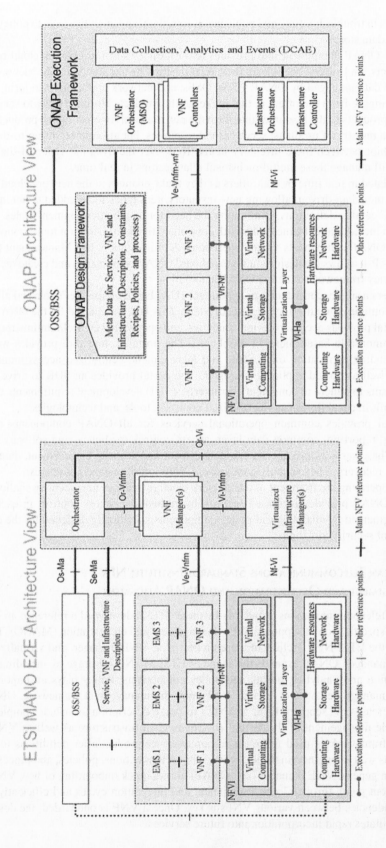

FIGURE 7.3 Comparison of ETSI MANO and ONAP architectures.

ONAP can be considered as a very full-featured, enhanced generic VNF manager as described by MANO NFV MANO architectural options (see Figure 7.3).

In addition, ONAP subsumes the traditional FCAPS functionality as supported by EMSs in the MANO reference architecture. This is critical to implementing analytic-driven closed-loop automation across the infrastructure and VNFs, analyzing cross dependencies and responding to the root cause of problems as quickly as possible. To successfully implement ONAP, the Ve–Vnfm–vnf interface (as well as Nf–Vi) is critical. It is expected that the Ve–Vnfm–vnf interface is standard interface defined by ETSI. ONAP's approach is to document detailed specifications for such interface(s) to collect rich real-time data in a standard format from a variety of VNFs and quickly integrate with ONAP without long custom development work.

7.2 MASTER SERVICES ORCHESTRATOR

In general, orchestration can be viewed as the definition and execution of workflows or processes to manage the completion of a task. The ability to graphically design and modify a workflow process is the key differentiator between an orchestrated process and a standard compiled set of procedural code. Orchestration provides a "template" approach, allowing for adaptability and improved time to market; it provides ease of definition and change without the need for a development engagement. As such, it is a primary driver of flexibility in the architecture. Interoperating with policy, the combination provides a basis for the definition of a flexible process that can be guided by business and technical policies and driven by process designers.

Orchestration exists throughout the ONAP architecture and should not be limited to the constraints implied by the term "workflow," as it typically implies some degree of human intervention. Orchestration in automated network clouds will not involve human intervention/decision/guidance in the vast majority of cases. The human involvement in orchestration is typically performed up front in the design process although there may be processes that require intervention or alternate action such as exception or fallout processing.

To support the large number of orchestration requests, the orchestration engine is exposed as a reusable service. With this approach, any component of the architecture can request execution of process recipes. Orchestration services are capable of consuming a process recipe and executing against it to completion. The service model maintains consistency and reusability across all orchestration activities and ensures consistent methods, structures, and versioning of the workflow execution environment.

Orchestration services expose a common set of APIs to drive consistency across the interaction of ONAP components. To maintain consistency across the platform, orchestration processes interact with other platform components or external systems via standard and well-defined APIs.

The MSO's primary function is the automation of end-to-end service instance provisioning activities. The MSO is responsible for the instantiation/release, and migration/relocation of VNFs in support of end-to-end service instantiation, operations, and management. The MSO executes well-defined processes to complete its objectives and is typically triggered by the receipt of "service requests" generated by other ONAP components or by order lifecycle management in the BSS layer. The orchestration "recipe" is obtained from the SDC component of the ONAP where all service designs are created and exposed/distributed for consumption.

Controllers (infrastructure, network, and application) participate in service instantiation and are the primary players in ongoing service management, for example, control loop actions, service migration/scaling, service configuration, and service management activities. Each controller instance supports some form of orchestration to manage operations within its scope.

Figure 7.4 illustrates the use of orchestration in the two main areas: service orchestration embodied in the MSO and service control embodied in the infrastructure, application, and network controllers. It illustrates the two major domains in a network cloud that employ orchestration, network service, and network infrastructure. Although the objectives and scope of the domains vary, they both follow a consistent model for the definition and execution of orchestration activities.

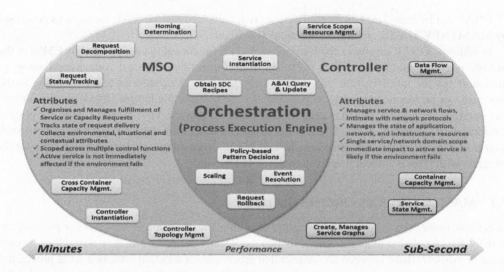

FIGURE 7.4   Comparison of MSO and controllers (ONAP view).

Depending on the scope of a network issue, the MSO may delegate, or a controller may assume, some of the activities identified above. The MSO's job is to manage orchestration at the top level and to facilitate the orchestration that takes place within the underlying controllers; it also marshals data between the controllers such that they have the "process steps" and all the "ingredients" to complete the execution of their respective recipes. For new services, this may involve determination of service placement and identification of existing controllers that meet the service request parameters and have the required capacity. If existing controllers (infrastructure, network, or application) do not exist or do not have capacity, the MSO will obtain a recipe for instantiation of a new controller under which the requested service can be placed.

SDC is the module of ONAP where orchestration process flows are defined. These process flows will start with a template that may include common functions such as homing determination, selection of infrastructure, network and application controllers, consultation of policies, and interrogation of A&AI to obtain necessary information to guide the process flows. The MSO does not provide any process-based functionality without a recipe for the requested activity regardless of whether that request is a customer order or a service adjustment/configuration update to an existing service.

MSO will interrogate A&AI to obtain information regarding existing network and application controllers to support a service request. A&AI will provide the addresses of candidate controllers that are able to support the service request. The MSO may then interrogate the controller to validate its continued available capacity. The MSO and the controllers report reference information back to A&AI upon completion of a service request to be used in subsequent operations.

7.3   SDC ENVIRONMENT

7.3.1   METADATA-DRIVEN DESIGN TIME AND RUNTIME EXECUTION

Metadata is generally described as "data about data" and is a critical architecture concept dealing with both abstraction and methodology. Metadata expresses structural and operational aspects of the virtualized elements comprising products, services, and resources as expressed in terms of logical objects within a formally defined model space. The attributes of these objects, and the relationships among them, embody semantics that correspond to real-world aspects of the modeled elements. The modeling process abstracts common features and internal behaviors in order to drive architectural consistency and operational efficiency. The logical representations of underlying elements can be manipulated and extended by designers in consistent ways, and uniformly consumed by the runtime execution framework. Alignment between elements in the model space and those in the real world is continuously maintained through tooling that resolves dependencies, handles exceptions, and infers required actions based on the metadata associated with the modeled elements.

One of the key benefits in a virtualized network architecture is a significantly decreased time from service concept to market deployment. The need to operationalize on a service-specific basis would be a major obstacle to this goal. Thus, network service operators have to manage their network cloud and on-demand services via the execution environment driven by a common (service independent) operations support model populated with service-specific metadata. In conjunction with this, these support systems will provide high levels of service management automation through the use of metadata-driven event-based control loops.

Such an approach achieves another benefit of driving down support systems costs through the ability to support new services with changes only in the metadata, and without any code modifications. In addition, a common operational support model across all offered services, coupled with high automation, drives down operations support costs (OpEx). ONAP implements a metadata-driven methodology in its network cloud by centralizing the creation of rules, analytic definitions, and policies in SDC. All ONAP applications and controllers will ingest the metadata that governs their behavior from SDC. Implementation of the metadata-driven methodology requires an enterprise-wide paradigm change in the development process. It demands an upfront agreement on the overall metadata-model across the business (e.g., product development) and the software developers writing code for ONAP, BSS, OSS, and the network cloud. This agreement is a behavioral contract to permit both the detailed business analysis and the construction of software to run in parallel. The software development teams focus on building service-independent software within a common operations framework for runtime automation, while the business teams can focus on tackling unique characteristics of the business needs by defining metadata models to feed the execution environment. The metadata model content itself is the glue between design time and runtime execution frameworks, and the result is that the service-specific metadata models drive the common service-independent software for runtime execution.

The metadata model is managed through a design environment (SDC) that guides a series of designers through third-party resource onboarding, service creation, verification, and distribution. The modular nature of the SDC metadata model provides a catalog of patterns that can be reused in future services. This provides a rich forward-engineering approach to quickly extend resource functions to manageable services and sellable products, further realizing benefits in time to market.

SDC is the ONAP component that supports metadata model design. Many different models of metadata exist to address different business and technology domains. SDC integrates various tools supporting multiple types of data input (e.g., YANG, HEAT, TOSCA, YAML, Business Process Management Notation [BPMN]/BPEL, etc.). It automates the formation of this metadata format to drive end-to-end runtime execution. SDC provides a cohesive and collaborative environment for design, test, certification, version control, and distribution of metadata models across the resource, service, product, and offer development lifecycle.

In SDC, the resource metadata is created in the description of the network cloud infrastructure and the configuration data attributes to support the service implementations on that network cloud. Subsequently, the resource metadata descriptions become a pool of building blocks that may be combined in developing services, and the combined service models to form products.

In addition to the description of the object itself, modeling the management *needs* of an object using metadata patterns in SDC may be applied to almost business and operations functions. For example, metadata can be used to describe the mapping of resource attributes to relational tables, through the definition of rules around the runtime session management of a group of related resources to the signatures of services offered by a particular type of resources. Such rules form the policy definitions that can be used to control the underlying behavior of the software function.

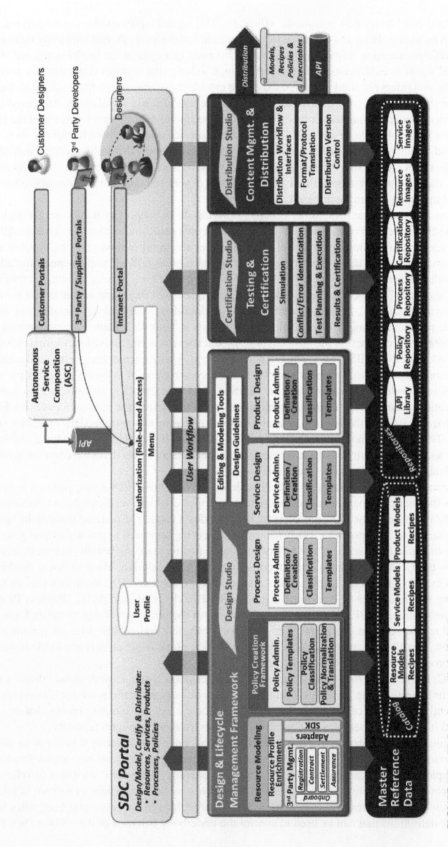

FIGURE 7.5   SDC design studio in ONAP.

Another example is to describe the workflow steps as a process in the metadata that can be used by ONAP orchestration to fulfill a customer service request.

Network cloud policies are one type of metadata, and will eventually be quite numerous and support many purposes. Policies are created and managed centrally so that all policy actions, both simple and complex, are easily visualized and understood together, and validated properly prior to use. Once validated and corrected for any conflicts, the policies are precisely distributed to the many points of use/enforcement; the decisions and actions taken by policy are distributed, but are still part of the policy component of ONAP. In this manner, policies will already be available when needed by a component, minimizing real-time requests to a central policy engine/policy decision point (PDP), or to policy distribution. This improves scalability and reduces latency.

SDC and policy creation exist in close relationship. Policies are created by many user groups (e.g., service designers, security personnel, operations staff, etc.). Various techniques are used to validate newly created policies and to help identify and resolve potential conflicts with preexisting policies. Validated policies are then stored in repositories. Subsequently, policies are distributed in two ways: (1) service-related policies are initially distributed in conjunction with the distribution of recipes (created via SDC), for example, for service instantiation and (2) other policies (e.g., some security and operations policies) are unrelated to particular services, and therefore, are independently distributable. In any case, policies are updatable at any time as required.

SDC employs a set of "studios" to design, certify, and distribute standardized models including the relationships within and across them. From a bottom-up view, SDC provides tools to onboard resources such as building blocks and to make them available for enterprise-wide composition. From a top-down view, SDC supports product managers who compose new products from existing services/resources in the SDC catalog or acquire new resource capabilities from internal or external developers. The components of SDC are shown in Figure 7.5.

The SDC "Design Studio" consists of a portal and back-end tooling for onboarding, iterative modeling, and validation of the network cloud assets. The design studio includes a set of basic model templates that each project can further configure and extend with additional parameters, parameter value ranges, and validation rules. The configured project-specific templates are used by the design studio graphical user interface (GUI) (as drop-down menu or rule-based validation) to validate the designs. This ensures the models contain the necessary information and valid values based on the type of model and the project requirements.

SDC provides access to modeling tools that can be used to create executable process definitions that will be used by the various network cloud components. Processes can be created with standard process modeling tools such as BPMN tools. The processes created are stored in the process repository and asset models in the catalog may refer to them.

The models in the master reference catalog can be translated to any industry-standard or required proprietary format by the distribution studio. The modeling process does not result in any instantiation in the run-time environment until the MSO receives a request to do so.

SDC will integrate with third-party suppliers, leveraging models for their management functions as needed for onboarding, configuration, analysis, and cataloging of software assets along with their entitlement and license models as part of the design studio.

7.3.2 DATA REPOSITORIES

SDC Data Repositories maintain the design artifacts and expose content to the designers, testers, and distributors. The repositories include the following:

1. Master reference catalog is the data store of designed models, including resources, services, and products, the relationships of the asset items, and their references to the process and policy repositories
2. Process repository is the data store of designed processes

3. Policy repository is the data store of designed policies
4. Resource images is the data store that contains the resource executables
5. Service images is the data store that contains the executables for the service
6. Certification repository is the data store of testing artifacts

7.3.3 CERTIFICATION STUDIO

SDC provides a certification studio with expanded use of automated simulation and test tools along with access to a shared, virtualized testing sandbox. The model-driven approach to testing enables a reduction in overall deployment cost, complexity, and cycle time. The studio provides the following capabilities:

1. Allows reuse and reallocation of hardware resources as needed for testing rather than dedicated laboratory environments
2. Supports test environment instantiation of various sizes when needed using a mix of production and test components, using a standardized and automated operations framework
3. Provides expanded use of automated test tools beginning in design through deployment for simulation/modeling behaviors, conflict and error identification, test planning and execution, and results/certification

7.3.4 DISTRIBUTION STUDIO

The models are stored in an internal technology independent format, which provides users greater flexibility in selecting the execution engines that consume the data. In a model-driven, software-based architecture, controlling the distribution of model data and software executables is critical. This studio provides a flexible, auditable mechanism to format, translate when needed, and distribute the models to the various network cloud components. Validated models are distributed from the design time environment to a runtime repository. The runtime distribution component supports two modes of access: (1) models can be sent to the component using the model, in advance of when they are needed or (2) models can be accessed in real time by the runtime components. This distribution is intelligent, such that each function automatically receives or has access to only the specific model components, which match its needs and scope.

7.4 SOFTWARE-DEFINED CONTROLLERS

7.4.1 APPLICATION, NETWORK, AND INFRASTRUCTURE CONTROLLER ORCHESTRATION

As previously stated, orchestration is performed throughout the ONAP architecture by various components, primarily the MSO and the application, network, and infrastructure controllers. Each will perform orchestration for one or more of the following:

1. Service delivery or changes to existing service
2. Service scaling, optimization, or migration
3. Controller instantiation
4. Capacity management

Regardless of the focus of the orchestration, all recipes will include the need to update A&AI with configuration information, identifiers, and IP addresses.

7.4.2 INFRASTRUCTURE CONTROLLER ORCHESTRATION

Like the MSO, controllers will obtain their orchestration process and payload (templates/models) from SDC. For service instantiation, the MSO maintains overall end-to-end responsibility for ensuring that a request is completed. As part of that responsibility, the MSO will select the appropriate controllers (infrastructure, network, and application) to carry out the request. Because a service request often comprises one or more resources, the MSO will request that the appropriate controllers obtain the recipe for the instantiation of a resource within the scope of the requested controller. After service placement is determined, the MSO may request the creation of a VM at one or more locations depending on the breadth of the service being instantiated and whether an existing instance of the requested service can be used. If new VM resources are required, the MSO will place the request to the infrastructure controller for the specific network cloud location. Upon receipt of the request, the infrastructure controller may obtain its resource recipe from SDC. The infrastructure controller will then begin orchestrating the request. For infrastructure controllers, this typically involves execution of OpenStack requests for the creation of VMs and for the loading of the VF software into the new VM container. The resource recipe will define VM sizing, including compute, storage, and memory. If the resource-level recipe requires multiple VMs, the MSO will repeat the process, requesting each infrastructure controller to spin up one or more VMs and load the appropriate VFs, again driven by the resource recipe of the infrastructure controller. When the infrastructure controller completes the request, it will pass the virtual resource identifier and access (VRID) information back to the MSO to provide to the network and application controllers. During the process, the MSO writes identifier information to A&AI for inventory tracking.

7.4.3 NETWORK CONTROLLER ORCHESTRATION

Network controllers are constructed and operate in much the same manner as application and infrastructure controllers. New service requests will be associated with an overall recipe for instantiation of that service. The MSO will obtain compatible network controller information from A&AI and will in turn request LAN or WAN connectivity and configuration to be performed. This may be done by requesting the network controller to obtain its resource recipe from SDC. It is the responsibility of the MSO to request (virtual) network connectivity between the components and to ensure that the selected network controller successfully completes the network configuration workflow. A service may have LAN, WAN, and access requirements, each of which will be included in the recipe and configured to meet the instance specific customer or service requirements at each level. Physical access might need to be provisioned in the legacy provisioning systems prior to requesting the MSO to instantiate the service.

7.4.4 APPLICATION CONTROLLER ORCHESTRATION

Application controllers will also be requested by the MSO to obtain the application-specific component of the service recipe from SDC and execute the orchestration workflow. The MSO continues to be responsible for ensuring that the application controller successfully completes its resource configuration as defined by the recipe. As with infrastructure and network controllers, all workflows, whether focused on instantiation, configuration, or scaling, will be obtained or originate from SDC. In addition, workflows also will report their actions to A&AI as well as to MSO.

Note that not all changes in network or service behavior are the result of orchestration. For example, application VFs can change network behavior by changing rules or policies associated with controller activities. These policy changes can dynamically enable service behavior changes.

7.5 PORTAL, REPORTING, GUI, AND DASHBOARD FUNCTIONS

SDNs and virtualized network functions (VNFs) are changing the way networks and services are designed and consumed. Imagine having on-demand access for designing, testing, and deploying

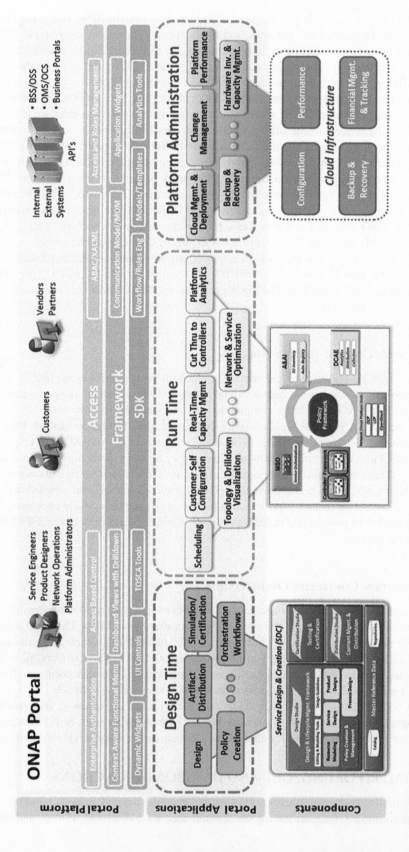

FIGURE 7.6 ONAP portal framework.

virtualized services, with the ability to configure, manage, and administer them in real time. The ONAP software is the VNF automation platform implementation of this vision, and ONAP portal is the access to manage these virtualized networks and services.

The ONAP portal provides a uniform web-based interface to all of the applications within ONAP. Think of the portal as a control panel or framework, which the individual ONAP applications "plug into," in a way that is transparent to the users of the portal.

As shown in Figure 7.6, the ONAP portal serves many different communities: those who administer, manage, and configure the network; those who operate and monitor it—eventually, even end users, who can use the portal for self-service via the web. It provides a uniform, consistent interface to the users of ONAP applications, and at the same time, it frees the developers of these applications from having to worry about applications other than their own. Because there are many different kinds of users with different needs, privileges, and responsibilities, the portal adapts itself based on who is using it, presenting just those functions and applications to which the particular user is allowed access.

The ONAP portal is actually more than just a framework or control panel; it is also a toolkit of software that makes it easy to create ONAP applications. Industry-standard open-source software is used—Java, common relational databases, modern JavaScript frameworks—all tools and languages that most developers already understand. Extensive support is provided to enable development of web-based applications with a consistent digital experience.

The portal, along with its software development toolkit (Figure 7.7), takes care of many functions needed by any web-based ONAP application, relieving the application developers of the burden of creating these. Among these functions are the following: single sign-on authentication; role-based authorization; security functions; required logging and auditing; and dashboard capabilities. For applications that are being newly built, the toolkit provides many potentially useful features that developers may wish to use in building their applications, such as the following: analytical and reporting engines, visualization and mapping tools (graphical information systems [GIS]), workflow and rules engines, network simulation, chat/video chat/screen sharing, etc. Not every application will need every component in the toolkit, of course, but these software components will be useful in many applications. These components have been hardened in production web applications, but they have been modernized and integrated into the ONAP portal toolkit for use by developers.

One of our goals in creating the ONAP portal is to make the application onboarding process—where applications are made accessible via the portal—as smooth and painless as possible. Existing mature applications—even required third-party applications—can be integrated quickly as plug-ins to the control panel. Over time, the level of integration of the applications with the portal can be increased, as business needs dictate.

While the ONAP portal serves as the control panel, the ONAP applications themselves that plug into the portal are not hosted by the portal. This allows application developers the freedom to work independently, eliminating development bottlenecks, while at the same time providing a seamless experience to users of the portal.

To summarize, the ONAP portal addresses the challenge of high growth in applications and functions that need to be developed and onboarded. It uses a flexible architecture that delivers variety of information, tools, applications, and dashboards. By leveraging a single mechanism, a thin-client, device-agnostic, browser-based web portal system, authorized users can configure, manage, and administer virtualized network and service functions on a network cloud. While this is a challenge, it is both a metric for success for ONAP portal architecture, as well as an opportunity to normalize applications, functions, and UX with scale.

7.6 DATA COLLECTION ANALYTICS AND EVENTS

In the virtualized network cloud vision, virtualized functions across various layers of functionality are expected to be instantiated in a significantly dynamic manner; this requires the ability to provide real-time responses to actionable events from virtualized resources, like ONAP applications,

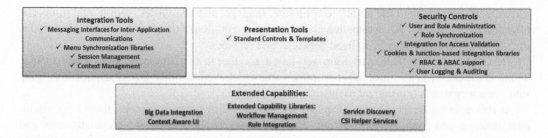

FIGURE 7.7   ONAP portal software development toolkit.

as well as requests from customers, partners, and other service providers. In order to engineer, plan, bill, and assure these dynamic services, DCAE within the ONAP framework gathers key performance, usage, telemetry, and events from VI. It does this to compute various analytics and respond with appropriate actions based on any observed anomalies or significant events. These events include application events that lead to resource scaling, configuration changes, other activities, as well as faults and performance degradations that require healing. The collected data and computed analytics are stored for persistence as well as use by other applications for business and operations (e.g., billing and ticketing). More importantly, DCAE has to perform a lot of these functions in real time. One of the key design patterns expected of this component is to realize the following:

"Collect Data → Analyze and Correlate → Detect Anomalies → Publish need for Action"

This pattern will take various forms and be realized at multiple layers (e.g., infrastructure, network, and service). The data are collected once and made available to multiple applications (network operators, vendors, partners, and others) for consumption. Applications supporting various operational and business functions are key consumers of the data and events made available by DCAE. DCAE is an open analytic framework, allowing network operators to be able to extend and enhance their capabilities, alter behavior and modify scale, supporting the evolution of the various network functions that are virtualized over time.

7.6.1   DCAE's Four Major Components

1. *Analytics framework*—a data development and processing platform with a catalog of micro-services—stand-alone, policy-enabled, elementary analytics functions that support structured and unstructured data processing.
2. *Collection framework*—a set of streaming and batch collectors that support virtualized network, device, and infrastructure data.
3. *Data distribution bus*—It enables different components within DCAE and ONAP to publish and subscribe to data as well as move data from the edge of the network downstream for processing.
4. *Persistence storage framework*—a data-storage platform for short- and long-term consumption.

Besides these four components, DCAE includes an orchestrator and a set of controllers that manage the data bus and micro-services and interprets different service designs and workflows.

DCAE, commonly referred to as "Open DCAE," adopts an open source philosophy. Shown in Figure 7.8, it fosters a data marketplace and a set of micro-service-based collectors and analytics functions. With this approach, a secure ecosystem is created where developers, third-party providers, data scientists, and engineers can contribute to the platform. The platform also supports analytics, collections, policy enablement, compute, and storage in three places: at the edge, in the

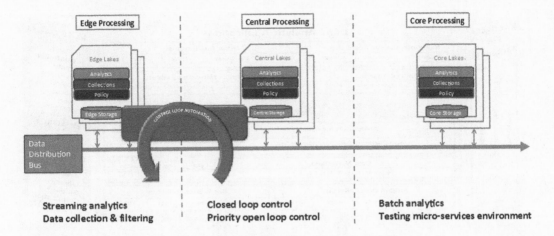

FIGURE 7.8 Open DCAE and control loop automation.

central sites (where there are regional cloud zones), and in the core (where batch-mode processing is enabled) of the network. Compute, data distribution, and storage resources can be optimized to achieve desirable scale, performance, reliability, and cost efficiencies.

Open DCAE enables a new generation of intelligent services. By capturing data from the edge of the network cloud, data can be analyzed securely and efficiently to identify and eliminate potential security threats. By taking data from VMs supporting virtual firewalls and routers, machine learning and advanced analytics can be applied to perform closed-loop automation (the automatic identification and resolution of failures). Those actions may include shifting traffic to new VMs or rebooting those machines during off-peak times.

There are some key technical challenges with Open DCAE development:

1. Scalability and movement in processing and management of heterogeneous and large-scale data, from structured to unstructured, and messages to events and files.
2. Development, certification, and automated onboarding of micro-services. These functions are typically developed through different environments, teams, languages, and styles, which means automated onboarding and certification must be consistent.
3. Optimization and orchestration of services by composing collectors and analytics micro-services at the edge, central, and/or core of the network to attain efficiency, cost effectiveness, security, and reliability.

7.6.2 Platform Approach to DCAE

It is essential that a platform approach be taken in order to fulfill DCAE's goals. The notion of platform here is a reference to the DCAE capabilities that help define how data are collected, moved, stored, and analyzed within DCAE. These capabilities can then be used as a foundation to realize many applications serving the needs of a diverse community. Figure 7.9 is a functional architecture rendition of the DCAE platform to enable analytic applications within the ONAP/DCAE environment.

As Figure 7.9 suggests, the DCAE platform requires a development environment with well-documented capabilities and toolkit that allows the development and onboarding of platform components and applications. The DCAE platform and applications depend on the rich instrumentation made available by the underlying cloud infrastructure and the various virtual and physical elements present in that infrastructure to enable the collection, processing, movement, and analysis necessary for elastic management of the infrastructure resources. In addition, it relies on robust interfaces with key ONAP

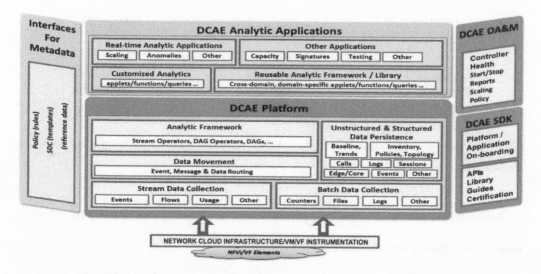

FIGURE 7.9 DCAE functional architecture approach to analytic applications within ONAP.

components for the reference information about the managed elements (A&AI), the rules (policy), and templates (SDC) that govern the behavior of the managed elements and the ONAP response.

7.6.3 DCAE PLATFORM COMPONENTS

The DCAE platform consists of multiple components: common collection framework, data movement, edge and central lake, analytic framework, and analytic applications. These are described as follows:

1. *Common collection framework*—the collection layer provides the various collectors necessary to collect the instrumentation made available in the cloud infrastructure. The scope of the data collection includes all of the physical and virtual elements (compute, storage, and network) in the cloud infrastructure. The collection includes the types of events data necessary to monitor the health of the managed environment, the types of data to compute the key performance and capacity indicators necessary for elastic management of the resources, the types of granular data (e.g., flow, session and call records) needed for detecting network and service conditions, etc. The collection will support both real-time streaming as well as batch methods of data collection.
2. *Data movement*—this component facilitates the movement of messages and data between various publishers and interested subscribers. While a key component within DCAE, this is also the component that enables data to move between various ONAP components.
3. *Edge and central lake*—DCAE needs to support a variety of applications and use cases ranging from real-time applications that have stringent latency requirements to other analytic applications that have a need to process a range of unstructured and structured data. The DCAE storage lake needs to support all of these needs and must do so in a way that allows for incorporating new storage technologies as they become available. This will be done by encapsulating data access via APIs and minimizing application knowledge of the specific technology implementations. While there may be detailed data retained at the DCAE edge layer for detailed analysis and trouble-shooting, applications should optimize the use of precious bandwidth and storage resources by ensuring they propagate only the required data (reduced, transformed, aggregated, etc.) to the core data lake for other analyses.
4. *Analytic framework*—the analytic framework is an environment that allows for development of real-time applications (e.g., analytics, anomaly detection, capacity monitoring,

congestion monitoring, alarm correlation, etc.) as well as other non-real-time applications (e.g., analytics, forwarding synthesized or aggregated or transformed data to Big Data stores and applications); the intent is to structure the environment that allows for agile introduction of applications from various developers. The framework should support the ability to process both a real-time stream of data as well as data collected via traditional batch methods. The framework should support methods that allow developers to compose applications that process data from multiple streams and sources. Analytic applications are developed by various organizations; however, they all run in the DCAE framework and are managed by the DCAE controller. These applications are micro-services developed by a broad community and adhere to ONAP framework standards.

5. *Analytic applications*—the following list provides examples of types of applications that can be built on top of DCAE and that depend on the timely collection of detailed data and events by DCAE:

   a. *Analytics*—these will be the most common applications that are processing the collected data and deriving interesting metrics or analytics for use by other applications or operations. These analytics range from very simple ones (from a single source of data) that compute usage, utilization, latency, etc. to very complex ones that detect specific conditions based on data collected from various sources. The analytics could be capacity indicators used to adjust resources or could be performance indicators pointing to anomalous conditions requiring response.

   b. *Fault/event correlation*—this is a key application that processes events and thresholds published by managed resources or other applications that detect specific conditions. Based on defined rules, policies, known signatures, and other knowledge about the network or service behavior, this application would determine root cause for various conditions and notify interested applications and operations.

   c. *Performance surveillance and visualization*—this class of application provides a window to operations notifying them of network and service conditions. The notifications could include outages and impacted services or customers based on various dimensions of interest to operations. They provide visual aids ranging from geographic dashboards, virtual information model browsers, detailed drilldown, to specific service or customer impacts.

   d. *Capacity planning*—this class of application provides planners and engineers the ability to adjust forecasts based on observed demands as well as plan-specific capacity augments at various levels, for example, cloud infrastructure level (technical plant, racks, clusters, etc.); network level (bandwidth, circuits, etc.); service or customer levels.

   e. *Testing and trouble-shooting*—this class of application provides operations the tools to test and trouble-shoot specific conditions. They could range from simple health checks for testing purposes, to complex service emulations orchestrated for trouble-shooting purposes. In both cases, DCAE provides the ability to collect the results of health checks and tests that are conducted. These checks and tests could be done on an ongoing basis, scheduled or conducted on demand.

   f. *Other*—the applications that are listed above are by no means exhaustive and the open architecture of DCAE will lend itself to integration of application capabilities over time from various sources and providers.

7.7 POLICY ENGINE

ONAP is responsible for the design, creation, deployment, and lifecycle management of virtualized network functions (VNFs). One of its goals is to do this in a flexible, dynamic, metadata, and policy-driven manner—letting users dynamically control ONAP's behavior without changing the system software. ONAP's policy component allows users to express, interpret, and evaluate policies, and then pass them on to other ONAP components or network elements for enforcement.

ONAP policies capture the service provider's network operational, security, and management intelligence—including proprietary domain knowledge related to how a service provider manages networks and services. One of ONAP's goals is to automate the life cycle of network-based services, including fault, performance, and service management, via control loops. Control loops define how VMs, VNFs, and service impairments are detected, isolated, and resolved, either automatically (closed-loop control) or through manual resolutions. As an example, a control loop policy might specify whether a VM is down or not responding (the signature it is exhibiting), then the VM needs to be restarted (the response required). The signatures and responses that define a control loop are specified as policies within ONAP; these policies capture the operational domain knowledge regarding the automation that is to be enabled.

Policies oversee various aspects of ONAP's behavior, include the following:

1. Service design—for example, constraints regarding where to place VNFs
2. VNF change management—for example, how software rollouts should be scheduled across VNFs and what health checks should be performed to validate changes
3. Management of the behavior of ONAP components—for example, how and when ONAP should collect data, and how long ONAP should retain that data

ONAP's policy framework consists of the following modules (shown in Figure 7.10):

1. Policy creation: Enables policies to be specified by services designers and network operators via a user interface and APIs
2. Policy evaluation: Evaluates/executes applicable policies reactively (responds to a query) and proactively (evaluates a policy and triggers actions)
3. Policy decision distribution: Distributes policy decisions to be enforced in distributed policy elements adjacent to other ONAP components
4. Policy validation: Validates policies, designed to minimize the risk of introducing bad policies into ONAP

There are many policy engine technologies in the industry, addressing different types of policies. For example, Extensible Access Control Markup Language (XACML) enables access control,

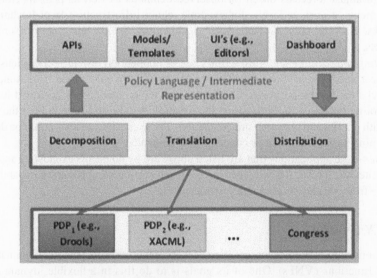

FIGURE 7.10   ONAP policy component framework.

while Drools supports the construction, maintenance, and enforcement of business policies. Given the diversity of ONAP's policy needs, the ONAP policy framework uses multiple policy engines. However, supporting various policy engines requires these policies be decomposed and distributed from the common policy creation layer to the different policy engines, providing translation to the policy engine-specific languages along the way.

ONAP's advanced automation facilitates operational efficiencies and faster and more consistent responses to network and service conditions. However, with automation comes risk. The policy framework thus uses a range of techniques to ensure "safe policies"—minimizing the risk associated with introducing policies that govern the automation that ONAP enables. ONAP uses advanced analytics to validate individual and combinations of policies before field deployment. Other mechanisms enable policies to be safely introduced into the system in a controlled fashion, and to retire expired policies. During runtime, "guard rails" limit the potential impact of inappropriate policies. For example, a guard rail may prevent too many VNFs from being taken out of service at a time, which would impact the network's ability to carry traffic.

The policy framework empowers operators and designers to control ONAP's behavior. However, the domain knowledge captured in ONAP policies is often distributed across many network operators and engineers, making it a challenge to collate. Machine learning techniques come to the rescue here by helping service providers automatically learn policies. In the case of control loops, machine learning enables signatures and responses to be automatically captured.

The policy platform plays an important role in realizing the network cloud vision of closed-loop automation and lifecycle management. The policy platform's main objective is to control/affect/modify the behavior of the complete network cloud environment (NFV infrastructure, VNF, ONAP, etc.) using field configurable policies/rules without always requiring a development cycle. Conceptually, "policy" is an approach to intelligently constrain and/or influence the behaviors of functions and systems. Policy permits simpler management/control of complex mechanisms via abstraction. Incorporating high-level goals, a set of technologies and architecture, and supportive methods/patterns, policy is based on easily updateable conditional rules, which are implemented in various ways such that policies and resulting behaviors can be quickly changed as needed. Policy is used to control, influence, and help ensure compliance with goals.

"A policy" in the sense of a particular network cloud policy may be defined at a high level to create a condition, requirement, constraint, or need that must be provided, maintained, and enforced. A policy may also be defined at a lower or "functional" level, such as a machine-readable rule or software condition/assertion, which enables actions to be taken based on a trigger or request, specific to particular selected conditions in effect at that time. This can include XACML policies, Drool policies, etc. Lower level policies may also be embodied in models such as YANG, TOSCA, etc.

7.7.1 POLICY CREATION

The ONAP policy platform has a broad scope supporting infrastructure, product/services, operation automation, and security-related policy rules. These policy rules are defined by multiple stakeholders (network/service designers, operations, security, customers, etc.). In addition, input from various sources (SDC, policy editor, customer input, etc.) should be collected and rationalized. Therefore, a centralized policy creation environment will be used to validate policies rules, identify and resolve overlaps and conflicts, and derive policies where needed. This creation framework should be universally accessible, developed, and managed as a common asset, and provides editing tools to allow users to easily create or change policy rules. Offline analysis of performance/fault/closed-loop action data are used to identify opportunities to discover new signatures and refine existing signatures and closed-loop operations. Policy translation/derivation functionality is also included to derive lower level policies from higher level policies. Conflict detection and mitigation are used to detect and resolve policies that may potentially cause conflicts, prior to distribution. Once validated and free of conflicts, policies are placed in an appropriate repository.

7.7.2 POLICY DISTRIBUTION

After completing initial policy creation or modification to existing policies, the policy distribution framework sends policies (e.g., from the repository) to their points of use, in advance of when they are needed. This distribution is intelligent and precise, such that each distributed policy-enabled function automatically receives only the specific policies that match its needs and scope. Notifications or events can be used to communicate links/URLs for policies to components needing policies; so, components can utilize those links to fetch particular policies or groups of policies as needed. Components in some cases may also publish events indicating they need new policies, eliciting a response with updated links/URLs. Also, in some cases, policies can be given to components indicating they should subscribe to one or more policies; so, they receive updates to those policies automatically as they become available.

7.7.3 POLICY DECISION AND ENFORCEMENT

Runtime policy decision and enforcement functionality is a distributed system that can apply to the various ONAP modules in most cases (there could be some exceptions). For example, policy rules for data collection and their frequency are enforced by DCAE data collection functionality. Analytic policy rules, anomalous/abnormal condition identification, and publication of events signaling detection of such conditions are enforced by DCAE analytic applications. Policy rules for associated remedial or other action (e.g., further diagnosis) are enforced by the right actor/participant in a control loop (MSO, controller, DCAE, etc.).

Policy decision/enforcement functionality generally receives policies in advance, via policy distribution. In some cases, a particular runtime policy engine may be queried in real time for policies/guidance, as indicated in the previous section. Additional unifying mechanisms, methods, and attributes help manage complexity and ensure that policy is not added inefficiently. Attribute values may be defined at creation time. Examples include policy scope attributes, described in the following section ("Policy Unification and Organization"). Note also that policy objects and attributes will need to be included in a proper governance process to ensure that correct intended outcomes are achieved for the business. Policy-related APIs can provide the ability to do the following:

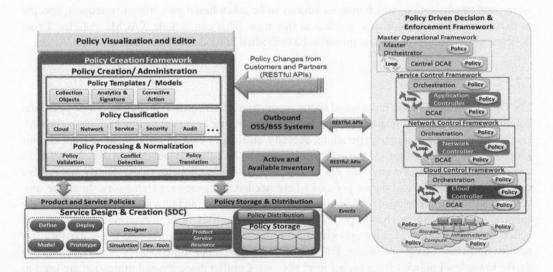

FIGURE 7.11   ONAP policy architecture framework.

1. Get (read) policies from a component, that is, on demand
2. Set (write) one or more policies into a component, that is, immediately pushed/updated
3. Distribute a set of policies to multiple components that match the scope of those policies, for immediate use (forced) or later use (upon need, e.g., time determined) by those entities

Figure 7.11 shows policy creation on the left, policy repository and distribution at the bottom, and policy use on the right (e.g., in control loops, or in VNFs). As shown in Figure 7.11, policy creation is in close association with SDC. When fully integrated, policies will be created either in conjunction with products and services (for policy scopes that are related to these), or separately for policies of scopes orthogonal to these (i.e., unrelated to particular products and services). Orthogonal policies may include various policies for operations, security, infrastructure optimization, etc.

Note that the architecture shown is a logical architecture, and may be implemented in various ways. Some functions in whole or in part may be implemented either as separate virtualized elements or within other (nonpolicy) functions.

7.7.4 POLICY UNIFICATION AND ORGANIZATION

In an expandable, multipurpose policy framework, many types of policy may be used. Policy may be organized using many convenient dimensions in order to facilitate the workings of the framework within a network cloud. A flexible organizing principle termed policy scope will enable a set of attributes to specify (to the degree/precision desired, and using any set of desired dimensions) the precise scope of both policies and policy-enabled functions/components. Useful organizing dimensions of policy scope may include the following:

1. Policy type or category, for example, taxonomical
2. Policy ownership/administrative domain
3. Geographic area or location
4. Technology type and/or specifics
5. Policy language, version, etc.
6. Security level or other security-related values/specifiers/limiters
7. Particular defined grouping
8. Any other dimensions/attributes as may be deemed helpful, for example, by operations. Note that attributes can be defined for each dimension.

By then setting values for these attributes, policy scope can be used to specify the precise policy scope of the following:

1. Policy events or requests/triggers to allow each event/request to self-indicate its scope, for example, which can then be examined by a suitable function for specifics of routing/delivery
2. Policy decision/enforcement functions or other policy functions to allow each policy function to self-indicate its scope of decision making, enforcement, or other capabilities
3. VFs of any type for auto-attachment to the appropriate policy framework and distribution mechanism instances, and most importantly
4. Individual policies to aid in management and distribution of the policies

7.7.5 POLICY TECHNOLOGIES

Policy in the network cloud will utilize various technologies (samples shown in Figure 7.12). These will be used, for example, via translation capabilities, to achieve the best-possible solution that takes advantage of helpful technologies while still providing in effect a single policy system.

Technology	Description	(Initial) Scope
Policy Applications	Plug-ins (in some fashion) to the Policy Platform, providing needed functions.	Additional functionality, e.g. for Conflict Detection
XACML++	XACML 3.0, extended for purposes beyond XACML's traditional access control focus.	1. Overarching/core policies 2. Policies not handled by DS technologies
OpenStack Congress	OpenStack policy-as-a-service	Detect policy violation for OpenStack resources
OpenStack Heat	OpenStack cloud orchestration	Resource orchestration policies, delegated
OpenStack GBP	Group-Based Policy (GBP) for OpenStack-managed resources	Neutron (networking), followed by Nova (compute), Swift (storage), etc
OpenDaylight GBP	Group-Based Policy (GBP) for network resources under SDN control	Service Function Chaining (SFC)
ASTRA	Policy-enabled firewall control, etc.	Security policies, firewall etc., delegated.
IAM / IDAM	Identity and Access Management	Security policies, Identity/Access, delegated.
YANG / TOSCA	Modeling approaches, SDN and higher level	Significant portion of SDN control
Drools	Business rules management system	Attribute/model based rule evaluation

FIGURE 7.12 ONAP sample policy technologies.

7.7.6 POLICY USE

At runtime, policies that were previously distributed to policy-enabled components will be used by those components to control or influence their functionality and behavior, including any actions that are taken. In many cases, those policies will be utilized to make decisions, where these decisions will often be conditional upon the current situation.

A major example of this approach is the feedback/control loop pattern driven by DCAE. Many specific control loops can be defined. In a particular control loop, each participant (e.g., orchestrator, controller, DCAE, VF) will have received policies determining how it should act as part of that loop. All of the policies for that loop will have been previously created together, ensuring proper coordinated closed-loop action. DCAE can receive specific policies for data collection (e.g., what data to collect, how to collect, and how often), data analysis (e.g., what type and depth of analysis to perform), and signature and event publishing (e.g., what analysis results to look for as well as the specifics of the event to be published upon detecting those results). Remaining components of the loop (e.g., orchestrators, controllers, etc.) can receive specific policies determining actions to be taken upon receiving the triggered event from DCAE. Each loop participant could also receive policies determining the specific events to which it subscribes.

7.8 ACTIVE AND AVAILABLE INVENTORY (A&AI) SYSTEM

A&AI is the ONAP component that provides real-time views of the network cloud resources, services, products, and customer information virtual network services. Figure 7.13 provides a functional view of A&AI. The views provided by A&AI relate data managed by multiple ONAP, BSS, OSS, and network applications to form a "top to bottom" view, ranging from the products customers buy to the services and resources used to compose the products. A&AI not only forms a registry of products, services, and resources, but also maintains up-to-date views of the relationships between these inventory items across their lifecycles. To deliver the vision of a dynamic, fungible network cloud, A&AI will manage these multidimensional relationships in real time.

A&AI maintains real-time inventory and topology data by being continually updated as changes are made within the network cloud. It uses graph data technology to store relationships between inventory items. The graph traversals can then be used to identify chains of dependencies between items. A&AI data views are used by homing logic during real-time service delivery, root cause analysis of problems, impact analysis, capacity management, software license management, and many other required functions.

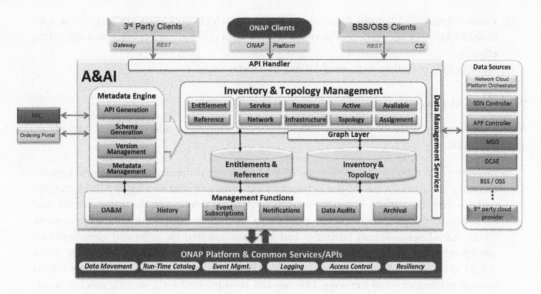

FIGURE 7.13 A&AI functional view in ONAP.

The inventory and topology data includes resources, service, products, and customer subscriptions, along with topological relationships between them. Relationships captured by A&AI include "top to bottom" relationships such as those defined in SDC when products are composed of services, and services are composed of resources. It also includes "side to side" relationships such as end-to-end connectivity of virtualized functions to form service chains. A&AI also keeps track of the span of control of each controller, and is queried by MSO and placement functions to identify which controller to invoke to perform a given operation.

A&AI is metadata driven, allowing new inventory item types to be added dynamically and quickly via SDC catalog definitions, reducing the need for lengthy development cycles.

7.8.1 KEY A&AI REQUIREMENTS

The following list provides A&AI key requirements:

1. Provide accurate and timely views of resource, service, and product inventory and their relationship to the customer's subscription
2. Deliver topologies and graphs
3. Maintain relationships to other key entities (e.g., location) as well as legacy inventory
4. Maintain the state of active, available, and assigned inventory within ONAP
5. Allow introduction of new types of resources, services, and products without a software development cycle (i.e., be metadata driven)
6. Be easily accessible and consumable by internal and external clients
7. Provide functional APIs that expose invariant services and models to clients
8. Provide highly available and reliable functions and APIs capable of operating as generic cloud workloads that can be placed arbitrarily within the network cloud infrastructure capable of supporting those workloads
9. Scale incrementally as ONAP volumes and the network cloud infrastructure scales
10. Perform to the requirements of clients, with quick response times and high throughput
11. Enable vendor product and technology swap-outs over time, for example, migration to a new technology for data storage or migration to a new vendor for MSO or controllers

12. Enable dynamic placement functions to determine which workloads are assigned to specific ONAP components (i.e., controllers or VNFs) for optimal performance and utilization efficiency
13. Identify the controllers to be used for any particular request

7.8.2 A&AI FUNCTIONALITY

A&AI functionality includes inventory and topology management, administration, and reporting and notification.

1. *Inventory and topology management*—A&AI federates inventory using a central registry to create the global view of the network cloud inventory and topology. A&AI receives updates from the various inventory masters distributed throughout the infrastructure, and persists just enough to maintain the global view. As transactions occur within the network cloud, A&AI persists asset attributes and relationships into the federated view based on configurable metadata definitions for each activity that determine what is relevant to the A&AI inventory. A&AI provides standard APIs to enable queries from various clients regarding inventory and topology. Queries can be supported for a specific asset or a collection of assets. The A&AI global view of relationships is necessary for forming aggregate views of detailed inventory across the distributed master data sources within this environment.
2. *Administration*—A&AI also performs a number of administrative functions. Given the model driven nature of ONAP, metadata models for the various catalog items are stored, updated, applied, and versioned dynamically, as needed, without taking the system down for maintenance. Given the distributed nature of A&AI as well as the relationships with other ONAP components, audits are periodically run to assure that A&AI is in sync with the inventory masters such as controllers and MSO. Adapters allow A&AI to interoperate with legacy systems as well as third-party cloud providers.
3. *Reporting and notification*—consistent with other ONAP applications, A&AI produces canned and ad hoc reports, integrates with the ONAP dashboards, publishes notifications other ONAP components can subscribe to, and performs logging consistent with configurable framework constraints.

7.9 CONTROL LOOP SYSTEMS: WORKING TOGETHER

Given the network cloud vision described in this section, network elements and services will be instantiated by customers and providers in a significantly dynamic process with real-time response to actionable events. In order to design, engineer, plan, bill, and assure these dynamic services, there are three major requirements:

1. A robust design framework that allows specification of the service in all aspects—modeling the resources and relationships that make up the service, specifying the policy rules that guide the service behavior, specifying the applications, analytics, and closed-loop events needed for the elastic management of the service.
2. An orchestration and control framework (MSO and controllers) that is recipe/policy driven to provide automated instantiation of the service when needed and managing service demands in an elastic manner.
3. An analytic framework that closely monitors the service behavior during the service lifecycle based on the specified design, analytics, and policies to enable response as required from the control framework, to deal with situations ranging from those that require healing to those that require scaling of the resources to elastically adjust to demand variations.

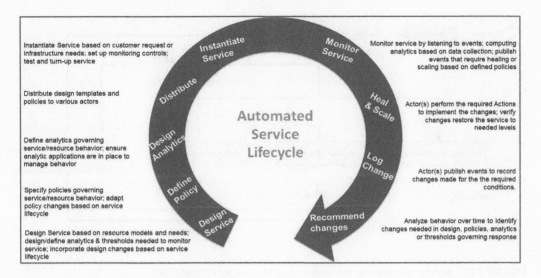

FIGURE 7.14 ONAP closed-loop automation.

The following sections describe the ONAP frameworks designed to address these major requirements. The key pattern that these frameworks help automate is the following:

"Design → Create → Collect → Analyze → Detect → Publish → Respond"

This automation pattern is referred to as "closed-loop automation" in that it provides the necessary automation to proactively respond to network and service conditions without human intervention. A high-level schematic of the "closed-loop automation" and the various phases within the service lifecycle using the automation is depicted in Figure 7.14.

The various phases shown in the service lifecycle above are supported by the design, orchestration, and analytic frameworks described below.

7.9.1 Design Framework

The service designers and operations users must, during the design phase, create the necessary recipes, templates, and rules for instantiation, control, data collection, and analysis functions. Policies and their enforcement points (including those associated with the closed-loop automation) must also be defined for various service conditions that require a response, along with the actors and their roles in orchestrating the response. This upfront design ensures that the logic/rules/metadata is codified to describe and manage the closed-loop behavior. The metadata (recipes, templates, and policies) is then distributed to the appropriate orchestration engines, controllers, and DCAE components.

7.9.2 Orchestration and Control Framework

Closed-loop automation includes the service instantiation or delivery process. The orchestration and control framework provides the automation of the configuration (both the initial and subsequent changes) necessary for the resources at the appropriate locations in the network to ensure smooth operation of the service. For closed-loop automation to occur during the lifecycle management phase of the service, the instantiation phase must ensure that the inventory, monitoring, and control functions are activated, including the types of closed-loop control related to the participating VFs and the overall service health.

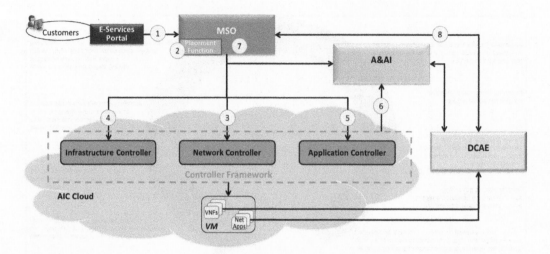

FIGURE 7.15   Service instantiation use case on a network cloud within ONAP.

Figure 7.15 illustrates a runtime execution of a service instantiation high-level use case. As requests come into ONAP (❶), whether they are customer requests, orders, or an internal operation triggering a network build out, the orchestration framework will first decompose the request and retrieve the necessary recipe(s) for execution.

The initial steps of the recipes include a homing and placement task (❷) using constraints specified in the requests. "Homing and placement" are micro-services involving orchestration, inventory, and controllers responsible for infrastructure, network, and application. The goal is to allow algorithms to use real-time network data and determine the most efficient use of available infrastructure capacity. Micro-services are policy-driven. Examples of policy categories may include geographic server area, serving area/regulatory restrictions, application latency, network resource and bandwidth, infrastructure and VNF capacity, as well as cost and rating parameters.

When a location is recommended and the assignment of resources are done, orchestration then triggers the various controllers (❸) to create the internal datacenter and WAN networks (L2 VLANs or L3 virtual private network [VPNs]), spin up the VMs (❹), load the appropriate VNF software images, connect them to the designated data plane and control plane networks. Orchestration may also instruct the controllers to further configure the VFs for additional layer 3 features and layer 4 and above capabilities (❺). When the controllers make the changes in the network, autonomous events from the controllers and the networks themselves are emitted for updating the active inventory (❻). Orchestration (MSO) completes the instantiation process by triggering the "test and turn-up tasks" (❼); this includes the ability for ONAP to collect service-related events (❽) and policy-driven analytics that trigger the appropriate closed-loop automation functions. Similar orchestration/controller recipes and templates as well as the policy definitions are also essential ingredients for any closed-loop automation, as orchestration and controllers will be the ONAP components that will execute the recommended actions deriving from a closed-loop automation policy.

7.9.3   ANALYTIC FRAMEWORK

The analytic framework ensures that the service is continuously monitored to detect both the anomalous conditions that require healing as well as the service demand variations to enable scaling (up or down) the resources to the right level. Once the orchestration and control framework completes the service instantiation, the DCAE analytic framework begins to monitor the service by collecting the various data and listening to events from the VFs and their agents. The framework processes the data, analyzes the data, stores it as necessary for further analysis (e.g., establishing a baseline, perform trending, look for a signature), and provides the information to the application controller. The applications in the framework look for specific conditions or signatures based on the analysis. When a condition is detected, the application publishes a corresponding event. The subsequent steps will depend on the specific policies associated with the condition. In the simplest case, the orchestration framework proceeds to perform the changes necessary (as defined by the policy and design for the condition) to alleviate the condition. In more complex cases, the actor responsible for the event would execute the complex policy rules (defined at design time) for determining the response. In other cases, where the condition does not uniquely identify specific response(s), the responsible actor would conduct a series of additional steps (defined by recipes at design time, e.g., running a test, query history) to further analyze the condition. The results of such diagnosis might further result in publishing of a more specific condition for which there is a defined response. The conditions referred to here could be ones related to the health of the virtualized function (e.g., blade down, software hung, service table overflow, etc.) that require healing. The conditions could be related to the overall service (not a specific VF, e.g., network congestion) that requires healing (e.g., reroute traffic). The conditions could also relate to capacity conditions (e.g., based on service demand variation, congestion) resulting in a closed-loop response that appropriately scales up (or down) the service. In the cases where anomalous conditions are detected, but specific responses are not identifiable, the condition will be represented in an operations portal for additional operational analysis.

The analytic framework includes applications that analyze the history of the service lifecycle to discern patterns governing usage, thresholds, events, policy effectiveness, etc. and enable the feedback necessary to effect changes in the service design, policies, or analytics. This completes the service lifecycle and provides an iterative way to continuously evolve the service to better utilization, better experience, and increased automation.

7.10   LEGACY BSSs INTERACTIONS WITH ONAP

Network cloud offers/products will be designed, created, deployed, and managed in near real time, rather than requiring software development cycles. ONAP is the framework that provides service creation and operational management of the network cloud; this software automation platform enables significant reductions in the time and cost required to develop, deploy, operate, and retire products, services, and network elements. The BSSs, which care for such capabilities as sales, ordering, and billing, will interact with ONAP in new network cloud architectures, thus will need to pivot to this new paradigm.

While BSSs exist for today's network, they will need to be enhanced in order to work and integrate with ONAP. These enhancements will need to be made with the assumption that the BSSs must also support existing products (perhaps in a new format) and enable faster time-to-market for new products and services in the network cloud.

The BSS transformation to support the dynamic network cloud environment is based on the following:

1. *Building blocks*—BSS migration from monolithic systems to a platform building block architecture that can enable upstream UX changes in how products are sold, ordered, and billed.
2. *Catalog-driven*—BSSs will become catalog-driven to enable agility, quick time to market, and reduce technology development costs.
3. *Improved data repositories*—support accuracy and improved access to dynamically changing data (e.g., customer subscriptions).
4. *Real-time APIs*—BSS platforms must expose functionality via real-time APIs to improve flexibility and reduce cycle time.

5. *New usage data and network events*—BSSs that have a need to know about network events (e.g., billing) will be retooled to support new information from DCAE and A&AI.
6. *Expose BSS functions*—provide BSS functions directly to our customers to streamline processes and allow new distribution channels.

7.10.1 BSS Scope

Figure 7.16 shows the BSS scope that includes customer management, sales and marketing, order lifecycle management, usage and event management, billing, customer finance, UX, and end-to-end BSS orchestration.

The BSS scope includes the following areas:

1. *Customer management*—focuses on customer information, retention of the customer, insight about the customer, managing customer service-level agreements, and building customer loyalty. Key new consolidated data stores of customer information for new virtualized services are the customer profile, customer subscription, and customer-interaction history.
2. *Sales and marketing*—provides all of the capabilities necessary to attract customers to the products and services being offered by a service provider, create solutions that meet customers' specific needs, and contract to deliver specific services. Sales and marketing hands off contracts to billing for implementation and solutions to ordering for service provisioning.
3. *Order lifecycle management*—provides the capabilities necessary to support the end-to-end handling of a customer's order. Orders may be for new service, or may be for moves, changes, or cancellation of an existing service. The experience of ordering will change in services built on the network cloud, as customers will experience provisioning in near real-time.
4. *Usage and event management*—focuses on end-to-end management of network cloud usage and events, including transforming from the traditional vertically oriented architecture to a decomposed function-based architecture. This will drive common collection, mediation, distribution, controls, and error processing. The scope of this includes usage

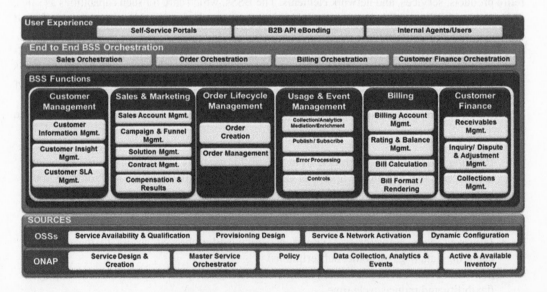

FIGURE 7.16 BSS scope in the network cloud environment (ONAP view).

and events that are required for real-time rating and balance management, offline charging, configuration events, customer notifications, etc.

5. *Billing*—focuses on providing the necessary functionality to manage billing accounts, calculate charges, perform rating, as well as format and render bills. Billing will evolve becoming more real time, and decomposed into modules that can be individually accessed via configurable reusable APIs.

6. *Customer finance*—manages the customer's financial activities. It includes the accounts receivable management functions, credit and collections functions, journals and reporting functions, as well as bill inquiries, including billing disputes and any resulting adjustments. As in the case of billing, customer finance will evolve under the network cloud to expose more API-based components that are reusable across platforms.

7. *UX*—provides a single presentation platform to internal and external users each of which receives a customized view based on role. The UX is divided into the self-service view for external customers, sales agents, care center agents, and other internal users.

8. *End-to-end BSS orchestration*—these functions identify and manage the business level processes and events required to enable a customer request and interactions between domains. They trigger activities across domains and manage status, provide a tool for customer care to holistically manage the end-to-end request, as well as the customer relationship for the duration of the enablement/ activation of the request.

8 Network Data and Optimization

Mazin Gilbert and Mark Austin

CONTENTS

8.1 Network Data and the Analytics Layer 138
8.2 Big Data 140
   8.2.1 Dealing with Big Data's 7Vs 140
   8.2.2 Data Quality 141
   8.2.3 Data Management: Lambda Architecture and Policies 142
   8.2.4 The Hadoop Ecosystem 143
      8.2.4.1 Storage: The HDFS 144
      8.2.4.2 Data Ingestion (Kafka, Flume, Sqoop) 144
      8.2.4.3 YARN (Yet Another Resource Manager) 144
      8.2.4.4 Storage: Nonrelational Databases, NoSQL, Not Only SQL 145
      8.2.4.5 Search (SOLR) 145
      8.2.4.6 Batch Processing (MapReduce, HIVE, PIG) 146
      8.2.4.7 Real-Time and Micro-Batch Processing Using STORM and SPARK 146
      8.2.4.8 Coordination and Workflow Management 147
   8.2.5 Analytics and ML 148
      8.2.5.1 Descriptive versus Reactive versus Predictive versus Prescriptive Analytics 149
      8.2.5.2 Supervised versus Unsupervised Learning 149
      8.2.5.3 Deep Learning 150
      8.2.5.4 Open Source Distributed Processing Toolkits for ML 151
8.3 Big Data Meets Network Cloud 151
   8.3.1 Data Collection, Analytics, and Events 153
   8.3.2 DCAE Functional Components 154
      8.3.2.1 Common Collection Framework 154
      8.3.2.2 Data Movement 155
      8.3.2.3 Edge, Central, and Core Lakes 155
      8.3.2.4 Analytics Framework 155
   8.3.3 MS Design Paradigm 156
   8.3.4 Control-Loop Automation 157
      8.3.4.1 Closed-Loop Modeling and Template Design 158
      8.3.4.2 CLAMP System Architecture 159
   8.3.5 Machine Learning for Closed-Loop Automation 160
   8.3.6 Deep Learning and SDN 161
8.4 Network Data Applications 162
   8.4.1 Self-Optimizing Networks 162
   8.4.2 A Customer Configurable Policy for a Content Filtering Smart Network 163
   8.4.3 Traffic Shaping Based on Historical and Current Estimated Network Congestion 163
   8.4.4 Utilizing SDN to Minimize Robocalling 167

8.4.5   Emerging Applications of SDN and NFV Including IoT, M2X,
               Ambient Video, Virtual Reality, Security, and Intelligent Agents 167
References 168

Collecting and analyzing network event data can be a daunting processing challenge for today's larger networks. For example, AT&T's network in 2016 passes more than 118 petabytes (PB) per day; if each byte was to be treated as an "event," it would imply a need to process, analyze, and act on approximately 1.4 trillion events per second. This sort of scale implies a need for a big data framework for event processing. This chapter describes some of the concepts for such data collection and analytics.

8.1   NETWORK DATA AND THE ANALYTICS LAYER

Having the ability to collect and analyze data from various points in the "end to end" dataflow is fundamental to being able to diagnose and optimize networks. While SDN separates the control plane and data plane, it is important to add an "instrumentation" layer to be able to collect this data as depicted in Figure 8.1.

Notice that this "data instrumentation layer" can collect data from the modules (endpoints), the virtualized network functions (VNFs), as well as the control plane managing the VNFs. Nevertheless, beyond the additional data collected in the Network Cloud, it is important to keep in mind the "data stack" needed to characterize the full end-to-end experience. This is shown in Table 8.1.

Device data are important as they provide the best view of the overall experience from the customer standpoint. For instance, it is the only data that covers the performance for when the device is on the network, off the network, or which network; it also captures device-specific attributes such as "resets," "battery," "memory," "software configuration," etc. issues. Collecting some of this data requires software on the device interfacing with either public or private APIs such as the Android APIs [Andr16], or the iOS developer library [iOS16].

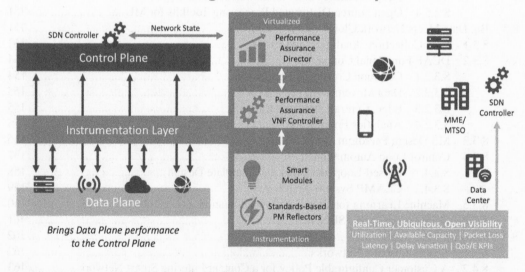

FIGURE 8.1 The instrumentation layer of big data and advanced analytics in SDN/NFV. (From Light Reading's Big Telecom Event, The Role of Big Data and Advanced Analytics in SDN/NFV, https://accedian. com/wp-content/uploads/2015/06/BTE15_The-Role-of-Big-Data-and-Advanced-Analytics-in-SDN-NFV. pdf.)

TABLE 8.1

Unique Data Contributions per Data Stack Layer with Traditional Network Data Collection and Network Cloud Data Collection Examples

Data Collection Layers	Unique Data Contribution to End-to-End View	Traditional Data Collection Examples	Additional "Network Cloud" Data Collection Examples
Application data	Application perspective: • Access stats across all networks (wireless, wireline, multiple operators, etc.)	Website logs	
Internet cloud data	Internet perspective: • Internet site logs, accessibility	DNS, CDN logs, etc.	
Core network data	Core network perspective: • Core network performance aspects per network function, e.g., latency, congestion, signaling	Deep packet inspection data Billing/call detail records	VNF control stats
Radio access data	Radio perspective: • User/device/application performance vs. radio characteristics, and location	RAN vendor-specific collected data	
Device data	User perspective: • Collect across multiple radio access networks (e.g., WiFi, cellular) • Device-dependent performance captured (applications, resets) • Collects no-service, and other network access issues	Device diagnostic collection software	

The radio access data contain a wealth of device reported data containing the interactions of the device and the radio network, as well as corresponding radio access node (RAN) data. These data help characterize network optimization and performance issues as a function of the radio link and node itself (e.g., good/bad coverage, good/bad interference, radio congestion, congestion between the radio node and core node). In addition, devices traditionally provide the radio network data in "Network Measurement Reports (NMRs)" related to their location—from coarse location data such as "what cell site a device is connected to," to time delay of arrival data from multiple cell sites that can help triangulate a given device to a much finer location area. Cell change updates (CCU) record data from devices as devices move from cell to cell, regardless as to whether a call or data session is in place, which makes these data particularly useful for visibility of all devices. Since the RANs are traditionally proprietary, access to these data is based on what the vendors will provide, or, alternatively what can be "probed" between the radio and core network connections.

The core network data contain data related to all nodes/functions within the more central core network nodes. Tapping these data allows visibility into not only core network node performance but also call and data interactions among other devices and endpoints. The interaction data allow an understanding of social graphs, and interests such as may be implied from websites or applications being accessed. Some of the most common data collected in the core network are call/data detail records, which are primarily recorded for billing purposes, but also have per device call performance data (e.g., calls blocked, dropped), as well as timestamps and cell-site locations. Other session data such as throughput, latency per device, per endpoint, are another common data source. In the Network Cloud, these traditional data sources are supplemented with VNF performance and control analytic data.

Internet cloud data consist of the interaction of the core network data with the Internet. These data are traditionally collected from external servers such as domain name servers (DNS) lookups, and/or content delivery networks (CDN), which can provide visibility into the volume of access for a given domain, the domain's accessibility content volumes from CDN. To the extent that content filtering or shaping is done, such as with "Open DNS" [Open16], per user per endpoint visibility can also be collected.

Finally, application data are important if they can be acquired as they provide full visibility for a given application—which devices from which networks, from which locations are accessing a given application. A simple example of this comes from weblogs, although more diverse application-specific data can be envisioned.

8.2   BIG DATA

Big data is a term that attempts to define data that are so large that traditional approaches toward analyzing or processing them are impractical. Nevertheless, as computers and applications become more powerful, exactly how large a data set needs to be to challenge traditional processing is a moving target. Beyond considering only the size of the data as the defining characteristic for big data, Doug Laney from Gartner [Lane01] has offered some further attributes of big data by introducing the "3Vs" of volume, variety, and velocity. Since then, others [Livi13, Data14, Wikibig] have extended the 3Vs to 5Vs or 7Vs to include other characteristics such as variability, validity/veracity, visibility/visualization, and value. These attributes describe a set of challenges when dealing with big data as described in the next section.

8.2.1   DEALING WITH BIG DATA'S 7VS

1. *Volume*: As noted above, the most fundamental element of big data is its size, or volume. The challenge for many businesses is the amount of data that is possible to store or process, which is overwhelming. To illustrate the size of today's digital data universe, it has been estimated that in 2015 the total volume of world's data was 7.9 zettabytes (1 zettabyte = 10^21 bytes) [Impa16]. Today, it is not uncommon for many businesses to store tens (or higher) PB of data. Storing and processing this amount of data cost effectively has led to new approaches such as the Hadoop ecosystem [Hado], described in subsequent sections.
2. *Velocity*: Perhaps even more daunting than volume, is how fast the volume of data is growing, which is known as it's velocity. Indeed, data are growing so fast that 90% of the world's data has been produced in the last 2 years. In addition, while the world has 7.9 zettabytes of data today, by 2020 the total digital data universe is expected to grow to 44 zettabytes, with each person generating 1.7 megabyte every second [Entr16]. At some point, the data become so large for a business that it becomes necessary to typically either limit its' retention or only keep summarized older data. Recently, new near real-time data processing technologies in the Hadoop ecosystem also provide the ability to process data in real time or micro-batches to extract the most important features of the data, which lessens the need to store the raw data at all. Organizationally, some [Entr16] have noted that the speed of data also suggests that it is impractical to have any single "data insight team" in a company, and that data analyses and processing needs to have some decentralization to keep up with the ever changing and new data.
3. *Variety*: Data also comes in different varieties. Some data are structured; having a known format such as might be produced from IOT devices emitting a known set of data elements in a predefined format. Some of it is unstructured, as might come from the content of a Twitter feed, social media, texting data, etc. Big data analyses often combine both types

of data to get the full picture of whatever is being analyzed. For instance, to better solve a customer's problem, a customer care big data application may utilize structured data having the account number, name, tenure, product type information, as well as unstructured data containing the "verbatim of the customer's problem description," any "error log files produced from the product" and any previous customer care "chat logs." Often, however, unstructured data are processed with machine learning (ML) and natural language processing (NLP) to extract structured features that can be further processed for insights and actions.

4. *Variability*: Raw data and derived data often have variability and changes over time. This can be a challenge for big data applications, which need to adapt based on the changing data. For instance, a "recommender engine" for movies needs to adapt and update its recommendations as a user's tastes change or new movies come out. In addition, some actions triggered from analytics on the data may utilize previously computed batch processed data, as well as real-time data. Thus, it is important that any big data architecture facilitate both batch and real-time processing. One such architectural approach to accomplish this is referred to as the "lambda architecture" as will be described in Section 8.2.3.

5. *Validity/veracity*: Most have heard the phrase "garbage in equals garbage out" and managing the quality or validity/veracity of the data is one of the largest challenges in big data. For instance, how much of the data has "missing values"? Is the data in the same format? (e.g., Do some set of phone numbers have the country code, area code, phone number, while others only have area code, phone number) Does the data have expected format? (e.g., Does a phone number have at least 10 numbers? Do names have both first name and last name?) Any big data processing system needs to have processes and checks in place to check, resolve if possible, or at minimum flag any data quality issues. Data quality issues are dealt with in a subsequent section.

6. *Visibility/visualization*: Understanding some of the basic attributes and summaries of a big data set is typically very hard, and having the ability to visualize the data via charts, graphs, heat maps, etc. is often very useful. Nevertheless, visualizing and manipulating large sets of data is a challenge for which new tools such as AT&T's nano-cubes [Att13] were invented. This tool allows visualizing and summarizing billions of data points in a common browser.

7. *Value*: Perhaps, the attribute that drives many of the basic decisions on the previous 6Vs is the value of the data. For instance, what is the incremental value in storing the last 2 years, 5 years, or 10 years of a given data set? Could the same value be garnered with storing summary data after 2 years? The challenge in deciding whether to keep or discard any data is to assess the value of its' use for applications and insights known today compared to those that could become possible in the future.

8.2.2 DATA QUALITY

Data quality has been recognized by the International Standardization Organization as an important topic from the early 1990s. While the rigor around what level of data quality is needed ultimately depends on the application, a variety of authors have written guidelines around data quality such as the 2011 ISO 8000 data quality standards [Wang10], and those summarized by Cai and Zhu [Cai15]. Table 8.2 summarizes some of the attributes that should be considered around data quality.

While the elements in Table 8.2 may not be exhaustive, it is important to address these areas throughout all phases of a big data project, to understand how they need to be treated to ensure useful actions and insights are realized.

TABLE 8.2

Important Data Quality Attributes for Big Data Systems

Availability	Usability	Reliability	Relevance	Presentation/Quality
Accessibility—Can you get the data?	Documentation—Is it understood how the data are collected?	Accuracy—Is the data accurate enough for the use intended?	Fitness—Can enough data be accessed, and is it useful for the intended purposes (e.g., indicators, prediction, classification)?	Readable—data content and format are clear and understandable
Timeliness—Is it the time period needed?	Credibility—Has a reliable person documented?	Integrity—Do the data values follow a standardized data model/type?		
Authorization—Is one authorized to use the data for the intended purposes?	Metadata—Are the data fields defined?	Completeness—Does the data have missing values?		
		Auditability—Is the data checked regularly against these aspects?		

8.2.3 Data Management: Lambda Architecture and Policies

As partially described in the previous sections, there are a variety of data management and policy issues that must be decided for a big data system, such as what type of data is collected, how long data is retained, whether data is summarized or anonymized, and whether data is processed and accessed in batch versus real time. Many big data systems utilize batch and real-time processing of data as summarized in the lambda architecture shown in Figure 8.2.

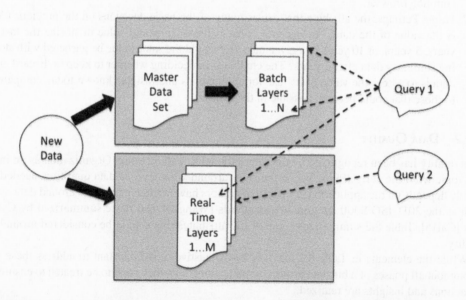

FIGURE 8.2   Lambda architecture. (From Lambda Architecture, http://lambda-architecture.net/.)

In many circumstances, other factors beyond the value of the data, such as policy and ethics, govern whether certain data are collected, utilized, retained, or shared. For example, the Federal Trade Commission (FTC), sets data collection and use policy for Internet companies, such as Google and Facebook, with respect to consumer's privacy. The Federal Communications Commission (FCC) established rules governing data collection and use by ISPs in November 2016, but those rules were based on the premise that the Internet Service Providers (ISPs) are uniquely "able to develop highly detailed and comprehensive profiles of their customers" [NPRM16], which is practically not the case. For example, the following "Future of Privacy Forum" blog [FPOF16], illustrates that a single visit to WebMD.com connects to 24 third-party sites, after which visiting another four websites, a total of 119 third-party sites were connected to. Since many of these third-party sites are advertising related, the data are naturally shared in data exchanges and a comprehensive view of the customer is formed. Further descriptions of various privacy policies and issues are described in this wiki [WikiPrivacy]. Because the asymmetry embodied in the FCC rules was not justified, Congress voted to repeal those rules in March 2017.

8.2.4 THE HADOOP ECOSYSTEM

The Hadoop ecosystem emerged primarily to deal with the large volumes of data, and ultimately with many aspects of the 7 Vs. Hadoop's originations come from Google, who in October 2003 published the "Google file system" article [Goog03], which outlined how to store large amounts of data cost effectively using common off-the-shelf hardware. This article was subsequently followed by another Google research article entitled "MapReduce: Simplified data processing on large clusters," which outlined how to process large amounts of data in a parallel fashion using many servers. From these background articles, the open source Apache Nutch project was born, which ultimately was named "Hadoop" by Doug Cutting who was working at Yahoo at the time [WikiHadoop]. Since then, the Hadoop ecosystem has evolved into many open source components covering the Hadoop distributed file system (HDFS), data ingestion, resource management, data processing approaches for analytics, batch, real time, search, ML, database storage approaches, and coordination and workflow management approaches as shown in Figure 8.3.

Today, the Apache software foundation [Apac] manages the updates to the Hadoop ecosystem components. There are several popular supported distributions of the Hadoop ecosystem from

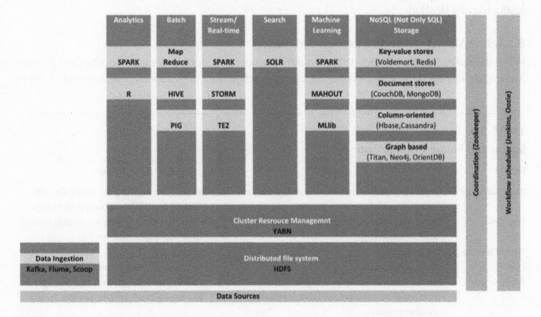

FIGURE 8.3 The Hadoop ecosystem and its components.

Hortonworks [Hort], Cloudera [Clou], and Map-R [Mapr]. The following subsections outline each of these components, and their importance.

8.2.4.1 Storage: The HDFS

The HDFS was a breakthrough in 2003, as it allowed the use of commodity hardware to store, access, and process large amounts of data by distributing the storage of data across multiple nodes. Specifically, HDFS achieves its fault tolerance by storing three or more copies of the data across different nodes, where the mapping of the data to nodes is contained in the *NameNode,* which is redundant itself. Nevertheless, one of the tradeoffs of HDFS is that it utilizes a write-once, read-many paradigm so that if changes are made on a single data point, the entire file needs to be rewritten. As per some reports in 2016, Yahoo has the largest Hadoop cluster in the world with 35,000 Hadoop servers, hosted across 16 countries, with a combined 600 PB of storage, that execute close to 34 million jobs per month [Dezi16].

8.2.4.2 Data Ingestion (Kafka, Flume, Sqoop)

Before data can be used, it must be ingested from the data sources, which could be structured or unstructured, and processing needs could be batch, micro-batch, or real time as described in Section 8.2.1. The Hadoop ecosystem has a variety of approaches for data ingestion such as Flume, Sqoop, and Kafka. Sqoop is particularly useful for extracting batch data from structured relational databases such as Teradata, Oracle, MySQL, SAP, etc. and Flume is particularly adept at ingesting data from multiple sources, which can be unstructured. A typical use for Flume is ingesting streaming log data. Kafka [Kafka], originally developed at LinkedIn, is like Flume; however, it is also a general purpose publish-subscribe model messaging system, which can also ingest data from multiple sources, and simultaneously have multiple consumers of the data. Dezyre [Dezy15] has a nice comparison of Sqoop and Flume, which is resummarized in Table 8.3.

While not specifically designed for Hadoop, Kafka is used by some for its ability to add multiple consumers of the data without affecting performance. Kafka achieves this scalability since it does not track which consumers have consumed the messages; it only persists the message for a defined period of time, and it is the responsibility of the consumer to track messages. Kafka is also particularly useful for ingesting streaming data across a variety of use cases as described on the Apache Kafka website [Kafka].

8.2.4.3 YARN (Yet Another Resource Manager)

When the initial version of Hadoop was rolled out, it only supported batch processing, and a single *NameNode* managed the entire cluster for both job scheduling and resource management. A more

TABLE 8.3
Sqoop and Flume Comparison

Feature	Sqoop	Flume
Basic difference	Useful for ingesting data from relational databases	Designed for ingesting streaming logs into Hadoop (e.g., JMS or spooling directory)
Data flow	Works well with any relational databased having JDBC connectivity	Works well for streaming logfile data sources from multiple sources
Type of loading	Not event-driven	Event-driven
Destination	HDFS, HBase, HIVE	From multiple sources into HDFS
Where to use?	Parallel data transfer, helps mitigate excessive loads to external systems, provides data interaction programmatically	Flexible data ingestion tool, high throughput, low latency, fault tolerant, linearly scalable

flexible approach was later introduced with YARN [YARN], which separates resource management and job scheduling to allow different applications to separately negotiate resources from a global resource manager. YARN thus allows multiple diverse user applications to run on a multitenant platform, so that users are no longer confined to the I/O intensive, high latency MapReduce framework, and alternate processing frameworks supporting near real-time processing (such as that described in Section 8.2.4.6) can be utilized within the same data lake.

8.2.4.4 Storage: Nonrelational Databases, NoSQL, Not Only SQL

In addition to HDFS, the Hadoop ecosystem defines several nonrelational "not only Structured Query Language (SQL)" (NoSQL) storage approaches. Table 8.4 illustrates a comparison approach of different types of NoSQL databases [Acci14].

As might be expected, which particular database storage solution is used depends on one's needs.

8.2.4.5 Search (SOLR)

Searching on Lucene Replication (SOLR) is an open source platform for searching data in HDFS. Searches can be across tabular, text, geo-location, or sensor data after "indexing" of the data is done via XML, JSON, CSV, or via binary. SOLR has the following attributes:

- Full text search
- HTML admin interfaces

TABLE 8.4
A Comparison of NoSQL Databases in Hadoop

NoSQL Database Type	Capabilities	Applications	Limitations
Key-value (BerkleyDB, MemcacheDB, Redis, DynamoDB)	Each value is retrieved by a unique key In-memory storage for fast retrieval	Applications requiring fast lookup (e.g., mobile, gaming, online)	Cannot update subset of a value Assigning keys may become challenging as data set grows
Document-oriented (MongoDB, CouchDB, Apache Solr, Elastic Search)	Value is a structured document Hierachical data structures possible, without predefining schema Supports queries on structured documents Common approach for search	Can manage a large variety objects with different data structure Large product catalogs in e-commerce, customer profiles	No standard query syntax Join queries are challenging
Column-oriented Cassandra, BigTable, HBase, Apache Accumulo	Value is a set of columns Columns can have multiple time-stamped versions Can be generated at run-time	Good for storing multiple copies of a log file with different time stamps Good for analysis that works	No join queries
Graph-oriented Neo4J, OrientDB, Apache Giraph, AllegroGraph	Models graphs as nodes and edges Graph computations very fast	Applications like calling/ texting/social networks	Scaling for large graphs is typically a challenge
Relational MySql, PostgreSQL, MariaDB, Oracle, SQL Server	Conventional relational database management system (RDBMS) structure consisting of fixed schema	Traditional applications (CRM, banking, etc.)	Scalability (horizontally limited) Inefficiency of nested data Challenge for unstructured data

Source: Choosing a NoSQL Database—Technology Comparison Matrix, October 14, http://accionlabs.com/ choosing-a-nosql-database-technology-comparison-matrix/.

- Statistics for monitoring
- Linearly scalable
- Configurable, XML configuration

Often after utilizing any of the various batch/real-time/analytic Hadoop ecosystem functions to process/crunch the data. SOLR is used to index it and serve it up for consumption. SOLR can also be used as a key-value store via its support for range queries.

8.2.4.6 Batch Processing (MapReduce, HIVE, PIG)

As mentioned previously, Hadoop originated with batch processing of large datasets with MapReduce. MapReduce allows parallel processing of data by performing a "divide and conquer" technique by dividing the data up across multiple servers in the map phase into key/value pairs and then after merging, combining, and sorting the data it uses a reduce phase to processes the data in parallel. To illustrate how MapReduce works, consider the illustration in Figure 8.4, which performs the simple task of "how to count the number of times a given word is used in a document." The data from a document are first divided into two documents across two servers and words are separated and individually tallied. They are then combined and sorted, after which the tallies are summed in the reduce phase.

MapReduce, however, has somewhat fallen out of favor in the ML community given its high overhead costs, and YARN has allowed alternate processing approaches like SPARK to exist in conjunction with MapReduce. To overcome some of the challenges of breaking down a computation task into map and reduce function, many programmers utilize the programming language PIG Latin, which is normally simply referred to as PIG. PIG supports user-defined functions written in Java, Python, etc. and then translates into MapReduce jobs.

HIVE is another popular batch processing approach, which allows querying and summarizing of data stored in HDFS and NoSQL databases using HIVEQL, which is like MySQL.

8.2.4.7 Real-Time and Micro-Batch Processing Using STORM and SPARK

STORM is a real-time distributed processing system [STORM], which allows processing of an unbounded sequence of data tuples, such as from a Twitter feed, into a new set of data tuples or event triggers. STORM defines a topology framework consisting of the following:

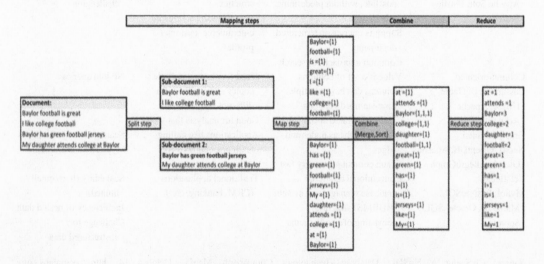

FIGURE 8.4 A simplified illustration of MapReduce to count the number of times a given word occurs in a document.

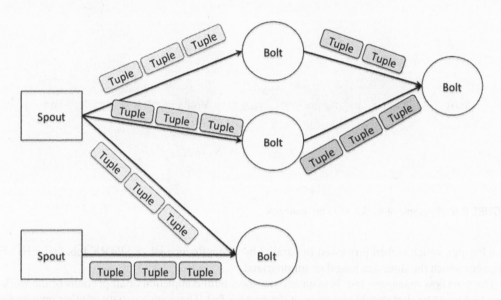

FIGURE 8.5   Illustration of an example of the topological data flow for STORM.

- Stream—the unbounded sequence of tuples
- Spouts—the sources of the stream (e.g., reading the API from the Twitter feed)
- Bolts—the process (e.g., functions, filters, aggregations, joins) that takes streams as an input and produces new streams as the output

The topological data processing of these framework elements is often illustrated as shown in Figure 8.5.

SPARK is an in-memory distributed data analysis and processing platform. SPARK has libraries for Machine Learning (MLlib), SQL, DataFrames, Graphs (GraphX), and Streaming. SPARK streaming allows micro-batches to be ingested from HDFS, Flume, Kafka, Twitter, and ZeroMQ in user-defined intervals into what is called resilient distributed datasets (RDD). RDDs are immutable distributed collections of objects, which can be operated on in parallel by storing the state of memory as an object, which can be shared across jobs. Consequently, this inherently allows speeding up batch processing jobs so that MapReduce jobs can essentially be done in-memory. SPARK as such has been known to have speed ups over batch jobs by 10–100 times [Tutor16].

8.2.4.8   Coordination and Workflow Management

As with any system, administrative and coordination solutions are needed. This is especially true with a distributed system with hundreds of servers, and Zookeeper [Zoo16] is one solution for a coordination service in the Hadoop ecosystem. Zookeeper is a centralized repository where distributed applications can place and extract data, which is used to coordinate processing. For example, with a huge number of machines on a network, race conditions can otherwise occur where a machine tries to perform multiple operations at a time, or deadlocks can happen when two or more machines try to access the same data simultaneously. Zookeeper acts as the coordinator and can serialize operations to avoid race conditions and it provides synchronization to avoid deadlocks.

In addition to Zookeeper, Oozie and Jenkins are two examples of open source workflow managers that help track the progress and status of an end-to-end data processing flow. For example, it is not uncommon that a given data result that needs to be updated regularly is the result of many different processes that need to be run. Figure 8.6 provides an example workflow using a directed acyclic graph, which shows the flow of data, starting with the ingesting of data with Kafka, followed

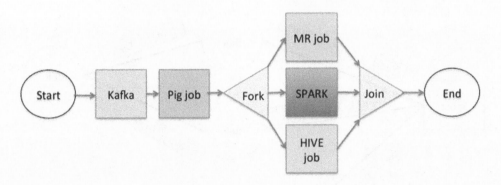

FIGURE 8.6   Example of workflow to be managed.

by a Pig job, which is then processed in parallel by a MapReduce job, a SPARK job, and a HIVE job, after which the three are joined or summarized.

The workflow manager's task is to ensure the successful completion of all portions of the workflow, and to act or alert should any stage of the process fail. There are a variety of other options for workflow managers, which could be considered—Azkaban, Airflow, Luigi, Cascading, etc.

8.2.5   ANALYTICS AND ML

We define ML as a computer program (or software agent) that can learn from data to predict a future state or condition. Traditionally, ML, like NLP, is considered a subfield of artificial intelligence (AI). We define an AI system as a more advanced form of an ML system that is able to act and continuously optimize its actions over time and context. One analogy to realize the relationship between ML and AI is the game of chess. ML is equivalent to predicting the next best move. This is an instant change. AI is equivalent to formulating a strategy to win the game. This typically involves strategizing and optimizing a sequence of best steps. The strategy is dynamic and updates on every move.

The foundation for ML and AI was created over many decades by numerous pioneers including Alan Turing, Claude Shannon, Arthur Samuel, Frank Rosenblatt, among others. The "Turing Test" was a measure of machine intelligence that was invented by Alan Turing in 1950 [Tur50]. It computed the length of time or the number of interactions a machine can fool a human in believing it is a human. While at Bell Labs, Claude Shannon advanced the field of AI by creating a memory mouse, named as "Theseus," for traversing a maze [Sha48]. The mouse is able to learn to strategize an optimal path from its source to its destination. Arthur Samuel from IBM wrote a program for the game of checkers that developed winning strategies and moves. Frank Rosenblatt is regarded as one of the fathers of neural networks, which attempts to simulate the operation of the human brain. The theory and the practical experiments developed by these pioneers and many others, has led to new fields such as deep learning, which is making its way into smart automation for industrial applications and services.

Over the past three decades, ML technologies have shown to be superior for applications including object detection, speech recognition, and NLP. In speech recognition, ML models, known as deep learning, are used to estimate the posterior probabilities of context-dependent linguistic units, such as phonemes and words [Den13]. Those probabilities are applied in a Markov hidden model to discover the best sequence of units. Training these ML models on a large variety and volume of data helps to make speech recognition systems speaker independent and robust to adverse environments and conditions. A form of ML, known as artificial neural network, has been used to estimate the articulatory parameters of a vocal tract model for generating synthesized speech [Rah91]. Other types of ML models, known as convolutional neural networks [Lec98], have been used to spot objects (e.g., characters, cars, dress, shoes) in an image enabling businesses to perform targeted advertising. ML has also been used to predict network failures. This proactive action can result in increased cost saving and improved customer experience [Gil16]. These successful demonstrations of ML applications continue to raise interest in academia and industry, a precursor for enabling smarter automated systems.

8.2.5.1 Descriptive versus Reactive versus Predictive versus Prescriptive Analytics

There are different forms of analytics that vary in their level of sophistication. Descriptive analytics is the simplest form of analytics that involves summarizing the data to generate a report. This is very commonly used today in commercial applications. Reporting can be in the form of tables or more sophisticated visualization methods. Descriptive analytics provides a useful aggregation of key performance indicators (KPIs) to monitor a system or a service.

Reactive analytics is also a commonly used approach in commercial applications. It involves raising an alarm based on processing data and events; reacting to a situation after the fact. For example, raising an alarm after an intruder burgled a house when a system shuts down due to a security threat or hardware failure. Although reactive analytics can help to explain what happened to an event, it does not avoid the accident or failure. Traditional rule-based or statistical methods such as clustering can be used for reactive analytics.

Predictive analytics is a more sophisticated form of analytics that involves predicting an event or a future state of a system based on historical data or observations. Predictive analytics can be unsupervised, supervised, or semi-supervised as will be explained in the next section. Algorithms for predictive analytics are either linear in nature, such as those using regression methods, or nonlinear, based on ML methods, such as neural network, support vector machines, or boosting [Kot07]. Examples of applications using predictive analytics include estimating the future value of an investment portfolio and the sentiment, emotion, or retention rate of users based on their historical interactions.

Prescriptive analytics is closer to AI where the goal is to identify an optimized sequence of steps or actions to take. Prescriptive analytics can be data-powered and may involve advanced ML models to perform optimization of the best-possible outcome. The nature of this analytics also requires the system to continuously be dynamic and reexamine its sequence of decisions as the context changes. An example of prescriptive analytics is applied in autonomous vehicles where the system will need to continuously optimize its decisions (turns, signals, etc.). Another example that is relevant to SDN and 5G is optimizing data routing across all network nodes and end devices as traffic changes rapidly.

8.2.5.2 Supervised versus Unsupervised Learning

Predictive and prescriptive analytics can be either supervised or unsupervised. Supervised learning implies that the training process has access to the correct targets or desired outputs, while unsupervised learning has only access to the observations or historical data.

Clustering is commonly used in unsupervised learning, where the goal is to separate data into "similar" groups based on a given set of attributes. An example is clustering images of people, which may lead to groups that have similar faces, identity, backgrounds, clothing, etc.

Regressions and classification are commonly used methods for supervised learning. Regressions enable predictive analytics of continuous data, while classification enables predictive analytics of data into discrete or finite groups. Supervised learning usually results in better performance than unsupervised learning, although it requires extensive labeling of data to identify the "truth" state. It is commonly used in speech recognition although it requires audio files to be manually transcribed.

Semi-supervised techniques provide a tradeoff between supervised and unsupervised learning. It involves labeling a subset of data that is used to create an initial model; applying the model to label a new data set, which is then included in training the supervised model. The process continues until all data is employed in training.

In the Network Cloud, supervised, semisupervised, and unsupervised learning are very relevant for closed-loop automation as will be shown in a later section. These methods are also relevant for 5G and next-generation access for traffic optimization. An excellent overview of supervised and unsupervised training can be found in Ng's Coursera course on ML [Ng16].

8.2.5.3 Deep Learning

The basic premise of more recent research in ML, referred to as deep learning, is that you can train a large number of nonlinear processing units with deep layers of the so-called neurons to learn any translation of data or patterns. This has attracted practitioners in areas of web search, language translation and chatbots, among others. The premise to transform and process arbitrary data, has significant impact to business applications requiring operational saving or driving new revenue growth.

Deep learning is based on the perceptron model. The perceptron model, shown in Figure 8.7, was originally proposed in 1969 by Minsky and Papert [Min69]. A perceptron is a nonlinear computational unit, which is capable of learning to partition simple patterns. It can accept both binary and continuous values.

A single perceptron is represented by a set of parameters that include weights, w, and a bias. It is able to classify simple patterns into two classes by forming a half-plane decision region. A single perceptron, however, is unable to separate input patterns that require more complex decision boundaries. For example, an exclusive OR problem cannot be solved by a single-layer perceptron. To overcome this restriction, one may introduce a hidden layer of neurons as shown in Figure 8.8. The three-layered network, which includes a set of M input features, two hidden layers, and an output layer of N nodes, is able to resolve arbitrary complex problems. Different applications may require adjusting the number of hidden layers and the number of respective neurons to attain a desired precision level.

Multilayer perceptrons are feed-forward nets with neurons that are fully interconnected between adjacent layers, but which are separate within one layer. The weights associated with the arcs across all layers can be trained with a back-propagation algorithm, hence making these networks very attractive for classification problems. By expanding the number of neurons and layers used, one can classify very complex signals such as those used in speech, language, and video processing. For multidimensional signals, such as images, more advanced deep learners, known as convolutional networks [Lec98], have been commonly adopted. Those networks support multiple dimensions across each layer.

One of the key advancements in deep learning is attributed to the discovery of the backpropagation algorithm by Rumelhart et al. [Run86]. The algorithm enabled supervised training of multilayer perceptrons. It is essentially a gradient search technique that minimizes a cost function between the desired outputs and those generated by the net. The aim is to establish a functional relationship for a given problem by adjusting the weightings between neurons. After selecting some initial values for

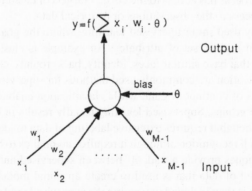

$$y = f\left(\sum x_i \cdot w_i - \theta\right)$$

FIGURE 8.7 A perceptron.

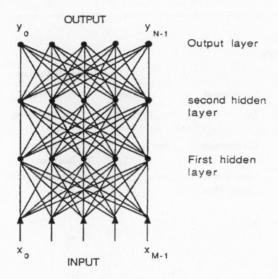

OUTPUT

y_0 y_{N-1}

Output layer

second hidden
layer

First hidden
layer

x_0 x_{M-1}

INPUT

FIGURE 8.8 Basic deep learner for event prediction.

the weights and internal thresholds, input/output patterns are presented to the network repeatedly and, on each presentation, the states of all neurons are computed starting from the bottom layer and moving upward until the states of the neurons in the output layer are determined. At this level, an error is estimated by computing the difference in the values between the outputs of the network and the desired outputs. The variables of the network are then adjusted by propagating the errors backwards from the top layer to the first layer. The application of deep learning to Network Cloud will be described in a later section.

8.2.5.4 Open Source Distributed Processing Toolkits for ML

There are a variety of open source toolkits available today that attempt to provide all types of ML under one umbrella (supervised, unsupervised, deep, etc.). One of the earliest toolkits originally produced with MapReduce was Mahout, whereas SPARK now has MLlib, and some of the newer toolkits like H2O and Deeplearning4j, start to incorporate deep learning approaches in their libraries as well. These are not exhaustive, and no doubt, further developments will be coming. Table 8.5 provides a summarized view of various functionalities of these toolkits at the time of this writing.

Depending on one's application, a different ML toolkit may be preferable; however, the newer and more active communities are currently in the SPARK and deep learning areas.

8.3 BIG DATA MEETS NETWORK CLOUD

While the Network Cloud is about virtualizing the network by having network functions run as software on commodity cloud infrastructure, big data, on the other hand, is about collecting and analyzing data for improving automation and extracting intelligence and information. The marriage of Network Cloud and big data is like peanut butter and jelly—it enables us to develop a network that is resilient, self-healing, and self-learning. It also provides us the ability to make rapid decisions through control loop services by processing and acting on data. This section describes a platform approach to collecting, ingesting, storing, and analyzing data to support virtualized services that can run reliably and securely in the Network Cloud.

Designing a data platform for the Network Cloud needs to adhere to five key design principles:

1. Supporting well-defined API's both within the platform components and when interfacing with external components, such as a policy engine or controllers.

TABLE 8.5
ML Toolkit Comparisons

Features/Attributes	Distributed ML Toolkits			
	Mahout	MLlib	H2O	Deeplearning4j
Languages supported	Java	Java, Python, Scala	Java, Python, R, Scala	Java, Scala, Clojure
Framework/library	Environment/ framework	Library/API	Environment/ language	Framework
Associated platform	MapReduce, Spark, Flink	Spark, H2O	Spark, Mapreduce, H2O	Spark, Hadoop
GPU support?	–	–	–	Yes
Current version	0.12.2	2.0.1	3.11.0.3652	
Graphical user interface?	–	–	Yes	–
Supervised Learning				
Generalized linear modeling (GLM), linear/logistic	–	Yes	Yes	–
Gradient boosting machine (GBM)	–	Yes	Yes	–
Distributed random forest	Yes	Yes	Yes	–
Support vector machine (SVM)	–	Yes	–	–
Naïve Bayes	Yes (Spark)	Yes	Yes	–
Ensembles (stacking)	Yes	Yes	Yes	–
Unsupervised Learning				
Generalized low rank models (GLRM)	–	Yes	Yes	–
K-means clustering	–	Yes	Yes	–
Streaming K-means	–	Yes	–	–
Spectral clustering	–	–	–	–
Principal components analysis (PCA)	Yes (Spark)	Yes	Yes	–
QR decomposition	Yes (Spark)	Yes	–	–
Singular value decomposition	Yes (Spark)	Yes	–	–
Chi squared	Yes (Spark)	–	–	–
Additional Algorithms				
Collaborative filtering algorithm	Yes (Spark)	Yes (ALS)	–	–
Topic modeling	–	Yes	–	Yes
Deep learning	–	Yes	Yes	Yes
Most recent version link	https://mahout. apache.org/	http://spark.apache. org/mllib/	http://www.h2o.ai/	https://deeplearning4j. org/

2. Creating reusable platform resources (such as data collectors and analytics capabilities) as plug-n-play components. These components, which are referred to as microservices (MSs), are stand-alone, model-driven and policy-enabled applications. (A deeper dive into MSs is provided later in this chapter.)

3. Allowing resources to be automatically configured and activated/deactivated to optimize performance and minimize cost.

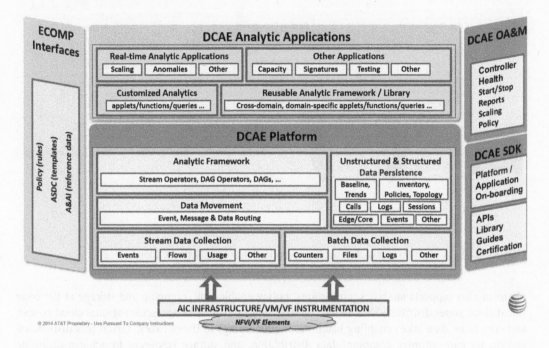

FIGURE 8.9 A platform approach to analytics applications in Network Cloud.

4. Orchestrating a model-driven service design dynamically.
5. Employing standardized data models to enable service chaining.

Figure 8.9 shows a high-level architecture of a big data platform supporting the Network Cloud. There are essentially four basic functionalities in the data platform:

1. Analytics framework—a data development and processing platform with a catalog of micro-services that can support structured and unstructured data processing.
2. Collection framework—a set of streaming and batch collectors that support virtualized network, device and infrastructure data.
3. Data distribution bus—this is the blood stream of the platform. It enables components within and outside the platform to publish and subscribe to data as well as to move data from the edge of the network downstream for processing.
4. Persistence storage and compute framework—a multitiered data-storage platform for short- and long-term consumption, as well as a distributed computing environment for supporting streaming and batch analytics.

Besides these four components, the platform would need to have a controller system to automate many of the operational and services configurations of the MSs and platform resources. More details on this platform are covered in Chapter 7.

8.3.1 Data Collection, Analytics, and Events

Figure 8.10 depicts a high-level conceptual view of data collection, analytics, and events (DCAE)—a big data platform within ONAP (see Chapter 7). DCAE, which is commonly referred to as "Open DCAE," adopts an open source philosophy. It fosters a data marketplace and micro-services of collectors and analytics functions. With this approach, we can create a secure ecosystem where developers, third-party providers, data scientists, and engineers can contribute to the platform. The

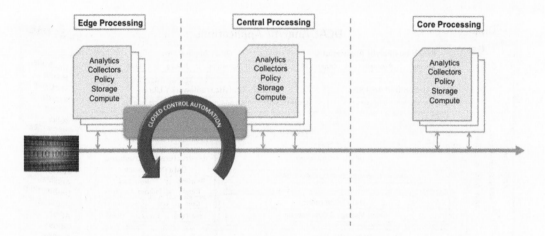

FIGURE 8.10 A platform approach to analytics applications in the Network Cloud.

platform also supports analytics, collections, policy enablement, compute and storage at the edge (local cloud zones distributed across the country), central (limited large-scale regional cloud zones), and core (core data lakes enabling batch-mode processing) of the network. Given this distributed design, we can optimize compute, data distribution, and storage resources to achieve desirable scale, performance, reliability, and cost efficiencies.

Open DCAE enables intelligent services to be dynamically instantiated or updated automatically. By capturing data from the edge of the network, we can perform both primitive and complex data processing ranging from data aggregation to advanced ML and predictive analytics. For example, by collecting performance and application data from virtual machines supporting virtual firewalls and routers, we can run advanced analytics at the edge to perform closed-loop automation. This may involve automatic identification of cyber-attacks and taking actions to terminate. Those actions may include shifting traffic to new virtual machines, deactivating IP addresses, or rebooting those machines (see Chapter 9).

In the Network Cloud, virtualized functions are expected to be instantiated in a significantly dynamic manner that requires the ability to provide real-time responses to actionable events from virtualized resources to applications and services. The big data platform is expected to gather key performance usage, telemetry, and events to compute various analytics and respond with appropriate actions based on observed anomalies or significant events. These significant events sometimes referred to as signatures, enable applications to dynamically perform resource scaling, configuration changes, traffic optimization, and detection of faults and performance degradation. Open DCAE enables collection and analysis of data for supporting such control systems at the edge of the network at a centralized location.

8.3.2 DCAE Functional Components

8.3.2.1 Common Collection Framework

The collection framework supports virtualized instrumentation for gathering data. The scope of the data collection would include all the physical and virtual elements (compute, storage, and network) and the management control functions in the cloud infrastructure. The collection data would include the types of events data necessary to monitor the health of the managed environment, the types of data to compute the key performance and capacity indicators necessary for elastic management of the resources, the types of granular data (e.g., flow, session, and call records) needed for detecting network and service conditions, etc. The collection will support both the real-time streaming as well as the batch methods of data collection.

8.3.2.2 Data Movement

This component facilitates the movement of messages and data between various publishers and interested subscribers. There are two types of data in the Network Cloud. Those that are frequent and in short bursts, such as messages and events, and those that are file-based and batched. Open DCAE adopts two technologies one is a KAFKA-based open source for message routing, and the other is a data router for enabling file transmissions. Those two technologies enable data to be streamed and moved within DCAE from edge to core, and enable movements of events across the Network Cloud platform to support services such as change management and closed-loop automation.

8.3.2.3 Edge, Central, and Core Lakes

DCAE needs to support a variety of applications and use cases ranging from real-time applications that have stringent latency requirements as well as other analytic applications that have a need to process a range of unstructured and structured data. The DCAE storage lake needs to support these needs and must do so in a way that allows for incorporating new storage technologies as they become available. This is done by encapsulating data access via APIs and minimizing application knowledge of the specific technology implementations.

Given the scope of requirements around the volume, velocity, and variety of data that DCAE needs to support, the storage will leverage technologies that a big data framework must offer, such as support for NoSQL technologies, including in-memory repositories, and support for raw, structured, unstructured, and semi-structured data. While there may be detailed data retained at the DCAE edge layer for analysis and trouble-shooting, applications should optimize the use of precious bandwidth and storage resources by ensuring they propagate only the required data (reduced, transformed, aggregated, etc.) to the core data lake for other analyses. Open DCAE is configured by the controller so that software-defined storage can be instantiated at different AIC cloud zones on an as-needed basis.

8.3.2.4 Analytics Framework

The analytics framework is an open ecosystem that is capable of onboarding MSs that are developed by analytics developers, data scientists, VNF analytics solution engineers, and business analysts. This concept is key for driving wide adoption of analytics in the Network Cloud, addressing trouble-shooting, service quality management, and DevOps.

The open ecosystem would need to support a full spectrum of analytics capabilities:

- Descriptive analytics (what is happening)
- Diagnostic analytics (why did it happen)
- Predictive analytics (what is likely to happen)
- Prescriptive analytics (what should I do about it)
- Exploratory analytics (what can I learn from the data and how do I use it to drive value to the business)

The analytics framework provides access to production data sources and enables automatic testing and certification of MSs. The framework includes

- APIs to
  - Expose a catalog of analytics MSs and/or applications to SDC
  - Support publishing and dynamic instantiation of those MSs and/or applications
  - Support interfaces with the base DCAE analytics framework (runtime, service based, pluggable)
  - Allow publishing and maintenance of the DCAE data catalog
- A runtime DCAE analytics platform for building, testing, and deploying batch and real-time streaming data solutions on Hadoop. In addition, the following key enablers will be addressed:

- An initial set of core MSs to jumpstart developers in using the framework
- Adaptors or modules (or flowlets in the Cask Data Application Platform [CDAP] parlance [CDAP16]) for common functions (e.g., reading records from the DCAE RDBMS)
- Criteria for defining the minimal set of interface (aka API) parameters necessary to enable orchestration
- A specification of what requirements a developer must meet to onboard their MS (i.e., analytics application) to the runtime platform
- A DCAE catalog:
  - The metadata in the DCAE catalog will enable users of the DCAE analytics framework to rapidly discover data elements/metrics/datasets and will allow them to define and share datasets with other analytics users; this component may also provide metadata necessary to enable orchestration of DCAE analytics framework (AF) MSs

A goal of this solution is to ease development and deployment of real-time applications (e.g., closed-loop analytics, anomaly detection, capacity monitoring, congestion monitoring, alarm correlation, etc.) as well as other nonreal-time applications (e.g., analytics, forwarding synthesized or aggregated or transformed data to big data stores and other sorts of applications for monitoring network, VNF, and service health) to support the VNFs and services deployed in the network. Further, the intent is to structure the ecosystem such that agile introduction of applications from various providers can shorten the time required to move software from development through test and into production. It is expected that analytic applications are developed by various organizations, however, they all run in the DCAE analytics framework and are managed by the DCAE controller (see Chapter 7).

8.3.3   MS Design Paradigm

MSs is a software architectural style in which large applications are composed from small loosely coupled services with independent lifecycles. This term has been popularized in the public domain by thought leaders such as Martin Fowler whose initial writings on MS architecture* suggest that the goal of such service design is to build the system using a collection of services, using common core components, with strong public APIs to achieve these benefits:

- Easy to deploy new capabilities—often just a matter of instantiating a new MS
- Easy maintenance—services can often be independently upgraded with no impact of other services
- Easy decommissioning—tear down the service (assuming no further subscribers)
- Language agnostic—services can be written in the language most appropriate for the service and programmer

We define four architectural styles for MSs and 10 principles [Wya16] (Figure 8.11):

1. Single capability focused: the MS architectural style composes applications from separate, independent services that provide a single-domain capability. This allows parts of the application to be changed and evolve over time without impacting other parts of the application.
2. Independence: this principle applies all aspects of MS delivery. MS must be independently developed, tested, deployed, configured, upgraded, scaled, monitored, and administered.

---

* http://martinfowler.com/articles/microservices.html

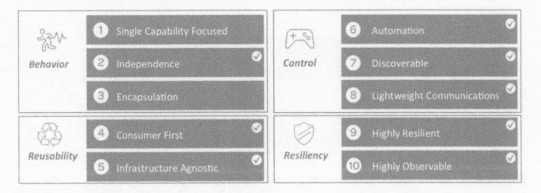

FIGURE 8.11 MSs principles.

3. Encapsulation: MS must hide implementation details. This insulates MS consumers from changes either within or downstream from the services they consume.
4. Consumer first: consumer first is about making it as simple and easy as possible for the consumers of an MS to successfully use it.
5. Infrastructure agnostic: MS must be portable and elastic. Moving a service across hosts, datacenters, or clouds should not be a problem. They should not be dependent on static addresses or ports when possible.
6. Automation: automation applies to all aspects of an MS. Automation means embracing DevOps and continuous integration and delivery processes (CI/CD).
7. Discoverable: since MS are auto deployed and infrastructure agnostic, they are required to be dynamically discoverable. Consumers must have the ability to query a service registry to lookup available instances and then route accordingly.
8. Lightweight communications: MS must follow a dumb pipe, smart endpoint approach for communication protocols. The core requirement is that no business logic be implemented within the protocol or broker and all domain logic remain within the service endpoint implementations themselves.
9. Highly resilient: building applications as a cohesive set of services requires the ability to tolerate and gracefully handle failures of all kinds.
10. Highly observable: provides insight and understanding of how request paths are processed and flow across services. Since large applications consist of many distributed services there is a need to collect, aggregate, and analyze logs and metrics across all services.

These MSs principles can be measured using different paradigms. One example being a metal rating (platinum, gold, silver, and Bronze) in which the rating reflects the percentage of match of the MS to the principles described above.

Within the analytics framework, we leverage the term "microservice" to describe a *reusable*, *open* analytic solution that provides useful output(s) and offers input APIs to control common data, computation, and control parameters. The DCAE analytics and collection framework described previously includes a controller that enables management and health checks of both analytics and collector MSs.

8.3.4 CONTROL-LOOP AUTOMATION

An intelligent service collects information about resources (users, devices, networks, and applications) used to deliver the service and about the environment in which it operates. It then makes decisions based on this information and domain knowledge, which includes adapting and personalizing

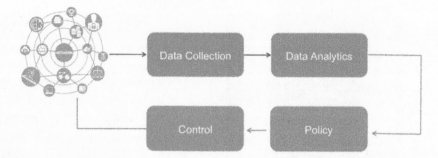

FIGURE 8.12   Basic components of a control-loop automation system.

the service for the users consuming it. An intelligent service receives feedback on its performance and learns. There are primarily three attributes that characterize an intelligent service, known as personalized, adaptive, and dynamic (PAD). A predictive personalized service is one that anticipates a user's need and proactively takes intelligent actions and recommends valuable and timely personalized information. An adaptive service learns from its past actions of all users and adjusts its behavior to provide superior service quality. A service that is dynamic is one that is robust and can survive and self-repair or self-organize itself from possible service interruptions.

In this section, we describe how intelligent services can be described essentially in the form of control-loop systems. These systems have been widely used in system design. The basic concept is to compute an error signal between an actual target and measurements computed from sensors and through a feedback control loop to attain a desired state where the error is zero. One example is a home thermostat that is continuously controlling the heating/cooling system to maintain a desired room temperature. The basic components of such a system, as shown in Figure 8.12, includes data collection for measuring the current temperature, data analytics to compute the variation between the desired and actual readings, policy for recommending the next best action, and a controller for performing the action. The action may include turning the heating on/off, cooling on/off, or do nothing.

In network management, control-loop systems play a vital role in delivering operational cost savings. Control-loop automation can be categorized into open or closed-loop systems. Open-loop systems capture telemetry and diagnostics information from the underlying cloud infrastructure (e.g., syslog, simple network management protocol (SNMP), fault, and network performance management events), perform a set of analytics, and provide reporting or alarms to the operations team. Closed-loop systems continuously monitor the system for fault, performance, security, etc. related problems and compute a set of signatures based on the detected anomalous condition. A policy engine then interprets these signatures and appropriate corrective actions are recommended to repair the system. Once the system has been repaired, a monitoring application checks the status to see whether the system responded to alleviate the detected problem. The goal is to attain zero downtime or minimal interruptions.

An implementation of the control-loop system in ONAP is shown in Figure 8.13. CLAMP, which stands for Control Loop Automation Management Platform [Gil16], includes three basic components. (a) A portal, which is essentially a web browser that enables authentication, construction, configuration, certification, testing, governance approval, and distribution of control loop templates. (b) A workflow engine that enables translation of the design template into an executable data model. The workflow engine communicates with ONAP through a set of well-defined ONAP-specified APIs. (c) A monitoring dashboard that enables telemetry data capture relevant to the performance of the control loop, status update, and diagnosis of failures.

8.3.4.1   Closed-Loop Modeling and Template Design

During the design phase, models for network functions virtualization (NFV) and ONAP platform elements are retrieved from a catalog and composed to create a service chain. This facilitates rapid

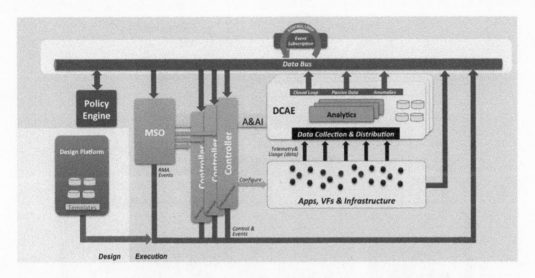

FIGURE 8.13 Control loop automation for network management in the Network Cloud. (From ECOMP [Enhanced Control, Orchestration, Management and Policy] Architecture White Paper, 2016, https://www.google.com/#q=ecomp+white+paper.)

service composition in a manner that is verifiable at different stages of the software lifecycle. The goal is to create a catalog, which will provide readymade templates (e.g. VM restart, VM migrate, VM rebuild, VM evacuate, etc.) that can be customized for different services through a model-driven approach.

8.3.4.2 CLAMP System Architecture

The CLAMP architecture uses all the relevant ONAP components to realize a closed loop in action. Once an application design is developed and distributed to ONAP, the following operations take place (Figure 8.14):

1. DCAE local (at the edge of the network) collects data from the relevant virtual functions and publishes the data into the data bus.
2. Analytics MSs subscribing to the data are then applied to compute events. A MS may involve a simple anomaly detector that looks for changes in the signal. The MS may be placed in local or regional cloud centers depending on the level of performance required and the resources available. MSs may also be chained together by DCAE.
3. Per rules acquired and interrupted by the policy engine, the DCAE platform outputs a signature once it encounters an anomaly and publishes that to the data bus.
4. The policy engine subscribes to that signature and recommends the next best action. The action is typically predefined by the designer of the service, and may include a simple notification to the operation team or some device reboot. The recommended action is published back to the data bus.
5. Per service design, the relevant controller (application, network, or infrastructure) picks up the action and performs the execution. In the case of a virtual function reboot, the service controller will communicate through a standardized protocol (such as OpenStack) and completes the action. It will also publish feedback information to the data bus to describe whether the action is a success or not. In the case of failure, subsequent recovery procedures may be performed automatically or manually per service design. If automation ultimately fails, then a ticket is issued to the appropriate operations center to do a manual override.

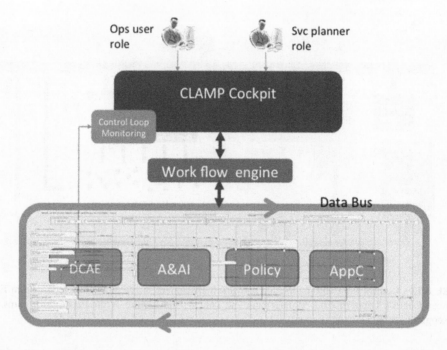

FIGURE 8.14 CLAMP execution engine. (From Gilbert et al. 2016. Control loop automation management platform, *IEEE GlobalSip 2016*.)

As described above, CLAMP supports a model-driven approach in which service design is encapsulated as a data model and distributed to ONAP. This enables rapid development and scalability of closed-loop services.

8.3.5 MACHINE LEARNING FOR CLOSED-LOOP AUTOMATION

Although there are numerous publications on the use of ML technologies for multimodal and multimedia applications [Mit97], there is far less research being done on the application of ML and AI for software-defined network. Some of the key opportunities we have identified include

1. Self-optimizing networks (SONs)
2. Cybersecurity and threat analytics
3. Fault management
4. Customer experience improvement
5. Traffic optimization

We will describe these applications in a little more detail in the next section. What is common across those applications is that ML and AI help to provide more automation by leveraging the closed-loop automation paradigm we described earlier. Rather than performing reactive analytics, which is typically employed in unsupervised learning models, ML helps to predict anomalies or events well in advance, based on time-varying signals. Similarly, instead of applying hard-wired policies, or rules, to guide the system on what best next action to take, AI can make that recommendation more intelligent based on data, context, previous patterns, and outcomes. Moreover, as the system continues to iterate and more data are collected, ML and AI can learn from successes and failures, and adjust their models accordingly to minimize errors. ML and AI help to transform the system from being static to being dynamic and from being reactive to being proactive. This is depicted in Figure 8.15.

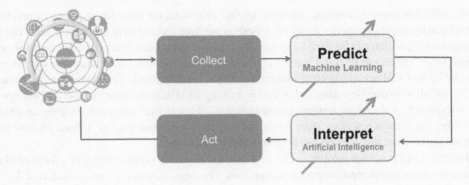

FIGURE 8.15 ML and AI for control-loop automation.

TABLE 8.6

Predicting Network Failure T Times in Advance

Predicting *T* Time Instances before the Event Happens. Values of T Below	Accuracy (%)
10	81.1
9	83.3
8	83.8
7	85.5
6	87.3
5	92.7
4	93.8
3	94.2
2	94.9
1	95.2
	98.0

Note: Every time instance is 5 min for this data.

The following is a preliminary experiment of using ML for predicting network failures in a closed-loop system [Rag16]. The goal is to predict as accurately as possible a network failure well in advance, that is, we are interested in predicting the occurrence of a network failure T time instances before the event has occurred. Around 50 signals were used that provide KPIs of a network performance. The failure translates to a CPU overload and system shut down (Table 8.6).

8.3.6 DEEP LEARNING AND SDN

In the Network Cloud, deep learning can be applied for classifying network data for identifying security threats or detecting performance degradations of virtual functions. For example, one can capture memory and CPU signals from a virtual machine supporting a virtual firewall to monitor the health check of the firewall. A set of deep learners can be trained to predict the health of this firewall over a set period of time. Detected failures can then be addressed by either automatically rebooting the firewall, migrating traffic to another firewall, or issuing a ticket for manual intervention.

The same concept of deep learning can be applied for interpretation when dealing with control loop systems. A deep learner can be trained to map analytics signals and context into a predefined set of policies/actions. At runtime, the deep learner can select the most likely policy or action. The learner can dynamically update its model based on positive and negative reinforcements. The back-propagation algorithm can be used to update the *a posteriori* probabilities to best reflect the correct action/policy. For example, if the learner predicted a no-action policy, and an action was necessary then the learner will negatively emphasize a "no action" and positively emphasize the correct action.

The pitfall with deep learning is that one is making an instant decision in time and not optimizing a sequence of decisions toward a common goal. This is fine when recognizing an object in an image, but present a serious issue when needing to recommend a set of actions on how best to migrate traffic from one network node to another, for example.

There are two potential solutions. The first is to use dynamic programming or a form of Markov model, such as that adopted in speech recognition. The basic concept is a set of different decisions with associated probabilities (or costs) computed at each time instant. Dynamic programming is then applied to find the best path based on minimizing a cumulative objective function.

The second approach is reinforcement learning, which is a form of ML [Mit97]. The learner receives reinforcement about its action but not being told the exact action. Reinforcement learning is typically formulated based on a Markov decision process. The objective is to maximize the credit assignment, or reward. A reinforcement learner can take an action based on its current state and reward input. The combination of deep learning and reinforcement learning to form a more holistic and powerful AI system has been explored most recently for gaming. The DeepMind's AlphaGo program beat world champion Lee Sedol in four out of five games of Go, a complex board game [Gog16F]. The system combined Monte Carlo tree search with deep neural networks that have been trained by supervised learning, from human expert games, and by reinforcement learning from games of self-play.

8.4 NETWORK DATA APPLICATIONS

8.4.1 Self-Optimizing Networks

SON concepts have been around for some time, with some of the earliest drivers coming from the "Next Generation Mobile Network Alliance" [Ngmn08] that documented a variety of use cases. In 2012, AT&T launched a nationwide deployment of SON for automatic neighbor relations (ANR), load balancing, and self-configuration features, which is a subset of the major typical SON features as shown in Table 8.7.

Table 8.7 also illustrates some of the types of radio access data being collected for SON, as well as the typical update rates for SON actions taken with the data. To put the data collected and action update rate a bit in perspective, AT&T currently collects 79 billion network measurements a day, resulting in approximately 1.4 million adjustments of the network. In comparison to typical "human adjustments" to tune the network, a good engineer may be able to make on average 10 adjustments per day; this would take approximately 140,000 engineers to accomplish the same volume of changes, which of course is not even practical.

Nevertheless, the ability to collect, process data, and make network changes is only useful if those changes have a positive impact on network quality improvement. AT&T's SON implementation when implemented resulted in a 10% improvement in call retainability, a 10% improvement in throughput, and a 15% reduction in overloading [Cell12]. Perhaps, even more impressive, when analyzing the improvement in call retainability, was the 50%–75% reduction of dropped calls due to "missing neighbor cells" which was achieved using the ANR SON feature as shown in Figure 8.16.

Similarly, Figure 8.17 illustrates the typical improvement seen when the load-balancing SON feature was introduced into a cluster of cells, which ultimately can be realized as increased network capacity.

TABLE 8.7
Popular SON Applications

Traditional SON Applications	Description	Primary Benefits/ Secondary Benefits	Primary Data Used for SON Decisions	Typical Update Frequencies
ANR	Dynamically assigns the neighbor list, and priority list priority per cell	Minimize dropped calls	Mobile reported neighbor cell measurement results	Minutes
Load balancing	Measure the load across cells/ radios/sites and balancing when near congestion	Minimize congestion, lower Capex	Radio network reported load per radio	Seconds
Antenna optimization	Dynamically adjust antenna tilts to optimize coverage/lower interference	Coverage/quality/ self-healing from site outages	Mobile reported quality cell measurement results	Hours-days
Energy savings	Turn down active transmitters during off-peak times	Lower Opex	Radio network reported load per radio	Hours
Self-configuration	Self-populate initial site parameters and neighbors for new sites/radios	Lower Opex/quality	Network topology, closest cells	NA

The above historical SON description was only an illustration of the benefits on what were 3G UMTS networks at the time, and as networks become more dense (more cells/area), have higher bandwidths deployed, consist of multiple technologies, the scale/complexity and opportunity will increase. It is expected that in 5G, SON will "not only apply to physical networks but to balance load across multiple networks/vendors," and as before it will apply to "traffic steering" and "dynamic spectrum allocation." Finally, with respect to the "data layers," this SON was only deployed for the radio access layer; and future SON's will have access to the full data layer stack giving the ability for much greater scale of optimization.

Implementing SON closed-loop automation applications in the ONAP framework could conceptually be done by designing and implementing an MS that defines the data collection from the network, analytics processing, and actions to be taken utilizing DCAE and ONAP implementing the various SON functions. As an example, Figure 8.18 and Table 8.8 illustrate how MS functions can be used to implement the ANR SON feature in the ONAP framework.

8.4.2 A CUSTOMER CONFIGURABLE POLICY FOR A CONTENT FILTERING SMART NETWORK

The SON application above was a good example of how radio access data are ingested to improve network quality and capacity. In addition, there may be "customer preferences" that could also drive network reconfiguration. A simple example of a customer preference could be to "block" some websites based on parental control attributes. As in SON, one can envision this implemented via "real-time" data collection of website IP addresses, accompanied by a "customer specific" policy as to whether parental controls have been enabled for a given website.

8.4.3 TRAFFIC SHAPING BASED ON HISTORICAL AND CURRENT ESTIMATED NETWORK CONGESTION

Traffic shaping is another Network Cloud application that can be used to improve the average "time to first byte," throughput, and latency across active users of a radio base station. This is generally achieved at the expense of high bandwidth users on the same radio during times of radio congestion.

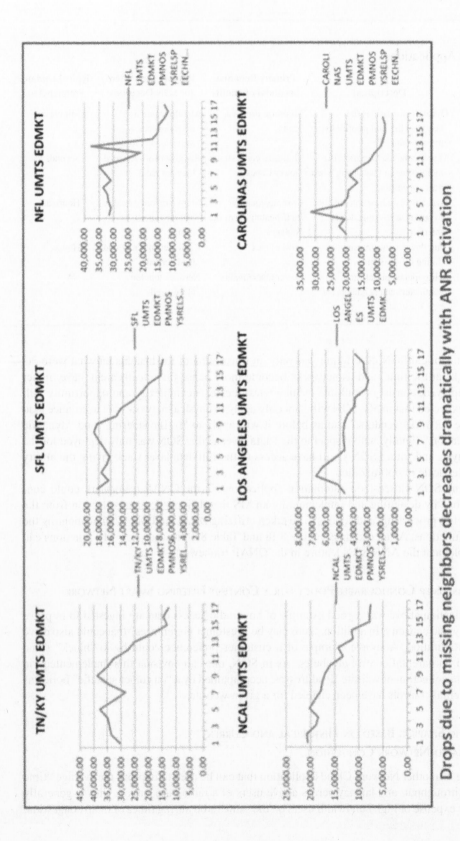

Drops due to missing neighbors decreases dramatically with ANR activation

FIGURE 8.16 Reduction of drops over a 17-day period due to missing neighbors (TN/KY UMTS ED MKT denotes the Tennessee/Kentucky UMTS market, and likewise SFL = South Florida, NFL = North Florida, NCAL = Northern California).

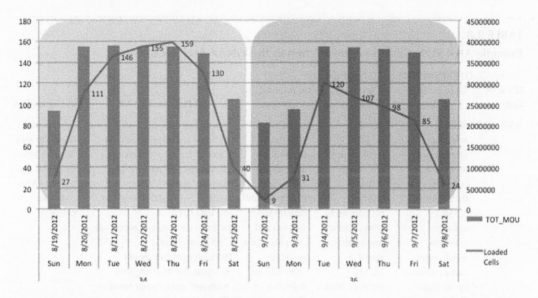

FIGURE 8.17 Typical loaded cell reduction seen with load balancing (~30%–40%), MOU = minutes of use. (Before load balancing implementation [August 19, 2012–August 25, 2012], and after load balancing implementation [September 2, 2012–September 8, 2012].)

The process of traffic shaping inserts time delays in a TCP/IP data flow for selected active sessions corresponding to high bandwidth usage. Those active sessions are then allocated fewer physical resource blocks (PRBs) on the base station radio, thereby freeing them for other active sessions. Traffic shaping only needs to occur on those base station radios where total usage, as measured by perhaps >80% average PRB usage, indicates heavy radio congestion.

The first steps for such a functionality are to determine whether a radio is in congestion and then which mobile devices are actively using that radio. Next, we must understand what type of data

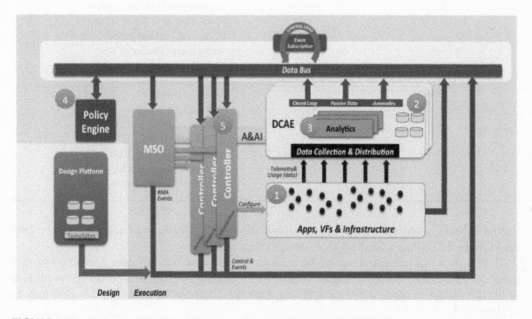

FIGURE 8.18 Reference DCAE architecture, referencing items 1–5 in Table 8.8.

TABLE 8.8

Example ANR SON Closed-Loop Function in the ONAP Architecture

D2 SON Feature	(1) Primary Data Used for SON Inputs	(2) Network Topology, Current State	(3) Analytics to Compute	(4) Policies for ANR	(5) Controllers
ANR	Mobile reported neighbor cell measurement results are collected with the DCAE "batch or real-time collectors" which collect data as defined by the Data Collector Policy= "ANR_collect" (see far right column)	Current neighbor relationships (i.e., which cells can handoff between each other). These results are stored in DCAE topology layer	The DCAE analytics MS for ANR analyzes the "nbhr_ measurements" data from the data bus and computes "best neighbor relations" the Analytics Computation Policy rule "ANR"	*Data Collector Policy "ANR_ collect"*: collectors should get "mobile reported neighbor cell measurements" data from network node XYZ every X seconds, and publish results to the data bus *Analytics computation policy rule "ANR_analytics"*: The analytics layer should compute the "best neighbor relations" per cell-pair based on the number of times a given cell was seen for each serving cell. It stack ranks the cell pairs, and performs a cutoff of the top N cell pair. If these are different than the existing neighbor list stored in the topology function (retrieved via API), then the analytics layer publishes the new "new_best neighbor_list" to the data bus *Actions policy "ANR_actions"*: the policy function shall inspect the data bus for the "new_best_neighbor_list-1234," and publish "implement_nbhr_list-1234" to the data bus intended for the "radio_access_Controller" to take action per *ANR Controller policy "ANR_controller"* Controller upon seeing "implement_nbhr_list-1234" publishes "new_nbhr_list" to the radio access network and if successful, updates topology with new neighbor relations	Radio access controller sees "change_nbhr_ list-1234" and follows "ANR controller" policy

session is being conducted by each of the mobile devices. Lastly, we must selectively insert time delays in the TCP/IP data flows for those device's data sessions.

To the extent that congestion cannot be measured in real time but only after a small delay, then historical data can be used to compare relative radio loading to estimate whether the radio is in or likely to be in congestion. Figure 8.18 and Table 8.9 illustrate how this SDN feature may be implemented.

TABLE 8.9
Traffic Shaping Feature Functional Mapping to DCAE

D2 Traffic Shaping Feature	(1) Primary Data Used For Throttling Inputs	(2) Network Topology, Current State	(3) Analytics to Compute	(4) Policies	(5) Controllers
High bandwidth traffic shaping	PRB utilization counts, from VNF edge nodes going into DCAE Data rates per Internet domain (bytes/time) generally observed at a VNF packet gateway or multiservice proxy (MSP) also going into DCAE	Current networks assert QoS-based allocation of resources for requested session (MIME) types on proprietary nodes PRB utilization is sent directly from proprietary nodes to DCAE	The DCAE microanalytics for throttling determines historical PRB utilization per base station radio and the associated session count per hour, per day of week. It also determines current number of active data sessions and estimate likely current load All results are placed onto data bus	Policy engine pulls down throttling microanalytics results from data bus and applied specific policies as follows. For those base station radios having high estimated current PRB utilization, insert TCP/IP packet time delays at the "controllers," which are the VNF packet gateway or VNF multiservice proxy as a function of session or MIME type for that radio. For instance, if a radio is heavily loaded, we can set up policies to throttle adaptive bit rate video, but not a*.jpg file transmissions Policy results/ commands are placed on to data bus	VNF packet gateway or VNF MSP receives a command via the data bus with radio ID and throttling level (based on PRB loading level) to invoke throttling e.g., "radio_X, throttle class Y"

8.4.4 Utilizing SDN to Minimize Robocalling

Combatting robocalling is an ever-increasing task in wireless networks whose solution also lends itself to software-defined networking. In this case, the SDN needs to identify robocallers using a "seeded" list of known robocaller phone numbers, and call detail attributes, and then subsequently by using ML to predict other robocallers dynamically. After detection, the suspected robocallers can either be sent to an operator for confirmation, or subsequently reprogram the network to block calls from suspected robocaller numbers.

8.4.5 Emerging Applications of SDN and NFV Including IoT, M2X, Ambient Video, Virtual Reality, Security, and Intelligent Agents

The expected data generated by Internet of things (IOT) continues to grow. Indeed, as per a recent article [Sdxc16], "The average person generates 600 to 700 MB of data per day; that's going to

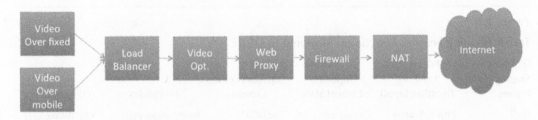

FIGURE 8.19 Example of traditional service chaining.

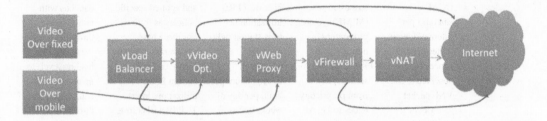

FIGURE 8.20 Example of dynamic service chaining.

increase to 1.5 GB by 2020, …. but that's paltry next to the expected numbers out of IoT: 40 TB per day from airplane engines, or 1 million GB per day for a smart factory." Consequently, to deal with the data explosion coming from IOT, more adaptable dynamic networks using SDN will be required, of which several concepts have been suggested such as dynamic service chaining, dynamic load management, and bandwidth calendaring.

Service chaining is not a new function in itself; it describes the various network functions that a data packet traverses throughout its transmission and has been discussed in Chapter 3. For example, these functions may entail Firewalls, traffic optimizers, network address translation (NAT) devices, web proxies, load balancers, and VPN terminations to name a few. In today's networks, which may contain traffic coming from both fixed and mobile access, a common core network would likely be the default route for these functions (see Figure 8.19).

Whereas, with SDN, packets can be programmed with a "service identifier," which will also entail which type of services a packet is routed as shown in Figure 8.20.

Dynamic load management can be thought of as balancing the load of the network based on current traffic conditions. In an SDN-powered wireless network, this may entail dynamically balancing the radio load across different layers of spectrum, different radios, or adjacent cell sites when possible. Within the core network, this can amount to different routing schemes based on congestion, or dynamically enabling more compute and storage capacity when needed.

Bandwidth calendaring amounts to scheduling or steering network transmissions from devices (e.g., IOT), toward times or locations on the network that are not congested. In wireless networks, this may entail steering devices to upload over WiFi when available, or scheduling large nontime sensitive transmissions to off peak-hours such as at night. SDN in this case entails dynamically sharing network transmission preferences with devices and having them upload or download traffic during nonpeak hours.

REFERENCES

[Acci14] Choosing a NoSQL Database—Technology Comparison Matrix, October 14, http://accionlabs. com/2014/10/16/choosing-a-nosql-database-technology-comparison-matrix/

[Andr16] Android Developer Package Index, Android API's, https://developer.android.com/reference/packages.html

[Apac] The Apache Software Foundation, https://www.apache.org/

[Att13] Nanocubes: Fast Visualization of Large Spatiotemporal Datasets, http://nanocubes.net/

[Cai15] Cai, L., Zhu, Y. 2015. The challenges of data quality and data quality assessment in the big data era. *Data Science Journal*, 14, p. 2. DOI: http://doi.org/10.5334/dsj-2015-002

[CDAP16] Cask Data Application Platform, http://cask.co/products/cdap/

[Cell12] AT&T Deploying SON Technology as Part of Latest Network Upgrades, http://www.cellular-news.com/story/Operators/53297.php

[Clou] Cloudera, http://www.cloudera.com/

[Data14] Understanding Big Data: The Seven V's, May 22, 2014, http://dataconomy.com/seven-vs-big-data/

[Den13] Deng, L. et al. Recent advances in deep learning for speech recognition at Microsoft, *ICASSP 2013—2013 IEEE International Conference on Acoustics, Speech and Signal Processing*, May 26–31, Vancouver, BC, Canada.

[Desa16] Desai, A., Nagegowda, K.S., Ninikrishna, T. 2016. A framework for integrating IoT and SDN using proposed OF-enabled management device, *2016 International Conference on Circuit, Power, and Computing Technologies (ICCPCT)*, March 18–19, Tamilnadu, India.

[Dezi16] Apache Hadoop Turns 10: The Rise and Glory of Hadoop, February 2016, https://www.dezyre.com/article/apache-Hadoop-turns-10-the-rise-and-glory-of-Hadoop/211

[Dezy15] Sqoop vs. Flume Battle of the Hadoop ETL Tools, October 15, https://www.dezyre.com/article/sqoop-vs-flume-battle-of-the-Hadoop-etl-tools-/176

[ECO16] ECOMP (Enhanced Control, Orchestration, Management and Policy) Architecture White Paper, 2016, https://www.google.com/#q=ecomp+white+paper

[Entr16] "Big Data" is No Longer Enough: It's Now All about "Fast Data," May 2016, https://www.entrepreneur.com/article/273561

[Eric14] 5G What is it? http://www.ericsson.com/res/docs/2014/5g-what-is-it.pdf

[Fow14] Lewis, J., Fowler, M. 2014. Microservices, http://www.martinfowler.com/articles/microservices.html

[FPOF16] Comprehensive Online Tracking is Not Unique to ISPs, May 20, 2016, https://fpf.org/2016/05/20/14382/

[Gil16] Gilbert, M., Jana, R., Noel, E., Gopalakrishnan, V. 2016. Control loop automation management platform, *5th IEEE Global Conference on Signal and Information Processing—IEEE GlobalSip 2016*, Washington, DC.

[Gog16] DeepMind AlphaGo, 2016, https://gogameguru.com/tag/deepmind-alphago-lee-sedol

[Goog03] Ghemawat, S., Gobioff, H., Leung, S., 2003, *The Google File System.* https://static.googleusercontent.com/media/research.google.com/en//archive/gfs-sosp2003.pdf

[Goog04] Dean, J., Ghemawat, S. MapReduce: Simplified data processing on large clusters, *OSDI'04: Sixth Symposium on Operating System Design and Implementation*, December 2004.

[Info2013] SDN, Big Data, and the Self-Optimizing Network, InfoWorld|October 23, 2013, http://www.infoworld.com/article/2612727/sdn/sdn--big-data--and-the-self-optimizing-network.html

[Impa16] The 7 V's of Big Data, https://www.impactradius.com/blog/7-vs-big-data/

[Hado] Apache Hadoop, Hadoop.apache.org/

[Hort] http://hortonworks.com/

[iOS16] iOS Developer Library, https://developer.apple.com/library/index.html

[Kafka] Apache Kafka: A Distributed Streaming Platform, https://kafka.apache.org/

[Kot07] Kotsiantis, S.B. 2007. Supervised machine learning: A review of classification techniques, *Informatica.* 31, 2007, 249–268. http://www.informatica.si/index.php/informatica/article/viewFile/148/140

[Lamb16] Lambda Architecture, http://lambda-architecture.net/

[Lane01] Laney, D. 2001. *3D Data Management: Controlling Data Volume, Velocity and Variety*, META Group, Stamford, CT. http://blogs.gartner.com/doug-laney/files/2012/01/ad949-3D-Data-Management-Controlling-Data-Volume-Velocity-and-Variety.pdf

[Lec98] Lecun, Y., Bengio, Y., Haffner, P. 1998. Gradient based learning applied to document recognition, *Proceedings of the IEEE*. 86, 11.

[Ligh15] Light Reading's Big Telecom Event, The Role of Big Data and Advanced Analytics in SDN/NFV, https://accedian.com/wp-content/uploads/2015/06/BTE15_The-Role-of-Big-Data-and-Advanced-Analytics-in-SDN-NFV.pdf

[Livi13] The 7 V's of Big Data, Livingstone Advisory, https://livingstoneadvisory.com/2013/06/big-data-or-black-hole/

[Mapr] https://www.mapr.com/

[Mit97] Mitchell, T. 1997. *Machine Learning*, Mcgraw-Hill Series in Computer Science. McGraw-Hill, Inc. New York, NY.

[Min69] Minsky, M., Papert, S. 1969. *Perceptrons: An Introduction to Computational Geometry*, MIT Press, Cambridge, Massachusetts.

[Ng16] Ng, A. 2016. Machine Learning, Coursera Course, https://www.coursera.org/learn/machine-learning

[Ngmn08] NGMN Use Cases Related to Self Organising Network, Overall Description, https://www.ngmn.org/uploads/media/NGMN_Use_Cases_related_to_Self_Organising_Network__Overall_Description.pdf

[NPRM16] FCC Releases Proposed Rules to Protect Broadband Consumer Privacy, Paragraph 4, https://www.fcc.gov/document/fcc-releases-proposed-rules-protect-broadband-consumer-privacy

[Open16] Web Filtering, https://www.opendns.com/enterprise-security/solutions/web-filtering/

[Rag16] Gopalan, R. 2016. Predictive Machine Learning for Software-Defined Networking Applications, AT&T Internal White Paper.

[Rah91] Rahim, M. 1991. *Artificial Neural Network for Speech Analysis/Synthesis*, Chapman and Hall, Cambridge.

[Rum86] Rumelhart, D.E., Hinton, G.E., Williams, R.J. 1986. *Learning Internal Representations by Error Propagation*, MIT Press, Cambridge, Massachusetts.

[Sdxc16] Intel Helps Developers Blend Machine Learning and IoT, https://www.sdxcentral.com/articles/news/intel-helps-developers-blend-machine-learning-iot/2016/08/

[Sha48] Shannon, C.E. 1948. A mathematical theory of communication, *Bell Systems Technical Journal*, 27, 3, 379–423.

[Tur50] Turning, A.M. 1950. Computing machinery and intelligence. *Mind*, 59, 236, 433–460. http://www.jstor.org/stable/2251299

[STORM] Apache STORM, http://storm.apache.org/releases/current/index.html

[Tutor16] Apache SPARK RDD, https://www.tutorialspoint.com/apache_spark/apache_spark_rdd.htm

[Wang10] Wang, J.L., Li, H., Wang, Q. 2010. Research on ISO 8000 series standards for data quality, *Standard Science*, 12, pp. 44–46.

[WikiBig] Big Data, https://en.wikipedia.org/wiki/Big_data

[WikiHadoop] Apache Hadoop, https://en.wikipedia.org/wiki/Apache_Hadoop

[WikiPrivacy] Privacy Policy, https://en.wikipedia.org/wiki/Privacy_policy

[Wya16] Wyatt, M. 2016. Microservices Principles, AT&T Internal White Paper.

[YARN] Apache Hadoop YARN, https://hadoop.apache.org/docs/r2.7.2/hadoop-yarn/hadoop-yarn-site/YARN.html

[Zoo16] Hadoop Zookeeper Tutorial, https://www.dezyre.com/hadoop-tutorial/zookeeper-tutorial

9 Network Security

Rita Marty and Brian Rexroad

CONTENTS

9.1 Introduction 171
9.2 Security Advantages of SDN and NFV 173
    9.2.1 Design Enhancements... 173
    9.2.2 Performance Improvements... 175
    9.2.3 Real-Time Capabilities 175
9.3 Security Challenges 176
9.4 Security Architecture 177
    9.4.1 Cloud Security 177
    9.4.2 AIC Security Evolution.. 180
    9.4.3 Network and Application Security 182
    9.4.4 Hypervisor and Operating System Security.. 185
    9.4.5 ONAP Security 187
9.5 Security Platforms 189
    9.5.1 Security Analytics 189
        9.5.1.1 Components of Security Analysis... 190
    9.5.2 Identity and Access Management 191
        9.5.2.1 Identity Ecosystem... 191
        9.5.2.2 End User Identities 192
        9.5.2.3 Client Application Identity... 193
        9.5.2.4 Resource Identity 193
    9.5.3 ASTRA 194
        9.5.3.1 ASTRA Overview... 194
        9.5.3.2 ASTRA System Overview 195
        9.5.3.3 ASTRA Defense in Depth 196
        9.5.3.4 Network Perimeter (Classic Network Protection)....................... 198
9.6 Topics for Further Study 199
Acknowledgments... 200
References.. 200

9.1 INTRODUCTION

To meet changing network demand driven by massive increase in data traffic, service providers need to transform their infrastructure into a software-enabled architecture. This relies on two enabling technologies: software defined networking (SDN) and network functions virtualization (NFV). These technologies will not only evolve the network to be more dynamic but also enable it to be more secure.

Security is built into the design of the SDN and NFV architecture day one. SDN and NFV deliver new security capabilities that make it easier to maximize responsiveness during a cyberattack and minimize service interruption to the customers. SDN enables the network to expand to handle additional traffic during an attack. The network is protected through virtualized on-demand defenses that provide security software as-a-service, built within a distributed cloud environment, integrated within the cloud provisioning process.

171

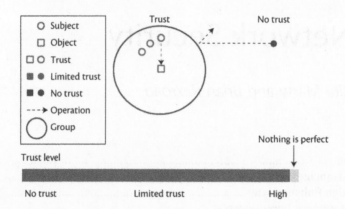

FIGURE 9.1 Initial enterprise perimeter state.

The above represents a major transformation to the traditional security architecture [1]. The traditional security approach was based on firewalls (FWs) protecting trusted entities inside the enterprise from untrusted access by outside entities via the public Internet. The intent of the perimeter view is to ensure that enterprise assets located inside the FWs are completely safe from untrusted access. Figure 9.1 shows a graphical representation of the trusted level typically associated with the perimeter approach.

This perimeter network model has been a pillar of protection design for security architects for nearly three decades. However, the overall enterprise trust in this setup has continually degraded as a results of connectivity decisions and evolving threats. As enterprise users began to covet access to emerging resources on the Internet and as email began to establish itself as the preferred medium of business communication, the enterprise was forced to allow connectivity through the trusted perimeter. This was done by creating FW rules allowing bidirectional connectivity to support the desired services as highlighted in Figure 9.2. This resulted in less-trusted external websites and email senders from untrusted external networks gaining access. As the Internet became more complex and varied, many new external connections began to appear. The perimeter architecture is no longer an effective security control resulting in critical infrastructure services being more vulnerable to cyberattacks.

This chapter will outline the security advantages of a software-enabled network including design enhancements such as security features embedded at design time, performance improvements such as reduced incident response cycle time, and real-time security capabilities such as resilience against a distributed denial of service (DDoS) attack. SDN and NFV introduce new attack vectors. A comprehensive security architecture will be required to mitigate the new attack vectors while realizing the security advantage of a software-driven network. This chapter will also outline the evolution of the security architecture from a traditional perimeter based model to a software-enabled dynamic security model.

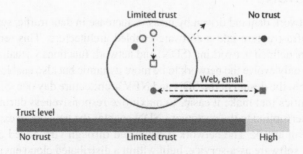

FIGURE 9.2 Perimeter model challenged with nontraditional access.

9.2 SECURITY ADVANTAGES OF SDN AND NFV

The industry transformation to SDN- and NFV-enabled ecosystem will provide several security advantages to service and cloud providers. Industry and standards organizations and security vendors are driving the realization of these security benefits for scalable carrier-grade implementation.

The benefits derived from SDN and NFV deployment can be divided into three categories: design enhancements, performance improvements, and real-time capabilities. Figure 9.3 expands on the benefits in each of these categories and are described in greater detail.

9.2.1 DESIGN ENHANCEMENTS

SDN and NFV provide flexibility and capabilities that enable security to be designed into the solution rather than being an add-on once the network and the rest of the infrastructure is architected, defined, and designed. Incorporating security capabilities as part of the design process allows for the security solution to be optimized. This will result in the security functions being deployed in locations and manners that will provide optimal security coverage while controlling costs better. NFV will enable the security components to be deployed in smaller less-expensive units in the locations where needed, while SDN adds to the flexibility as to where and how these capabilities can be deployed.

SDN and NFV allows for the security controls to be customized to the particular flows, enabling particular traffic to flow through a single security device or multiple security devices, depending on the security risks that need to be addressed. This is shown in Figure 9.4. This is referred to as a "Defense in Depth Architecture."

As an example, if there is risk of sensitive information being leaked, then the traffic can be directed to flow through a data leak protection (DLP) control as well as a FW. In Figure 9.4, Vendor 1 Security Tool could be the FW, Vendor 2 Security Tool could be the intrusion detection system (IDS), and Vendor 3 Security Tool could be the DLP protection. If this risk is not present, then the DLP control can be bypassed and just the FW and IDS controls are experienced. This use of the service-chaining ability provides greater efficiency both in optimizing the amount of DLP controls needed in the infrastructure, as well as the benefit of the nonsensitive information flowing more efficiently through the network (e.g., not being subject to the DLP control both in processing and routing).

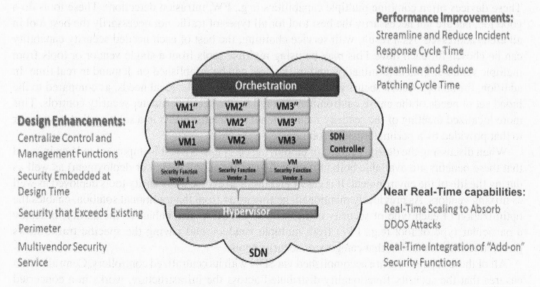

Design Enhancements:
Centralize Control and Management Functions

Security Embedded at Design Time

Security that Exceeds Existing Perimeter

Multivendor Security Service

Performance Improvements:
Streamline and Reduce Incident Response Cycle Time

Streamline and Reduce Patching Cycle Time

Near Real-Time capabilities:
Real-Time Scaling to Absorb DDOS Attacks

Real-Time Integration of "Add-on" Security Functions

FIGURE 9.3 Security advantages of SDN and NFV.

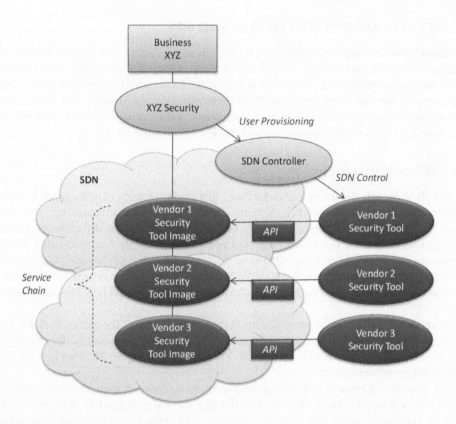

FIGURE 9.4 Defense in-depth architecture.

The flexibility that SDN and NFV bring also enables the optimization of the tools used. As compared to the traditional implementation, SDN offers the ability to mix and match the tools used. In a traditional environment, many of the security capabilities are deployed in a perimeter position. A perimeter control point enables and encourages the deployment of fewer large security appliances. These devices often combine multiple capabilities (e.g., FW, intrusion detection). These tools do a good job, but are not necessarily the best tool for all types of traffic nor necessarily the best tool in all areas addressed. As a result, with service chaining, the best of each needed security capability can be chosen for each flow. This may be using multiple tools from a single vendor or tools from multiple vendors. This determination and traffic flow can be established on demand in real time. In addition, the sizing of each security tool will be determined by the local needs, as compared to the broad set of needs of the entire environment, served by the set of perimeter security controls. This more localized meeting of the security requirements and needs results in a solution that is superior to that provided by a perimeter-based solution.

When discussing the design benefits of employing SDN and NFV, it is important to keep in mind that these benefits are available both at initiation (e.g., the initial design or deployment) as well as during the life of the environment. It is easier and faster to evolve the security tools deployed as well as mixing vendors. As previously mentioned, getting away from the traditional solution, enables the optimization of each type of security control. SDN and NFV also enhances the ability to deploy a particular type of tool (e.g., FW) from multiple vendors, and having the specific traffic flows directed to the best tool in that category of security tools.

All of these capabilities are accomplished via SDN with its centralized controllers. Centralization ensures that the security functionality distributed across the infrastructure, works in a concerted manner providing the appropriate and needed security.

9.2.2 Performance Improvements

From a performance perspective, SDN and NFV can enhance the security of the environment. Patching of virtual machines (VMs) is easier and simpler with SDN and NFV. SDN and NFV enables the patching systems by simply creating a new instance of a function with the patched version of the software and then redirecting the traffic to the new instance(s). The deployment of, as well as the transition to, the new instances is simplified with SDN and NFV. In addition to the patching process being more automated and having essentially no impact on the operation and performance of the application, the inventory is automatically updated to reflect the changes.

The incident response process is also improved in a number of ways. Assuming just one or a subset of the instances of the virtual function is impacted or involved in an incident, taking it down will not cause the entire function to be taken off-line thereby reducing the impact. In all cases, whether it is one or many instances of the VNF that is impacted, adding instances of the VNF and routing to the new instances is simplified with SDN. Being able to instantiate new instances of the VNF allows for freezing and grabbing a snapshot of the impacted VNF, which will provide better information for the forensics operation.

9.2.3 Real-Time Capabilities

As has been described earlier, SDN enables the swift response to a security incident. In the realm of security, this includes possibly adding additional instances of capacity, or routing traffic through additional or different security controls.

A prime example of real-time capabilities would be the resolution of DDoS attacks. There are services and capabilities to mitigate DDoS attacks today. However, there is often a delay in activating the control due to the risk associated with dropping valid or desired traffic. SDN and NFV, as depicted in Figure 9.5, offers the ability to scale out the impacted constrained resource to provide better performance until the traffic and the impact of other controls can be evaluated. When, for

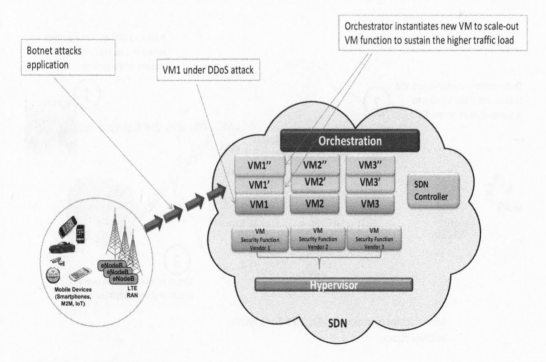

FIGURE 9.5 DDoS attack resilience.

example, a botnet attacks an application, the instance of the application is overwhelmed. However, when the overload condition is observed, the orchestrator can instantiate additional instances of the VNF to cover the increase in traffic. Once the analysis is performed and the other controls kick-in, the additional instances of the VNF can automatically be removed as the demand decreases.

9.3 SECURITY CHALLENGES

The deployment of SDN and NFV technologies in networks presents some security challenges in a number of different areas: architecture, technology, and operations. This section focuses on those challenges that are specifically related to SDN and NFV. There are a number of other challenges that are more appropriately considered cloud-deployment challenges. An example would be the ability for traffic to stay within a server and not being subject to some security controls. There are also a number of challenges that are shared with cloud deployments. For example, the existence of more software in the deployed solution—(both cloud infrastructure software, as well as SDN and NFV require more software). This creates a security challenge with the opportunity for bugs or malware to be introduced into the environment. Other examples include the need for new or modified processes, different data flows, different operations flows, more distinct separation between infrastructure and applications, and the resulting impact and coordination with operations teams. Another shared challenge is that components of the infrastructure may react to stresses differently—what may happen with a particular device or appliance (e.g., fail open); might work differently in the cloud environment than in a SDN and NFV implementation (e.g., fail closed).

From an architecture perspective, SDN both centralizes the control and distributes the implementation points for the networking. This provides the ability to provide more coordinated control; however, should the central controller have a security issue, the impact could be dispersed across the entire network. Establishing an architecture and implementation that provides a significant barrier for accessing the SDN controller is critical to help ensure the SDN infrastructure cannot be used as part of an attack.

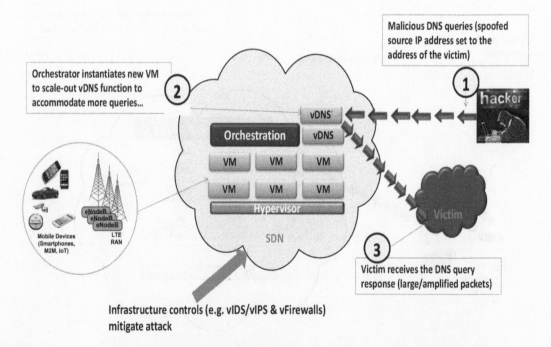

FIGURE 9.6 DNS amplification attacks enhanced by elasticity function.

Since the implementation is distributed, it is imperative that the controller and the centralized source of information have correct information on the structure of the network and what is distributed where. Sending commands to the incorrect virtual router or switch could result in unwarranted consequences. Recovery from this state would probably rely on backing out the latest commands and a subsequent analysis to determine what went wrong and related corrective actions. Figure 9.6 depicts an example of how the technology can provide a challenge.

In this scenario, what was deemed as a benefit when dealing with a DDoS attack (e.g., adding capacity) can work to our detriment when applied to an attack mechanism. An attacker sends malicious domain name system (DNS) queries from a spoofed source IP address. The infrastructure reacts to the increase in queries by standing up additional virtual DNSs (vDNSs), which allows more attack traffic to reach the target or the victim. The solution in this case, involves applying traditional controls (e.g., FW, intrusion detection and prevention) as well as processes to detect and limit the damage. These controls and processes work together to alarm on the unusual activity and limit the damage. It is the balance of the flexibility and the controls that will ultimately determine and constrain the damage imparted on the victim. This also demonstrates a case where the operations team and processes need to be aware of, and be in a position to react appropriately to, undesired traffic.

9.4 SECURITY ARCHITECTURE

Our approach to security of a software-enabled network is comprehensive. The security architecture addresses multiple security pillars or layers including (1) cloud, (2) network, (3) applications/tenants, and (4) Open Network Automation Platform (ONAP) security. Security controls will be embedded day one at each layer. The critical need is to have the security controls at each layer well coordinated so that they enhance the overall security framework instead of allowing for cracks in the implementation where signals or red flags might fall through and be missed.

Across each of these core components or layers of the architecture, three major platforms will be ingrained: security analytics, identity and access management, and ASTRA. These platforms enable a multilayer defense strategy. The security analytics platform provides a comprehensive threat detection and investigation capability. The identity and access management platform provides an adaptive security capability that enables effortless protection via authentication and authorization of hardware, systems, and applications in the cloud. ASTRA provides an adaptive security platform for the cloud infrastructure. Each of these three core components are described in detail in Section 9.5.

The cloud architecture was created to provide some core security protection. The separation between the control planes and the VNF on the data plane helps limit the security risks. If there is no direct connectivity between the planes, then attacks and attempts to steal information or data is severely curtailed.

9.4.1 CLOUD SECURITY

AT&T Integrated Cloud (AIC) refers to AT&T's common cloud environment, which is discussed in detail in Chapter 4. AIC deployments are based on OpenStack software and are deployed in multiple models. All the deployment models are based on the same OpenStack foundation and are tailored and sized to support different needs. The factors defining the deployment models include network topology, location, physical resource availability, and applications targeted for a given location. For example, some sites will be designed to support VNFs due to their geographic location while other sites will support more generic computing needs due to the network connectivity and cost of the infrastructure. Based on the expected needs, costs, etc., those sites designed to support VNFs will be smaller than those used for more generic computing. In all instances because of the common OpenStack base, there is the capability and flexibility to handle whatever VMs or applications that are needed at that location.

The AIC security framework is a composite of core AIC architecture and design decisions, which incorporates principles that ensure and achieve some level of security. For example, the architectural requirement that there be separation between management and bearer traffic helps provide an inherently secure architecture. The implementation and distribution of security functionality and capabilities within and on top of the infrastructure further builds out the security model. The AIC security framework is predominantly based on components outside of the security functionality provided by OpenStack as part of it base functionality. The ultimate goal is to have a security solution that meets and exceeds expected corporate security requirements.

AIC connects to many different types of networks. All of these networks have the potential to impact the AIC security framework. The specific security implementation for each AIC model and data flow is based on the functionality and data content provided by AIC in that instance. It is the functionality and data that defines which requirements and security controls apply in each case or flow, all facilitated by service chaining enabled via SDN.

Figure 9.7 depicts the areas that impact how, when, and where the security functions that comprise the AIC security framework are implemented. The top layer of the diagram illustrates various cloud implementations while the layers below it depicts the various approaches, technologies, controls, and organizations that impact the AIC security framework.

The top layer depicts the spectrum of designs. They range from (a) an architecture that is based on a design with a single instance of OpenStack at a site, (b) to an architecture that is based on multiple virtual instances of OpenStack within a site with a centralized control structure, and (c) to an architecture with a more distributed control structure and a more distributed tools structure.

These designs impact the AIC security framework in many ways. From an individual OpenStack component perspective, the movement of OpenStack components can require new network routes, which must be evaluated and secured. In addition, as the OpenStack components evolve, they may require new interactions with systems outside of AIC. From an architecture perspective, changing the design (e.g., dual virtual OpenStack instances) can create new security requirements as a result of there being new flows and new connectivity. Each of these implementation options has defined solutions as to which security functionalities apply to each implementation.

The next layer in the diagram highlights the fact that the network design has a significant role in the security framework. AIC is dependent upon many different types of networks, all with potential impacts to the AIC security framework. These networks include how AIC connects to external networks or services, internal routing, and the use of virtualization. Two examples of how the various networks impact security include the use of SDN inside the environment (SDN-Local) and how networks are connected. SDN-Local provides the ability to service chain capabilities. These capabilities can be security capabilities (e.g., FWs, data leakage protection) or could be application capabilities. Each of these will define or drive the security functionality needed and deployed to meet the goals of the security framework for the particular flow and functionality. The bridging of networks, whether they be virtual or physical, must be designed to maintain the security. Given the flexibility provided by the cloud implementation, this must be an integral part of the core AIC design.

The layer below network designs captures the introduction of new technologies and functions that are part of AIC and impact the security framework. The list of new technologies and functions is long and each brings its own impact on security. For example, containers bring a new model, which can provide new boundaries and structure for applications. Their role in achieving and preserving our security framework cannot be ignored. Similarly, the greater use of software, whether it be the hypervisor, VNFs, or other capabilities bring additional risks and interactions. Controls (e.g., hypervisor separation) and evaluations need to be performed to ensure a proper understanding of possible security issues and impact on the security framework. For example, if when a virtual FW (vFW) fails it allows all traffic to pass, then this would be a significant difference from a standard capability of FW appliances today.

The bottom layer in the diagram captures the external processes and the requirements that form the security framework. In addition to the security policy and requirements, there are the strategies,

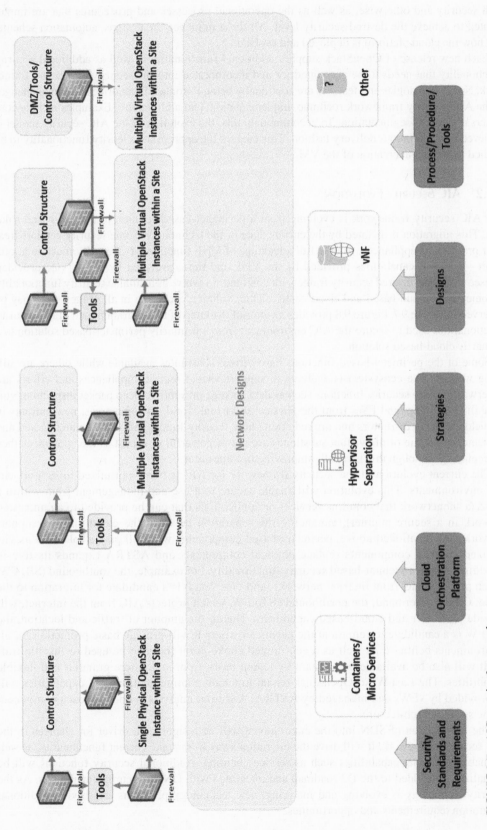

FIGURE 9.7 AIC security framework.

both security and otherwise, as well as the operational processes and procedures that are implemented to achieve the desired security level. All these influence the designs, automation schema, and how the cloud platform is deployed and evolved.

Each new release of OpenStack supplies additional functionality as well as additional security functionality that needs to be accounted for and incorporated into the security model and framework. New technologies and features are continually being introduced that drive additional changes to the AIC security framework roadmap, implementation, and architecture. Examples include containers and micro-segmentation. To accommodate this, the evolution of the AIC security model is achieved in a continuous delivery fashion. This enables the appropriate security functionality to be applied by default at creation of the VM.

9.4.2   AIC Security Evolution

The AIC security framework is evolving from a perimeter-based solution to a cloud-based solution. This migration is dictated by the current state of the technology. Some security vendors treat their products as appliances; to take full advantage of VNF functionality, security must be a joint effort—a mix of capabilities provided by the VNF and those provided by our service provider. Consequently, the desired security framework continues to evolve as available security functionality becomes more cloud-based and cloud-centric. This evolution occurring in all of the layers may be observed in Figure 9.7. Figure 9.8 provides a pictorial roadmap of the migration path of the security functionalities used to secure the AIC environment from a primarily perimeter-based solution to a primarily cloud-based solution.

Some of the perimeter-based functions have virtual substitutes available while others are still being worked. The ecosystem is evolving to support virtual security appliances and efforts are underway to move security functions such as data leakage prevention, deep packet inspection, vulnerability scanning, and FWs from the physical to virtual world. In some areas, new entrants in the field are being disruptors and are redefining how security functionality can be architected and implemented. Some of the virtual substitutes are not as powerful or robust as the appliances they are replacing, although this is rapidly improving in some areas.

The current evolution of the security framework for AIC is being optimized to support virtual environments. This evolution will enable secure tool use and management from within a DMZ (a subnetwork that contains services or applications that can be provided to an untrusted network in a secure manner), enhancing the separation between the cloud and the support networks. As mentioned above, perimeter-based components will still play a role even as virtualized security components replace physical components and ASTRA expands its role in providing enhanced tenant-based security functionality. For example, the southbound (SB) FW, which protects AIC from internal networks (and vice versa) is a candidate for migration to the cloud. On the other hand, the northbound (NB) FW, which protects AIC from the internet, will provide perimeter and cloud-based capabilities. Due to the amount of traffic and location, the NB FW is a candidate to remain at the perimeter where it will provide basic protection for all environments behind it as well as a centralized choke point that can be used to throttle traffic. It will also be available on a tenant-by-tenant basis, to provide more granular and flexible capabilities. The East-West (OpenStack tenant to tenant communication) FW capabilities will be provided by vFWs and managed by ASTRA. A similar analysis is done for each component of the security architecture.

The introduction of SDN into the environment will be a significant driver for changes in the AIC security framework. It will drive the migration away from some current functionality, as well as enable additional capabilities such as service chaining. Additional security functions will be virtualized and added to the D2 roadmap and integrated with ONAP security capabilities. As the security technology is evolving and maturing, new technology platforms are driving additional integration requirements and opportunities.

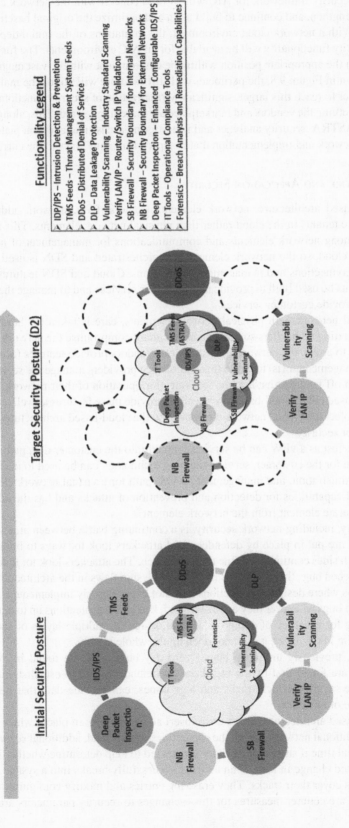

FIGURE 9.8 AIC security framework evolution.

Functionality Legend

Functionality Legend

IDP/IPS – Intrusion Detection & Prevention
TMS Feeds – Threat Management System Feeds
DDoS – Distributed Denial of Service
DLP – Data Leakage Protection
Vulnerability Scanning – Industry Standard Scanning
Verify LAN/IP – Router/Switch IP Validation
SB Firewall – Security Boundary for External Networks
NB Firewall – Security Boundary for Internal Networks
Deep Packet Inspection – Enhanced Configurable IDS/IPS
IT Tools – Operational Compliance Tools
Forensics – Breach Analysis and Remediation Capabilities

The target security framework for AIC will work in concert with the network cloud components to enhance, strengthen, and continue to build upon and optimize the original baseline set of security functionality. With a network cloud environment, the limitations of the initial deployments will be lifted and security functionality will be available for all AIC environments. The functionality will be implemented in the appropriate position within the environment with the vast majority being cloud-based. As shown in Figure 9.8, the perimeter will still exist but will not be the main tenant of cloud security. In order to reach this target, significant virtualization of security functionality will need to take place, something the vendors and marketplace are marching toward. This, along with the continued rollout of ASTRA, security analytics and identity, and access management platforms, will bring a security framework and implementation that is superior in many ways to the current environment.

9.4.3 Network and Application Security

In the cloud-based architecture, network elements (routers, FWs, network address translations [NATs], etc.) are tenants in the cloud rather than physical network elements. This means that communications among network elements and communications for management of network elements are part of the cloud, so the network elements are orchestrated and SDN is used to manage both customer data connections and management connections. Cloud and SDN features such as service chaining can thus be used both to provide new customer services and to manage the virtual network elements that provide customer services.

In a classical network with physical network elements, care is taken to "harden" the network element itself, setting up barriers so that outsiders cannot compromise the network element and use it as a platform to attack the network infrastructure. Vendors provide features (access control lists, controls on management ports) to aid in this and service providers manage the services available on routers and turn off features that are not necessary for operation of their network. The virtual network elements used in the cloud-based architecture provide these features as well, and they continue to be used to protect the virtual network elements. But the cloud-based architecture offers additional opportunities for security.

For example, just as a vFW can be service chained into the customer data path to provide additional protection for the customer, service-chaining technology can be used to insert a FW into the operations, administration, and management (OAM) path for a virtual network element. This provides additional capabilities for detection and prevention of attacks and has the added benefit that the FW is a separate element from the network element.

Cybersecurity, including network security, is a continuing battle between attackers and defenders. Protections are put in place by defenders, and attackers look for ways to break through those protections. Both sides continually refine their approach. The attackers look for holes in the system, including unpatched bugs that open gaps in defenses, design flaws in the architecture of a particular system, and gaps where designed protections have not been properly implemented. Defenders work to patch security bugs whenever they are discovered, keep their protections up to date, monitor their systems looking for evidence of compromise, and design in multiple layers of security so that if there is a problem in one layer, it does not open up the whole system.

At any time, previously unknown bugs ("zero day bugs") can be found in existing systems, and sometimes are found first by the attackers. So defenders need to establish multiple layers of defenses. Active security, which tracks and logs issues, can provide data that helps identify the problems in case of an attack.

The cloud-based architecture allows us to insert active security in places where it would not be feasible in a traditional network. Given the capabilities of the cloud, additional security features can be inserted in real time if suspicious activity is detected to help determine whether it is an attack or just an unexpected change in load. If an attacker successfully breaks into a system, one of the first things they do is cover their tracks. They erase log entries and modify configurations to hide their presence. There are counter measures for this—changes to security parameters are logged and all log entries are typically sent to a remote logging server so they cannot be erased if the system is compromised.

By service chaining security components into the OAM path of a virtual network element, we isolate the security system from the network element itself. The service-chained security component runs in a separate VM, so even if the network elements were completely compromised the attacker would not be able to disable or modify the security check in the security component. This provides an additional layer in the defense of the network. Figure 9.9 shows a security function being service chained into the OAM path of a VNF.

Various VNFs can be service chained into the management paths. FWs, intrusion detection and intrusion prevention systems, debugging tools, log analysis tools, and many other can all be used. The following basic principles leveraged in the design of MPLS VPN network also apply to cloud-based architecture [2].

Separation. SDN and cloud-based architecture allow computing resources as well as networking resources to be efficiently and securely shared across many customers. Those same technologies are used to separate management functions in the cloud architecture. The principle of *least privilege* states that users, be they customers or network managers, should only be given access to the resources they need to do their jobs. Cloud architecture allows us to set up separate management domains using the same technology used to set up different environments for different customers. For example, management of the hypervisors and other resources that create the cloud environment itself, is separate from the management of the VNF tenants that customers see.

Automation. ONAP orchestrates resources, allowing networking and computing resources to be adjusted as needed when the load changes. It also makes sure that all the necessary security functions are included; the security rules are stored as part of the description of a given VNF, so when a new copy of the VNF is instantiated, the appropriate security features are automatically included.

Monitoring. Physical and virtual elements are monitored and additional capacity can be added as needed to meet observed load. If a potential security issue is seen, automation can service chain additional security tools to investigate the situation. A copy of a misbehaving VNF can be sent for forensic evaluation and a new, clean copy started fresh.

Control. Operational excellence require 24×7 network operations support. Operational security is enhanced by use of separate networks for OAM and by requiring security gateways between networks that operate at different security levels. This is part of the principle of least privilege discussed above.

Testing. Using some of the best "ethical hackers" to probe, test, and try to find weaknesses have been very effective in the deployment of the MPLS network. Occasionally, they find an area in which improvements are necessary—and steps are taken immediately to address their findings.

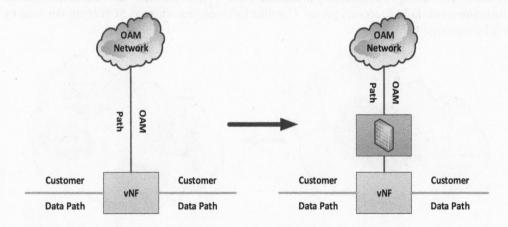

FIGURE 9.9 Service chaining a security function into the OAM path.

Response. Security incident response is performed using a tiered operations structure. AT&T utilizes a mature, global three-tiered 24/7 security operations team that is centrally coordinated in the Global Network Operations Center in Bedminster, NJ. Expert Tier 3 security analysts support this structure as incidents are escalated using well-defined security methods and procedures. At Tier 1, trained operations managers use mature monitoring tools to proactively identify conditions that might warrant response. A Tier 2 management interface oversees this activity and is used to tie together conditions that might be brewing in disparate locations.

Innovation. Tiers 1, 2, and 3 continuously evolve as security is an ongoing process. Potential attackers continue to look for new ways to attack and disrupt, defenders need to anticipate and continue to build up defenses. Beyond that, security in computing and computer networking remains an active field of research and new advances are frequently made. Service providers and network operators need to stay current with these advances.

In a "classical," appliance-based network, a security perimeter protects the network elements from attack. This may consist of access controls in routers, physical FWs, and systems for authentication and authorization of access to network elements. The perimeter is set up by carefully configuring the network elements, and auditing tools are used to periodically check that the proper protections are in place. Since setting up the perimeter requires changes to many different physical components from many different vendors, maintenance of the perimeter is a complex and time-consuming operational job (Figure 9.10).

The perimeter remains a useful concept in the cloud-based architecture, but the nature of the perimeter changes [3]. Rather than one perimeter that encloses all the critical resources, the cloud-based architecture allows critical resources to be distributed, which enables finer-grained protection rules to be applied. Where the appliance-based perimeter divided the world into "Management Team" and "Everyone Else," the cloud-based architecture allows us to partition critical resources into smaller groups and allows them to be in their own networks. By controlling access to those networks, we enforce least privilege and make it harder for a single attacker to compromise substantial parts of the network.

The same security best practices apply to the SDN controller. The SDN controller (or the VMs and servers hosting the SDN controller) will be protected with the same controls that protect the network from various types of intrusion and vulnerabilities. These security mechanisms include but are not limited to software scanning, security audit, secured access based on roles and responsibilities, and active DDoS protection.

A network cloud elastic infrastructure will enable dynamic security control for SDN controllers. A benefit of deploying SDN controllers in a cloud platform is that if compromised, the SDN controller will be isolated, shutdown, quarantined, and replaced by another instance of a secured SDN controller. This ability to dynamically instantiate a new VNF to maintain SDN controller function also improves the failure recovery period. Consider two scenarios where an SDN controller security may be compromised:

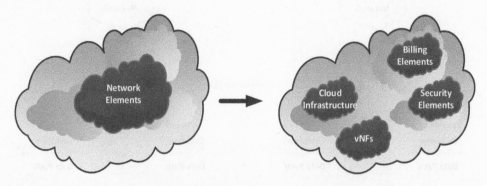

FIGURE 9.10 Evolution of the perimeter (gray line represents the perimeter).

- The SDN controller is hacked via a vulnerability in the Open Source code, for example, the hacker exploits a vulnerability in the Open Source code and infects the SDN controller with malware, then the compromised SDN controller VM can be quickly detected, isolated, shutdown, quarantined, and replaced by another instance of a secured SDN controller VM while the operations team quickly work on resolving the threat (e.g., apply security patches to fix the Open Source code vulnerability).
- The SDN controller is hacked via a vulnerability in the Open Source code, for example, the hacker exploits a vulnerability in the Open Source code and gains access to the SDN controller, then. There are several security best practices to consider to mitigate this risk. A role-based-access control function can limit the hacker to performing least-privileged activities (e.g., will not disrupt service) while we quickly work on resolving the threat. Security scans of network elements (i.e., SDN controller software) can be conducted to detect known vulnerabilities. A zero-day (unknown) vulnerability can be mitigated by quickly applying security patches.

The following paradigms are important architecture considerations:

1. Software-defined configurability of VM FW rules will help SDN controller security: SDN controller can dynamically modify the FW rules. For example, the SDN controller can dynamically modify the security gateway rules in response to a security incident. In this case, the SDN controller will modify the FW rules to dynamically block malicious traffic or a DDoS attack preventing service disruption. This provides a real-time mechanism to respond to security threats leveraging the SDN controller.
2. The network cloud shared infrastructure benefits from this security architecture: Using rigorous tools and processes, the network cloud shared infrastructure components are protected and continually monitored for security events.
3. The network cloud elastic infrastructure will enable dynamic security control for VNFs and network services: All the features of a network cloud on-demand elastic cloud-like platform that protect against a failure also protect VNFs and network services against a potential security breach or attack. A compromised VNF VM can be quickly isolated, shutdown, quarantined, and replaced by another instance of a secured VNF VM. If an entire site is compromised then all the VNFs at that site can be replicated to other sites (although, due to redundant distributed design this may not be necessary except in extreme cases).
4. Hypervisor technology will provide another layer of security: VNFs can run as software modules and processes under any modern multiprocess operating system (i.e., any version of Linux). Although using the same physical hardware, virtualization technology (via hypervisor) implements a strong partitioning of resources and isolation between VMs. It further implements strong VM quarantine or isolation capability for each software element on the network cloud shared infrastructure. Therefore, should a VM be compromised, the corrupted VM can be terminated and the VNF can be instantiated somewhere else.

9.4.4 Hypervisor and Operating System Security

NFV moves operations such as routing and FWs onto the hypervisor. Figure 9.11 depicts how infrastructure components and network traffic flows are expected to change with the introduction of NFV. The previous section highlights a traditional application stack where each component is on a physically separate hardware. Therefore, all the network traffic traverses physical connections. In the future, when the operating system has been virtualized and now the traffic between some components at the top of the stack may only traverse the virtual connections established on the hypervisor.

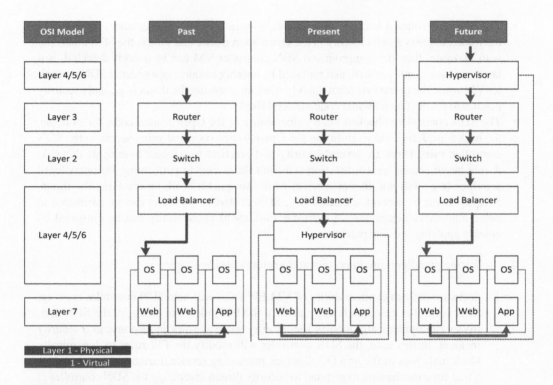

FIGURE 9.11   Hypervisor and operating system security.

The diagram depicts the example of a web application where the web server and application server are running as two separate VMs on the same hypervisor. The traditional networking functions—like the router, switch, and load balancer in the diagram—may also be VMs on the same hypervisor. The network is now traversing virtual links within a single hypervisor. While the end result maybe exactly the same, from a security monitoring perspective the ability to discern these components, the interrelationships between them, and the ability to place controls on them and in the network paths between them, requires a new way of operating. Mitigation strategies for the three new challenges:

- *Isolation*. Requires creating a baseline for secure configuration of the hypervisor and partitioning strategies such as the separation of workloads with internal- and external-facing workloads.
- *Relationships*. Requires correlating real-time asset inventory data from multiple sources to depict the nature of components in the stack and the network traffic channels between them. This understanding can then enable the placement of more effective security controls into the environment.
- *Uniqueness*. The process of capturing real-time asset inventory must account for the dynamic nature of assets in a cloud ecosystem. Typically, organizations use some sort of global identifier value to reliably track assets over time.

The introduction of network virtualization further accentuates the importance of maintaining comprehensive asset inventories. As the volume of network traffic originating and terminating within a single hypervisor increases, it will become critical to understand all the assets dependent upon that hypervisor. This type of understanding will only be possible with robust inventories that maintain adequate information about asset interrelationships.

There is noted improvement in asset inventories when device records are automatically generated via orchestration tools in cloud platforms. However, issues related to missing or incorrect inventory records can be exacerbated by the increased volume and more frequent changes to assets in a virtualized environment.

In summary, NFV introduces three new challenges relevant to continuous monitoring programs:

- *Isolation.* The adoption of network virtualization increases the importance of securing the hypervisor host. Vulnerabilities on the hypervisor may now affect the security framework of all associated network segments and devices.
- *Relationships.* The relationship between the device and other layers of architecture must be consistently captured in asset inventories.
- *Uniqueness.* Increased volatility in the asset lifecycle makes it much more important to uniquely identify a device and track its security status over time.

9.4.5   ONAP SECURITY

The ONAP software platform [4] delivers product and service independent capabilities for the design, creation, and lifecycle management of the network cloud environment for carrier-scale, real-time workloads. It consists of multiple software subsystems covering two major architectural frameworks: (1) a design time environment to define, and program the platform and (2) an execution time environment to execute the logic programmed utilizing closed-loop, policy-driven automation.

ONAP performs tasks such as provisioning of resources (VMs), applications, configuration management (of storage, compute, networking), security monitoring, and reporting. Misuse, either intentionally or unintentionally, of the powerful automation that ONAP provides could easily lead to disruption within the SDN and NFV environment. Security of the ONAP platform is critical to the stability and availability of the SDN and NFV environment.

There are two main aspects to security in relation to the ONAP platform: security of the platform itself and the capability to integrate security into the cloud services. These cloud services are created and orchestrated by the ONAP platform. This approach is referred to as security by design.

The enabler for these capabilities within ONAP is an application programing interface (API)-based security framework, depicted in Figure 9.12. The illustration highlights the security platforms work with ONAP to provide security functionality. The security platform is connected to ONAP through a set of security APIs. While the security platform could be specific to the service provider, the ONAP framework can be utilized by other security platforms. This framework allows for security platforms and applications existing outside of ONAP, to be leveraged for platform security and security by design for services it orchestrates.

Security of the platform begins with a strong foundation of security requirements with adoption of security best practices as an inherent part of the ONAP design. Some examples include

- Deployment of the platform on a secure physical and network infrastructure
- Adherence to secure coding best practices
- Security analysis of source code
- Vulnerability scanning
- Defined vulnerability patching process

Building upon this foundation, external security platforms that provide additional security capabilities such as identity and access management, micro-perimeter controls, and security event analysis are integrated onto the platform by leveraging the ONAP security framework.

The ONAP platform also enables security by design for services it orchestrates, by engaging a security trust model and engine. This begins with validation of security characteristics of resources as part of the SDC resource certification process. This assures service designers are using resource modules that have accounted for security. By leveraging the ONAP security framework to access an external security engine, additional security logic can be applied and enforced during service creation.

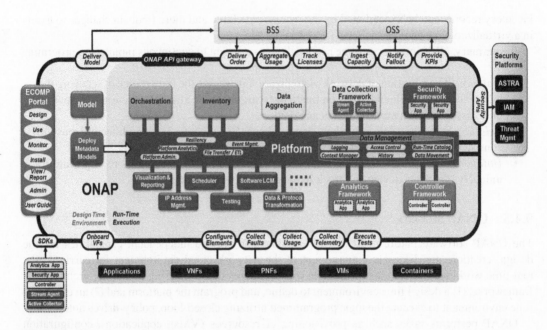

FIGURE 9.12 ONAP platform and interfaces.

ONAP is a platform for many types of services. Because of its inherent security, it is also a powerful means to provide security as a service. In many ways, security services are similar to introducing other services. However, security services must be provided via a platform and infrastructure that is inherently secure. The various types of security services can be access control, authentication, authorization, compliance monitoring, logging, threat analysis, management, etc. This is illustrated with the example of managed vFWs. Using a customer web portal, the customer provides the needed information to enable ONAP to determine and orchestrate the FW placement. In addition, the FW capabilities (e.g., rules, layer 7 FW) are instantiated at the appropriate locations within the architecture. If necessary, service chaining of many security controls and technologies including FWs, URL blocking, etc. can also be service chained. As part of an overall security architecture, the log data from the FWs can be captured by DCAE and used by the threat management application to perform security analytics. Should a threat be detected, various mitigation steps can be taken, such as altering intrusion prevention system (IPS) settings, change routing, or deploy more resources to better absorb an attack. This can be achieved by ASTRA working with ONAP to deploy the appropriate updates across the infrastructure, minimizing the service interruption caused by security threat.

SDN increases the dependency on application software, which brings with it new benefits and challenges. The benefits include new partitioning options like containers and the opportunity to automate via integration with continuous integration (CI), continuous deployment (CD), and defect management tools. The primary challenge is to maintain emphasis on traditional secure software development lifecycle (SDLC) measures such as threat modeling, security testing, and vulnerability remediation.

In AT&T's implementation of the network cloud, VNFs are onboarded to SDC module. Prior to distribution into the run-time environment, all VNFs undergo certification testing. The overall objective of this effort is to incorporate testing of a number of key security requirements and best practices into the VNF certification process in SDC where VNFs will be evaluated against (~35 security requirements). The goal is to automate this compliance function. Automating the evaluation of those security requirements is done in two parts:

- Part 1—UPSTREAM automation
  - The upstream processing involves locating data points about the VNF (resource module or service) to determine
    - Whether a particular security requirement is applicable, and if so
    - Notifying the VNF owner of the evaluation, which will be performed to determine whether the security requirement is effective
- Part 2—DOWNSTREAM automation
  - The downstream processing involves evaluating a particular VNF against the relevant security requirements identified in Part 1 including
    - Measuring the effectiveness of each security requirement and providing a pass/fail conclusion
    - Communicating to the VNF owner the results of each control measurement including instructions on remediation requirements in order to obtain approval for distribution

Both the upstream and downstream automation is performed by the ASTRA Security Certification Engine. This functionality acts as a standalone application controller, which oversees security aspect of the VNF onboarding, certification, and distribution processes.

9.5 SECURITY PLATFORMS

As mentioned earlier, security of AT&T's cloud-based architecture is based on a multilayered defense strategy and includes three critical platforms: (1) identity and access management, (2) security analytics, and (3) ASTRA.

It starts with a comprehensive identity and access management (IAM) approach. This approach goes beyond traditional methods such as self-created passwords. Strong multifactor protection includes user credentials, biometrics information, and a user's mobile device provide an effortless way to protect hardware, systems, and applications.

ASTRA is a granular security policy engine, which can create "rings" of protection at the cloud perimeter, tenants, and workload levels. These rings act as a micro-perimeter for every asset that needs to be protected. Security policies created for one asset can automatically apply to new instance of that asset. ASTRA enables customized protection for the cloud versus having a one-size-fits all approach to security. Leveraging ASTRA, it is possible to shrink-wrap security around each object in the cloud.

Security analytics is the practice of looking for abnormal behaviors or malicious activities on the network. The security analytics platform detects malicious activities and leverages ASTRA to automate the mitigation strategy. The security analytics platform uses event data from ASTRA and ONAP Data Collection Engine to quickly detect and characterize many types of attacks providing valuable information to mitigate attacks early on. A team of analyst leverage big data techniques to alert on anomalies. Automated data collection and virtualization helps analysts quickly characterize the attacks, and smart agents automatically update the policies in ASTRA. These new policies propagate quickly across all existing and new cloud assets.

Today's constantly evolving cyber threat landscape demands relentless focus on security. With strong identity and access management, the ASTRA policy engine, and security analytics platform are important part of what is needed to orchestrate a robust, automated, and multilayered cloud defense strategy.

9.5.1 SECURITY ANALYTICS

There are two fundamental functions of security analytics:

1. First and foremost, the purpose of security analytics is to identify potential misuse, abuse, or attack behavior that was not originally anticipated. For this, the objective is to define general analysis or tools that can be used to detect a wide variety of abuses.

These generally fall into two categories. (i) signature-based detection (use of identifiers to detect symptoms of known attacks and abuses) and (ii) behavioral analysis (detection of anomalies in activity, and determining whether the anomaly is relevant to security).

2. If or when indications or suspicions exist that there is a security issue, security analytics provides the means to investigate (i.e., hunt) for further evidence of the issue and to provide the means for learning what activities are needed to remediate the issue.

In a network service environment, there are two perspectives:

1. Verify the integrity of the platforms providing the service. In this perspective, no infection is acceptable. This includes focusing on the integrity of the cloud infrastructure that hosts applications, assuring tenant applications within the cloud infrastructure are secure, and assuring that any control functions associated with creating, removing, and configuring the infrastructure and applications are protected from unauthorized changes.

2. Verify activities on the service are not going to impact availability of the service. Generally, the responsibility of network services is to transport traffic from origin to one or more destinations. As a network service provider, confidentiality of data traversing the network is important. However, assuring service availability is also important. It is a fact that customers of any large network service will have security issues that are beyond the control or even influence of the network service provider—compromised machines, botnets, DDoS attacks to name a few. The objective of security analytics in this environment is not to police the network. The objective is to provide a mean to detect attack vectors, trends, and indicators that might ultimately impact the service that is being provided to customers. This is often more of a quantitative assessment of activity rather than a matter of detecting discrete events. For example, we do not want a security issue that one customer has to impact the service to any other customer.

9.5.1.1 Components of Security Analysis

Any analysis function to result in some sort of response action. Often, analysis algorithms focus on the concept of "anomaly detection." However, there are a number of other functions that are necessary to result in appropriate responses.

1. Data prefiltering and selection—there are endless sources and types of available data. It is important to select robust and concise data as well as data formats that will provide good indicators of events of interest on the network and infrastructure.

2. Anomaly detection is a means to flag situations that are of interest. This includes both signature-based detection as well as detecting abnormal behavior (e.g., volumetric events).

3. Details collection and characterization—this step may include correlation, but correlation is a subset of details collection. The objective is to find and collect any related data that might be relevant to interpreting and quantifying the event. For example, if there is a volumetric anomaly during a certain time period on a particular TCP port, identify what addresses appear to be sending the data and who are the recipients of the data. How long has this been taking place? Are there past cases? Are there other anomalies or is there threat intelligence associated with the addresses? Are there other events that have been identified as security relevant that have similar characteristics? Do the TCP sessions look legitimate and complete, or do they have indications of failed connections or perhaps no successful connections? Were there successful connections prior to the anomaly?

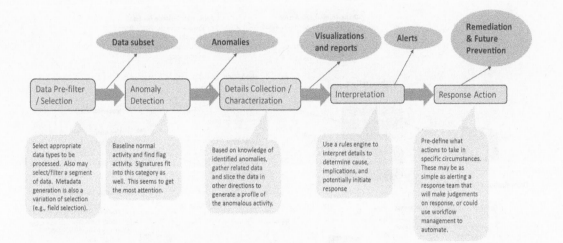

FIGURE 9.13 Generic security analysis algorithm template.

4. Interpretation is perhaps the most difficult step in the processes. Neither a system nor people can accurately interpret a situation with insufficient information. Often manual processes are introduced not because of lack of automation, but because the information needed to make a decision come from sources external to the analysis environment. If analysts need to consult with people to make an interpretation, then this generally cannot be automated until the information those people have can be incorporated into the processing environment (Figure 9.13).

9.5.2 Identity and Access Management

IAM services are a critical component of any virtualization strategy. As cloud infrastructures dynamically expand and contract, each new node has to be provisioned with appropriate machine and user identities and credentials. In fact, the maturity of cloud-based offerings can often be measured by the level of IAM capabilities and without them these offerings will struggle to be viable for serious business application. All major cloud providers have focused on evolving their IAM implementation.

Virtualization adds several additional layers to the IAM ecosystem, which creates both opportunities and potential threats. Hypervisors are able to make physical resources available to the VMs and software-defined networking capabilities dynamically create network segments for specific VMs and their virtual interfaces. With all of these layers of abstraction and dynamic configurability the hypervisor and orchestration tools become a prime target for attack.

Security modules such as the IAM platform provide critical security capabilities to the ONAP solution. Access management enhancements deliver preventive and detective access controls for the ONAP portal and related front ends. Options for fine-grained authorization capability also exist. For identity lifecycle management, IAM provides user provisioning, access request, approval, and review capabilities and have efficient role design to minimize administrative burden.

9.5.2.1 Identity Ecosystem

The identity ecosystem needs to address the security of user, client application, and resource identities (Figure 9.14).

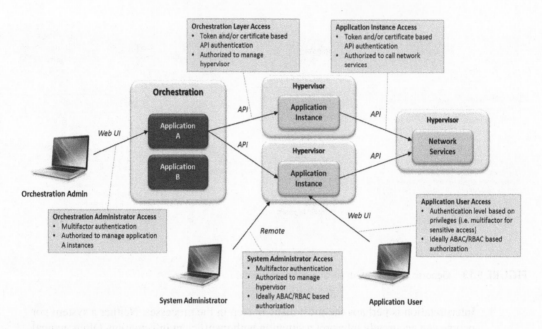

FIGURE 9.14 Sample network cloud identity scenario.

9.5.2.2 End User Identities

End users will interact with various business applications that require access to the control layer. It is critical that users with authorization for privileged access to the control layer are properly vetted from an IAM perspective. This is geographically represented in Figure 9.15 and entails

- Identity proofing prior to digital identity creation
- Strong authentication capabilities

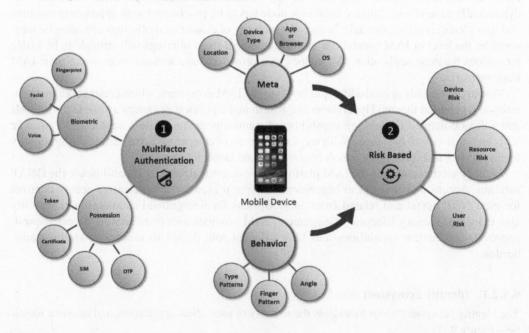

FIGURE 9.15 Authentication approach.

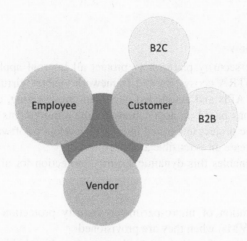

FIGURE 9.16 End user identity classes.

- Multifactor authentication with focus on biometric and possession-based factors
- Risk-based analysis to override static authentication policies as needed
- Attribute-based access control (ABAC) and risk-based access control (RBAC) to simplify management of entitlements and limit access creep.

Figure 9.16 highlights the usual user identity classes:

- *Employees and contractor.* This category is structurally very hierarchical and includes a broad range of entitlements including ABAC/RBAC policies.
- *Customer:*
  - *Consumer (Business to consumer [B2C]) identities* have limited hierarchies but are more complex from peering (friend concept) perspective
  - *Business customer (Business to business [B2B]) identities* range from small and simple for a small business to hierarchical and complex for a large signature customer
- *Vendor.* These identities allow a vendor to access and often support implementations of their solutions.

9.5.2.3 Client Application Identity

Each of the business applications needs to be approved for specific access (least privilege) into the control layer.

- For client credential access patterns, each application instance must have unique credentials, issued at installation and ideally via automated orchestration.
- The preferred access pattern for client applications that needs to access resource APIs, is the use of resource credentials that are issued by an authorization server as in OAuth 2.0 implementations. Ideally, this solution
  - Has authenticated the client application
  - If possible, authenticated the end user
  - Verified basic authorization and issued unique resource credentials for the specific requesting client application instance

9.5.2.4 Resource Identity

This identity is used to target a specific resource. Associated attributes such as IP address as the network identity, can be dynamic due to DNS-based global load balancing and also obfuscated due to data center level traffic management features such as load balancers (client is targeting a virtual IP).

9.5.3   ASTRA

9.5.3.1   ASTRA Overview

ASTRA is an innovative security platform to protect all internal applications within the cloud environment [5]. The ASTRA ecosystem and framework enables virtual security services to be delivered effortlessly via APIs and automated intelligent provisioning, creating micro-perimeters around specific applications based on application-specific requirements. Using an Agile software development approach, the project integrated internally developed software with both open source and vendor solutions to create an extensible architecture.

ASTRA's ecosystem enables this dynamic security protection for all computing resources by providing:

- Automatic orchestration of micro-perimeter security protection "rings" for workload objects (individual VMs) when they are provisioned
- Automatic orchestration of workgroup (collection of workload objects) protection "rings"
- Dynamic security policy provisioning using API communications to the workload, workgroup, and classic perimeter protection systems
- Security event collection and forwarding to authorized subscriber system
- User interface support for security administrators (Figure 9.17)

ASTRA implements these functions using a bot-like self-healing architecture that was designed from the start to use API services to enable a flexible and manageable environment that supports integration with leading marketplace security technologies and to be an open system for third-party solution integrators. ASTRA use of API enable effective communications and controls for security policies and provisioning as well as seamless integration with external business system-like asset inventory, threat management, and record retention services.

ASTRA leverages SDN to enable cloud or physical data center security protections and takes advantage of the new technology capabilities available in OpenStack and other cloud integration environments, providing north–south and east–west protections for the cloud. Security function virtualization (SFV) is the hallmark of ASTRA. SFV is a key enabler for ASTRA to enable dynamic and intelligent security protections required by the workload in a cost-effective and right-sized manner.

As explained earlier, part of today's security challenge stems from the fact that classic perimeter no longer truly exists or has become so porous that it often feels as though it no longer exists. The need to communicate information with partners, the use of mobile bring your own device (BYOD), and more sophisticated advanced persistent threat (APT) attacks are some of the key contributors to

FIGURE 9.17   ASTRA architecture features.

this breakdown. This already complicated problem is being further exasperated by corporate mandates to move their IT services into the cloud—where traditional security protections were never designed adequately to protect critical business systems.

ASTRA attempts to provide the tools and technologies to help solve this difficult challenge. This is explained in more detail in the next section.

9.5.3.2 ASTRA System Overview

The functional ASTRA architecture is depicted in Figure 9.18. The ASTRA architecture can be broken down into multiple layers loosely partitioned into ASTRA Core and ASTRA Satellite.

The ASTRA core environment is an integrated platform purposely architected to provide a rich set of command and control systems to manage enterprise and service provider sized cloud computing environments. The ASTRA core platform provides mission critical capabilities needed to effectively manage cloud-based security including

- Intelligent creation of security polices and controls
- Enabling auditing of security polices
- Acquiring and maintaining cloud object inventory
- Orchestrating the dynamic implementation of security policies
- Acquiring and managing massive volumes of security event logs
- Providing instant notification services to authorized administrators and system owners

ASTRA core presents these functions using a comprehensive user-interface portal. It also seamlessly interoperates with other key management systems such as operations support systems (OSSs),

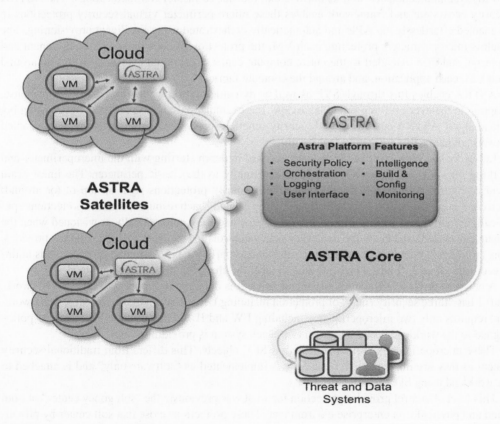

FIGURE 9.18   Functional ASTRA architecture.

business intelligence (BI) systems, and the ONAP, IAM, and security analytics systems discussed earlier in this chapter.

ASTRA satellites provide the ability for ASTRA to manage and control the security protection of distributed cloud environments. Lightweight ASTRA satellites work with ASTRA core to provide a complete set of command and control functions and telemetry services that ensures effective security management including distributed security event collection and managing the distribution of event data to authorized subscribers. ASTRA satellites can be extended to implement any service that operates in the distributed cloud environment.

One of the keys to the effectiveness of the overall ASTRA architecture is that it utilizes APIs to enable an open and effective integration strategy. The use of agile development techniques with a CI model provides dynamic and responsive software delivery capability to support new feature and function introduction. The next section will explore the multifaceted and multilayered security protection capabilities provided by ASTRA.

9.5.3.3 ASTRA Defense in Depth

AT&T's own experience over the years with the growing cyber security challenge of cloud computing and the erosion of the security perimeter, has led with an approach that addresses both the cloud security implementation issues as well as the dissolving perimeter problem. The answer is based on trust in layers of protection, but in today's ASTRA model, the controls will be implemented in new and novel ways where the security protection is dynamically aligned and provisioned to the business application using concentric rings to provide a seamless and intelligent security model for the enterprise.

Micro-perimeter controls are required to secure cloud platform environments and especially critical initiatives like network cloud. These micro-perimeters provide security in east–west (VM to VM) communications as well as north–south (outside to inside) communications. The ASTRA security ecosystem and framework enables these micro-perimeter virtual security protections to be enabled effortlessly via APIs and automatically orchestrated using intelligent provisioning. The result is microperimeters protecting each VM, the protection of collections of system (tenant and projects), and then extended to the entire compute center. In summary, security is enable around each VM, each application, and around the compute fabric.

ASTRA enables this through SVF as well as dynamic real-time security controls in response, adapting to the ever-evolving threat landscape. For example, based on security analytics using big data analysis, ASTRA enables virtual security functions on-demand, leveraging the SDN-enabled network, dynamically mitigating security threats.

Let us look at each layer of the security defense in depth starting with the microperimeter and working outward to tenant or project level and finally to the classic perimeter. The finest grain security control is to establish intelligent micro-security protections right on top of (or around) each computing resource—conceptually as "a ring around" each resource, with the protections specifically required for that resource. These microperimeters are automatically provisioned when the resource is created and carry the necessary policy controls for the work that the resource provides. This "workload" policy management is established as part of the provisioning process and is maintained through an API-based orchestration to meet the changing need of the workload.

A conceptual view of the microperimeter concept is depicted in Figure 9.19. These VM (workload) 1 has "three security rings" of protection including URL, FW, and IDS. In Figure 9.20, workload requires only two microperimeters including FW and IDS. Each of these have specific policy aligned to the workload based on the services each system is providing.

These microperimeters are implemented as SFV objects. This differs from traditional security system as they are implemented, since SFV is implemented as "software only" and is attached to the workload using SDN techniques.

This level of control provides protection for what was previously the "soft gooey center" of both cloud and physical iron enterprise environment. These protections close that soft center by providing east–west protections between VM's that previously did not exist.

FIGURE 9.19 Conceptual view of microperimeter on VM1 (three rings).

To extend the security protection technologies beyond the individual workload objects, the ASTRA security construct also provides additional layers of protection through the provisioning and orchestration of tenant-level or project-level protections. A tenant can be thought of as a logical grouping of workload objects (VMs) into a workgroup cluster. This workgroup or logical tenant grouping is not bound by physical networking and can be geographically distributed.

By provisioning and orchestrating a higher-level policy to the workgroup, an additional protection ring is available to shield the entire tenant group. The workgroup perimeter works in conjunction with the micro-perimeter protection while enabling alternative protection technologies and policies to be provisioned that protect the entire workgroup (Figure 9.21).

The workgroup protection model extends the access controls directly applied to the individual workload object and allows for policy controls that can protect the external access to all logical objects in the workgroup independent of network and system topology location of the objects. This extends the east–west protections as well as provides a strong higher-level north–south protection for the entire workgroup. The workgroup protections enable policy controls that are not as effectively implemented at the workload object level. The next security ring completes the protection model by adding perimeter protection.

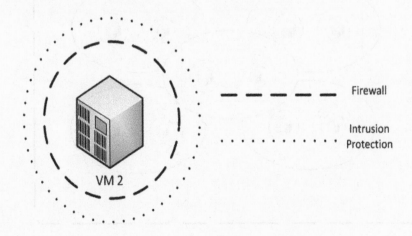

FIGURE 9.20 Conceptual view of microperimeter on VM2 (two rings).

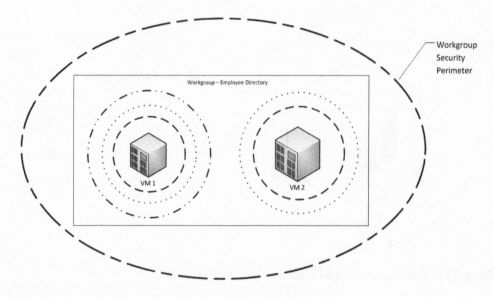

FIGURE 9.21 Conceptual view of workgroup/tenant-level perimeter.

9.5.3.4 Network Perimeter (Classic Network Protection)

As mentioned previously, the classic perimeter protection strategies are by themselves not capable of completely protecting today's enterprise and especially cloud-computing environments. However, used in conjunction with a multilayered protection strategy, the network perimeter is still an important capability to help protect today's enterprise data center environments. This gatekeeper function provided by traditional perimeter security technologies can still establish highly scalable and effective first level of defense capabilities when used in conjunction with intelligent and dynamic policy provisioning and orchestration. This is depicted for a single site in Figure 9.22, and collectively across multiple data center sites as shown in Figure 9.23, which extends the perimeter protection to the entire enterprise.

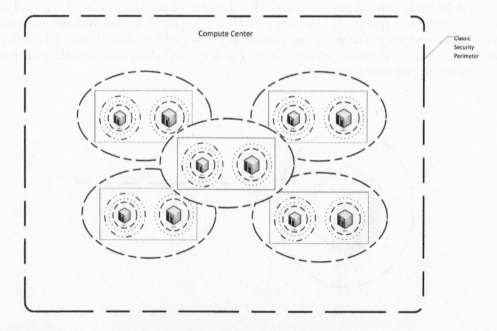

FIGURE 9.22 Conceptual view of classis security perimeter—single site.

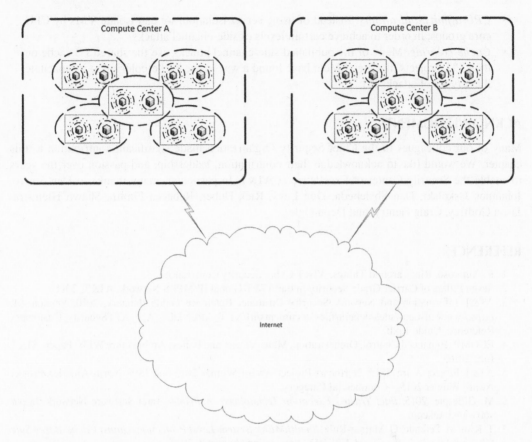

FIGURE 9.23   Conceptual view of classis security perimeter—multiple sites.

The combination of these security protection layers provides a comprehensive protection strategy ranging from the smallest work object, the work group, and to the enterprise data center. This multilayered protection model is dependent on intelligent and dynamic provisioning and orchestration to enable the cohesive and elastic security protections.

9.6   TOPICS FOR FURTHER STUDY

This section highlights the potential for future research and development to address hypervisor vulnerabilities. Hypervisor hardening and VM nonisolation are research and development domains required to address hypervisor vulnerabilities. Realizing that stranded resources could become an issue when isolating VMs to create separation of internal and externally exposed resources, we are preparing to shift to a hardened hypervisor to remove requirements for isolation. Key components of this hardening will be the use of trusted execution, expanded patching lifecycle, privilege escalation detection, CPU affinity, cache-coloring, among others.

- *Trusted execution.* VM validation/along with platform validation at time of instantiation will ensure the correct VMs are spun up, and spun up on the appropriate hypervisor [6].
- *Expanded patching lifecycle.* Utilizing smart patching techniques for only updating patches for packages installed, utilizing image patching with VM replacement, and other techniques we expect to harden our VMs just as much as the hypervisor.
- *Privilege escalation detection.* Auditing and logging of executed commands in user space will allow us to detect when programs have been run without proper authorization.

- *CPU affinity/anti-affinity.* Instead of using server isolation, we can isolate VMs by CPU core groups, in order to achieve certain levels of side-channel attacks.
- *Cache-coloring.* Many of the published side-channel attacks use the shared L3 cache on the CPU, projects like SteathMem have found a way to use cache-coloring to incapacitate these types of attacks [7].

ACKNOWLEDGMENTS

Many of our colleagues in the Chief Security Organization made significant contribution to this chapter. We would like to acknowledge their contribution, leadership, and passion over the years to enable the Security Center of Excellence at AT&T. In particular, many thanks to Dan Solero, Johannes Jaskolski, Dan Sheleheda, Don Levy, Rick Huber, Rebecca Finnin, Shawn Hiemstra, Jason Godfrey, Craig Gentry, and Deon Ogle.

REFERENCES

1. E. Amoroso. Rings around Things, AT&T Cyber Security Conference.
2. Seven Pillars of Carrier Grade Security in the AT&T Global IP/MPLS Network, AT&T, 2011.
3. AT&T Information and Network Security Customer Reference Guide, January, 2010, Version 4.1. https://www.nist.gov/sites/default/files/documents/itl/AT-T_APENDIX_A_-_ATTSecurity_Customer_Reference_Guide-1.pdf.
4. ECOMP (Enhanced Control, Orchestration, Management, and Policy) Architecture White Paper, AT&T Inc., 2016.
5. AT&T Project Astra, ISE® Northeast Project Award Winner 2015 and ISE® North America Project Award Winner 2015—Commercial Category.
6. M. Gillespie. 2015. *Intel Trusted Execution Technology—A Primer,* Intel Software Network: http://software.intel.com.
7. T. Kim, M. Peinado, G. Mainar-Ruiz. *StealthMEM: System-Level Protection against Cache-Based Side Channel Attacks in the Cloud*, USENIX Association, August 8, 2012.

10 Enterprise Networks

Michael Satterlee and John Gibbons

CONTENTS

10.1 Evolution of Network Complexity 202
10.2 Technology Innovations..203
    10.2.1 Virtualization of Network Functions...204
    10.2.2 Increasing Packet Processing Capabilities 207
    10.2.3 Optimizing the Virtual Environment 208
        10.2.3.1 Linux Bridge Approach 209
        10.2.3.2 Virtual Switch (OVS)-Based Approach... 210
        10.2.3.3 SR-IOV-Based Approach 211
    10.2.4 Optimizing VNF Performance... 213
        10.2.4.1 Processor Affinity 213
        10.2.4.2 Intel DPDK and Related Technologies 213
10.3 Memory and Storage Resources.. 214
10.4 Network Management and Orchestration.. 214
    10.4.1 Call Home Function... 215
    10.4.2 Data Modeling for Network Design 215
    10.4.3 VF Deployment and Management.. 216
10.5 DPI and Visibility 217
10.6 Network Functions on Demand 217
10.7 Universal CPE... 218
    10.7.1 uCPE Phone Home Process 221
10.8 Conclusion 221
References... 222

Today, virtualization is transforming the way traditional network functions are consumed within an enterprise. Router, firewall, and other network appliance companies are adopting software business models that allow them to leverage compute utilities for functions that once required a specialized hardware platform. Current network deployments within enterprise domains are static and nonprogrammable in nature. Many networks are provisioned by direct command line interface (CLI) access or via simple script-based tools. This prevents network engineers from meeting basic business requests to deliver dynamic network services, like bandwidth-on-demand, rate limiting, dynamic quality of service, security policy changes, etc. An SDN architecture allows for a fully programmable physical and virtual network where organizations can deploy high-value network services with speed and flexibility.

Enterprise networks are flooded with devices from many vendors with each device using proprietary hardware and providing unique functions such as routing, switching, firewall security, intrusion prevention system/intrusion detection system (IPS/IDS), caching, wide area networking (WAN) acceleration, distributed denial of service (DDoS) monitoring, traffic analysis, etc. In addition, most of these individual products come with separate management products and sometimes require additional products from a different vendor. For example, when managing multiple firewall appliances, a separate firewall management solution is used, for managing routers or switches. In essence, the widespread use of purpose-built hardware appliances along with lack of a complete solution approach has also contributed significantly to a complex, expensive, and inflexible networking environment for the enterprise. New technologies such as software defined networking (SDN) and network functions virtualization (NFV) provide an opportunity to simplify this networking environment.

The programmability aspect allows network administrators to dynamically instantiate services in the cloud and at the customer premise by creating concepts such as a Network Function App Store. These network functions can be offered as value-added service for customers. Programmability enables even more benefits for managed LAN and WAN services, whereby service providers can provide a unified multivendor solution for devices embedded in a customer's LAN environment.

This chapter will discuss how one can leverage these technologies for creating next-generation customer premise equipment (CPE) products and managed services for enterprise customers.

10.1 EVOLUTION OF NETWORK COMPLEXITY

The very basic role of a "network" of any size is to enable users to access services and resources in a reliable, timely, and secure manner.

However, over time, the functional requirement associated with a "network" has grown significantly. Many new functions are now considered core requirements along with enablement of user access. Examples of some of these functions include the following:

- Functions to mitigate security threats—firewalls, intrusion prevention system (IPS)/intrusion detection system (IDS), malware/virus protection, web proxies, etc.
- Functions to authorize and authenticate users
- Functions to enable transport-level security
- Functions to optimize traffic such as WAN acceleration solutions, content caching solutions, etc.
- Functions to load balance the traffic across different locations, different servers, and different networks
- Functions to optimize the cost for using the networks
- Functions to manage remote access and bring your own device (BYOD) and their security
- Functions to provide high availability of networks and services

In the past decade, the complexity of the network has grown significantly as each new function was developed and deployed using a hardware centric model. Each new functional requirement served as an innovation opportunity that resulted in the invention of a new box (sometimes referred to as middle box). These middle-box functions implemented a specific function using a "dedicated appliance" with proprietary hardware and/or software. Today's networking platforms are a Gordian knot of middle boxes and their support systems. Equipment vendors have delivered platforms built from proprietary hardware coupled tightly to proprietary software. Quite often, the cost of a device is proportional to the length of the datasheet, which describes the services it supports.

At the same time, operators lack the ability to selectively decompose and recompose individual service elements from these platforms. While this tight coupling of systems and services has a basis in sound engineering, it also originates from vendor business models looking to optimize their opportunities. On the one hand, the tight coupling allows for optimal design of scale, serviceability, and reliability. On the other hand, the closed nature of the platforms promotes both economic and architectural lock in.

Traditionally, network administrators relied heavily on management of hardware-centric network appliances located at the customer premise. This model required shipping, racking and stacking, test and turn up, configuration, monitoring and ongoing maintenance for the dedicated routing/switching, or middle-box platform. Many of these platforms use a CLI while others provide a graphical user interface (GUI) specific to that vendor. Also, some require use of a separate third-party management suite.

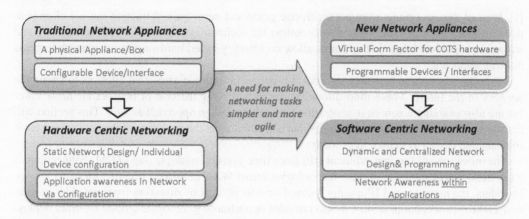

FIGURE 10.1 Shift to software for enterprise networking.

Enterprise IT managers are used to seeking the help of scripting languages and of system integration companies to deploy and manage these boxes. Many also tried adding new management platforms that provided a "single pane of glass" management interface. However, because new functions were being created in isolation, the customer was left with multiple "single pane of glass" management interfaces.

As enterprise IT continuously deployed additional middle boxes, the rules to manage traffic between the network elements became so complex that it often required highly skilled networking professional to plan and test a small change in network design. In addition to the middle-box and management issues, growing network demand due to the change in business environment has also been contributing to an increase in network complexity.

Whether a business is expanding or consolidating, it requires the capabilities to run demanding applications like streaming data, email, and video content. IT managers are looking for a network service that is fast, versatile, and can carry mission-critical traffic without any downtime. The adoption of cloud services and the need for user-mobility requires significant change in traditional network design.

From service-provider perspective, one needed a broader approach—a software-centric model for all networking needs. New virtual network appliances that met the requirements associated with software-based cloud services could also be used to address use-cases associated with CPE deployment. One can create a software-centric model for CPE devices that will provide managed service providers and their customers with an opportunity to evolve the enterprise networks to be more flexible, responsive, and future proof. This is depicted in Figure 10.1.

In this new software-centric model approach, the network becomes a collection of virtual network services. The services can be instantiated on-demand at the customer premise, within network nodes, or within the customer datacenter as needed. This also allows the possibility of creating a dynamic policy control environment where different policies or services can be inserted, based on traffic characteristics of individual flows.

In short, many technologies that were once limited to the datacenter space such as software-defined networks, server virtualization, overlay networking technologies, etc. can now be used to promote simplification of the end-to-end network environment across WAN boundaries. The next section describes some of these technologies as it pertains to enterprise applications.

10.2 TECHNOLOGY INNOVATIONS

Intel x8 6-based server technology has been around for ages and the ability to run network functions such as routers and firewalls has been there for some time, especially in the Linux environment

204   Building the Network of the Future

[1]. Most of these solutions were not enterprise grade and never gained traction outside of embedded consumer devices. This section will outline the technology advances in x86 technology and software programmability that will now allow commodity-based hardware to run enterprise grade network applications.

In recent years, there has been significant progress in multiple areas that allow us to create the network of the future. These innovations are not only driving the cost of network elements lower but are also enabling a new next-generation networking design approach as well. This section will discuss some of the enabling technologies that directly impact the evolution of the networking environment and design as known to us today.

The move toward Internet protocol (IP) from time division multiplexing (TDM) connections is a key trend that enables use of these technologies across WAN networks. Initially, for many service providers, the move toward IP mainly focused on cost savings by divesting from the older generation TDM network, but in reality, it also provides opportunity to innovate around Ethernet deployments. As an example, with Ethernet only access, the CPE edge device at customer locations will no longer require TDM interfaces.

With Ethernet only interfaces, CPE devices now look very similar to x86-based server platforms used in the data center but with more ports. While CPE devices may look like x86 on the outside it still is much more than a basic server platform on the inside. Technologies such as virtualization along with innovations in software provide the opportunity to transform the simplified hardware platform to develop a next-generation CPE that is capable of running any network function.

10.2.1 VIRTUALIZATION OF NETWORK FUNCTIONS

The initial efforts associated with virtual CPE projects were based on simple concepts of virtualization that have been used by cloud computing since 2010. The basic idea is if one can use an existing x86 server platform and instantiate "a network function" on it (instead of deploying physical appliances shipped by vendors), it will provide many benefits from cost, planning, logistics, deployment, and operational point of view.

As one delves further, it is easy to discover that theoretically, it is also possible to instantiate multiple virtual network functions on the same platform and service chain them together so that the traffic can be directed to individual functions based on rules. In the past, this capability could be achieved but acceptable performance required high performance compute platforms that were expensive. The key to enabling this capability is the performance optimizations that have been recently developed and are still emerging to allow multiple functions to run on the same platform, while keeping the platform cost at or below existing costs. The following illustration highlights the difference between the two concepts described above. The diagram on the left represents a limited use of virtualization with single function running on a server appliance. This model has been used by many in the industry for some time now, especially for applications that handle control traffic only, such as call-control application. However, the model on the right (whereby multiple functions are consolidated on the same box) provides the optimum use of hardware resource and true benefits of virtualization technologies to end customer (Figure 10.2).

It is clear that once this new model is developed, it would provide significant cost savings to customers. It not only eliminates the middle-box problem discussed in the previous section, but also provides a software-centric approach to solving networking problems in a premise environment. Customers will only be required to deploy limited hardware resources at the customer premise at the beginning. Subsequently, all the network use-cases can be implemented by instantiating and inserting the appropriate network function software as traffic passes through the CPE.

In addition, the concept can be expanded further to evaluate another model whereby service chains are extended between network functions, across the WAN network. In this model, one can instantiate multiple network functions on the CPE platform and/or a cloud platform. Irrespective of where the functions are instantiated, it could be deployed and managed with the same set of tools.

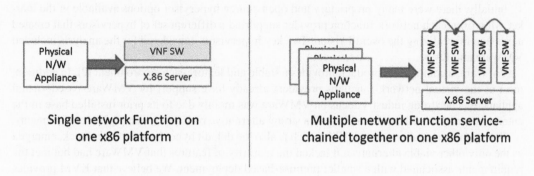

FIGURE 10.2 Shift to multiple VNFs.

The customer would choose where best to deploy a function, either locally on premise or in cloud for multiple branch locations (Figure 10.3).

With this vision, we embarked on this evaluation by putting together a proof of concept of a very simple topology, a single network function (virtual router) running on a x86 server platform. The very first question one had to answer was which virtualization environment to use for the project, that is, which hypervisor. The hypervisor is a program that enables hosting of several different virtual machines on a single physical server (see Chapter 3). In this form of virtualization, the underlying physical hardware is still working as an independent appliance while the hypervisor is providing the role of hardware virtualization and managing the hardware's resources across the virtual functions (VFs) (Figure 10.4).

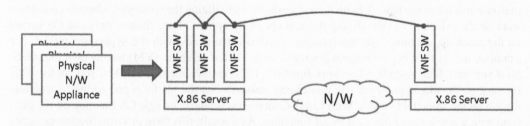

FIGURE 10.3 Service chains across WAN boundaries.

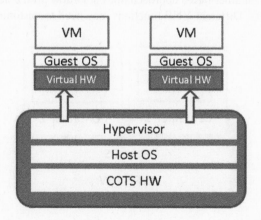

FIGURE 10.4 Hardware virtualization.

Initially, there were many proprietary and open-source hypervisor options available in the market. Moreover, each network function provider supported a different set of hypervisors that created a problem for realizing the overall vision. Two key hypervisors considered for the analysis included VMWare and KVM.

VMWare, a commercial product, provides a stable and feature-rich environment [2]. In addition, most of the virtual network functions providers already have support for VMWare-based virtual appliances. The strong industry focus on VMWare was mainly due to its prior installed base in the enterprise cloud environment and lack of a strong alternative platform for enterprise environments.

One open-source alternative, KVM, which is also the default hypervisor for OpenStack, emerged as the only other viable alternative. It lacked the maturity of features that VMWare had but met the requirements associated with a smaller premise-based deployment. We believe that KVM provides several technical and nontechnical benefits that makes it a very good hypervisor choice for office branch (remote office) environment. The key reasons are listed below:

- For service providers, the branch platform is generally considered an extension of its network-based cloud platform. Since most service providers are relying on Open Source technology such as OpenStack/KVM for their cloud initiatives, the ability to leverage the tooling for branch environments provides opportunities for significant cost benefits in future.
- The existing limitations of the KVM hypervisor were not applicable to a single server deployment in a branch environment. With the increased focus and adoption of Open Source initiatives, most vendors today already support both VMWare and KVM hypervisors for the network functions they offer.

The discussion on virtualization, however, cannot be concluded without exploring another form of virtualization that is getting a lot of traction, especially in a cloud environment called Linux Containers [3]. While Linux Container as a technology is not new, its application as a form of virtualization is relatively new. A hypervisor focuses on virtualizing the hardware, whereas containers work on the principal of virtualizing the host operating system. Since there is only one OS/kernel on the machine, containers do not have the overhead that exists with the hypervisor model. For example, in a typical hypervisor environment, each virtual machine (VM) has its own guest OS. So if you have four virtualized network functions (VNFs) implemented as VMs, you get five OSs running on the platform—each one consuming resources on the platform and trying to schedule the job on the same CPU cores. With Linux Containers, we have a single OS function on the platform with a single scheduler used by all functions. As a result, this form of virtualization provides significant boost to improve the packet-performance capability necessary for applications such as network functions (Figure 10.5).

However, today the Container-based approach does not allow all the needed flexibility required by different VNF vendors. Different VNF functions may need to customize the kernel, or depend

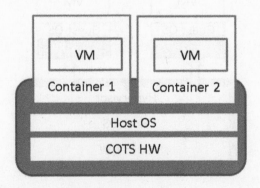

FIGURE 10.5 Linux containers.

on certain kernel versions, and this does not work well since all functions will need to share the same kernel. In particular, for an environment where the same machine is required to host software applications sourced from different vendors let alone different functions, these issues become much more relevant.

Multiple VNFs sharing a single kernel can also create security and stability issues that are yet to be solved. As an example, since every application uses the same kernel, it is possible that one guest operation can affect the functionality of other guest applications on the same machine. Hypervisor-based approaches do not pose this issue because guest VMs do not share any kernel code (they each have their own guest OS). More work is needed to find an optimal host OS configuration where multiple network functions can efficiently use the container-based environment.

As container technology becomes more mature and VNF vendors evolve to provide container-ized network functions, service providers will be in a good position to leverage the new container-based approach to further simplify and enable faster delivery of software services. With the existing state of container technology, nonpacket forwarding VNFs that do not have complex kernel require-ments, are suitable for containerization.

10.2.2 Increasing Packet Processing Capabilities

Once we figure out the mechanisms to virtualize a set of network functions via a hypervisor, the next step is to determine the overall performance of the VNF. Since the customer requirements (and expectations) for network elements do not change based on implementation (i.e., physical or virtual), the overall performance goal is to be as good as, or better than, the physical environment, especially to dispel the prevailing perception that x86 platforms cannot perform at the same level as purpose-built hardware.

The initial performance work focused on the most basic network function, the vRouter (virtual router). The vRouter is considered to be a necessary VNF building block for other higher-level VNFs such as vFirewall, vWANx, etc. The primary focus was to assess speed of processing packets while maintaining acceptable levels of packet loss, latency, and jitter.

At the initial stage of the investigation, baseline performance data were collected, with no x86 packet optimization techniques enabled on the platform. While the baseline results were consistent across multiple vendors, they were severely limited when compared with the modeled expectations and certainly were not at par with existing physical routers in relation to throughput, packet loss, and jitter.

This exercise provided valuable insight into the obstacles to be overcome before implementing packet processing VNFs on a generic server platform. Both the hypervisor within the server, and the virtual router code, suffered from bottlenecks that contributed to limited performance and was sus-ceptible to unpredictable packet loss and jitter. For example, running a canned traffic performance benchmark test on a physical router in a static state typically shows very similar results each time (e.g., with a 100 MB/s throughput and up to 0.01% packet loss, the results are usually consistent). But in the virtual implementation, with the initial server testing, the results could vary widely from test to test. While some of these impairments were significantly reduced when additional CPU resources were allocated to the VNF, the additional cost of CPUs makes it cost prohibitive to vir-tualize the function.

Within the server's hypervisor, a packet needed to go through additional processing that con-tributed to cost in terms of overall performance. The packet processing throughput of the hyper-visor itself was far less than what was typically observed with physical appliances, let alone when adding a VF on top of the hypervisor—the hypervisor itself became the first bottleneck to overcome.

By temporarily eliminating the bottleneck of the hypervisor by using more CPU resources, we then found that the throughput of the vRouter was also significantly lower. The data showed us that while a physical box provided well-known vendors support up to 1G of traffic, the virtual environment using the same amount of processing power (i.e., CPU) was severely limited. The packet-processing capability in the virtual environment ranged from 200 to 350 Mbps depending across various vendor vRouters.

In addition, it was unclear how and at what cost (in terms of performance), one can use SDN techniques (such as granular policy-based controls) to switch packets to different network functions within the same machine. In order to address these issues and gain a deeper understanding about these bottlenecks, we started by creating a proof-of-concept environment. The idea was to identify an ideal virtual environment for the CPE that would provide maximum flexibility and packet-processing capability on common-off-the-shelf (COTS) platform.

For improvement in processing capabilities of the VNF itself, we worked with engineering teams of several vendors independently to optimize their respective code-bases to achieve higher packet processing capability on COTS platform. We pushed for the use of latest (and evolving) technologies such as Data-Plane Development Kit (DPDK) and single root I/O virtualization (SR-IOV) (this is discussed in Chapter 3) in the proof-of-concept environment. Subsequent results showed that the use of these techniques, in addition to general software performance tuning that focuses on I/O optimization and CPU optimization, provided significant performance gains.

The next section provides more details on the various options that were considered for creating a virtual implementation on the CPE. We will also discuss key reasons for getting higher or lower performance in each option, along with the tradeoffs one has to be aware of before selecting a specific model for a virtual environment.

10.2.3 OPTIMIZING THE VIRTUAL ENVIRONMENT

When we look at I/O performance in the virtual world we have to look at two areas, physical network interface cards (pNICs) and virtual NICs (vNICs). In a physical machine, Ethernet network traffic is handled by the pNICs installed on physical machines. Similarly, a virtual machine handles the traffic as it arrives at its virtual Ethernet interface or vNIC. The virtual environment between the pNIC and vNIC is responsible for transferring the data between the physical network and the VM/VNF. To understand the bottlenecks within the virtual environment, we needed to look into the implementation details associated with the virtual environment. We need to understand the life of a packet as it traverses from the pNIC card on the machine to the VNF and vice versa (see Figure 10.6).

In fact, there are multiple ways to create this virtual environment. Depending on how the environment is created, the cost (for packet processing performance) and feature availability (for granular

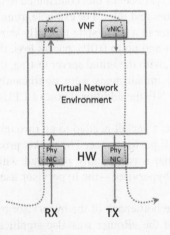

FIGURE 10.6 Virtual environment within CPE.

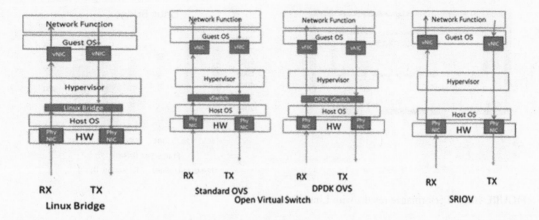

FIGURE 10.7 Virtual environment options.

traffic control), changes. The illustration below highlights the four different virtual environment options that were considered in the proof-of-concept evaluation:

- Linux bridge-based approach
- Use of standard virtual switch (open virtual switch [OVS])
- Use of DPDK-enabled virtual switch (DPDK OVS)
- SR-IOV (Figure 10.7)

Each option uses different mechanisms to transfer the packet to the VNF as it arrives on the pNIC. The same process is repeated in the reverse direction as the packet is transferred from the VNF to the NIC card. Let us look into each of these options in more detail.

10.2.3.1 Linux Bridge Approach

The "default" way to create a virtual networking environment in Linux is to use a bridge device. In reality, the Linux bridge [4] is a very simplified form of a virtual switch implemented inside the Linux kernel.

First, we create the Linux bridge on the platform and connect the physical interface to the bridge. Subsequently, each individual virtual network function is configured to connect to the same bridge, or multiple bridges can be created to allow for separation of traffic or service chaining between VNFs within the server.

As the packet arrives on the pNIC, it is copied on to the Linux bridge by the kernel. The packet is then switched, based on destination Media access control (MAC) address, to the appropriate bridge port—in our case the vNIC of the VNF. The VNF then receives the packet on its vNIC and copies the packet into its memory for further processing. The same steps are repeated for each packet as it arrives at the pNIC and once again in the reverse direction when packets go from the VNF to the pNIC card after processing.

Figure 10.8 provides the topology and performance results associated with this approach.

Early measurements indicated that the Linux-bridging environment was fairly limited in terms of what it could do. It could switch packets based on layer 2 information such as MAC address or VLAN tags, but that is about it. In essence, using the default Linux bridge approach does not provide the necessary capabilities to implement granular traffic policy controls as it traverses through the virtual environment.

This lack of granular matching/forwarding criteria also contributes to another problem with Linux bridges: service chain complexity. When creating a service chain of multiple VNFs, multiple Linux bridges need to be used in order to keep the traffic segregated.

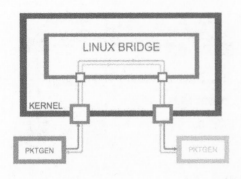

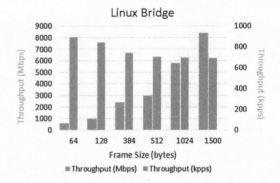

FIGURE 10.8   Performance results with Linux bridge.

The performance of the Linux bridge was good, but not as good as DPDK-based OVS. The Linux bridge is a kernel mode feature, so any packet that needs to go from one VM to another, or to/from VM to pNIC, needs to be copied between user space and kernel space. This causes a high amount of context switching in the kernel, which degrades performance.

10.2.3.2   Virtual Switch (OVS)-Based Approach

Another option for virtual networking is the virtual switch approach. This provides a better feature-set with support for features such as VXLAN, OpenFlow, GRE, QoS, NetFlow, and granular matching/forwarding criteria. Traffic flows can be statically or dynamically diverted, by match-action criteria, to different VNFs based on higher-level policy or input from an SDN controller. The matching criteria can look at layers 2–4, and also has the possibility to go up to layer 7 when integrating with a deep packet inspection (DPI) engine.

However, test results show that virtual switch implementations (such as standard open virtual switch) are inferior, from a performance perspective, to simple bridging environment. The reason for this is increased packet processing (deeper analysis done on each packet) that not only impairs switching performance but also incurs more.

Figures 10.9 and 10.10 provide the topology and performance results associated with OVS approach without and with DPDK.

To increase the overall packet performance, a DPDK-enabled virtual switch was introduced instead of standard OVS [5]. In short, the DPDK-enabled virtual switch is a newer version of OVS that was enhanced by Intel to use DPDK libraries designed to accelerate packet processing on x.86 hardware. Additional details on DPDK are discussed in a later section of this chapter.

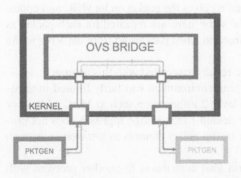

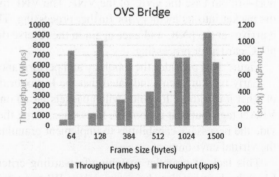

FIGURE 10.9   Performance results with OVS.

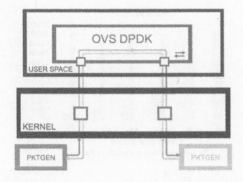
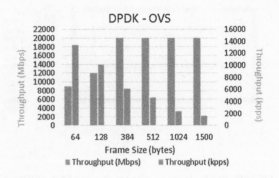

FIGURE 10.10 Performance results with OVS DPDK.

The performance gains associated for packet matching and forwarding the packet through the virtual switch, are significant. As long as the VNF implements a virtual driver to make use of DPDK-enabled virtual switching, it could copy the packet directly from the pNIC to its own memory space (and vice versa, in the reverse direction). While the DPDK-enabled virtual switch approach provides the best of both worlds, that is, good performance and an excellent feature-set, it comes at a cost. The processing associated with its feature-set and deeper analysis of packets requires dedicated CPU resources.

In order to get extremely high performance, DPDK uses a poll-mode driver instead of an interrupt mode driver. A poll-mode driver constantly checks the receive (RX) queues of the interfaces for packets arriving, whereas an interrupt mode driver will notify the CPU via an interrupt that a packet has arrived and needs processing. The interrupt-mode driver takes longer to complete and also causes additional CPU process of the interrupt. Because the poll mode driver is constantly checking the RX queues, the entire CPU can be consumed even if there is no traffic to process. This is not an issue for networking VNFs that constantly have a high rate of traffic to process, such as core or aggregation routers, for VNFs that have a light load of traffic this leads to CPU inefficiency—the wasted CPU cycles could have been used by other VNFs on the same server to process traffic. In order to address this, DPDK.org as well as vendors have started to introduce optimizations and tuning to create an "adaptive" poll-mode driver [6]. The adaptive poll mode driver can adjust the frequency of polling based on the observed traffic load, so that when lighter loads are seen, polling is less frequent.

10.2.3.3 SR-IOV-Based Approach

The Linux bridge and standard OVS approach focuses on switching the packets within the host to different VNFs based on destination MAC address or VLAN tag.

With both of these approaches, every packet transmitted or received has to be processed by the hypervisor as it gets copied from the NIC to the VNF memory space, and also processed by the Linux bridge (or virtual switch) in order to make the switching decision. Processing of each packet typically requires the CPU to service three interrupts—one to handle the packet as it arrives to the host pNIC and send it to the Linux bridge or OVS, a second to make a switching decision, and a third to take the packet from the vNIC to into VNF memory space for processing. As the packet arrival rate increases, so does the CPU workload and context switching overhead (i.e., overhead associated with storing and restoring the state or context of a process so that execution can be resumed from the same point at a later time) of CPU cores associated with serving the interrupts. This creates an I/O bottleneck and results in overall performance degradation of the system.

A new approach referred to as SR-IOV [7], solves the above issue (this is discussed in Chapter 3). In this approach, the VNF is granted direct access to hardware resources at the network interface level by the hypervisor. SR-IOV allows one to create multiple virtual hardware instances, called

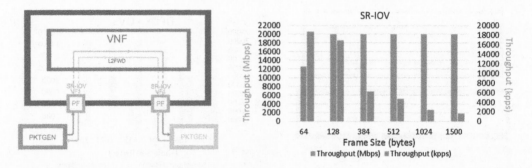

FIGURE 10.11 Performance results with OVS DPDK.

VFs of the pNIC (or in SR-IOV terminology, the physical function or PF) such that multiple VNFs can each be assigned a VF and share the same PF (or pNIC.)

As the packet arrives at the network interface, it is directly placed into the hardware (HW) queue associated with the specific VNF, based on destination MAC or VLAN tag. Subsequently, as the CPU tied to the VNF is ready to process the next packet, it can be directly accessed using direct memory access (DMA). SR-IOV essentially provides an optimized environment where packets are not copied multiple times before it can be processed by the VNF.

Figure 10.11 provides the topology and performance results associated with SR-IOV-based approach.

While SR-IOV provides better performance in terms of processing traffic at a higher rate, it may not be a suitable implementation for all cases. Implementation of SR-IOV is essentially a tradeoff between higher performance and feature-functionality. For example, since in SR-IOV, the VNF is directly tied to hardware, dynamic VNF mobility (i.e., moving a VNF to another server) is not possible. In addition, one also loses the capability to distribute traffic among different VNFs using granular policy based on L3 -L7 information within the server. In case the SR-IOV mechanism is to be used, such distribution must be performed prior to being sent to the server's pNIC.

The following chart shows the throughput comparison of the four approaches discussed for traffic at 384 bytes. As apparent from the chart, one can significantly gain performance with the use of newer performance optimization techniques such as SR-IOV and DPDK enabled OVS as compared to traditional approach of using standard OVS or Linux bridge for networking environment within the CPE (Figure 10.12).

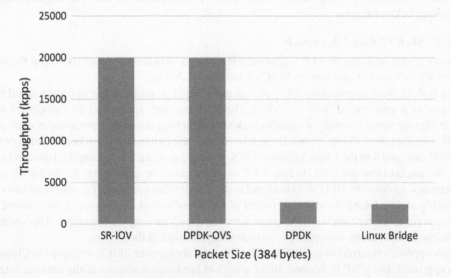

FIGURE 10.12 Comparing throughput results of optimization approach.

10.2.4 OPTIMIZING VNF PERFORMANCE

Another way to achieve higher overall performance is to focus on the efficiency at which each packet is processed by the VNF itself.

10.2.4.1 Processor Affinity

One simple and popular technique that often gets highlighted is processor affinity to a VNF.

Processor affinity (also known as CPU pinning) binds a specific process to a specific CPU or to a range of CPUs, so that the process will execute only on the designated CPU or CPUs. In addition, when a process is pinned to a certain CPU, CPU cache always remains associated with that process thus avoiding cache misses.

Although the basic approach of CPU affinity does not align well with the principles of virtualization, it sometimes can be used to achieve better performance depending on the environment.

As part of our proof-of-concept testing, this approach was put to test by assigning two vCPUs to the VNF and then retesting with multiple VNFs on the same machine. The objective was to determine whether CPU pinning improves the overall performance by reducing the context switching or faster access to data within CPU cache after context switching.

The test results suggested that pinning did not have better performance in our environment. One of the potential reasons for this could be because process affinity varies significantly depending on system environment and scheduler implementation. As an example, results could be different if two virtual CPUs assigned to a VNF are implemented on the same core (via hyper-threading) versus two cores on the same physical processor versus separate physical processor.

The key learning from this exercise was to not assume any performance gain without testing the specific VNF in a replica of the production system environment.

10.2.4.2 Intel DPDK and Related Technologies

Another method to improve the performance of VNFs is to leverage the Intel instruction set features that are specifically designed to accelerate packet processing on x86 platforms.

From a hardware point of view, today's network appliances (such as switches, routers, firewalls, etc.) are built using application-specific interacted circuit (ASIC), field programmable gate array (FPGA), and "network processors." These are "fixed-function" dedicated units that are used for offloading tasks such as crypto, compression, and packet processing.

In recent years, however, Intel has demonstrated that new multicore architecture processors along with specially designed set of software libraries, can perform as well as ASICs, FPGA, or network processors for access network use-cases.

One of the key technologies is DPDK. It is a set of libraries and drivers that essentially facilitates this task by providing application programming interfaces (APIs) for efficient memory management and packet handling as well as optimized poll-mode drivers. It offers

- Environment abstraction layer—provides access to low-level resources such as hardware, memory space, and logical cores using a generic interface that obscures the details of those resources from applications and libraries.
- Memory manager—allocates pools of objects in huge-page memory space; an alignment helper ensures that objects are padded, to spread them equally across dynamic random access memory (DRAM) channels.
- Buffer manager—significantly reduces the time the OS spends allocating and deallocating buffers; fixed-size buffers are preallocated and stored in memory pools.
- Queue manager—implements safe lockless queues (instead of using spinlocks), allowing different software components to process packets while avoiding unnecessary wait times.
- Flow classification—incorporates Intel Streaming SIMD Extensions (Intel® SSE) to produce a hash based on tuple information, improving throughput by enabling packets to be placed quickly into processing flows.

In addition to DPDK, Intel has also launched other technologies such as QuickAssist and Hyperscan that can be used to achieve higher x.86 performance for implementing functions such as firewall, IPS/IDS, malware/virus protection, etc. [8].

HyperScan is a software pattern matching library that can match large groups of regular expressions against blocks or streams of data and therefore is very useful for functions like firewall, IDS, and DPI.

QuickAssist on the other hand focuses on improving the processing capability for applications that heavily use encryption and compression features. This technology provides

- Symmetric cryptography functions including cipher and authentication operations
- Public key functions including RSA, Diffie–Hellman, and elliptic curve cryptography
- Compression and decompression functions

While DPDK and Hyperscan were initially introduced by Intel, these efforts now are maintained as Open Source projects and any VF provider can leverage these software libraries to get better performance for their code on an x.86 platform (as well as on other platforms such as ARM or PowerPC).

The idea of using general purpose hardware instead of proprietary hardware, creates significant opportunities as well as challenges for equipment vendors. While the vendors have the opportunity to significantly reduce the development cost and can do a quick transition to a virtualized product by mainly focusing on the software, they will be restricted to creating product differentiation (*vis-à-vis* other vendors) only via software. These solutions are also lowering the barriers to entry for new vendors in the market; new entrants can launch cost-effective products much more quickly since they do not have to worry about the development of hardware.

10.3   MEMORY AND STORAGE RESOURCES

In the sections above, we discussed evolving technology options for optimized use of I/O and CPU resources. In addition, we need to also look at memory and storage. In our use-cases, while memory and storage did not have huge impact on the throughput, they do prove to be a big cost component for the CPE. Since, many products or solutions came from a datacenter world where memory is plentiful (hard drives), many vendor designs do not pay much attention to optimized memory use. For the CPE, because of cost constraints, it is essential to implement a solution within the limitations of available memory and storage. This involves optimizations such as reducing the size of the images, the amount of storage allocated for writing log files, memory map files, and reducing or right-sizing memory reservations for the CPE platform. In particular, with a platform approach, where one wants to run multiple VNFs from different vendors on the same platform, the efficient use of memory and storage is critical [9].

10.4   NETWORK MANAGEMENT AND ORCHESTRATION

Network management and orchestration are most critical for the simplification of the network environment, both at the stage of deployment and over the life cycle. This is one area where we can see significant changes in the industry. The strong industry-wide focus on SDN brought concepts such as network programmability to center stage. This essentially reignited the discussion on how one should design a product for orchestration and management of network elements. In conjunction with an SDN approach, wide-spread adoption of network function virtualization and the cloud services model are the main driving forces that are shaping the management aspects of next-generation products and services. While this topic is explored in detail in Chapter 6, this section will discuss some of the key problem statements related to the enterprise network, from the standpoint of existing management models, and provide insight into new solutions.

10.4.1 CALL HOME FUNCTION

One of the biggest pain points for enterprise IT administrators today is the planning and logistics involved with the roll out of new network appliances or the replacement of older appliances. Today, not only is this process largely manual and complex but also takes many weeks and sometimes months to execute. As the industry started focusing on network simplification, a variety of solutions emerged to solve these issues. These solutions are often referred to as a Call Home service (also referred to as Phone Home Service, Zero Touch Provisioning, and Plug-n-Play). Each of these solutions is mainly focused on avoiding truck-rolls needed for appliance installation. While there are multiple solutions in the market, each solution often uses a proprietary approach. We do have Internet Engineering Task Force (IETF) drafts* that provide a secure and standard-based approach to connect with a newly deployed device, and configure it with the appropriate feature-set required for the service.

10.4.2 DATA MODELING FOR NETWORK DESIGN

Today, the network design process is centered on a set of documents—the network requirements and design documents that contain a variety of information such as high-level network architecture diagrams, low-level detailed designs, and specification of what features are enabled and sometimes, examples of how those features can be enabled. One of the common issues encountered is identifying the best way to configure the network and maintaining the configuration over a long time period. This is especially challenging in multivendor environments and sometimes with different products from the same vendor, mainly because of different configuration interfaces to configure individual features. This calls for developing variety of configuration tools or scripts and troubleshooting becomes extremely challenging. While this approach has served the networking industry for many years now, today network engineers are moving toward a service-model-based approach that uses the concept of abstraction to help modularize the network design and implementation process.

As shown in Figure 10.13, the network design gets abstracted into three different layers:

- The service abstraction layer essentially focuses on the service definition that is offered to the customer. Data models associated with this layer include service-specific objects, their parameters and constraints exposed to external customers and applications.
- The network abstraction layer captures a common informational model and data store that are used to abstract an up-to-date view of the network to the service layer. This facilitates powerful concepts such as centralized network control and centralized monitoring, and provides a real-time current view of network state on which different services can be instantiated. Data models for the network layer contain information about functions or services that are currently in use on the network elements and can apply specific engineering rules that allow or restrict certain functions based on available resources (like a bandwidth upgrade can be blocked if the underlying device does not support that bandwidth level).
- The device layer focuses on specific devices and is responsible for taking the information from the network data model and populating specific parameters using the specific device's configuration language. This layer essentially captures the attributes that must be configured on the device to enable a specific service, and how it should be configured for a given vendor.

The above approach enables designers to catch errors and oversights early in the cycle, when they are easy to fix. In addition, this basic abstraction approach also simplifies several tasks of the network design team. For example:

- Much of network design information that once was only captured in documents is now part of the data model. Design teams can simplify creation and update of network design documents by just referring to the data models created as part of service instantiation. The data

---

* https://tools.ietf.org/html/draft-ietf-netconf-call-home-17 and https://tools.ietf.org/html/draft-ietf-netconf-zerotouch-09.

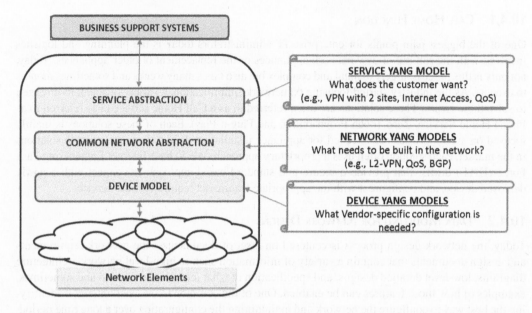

FIGURE 10.13 Abstraction layers for data modeling.

model also drives the agreement on vocabulary and jargon. In addition, a data model can automate some tasks—design tools can take a model as an input and generate the initial database structure and device configuration.

- Data models could serve as an important tool to estimate the potential complexity associated with a network design project. This estimation could eventually be useful to gain insight into the level of development effort and project risk.
- Enhancements to services as a result of new business requirements can assess the impact to the existing network design by identifying the scope very clearly. In addition, reuse of the data models enables network designers to shorten their development cycle considerably.
- Deployment and replacement of specific network devices becomes easy; this in turn reduces the time to market or time for resolution of issues.

Today, the industry has more or less standardized on using YANG as a data modeling language for networking environments. The use of this modeling language has significantly improved the process of network design. Use of the new process has allowed us to improve both cost and time associated with service development while reducing the typical errors seen in prior processes. Just like an architect creates a blueprint before constructing an actual building, the network designers can now create YANG-based blueprints before worrying about the choice of vendor device, capability of each device, etc. Once the service models and device model blueprints are ready, software tools are used to generate device-specific CLIs thus restricting human intervention only to the service design stage.

10.4.3 VF Deployment and Management

As mentioned in Chapter 4, many service providers and data centers have standardized on the use of OpenStack in the cloud environment. However, the use of OpenStack has been focused on deployment and monitoring of VFs in the data centers (and lately, central offices) but not for the branch environment. OpenStack has addressed problems pertaining to highly scalable data centers, where thousands of servers are interconnected via gigabit links, whereas in a branch environment a single low-end compute server is used to deploy multiple functions across a bandwidth constrained WAN network. OpenStack still needs to evolve in terms of efficient use of compute resources, network resources, and support of distributed WAN topologies. There are projects underway within OpenStack and it is expected to have the optimizations and enhancements needed for use in a highly distributed environment. Until then, AT&T's plan has been to use a home grown system (based on open standards like NETCONF and YANG) to manage the virtualized branch environment.

10.5 DPI AND VISIBILITY

DPI is commonly used in the security space for searching for protocol noncompliance, viruses, spam, intrusions, or defined criteria to decide whether the packet may pass (see Chapter 8). It is also used in the packet-forwarding space to allow for application layer routing, or for the purpose of collecting statistical information. Lately, DPI technology has been used in management platforms to provide a deeper understanding of current network traffic, which allows a simple graphical interface to control the network traffic on a per-application flow basis. Protocols such as the simple network management protocol (SNMP) that have been historically used for basic monitoring of network traffic, do not provide fine-grain traffic information. IT administrators were forced to deploy additional traffic analysis devices in the network (at a high cost) to get a deeper understanding of the actual traffic. Today, equipment vendors have implemented a DPI stack on the newer versions of their products. This allows them to provide the capability to gather and export, a deeper level of information as traffic traverses the equipment. However, in understanding the problem of better traffic visibility, the real value of DPI can only be realized with a complete solution whereby one has the ability to collect and analyze data and implement the appropriate control mechanisms to improve network performance under high traffic conditions.

10.6 NETWORK FUNCTIONS ON DEMAND

The vision for next-generation network service is simple—provide the ability for enterprise administrators to choose, deploy, and manage any network function, anywhere within their network. Based on this vision, AT&T started the Network Functions on Demand (NFoD) effort, which is illustrated in Figure 10.14.

It is about having a library of virtual network functions that can be instantiated on the service provider's cloud platform or at customer premise locations. To implement this, the following three core principles are used in the NFoD platform:

- Vendor neutral interfaces for the NFoD platform—the library of network functions not only provides a choice of various network functions but also a choice of vendors for each function.

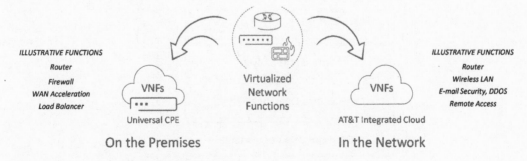

FIGURE 10.14 Network functions on demand.

- Choice of deployment models—network functions can be instantiated on compute resources within the Network Cloud (on AT&T Integrated Cloud [AIC]) or on premise using the universal CPE (uCPE) platform (discussed in the next section).
- Flexible management models—while the service provider takes on the responsibility for managing the underlying hardware and software platform that VNFs run on, in both the AIC or uCPE, customers can choose the VNF management model. They can outsource the complete management to the service provider (thus achieving the typical managed services model) or self-manage via tools provided by the service provider (thus achieving the typical cloud services model).

These three NFoD principles provide the much-needed flexibility and operational simplicity for enterprise networking use-cases. While moving functions to the cloud is an obvious trend in the industry, there is still a need to support functions at the customer premise location. To address this need, a new open hardware platform that is capable of running any function that exists in VNF software library, was created—this is the uCPE.

10.7   UNIVERSAL CPE

The uCPE is an open hardware platform that is capable of running different types of virtualized functions at the customer premise. Think of the uCPE as a small extension of the AIC that runs network functions and also directly integrates with the AIC. So if a customer requires some functions to run locally at the premise it is run on the uCPE, while it may make sense to run the other functions in the AIC. The service will service chain the premise into the AIC to provide a seamless service. Before we talk about the services that can be enabled at the customer premise, we have to talk about the uCPE hardware platform on which these services run. The uCPE is a generic hardware platform built with commercial-off-the-shelf (COTS) components. The platform will run Open Source software for the base system functions and allow VNFs to run as services on top, in the same manner as services that run on a cloud platform. Figure 10.15 outlines the uCPE architecture.

The goal is to create a generic and open hardware platform using cheap commodity components to drive down cost, continue to maintain performance, and add cloud-like flexibility for network services.

The initial release of uCPE will focus on an Intel-based x86 platform but the uCPE architecture does not dictate it has to be Intel-based. The uCPE architecture could run on other hardware in the future (such as ARM-based platforms). Today x86 is best suited to support the upper layer services required, but the architecture can include any future COTS-based solution that can drive down costs or add functionality.

The layer above the hardware is the operating system for the uCPE. This will be an open source operating system optimized for running VNFs. This will include functions to allow service activation and management interfaces to the Open Network Automation Platform (ONAP).

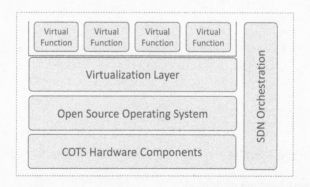

FIGURE 10.15   uCPE with VF.

One of the important benefits of the uCPE is the ability to run multiple VNFs on a single hardware platform. In today's customer premise implementation, multiple boxes are needed at premise locations to support different network functions. For example, a site may need a box for the router, another box for a firewall, a third for WANx, etc. With cloud-based virtualization technology in the hypervisor layer, multiple network functions can run on the same hardware, each in its own VM. Longer-term, this can be optimized with network functions running in Linux containers, which will eliminate the hypervisor overhead but maintain application separation.

The final piece of the architecture is the ability to orchestrate the uCPE using SDN. The uCPE will have orchestration APIs so the ONAP platform can orchestrate VNFs on the uCPE. This will include turning up a VNF VM, turning down a VM, and lifecycle management of a VM.

As mentioned, one of the key characteristics of the uCPE is to run multiple VFs on a single hardware device. Figure 10.16 outlines how a typical customer premise may be set up in today's model and how the uCPE can simplify this with running multiple functions on a single box.

In today's premise model, a typical site will have a premise-based router that will have a wide area circuit to a provider's network. The router is a purpose-built box that includes proprietary hardware that is running proprietary software provided by the router vendor. The hardware and the software are tightly integrated together; you cannot use one without the other. Routers have evolved to support higher-level services such as firewall and security functions but typically these functions are limited compared to purpose-built appliances, and typically the best of breed choice for each network function does not come from the same vendor.

Therefore, if the site needs additional functions such as an advanced firewall, that is provided as a separate box. The firewall will also have its own proprietary hardware and operating system, just like the router. There are many other network functions that follow the same "middle box" model including WANx, network analyzers, caching, or wireless LAN controllers to name a few. If the customer wants the best of breed for each function, they likely will be left with multiple hardware vendors with multiple proprietary network management interfaces. Also, if a customer wants to add a network function to a site, a new piece of hardware has to be purchased, configured, shipped to the site, and a truck roll to install and bring up that service.

The uCPE model will fundamentally change how premise-based services are delivered. The hardware will be the open hardware platform discussed above for all customer sites. There will be different size platforms for different site types, but the architecture will be consistent.

The VNF functions that run on the uCPE will be typically available in a VNF library, where the customer can decide what network functions they would like to download and run on the uCPE. Think of this like an App Store model for enterprises. All the major network hardware providers now offer their proprietary software as a software-only option that will run on x86. A good example is the Cisco Cloud Server Router (CSR), which is their traditional IOS software that can be purchased and run on an x86 platform. Vendors in the routing, security, analysis, and WANx areas have all done the same.

The library will have multiple software options within each category including routing, security, analysis, WANx, wireless, VoIP, and others where the customer can choose the vendor they prefer

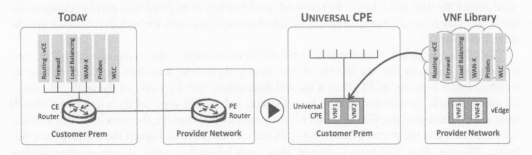

FIGURE 10.16 CPE and provider network with virtual services.

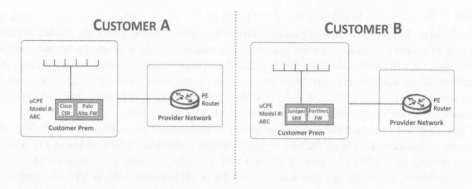

FIGURE 10.17   uCPE—same hardware with different services.

to run. Since the hardware is generic, all of these functions will run on the same hardware platform. So if customer A would like to order a Cisco router with a Palo Alto firewall and customer B would like to order a Juniper router with a Fortinet firewall they both have the same hardware but use SDN and NFV to customize that hardware to their own needs. This concept is depicted in Figure 10.17. Customer A orders uCPE Model ABC and then via a user portal requests that the uCPE run a Cisco CSR router and a Palo Alto firewall. Customer B also orders uCPE Model ABC but via the user portal requests a Juniper vSRX router and a Fortinet firewall. The same hardware is used in both cases, but by using SDN and NFV, each box is fundamentally different based on the customer's specific needs.

This is at the heart of the strategy to turn the network into a platform that is controlled by software. The underlying layer is not tied to any vendor so it can change and grow using a software model and avoids the need to upgrade hardware when new functionality is needed.

As a use-case let us assume a customer today has deployed a physical router at a site. If there is a security issue at that location and requires the customer to deploy a firewall, today it takes 30–45 days to acquire a new firewall, provision it, and to have it professionally installed at the site. With the uCPE model, if the same customer has the uCPE deployed with just a vRouter enabled, and if there is an urgent need for a firewall, it can be ordered via the portal, and within minutes that firewall can be up and running. Another use-case, with WANx, has the customer with a point-of-sale application that is getting very slow response time. The WANx VNF can be deployed in a matter of minutes and configured to optimize the point of sale flow.

This model also allows for new service and business models. An example is a try-and-buy model where the service provider can offer the customer the ability to try a new VF or a new vendor within a functional category. Today, if a service provider wanted a customer to trial a WANx physical appliance at a branch site, there is a lot of work involved for this to occur. The device has to be shipped to the site, the customer has to physically install it, which will likely cause the site connectivity to be lost during the install. In the event there is an issue with the WANx hardware during the trial and it has to be removed, someone has to be physically there to deinstall the hardware. This is not an attractive model to introduce new functions for both the customer and the service provider.

The uCPE can greatly improve this. If the uCPE is installed at the customer site and the customer would like to try the new WANx, all they have to do is order it from the portal, download the software image on to the uCPE, boot it up, and dynamically service chain on the uCPE. No physical changes are needed, nobody has to be at the site, and the site does not incur any down time. If there is a problem with the WANx trial, or if the customer does not see the value in it, it is a simple software change in the portal to remove it from the service chain and revert the site to what it was before. This can allow customers to almost "play" with VFs at their sites, till they get the combination of functions and the set of vendors that works best for them.

Finally, the open software model can drive real innovation into the enterprise network space with a lower barrier to enter. A new player can innovate faster and can optimize performance easier than an established network vendor, since their VNF software is built from the ground up to work in a cloud environment as compared to an established vendor who may be porting code from an existing hardware appliance.

The uCPE will start out by offering the same services and functions offered today, such as routers, firewalls, WANx, or network analyzers to name a few. But the uCPE is an x86-based cloud platform so any application that can run on x86 can run on the uCPE. Other functions like network attached storage (NAS) file servers, print servers, as well as micro applications will be able to run as well. The service provider can allow customers to "bring their own application" and load it as a VM.

10.7.1 uCPE Phone Home Process

We have talked about the flexibility a cloud model brings to the customer premise with the uCPE and the ability to dynamically turn up and turn down applications. But the service provider still has to get the physical uCPE to the site to allow for all of this to happen. As we continue to compare the uCPE to today's model, getting a physical device like a router to the site can take many days. Again the box has to be purchased, configured specific for the site, shipped, and a certified person needs to install it.

The uCPE will greatly simplify this process as well, using a Phone Home or Plug-and-Play (PnP) model. The uCPE can be overnight-shipped from a warehouse with no specific customer configuration settings. The customer can simply plug the uCPE into the network port, power it up, and the uCPE will use a Phone Home protocol to call back to the network management cloud, authenticate, and then download the site configuration for the device. In a matter of minutes, the site can be up and running.

This can open new distribution models for the hardware. Instead of the service provider having to warehouse a larger inventory of hardware, the hardware can be drop-shipped from the manufacturer directly to the customer site. Another model could leverage the service provider retail locations for stocking uCPEs and if a customer needs a site up the same day, they can get one at the nearby retail location.

In order for the Phone Home solution to be successful, it has to be secure as it will have to map the hardware unit to a specific secure customer domain. The first level of authentication will be hardware based, a burned-in item such as the unit's serial number or MAC address is passed from the uCPE to the Phone Home server. The service provider will know what MAC address or serial number was shipped to the customer site so once this comes in, they know that the device is attempting to obtain its initial configuration.

However, this is not secure enough; let us say the device was lost or stolen during shipment and someone tries to plug it in, there is a possibility a rogue entity could gain access to the customer's private network space. Therefore, a second level of authentication is required. The Phone Home server will then generate an email or text message to the customer's administrative contact with a unique authentication code that can be used to fully authenticate the uCPE. This is outlined in Figure 10.18.

10.8 CONCLUSION

The rise of mobility and cloud continues unabated, and both of these major trends are driving business requirements for simplified networking models. With virtualization and programmability concepts such as SDN, the industry as a whole has just started its journey toward the next-generation networking environment. As vendors, service providers, and customers focus on the need for simplicity, they will continue to accelerate the current momentum toward a software-centric networking model. Continued innovation in general purpose hardware, open source software and APIs, management simplicity, and performance optimization are the key to success of this evolution.

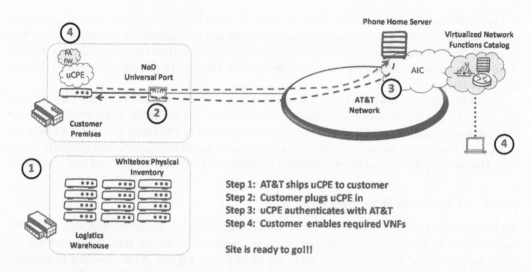

FIGURE 10.18 uCPE call-home process.

Step 1: AT&T ships uCPE to customer
Step 2: Customer plugs uCPE in
Step 3: uCPE authenticates with AT&T
Step 4: Customer enables required VNFs

Site is ready to go!!!

REFERENCES

1. http://www.nongnu.org/quagga
2. VMWARE Virtualization: http://www.vmware.com/solutions/virtualization.html
3. Linux Containers: https://linuxcontainers.org
4. Linux Bridge: https://wiki.linuxfoundation.org/networking/bridge
5. Open Virtual Switch: http://www.openvswitch.org/
6. DPDK Poll-mode: http://dpdk.org/doc/guides/prog_guide/poll_mode_drv.html
7. SR0-IOV: http://www.intel.com/content/www/us/en/pci-express/pci-sig-sr-iov-primer-sr-iov-technology-paper.html
8. Hyperscan: http://www.intel.com/content/dam/www/public/us/en/documents/solution-briefs/hyperscan-scalability-solution-brief.pdf
9. http://www.ijcaonline.org/research/volume131/number16/sarma-2015-ijca-9076010.pdf

11 Network Access

Hank Kafka

CONTENTS

11.1 Introduction 223
11.2 What Makes Access Networks Different 224
11.3 Extending NFV and SDN to Network Access...226
11.4 Wireline Access Technologies...229
    11.4.1 PON Technology 229
    11.4.2 G.fast Technology...232
    11.4.3 Wireline Access Hardware 233
    11.4.4 Merchant Silicon...234
    11.4.5 Wireline Access Hardware Standards and Open Specifications......................234
    11.4.6 Open vOLT Hardware Specifications 235
    11.4.7 Wireline Access Software 237
    11.4.8 Network Abstraction Layer 238
    11.4.9 SDN Access Controllers..240
    11.4.10 Open Access Network Software...240
11.5 Mobile Wireless Access Technologies 241
    11.5.1 LTE RAN Configurations 242
    11.5.2 5G Wireless 244
        11.5.2.1 Advanced Multielement Antenna Structures....................................245
        11.5.2.2 Dual Connectivity..246
        11.5.2.3 Ultrahigh Density and Self-Backhaul...247
        11.5.2.4 Flexible Carrier Configuration 247
Acknowledgments...248
References..248

This chapter will examine the general characteristics of access networks and how those characteristics create challenges for implementing network functions virtualization (NFV) and software-defined network (SDN). It will then describe the general solutions that the network cloud architecture uses to address those challenges. Finally, the chapter will provide more concrete examples of these concepts by describing Gigabit Passive Optical Network (GPON) access technology and wireless 5G access technology in more detail, and showing how the network cloud alters the evolution of access networks.

11.1 INTRODUCTION

In an all Internet protocol (IP) network, there are two predominant means of accessing the network from customer endpoints—wireline technologies and wireless technologies. Both of these technologies require a highly distributed physical infrastructure, shared and allocated across multiple customers and devices. Wireline technologies are exemplified by the GPON standard, which is evolving from fixed optical wavelengths to new technologies with higher speeds and tunable wavelengths [1]. Wireless technologies are exemplified by 4G long term evolution (4G-LTE), which is evolving into 5G technology with higher speeds, lower latencies, and support for vastly larger numbers of devices [2]. The move to the next generation of the underlying access technologies provides an opportunity for synergy with the move to network clouds, although the simultaneous evolution in both dimensions increases the complexity of the architectural roadmap.

Access technologies create another set of challenges for network clouds. Two initial principles of NFV are that virtualized functions can be transformed from physical boxes to software running on generalized compute platforms, and that these generalized compute platforms can be located in any physical location. Network access technology, by its very nature, violates both of these initial principles. As a result, the implementation of network clouds in access networks is less mature than the implementation in core networks. However, by creating an open architectural framework that disaggregates the components of network access nodes, and by broadening the concept of open compute platforms, the network cloud architecture extends the key benefits of NFV to network access elements and also brings them under SDN control.

11.2   WHAT MAKES ACCESS NETWORKS DIFFERENT

Since NFV is an extension of concepts introduced for cloud computing in centralized data centers, some fundamental characteristics of access networks make them different from cloud computing functions and core network functions; this prevents access networks from being able to run in their entirety on conventional data center hardware platforms.

As shown in Figure 11.1, access networks are inherently distributed and cannot have their functions completely centralized. The endpoints of the access network are not located in data centers—they are located at the customer premise, such as homes or businesses, and are incorporated into customer-owned devices such as mobile handsets. Standards define the protocols and behaviors that allow these endpoints to connect to the network. These standards support limited transmission distances, which means that network access elements that directly interface with the customer must be located close to customer locations and devices.

The distributed nature of the access network endpoints would not in and of itself stop the compute and networking components of the access network from being fully centralized. If the transport component of the access network provided a fully transparent constant bitrate pipe from each endpoint device to the centralized location, with each transparent pipe having negligible delay and operating at the full peak speed of the endpoint device, it would allow for a high degree of centralization. This type of transport would be a good way to characterize the connectivity between compute nodes and the top-of-rack switches in a data center, and it is the key to allowing the compute functions to be fully virtualized within a data center.

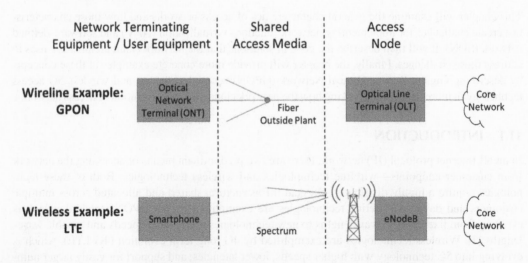

FIGURE 11.1   Access network architecture examples.

However, this type of transport is not available in wide-coverage broadband access networks, because the distances are much greater, requiring specialized and more expensive electronics, optics, and radio transceivers for access links of a given speed. Also, access networks require substantial physical infrastructure, such as buried or aerial cable with splices, cell site infrastructure, and spectrum. The costs of the electronics and infrastructure means that a fully transparent dumb pipe infrastructure would make broadband services unaffordable. Modern wireless and wireline broadband access networks are cost effective because they use specialized hardware and sophisticated network protocols to enable longer transport distances, automatic range detection with timing adjustments for devices at different distances from the network, and multiple access with statistical multiplexing on shared access media. These techniques make broadband access affordable by allowing the cost of one access fiber, or one cell site's radio frequencies, to be efficiently shared by multiple endpoints.

The software-driven functions that enable resource sharing in the access network must be distributed close to the access endpoints. Some functions such as radio interference mitigation require very low latency with distances limited by the speed of light. Other functions such as statistical multiplexing must take place close enough to the customer endpoints to enable effective sharing of the access network infrastructure. In practical terms, the allowable distances for performing a specific function will vary, depending on factors such as propagation limits of the access signals and speed of light latency requirements.

Certain software functions of access networks need to be within no more than 10–20 km of the customer endpoints. For high-speed broadband networks, the total capacity of the access medium (the fiber of a GPON or the wireless carrier of a cell site) needs to be statistically multiplexed across multiple customer endpoints and devices. This requires some of the resource optimization and scheduling functions of a SDN to be carried out at the access node that is closest to the customer. The combination of distance constrained functions and shared media statistical multiplexing means that some of the functionality and software of cost-effective access networks must be distributed throughout a network close to the customer endpoints and devices, rather than being centralized in large data centers.

Access networks also have functions that cannot be optimally run on normal mainstream compute and storage platforms, creating challenges for cost-effective virtualization. The input/output and communications protocols that are integrated into commercial compute platforms are designed for short-range high-speed point-to-point communications, such as connections from the compute node to the top-of-rack switch. Access networks require more complex and specialized real-time functions to support high-performance shared media, such as sophisticated transmit power control, variable range-compensating timing offsets, dynamic performance monitoring, dynamic error correction coding, dynamic modulation, and low latency request/grant resource scheduling. In wireline access networks, these functions are supported by dedicated system on a chip (SOC) integrated circuits with specialized hardware functions optimized for the specific requirements of each media and protocol.

The functions in a typical wireline access SOC are only a starting point for the specialized hardware functions required for high-performance broadband mobile wireless access networks. Mobile wireless base station equipment is built around baseband signal processing, which utilizes massive amounts of digital signal processing. As spectrum auctions and valuations have repeatedly demonstrated, the spectrum used by mobile wireless access networks is a scarce and valuable resource, and the explosion of mobile data traffic has continued to emphasize that value. The advancement of mobile wireless data protocols and standards has been a relentless pursuit to employ real-time algorithms and digital signal processing electronics to continually achieve higher spectral efficiency [3]. Over time, as Moore's law has increased the amount of digital signal processing available on dedicated integrated circuits optimized for wireless access network technology [4], wireless standards and network equipment have consumed that capacity to improve spectral efficiency and capacity. The mainstream compute platforms used in data centers do not provide the level of specialized digital signal processing that is needed to build a high-performance mobile broadband wireless access network.

The extension of SDN and NFV all the way to the end of the access network gives a complete end-to-end implementation of the SDN/NFV and open network automation platform (ONAP) architecture across all access technologies, enabling the creation of new features and services that can incorporate the access network and that can be written once and applied across multiple access technologies, even as those access technologies continue to evolve. Access technologies present a challenge for NFV and SDN. The virtualization of high-performance access networks does not directly fit with conventional centralized general-purpose compute and storage platforms, because components of the access network need to be distributed rather than centralized, and because access networks rely on specialized hardware to perform functions that cannot be efficiently executed on general purpose compute platforms. The access network is a critical and expensive part of the overall end-to-end network, and the full benefits of NFV and SDN control cannot be achieved without including the access network in the overall architecture. This chapter explores how NFV and SDN can be maximally applied to access networks, given their topology constraints.

11.3   EXTENDING NFV AND SDN TO NETWORK ACCESS

To extend the network cloud for access application, compute and storage resources need to be dispersed to remote locations. The new architecture needs to support cost-effective smaller compute and storage instantiations at these locations. In order to overcome the specialized hardware challenge, the network cloud architecture is extended to provide open specifications for specialized hardware with defined generic interfaces that allow access software to be implemented independently from the specifics of the access hardware. In order to overcome these challenges and implement these approaches, the network cloud architecture decomposes the integrated functions and features of a traditional access network element, allowing those functions to be executed at different locations and on different hardware platforms.

Large data center environments are optimized for supporting large numbers of compute and storage nodes, including thousands of processors interconnected with spine and leaf switches. Open Stack is well suited to provide orchestration at the scale required environments. However, minimal configurations of Open Stack still require multiple servers. The general purpose processing required for a distributed access node may include only a single processor integrated as part of an SOC, and that single processor at the access node will be performing media access control (MAC) and quality of service (QoS) functions that must be coordinated with SDN in the larger network. That single processor will also be generating traffic performance and usage statistics that must be integrated into the large-scale ONAP Data Collection, Analytics and Events (DCAE) platform. These requirements mean that the access network requires the network cloud architecture to run in an integrated fashion across small-scale distributed access nodes and large-scale centralized data centers. Several options are currently being explored to provide the integration between the large-scale centralized AT&T Integrated Cloud (AIC) nodes and the highly distributed access nodes.

The first option is to treat the access nodes as peripherals, which perform specific functions or provide specific customized interfaces. This can be an appropriate option for simple and stable access functions that do not need to be updated to provide new capabilities for new services. However, because this option relegates the software on the access node to its own compute structure, it removes that software from the benefits of being part of the overall network cloud compute structure. In a sense, this software essentially becomes a new form of embedded software—being different only in that its interfaces are designed to work with an ONAP SDN/NFV environment.

The second option would be to modify the overall network cloud software infrastructure to support smaller highly distributed environments. This requires expanding the configurability of all components so that they can scale down to support very small distributed "data centers," as measured by the number of compute, storage, and switching nodes. Ideally, this might enable the AIC environment to include distributed embedded processors as a native part of its architecture, allowing the compute resources embedded in a SOC to become part of a small cloud implementation. While current technology has not demonstrated a single orchestration architecture that can efficiently span the extreme dynamic range required to support this approach, new orchestrators or extensions of current orchestrators to support smaller configurations are under investigation and should be feasible. One example of use of this type of configuration would be the instantiation of a localized lightweight remote orchestration agent at the small cloud node that could allow remoting most of the ONAP capabilities to a larger cloud node. With the support of sufficiently small configurations, the combination of an orchestrator capable of supporting a wide range of cloud node sizes, along combination with the approach that treats extremely small access nodes as peripherals, may be an approach for realizing most of the benefits of NFV and SDN for distributed access nodes.

The third option provides an alternative to extending a large-scale orchestrator to support small-scale access nodes. In this option, lighter-weight orchestration environments can be developed that can manage local resources for small-scale cloud instantiations. Coordination between the orchestrators and associated ONAP components could potentially be established in a manner similar to the hierarchical controller and subcontroller architecture developed for SDN control functions. This approach could provide nearly all of the benefits of a single orchestrator capable of supporting both very large and very small configurations, at the cost of defining interfaces and behaviors across the orchestrator infrastructure and related systems such as ONAP.

Distributed access nodes generally require specialized hardware to carry out specific functions, usually with custom application-specific integrated circuits (ASICs) and/or field programmable gate arrays (FPGAs). In a conventional network architecture, these hardware functions are specific to each hardware supplier and require custom integrated software from the supplier. Large cloud computing data centers gain cost advantages by fully separating the hardware from the software. This provides cost advantages on the hardware side by allowing the use of generic hardware, which could be described in open specification and built by multiple suppliers. This approach provides cost advantages for software by allowing the software to be developed one time (often as open source) independent of the underlying hardware, rather than requiring multiple developments of customized software (and subsequent software upgrades) for each supplier's hardware.

The new architecture for access elements needs to capture these benefits to the largest extent possible, while still maintaining specialized access hardware as a component of the architecture. This can be accomplished by defining the detailed functions of the specialized hardware platform in an open specification, with interfaces and functions specified in a way that allows the hardware to be built by multiple manufacturers. Ideally, the specified interfaces and functionality would be sufficiently complete so that hardware manufacturers would have full flexibility in designing for optimal performance at the lowest cost, and software could be written to execute on any compliant hardware implementation from any hardware supplier. In practice, it is likely to take some time before enough experience is gained to enable writing complete and open specifications that achieve this ideal. As a first step toward this ideal specification, hardware specifications may be written that include some degree of dependence on specific hardware components, such as a hardware design specification based on selected SOC or FPGA components. As experience is gained with these implementations, their interfaces and behavior can be further abstracted from the underlying integrated circuit technology to drive toward the ideal state.

As previously described, mobile wireless network equipment requires intensive digital signal processor (DSP) implementation, and high-performance networks will continue to demand ongoing upgrades to maximum available DSP silicon horsepower as Moore's law continues to expand those capabilities. To achieve the maximum economic benefits, the architecture needs to allow software to remain unchanged while being able to take advantage of more powerful DSP engines as they become available over time. Furthermore, the specific calculations performed by these DSP engines need to be flexible to adapt to changing services, changing network conditions, and changing standards and features, over periods of time ranging from less than 1 s (for changing loads or conditions) to multiple years (for changing standards and features). This goes beyond the capabilities of application programming interfaces (APIs) and functional specifications.

This type of DSP functionality does not show up in cloud data center environments. However, there is a very successful example of a similar type of specialized function in general purpose computing—specifically, the graphics processor, used in PCs by consumers to support the generation of high-resolution video for high-end games. One key element that has made the graphics processor highly successful is OpenGL—a graphics language and library that abstracts the computations of the graphics processor from the details of the underlying hardware. OpenCL is a similar language developed for defining FGPA-based accelerators for parallel computing [5]. This concept can be extended to the specialized DSP engines needed to support high-performance mobile wireless services. In place of OpenGL or OpenCL, an open interface specification and language needs to be developed to support the specialized accelerators needed for access technology—it might be called the Open Access Language (OpenAL). Access accelerators executing OpenAL could then provide the specialized hardware functions needed by access networks, such as the DSP engines required to support mobile wireless networks. This enables the development of a range of hardware accelerator platforms in various sizes and configurations, suitable for deployment in small highly distributed nodes as well as large centralized nodes. Software can be written once, and run on specialized hardware accelerators in a variety of sizes and capacities. As Moore's law enables more power, new higher-performance accelerators could be introduced without requiring a change in the underlying hardware or software.

Figure 11.2 illustrates the architecture for access networks using AIC components in an extended configuration, including support for small distributed nodes and specialized hardware resource pools. This conceptual architecture overcomes the unique challenges presented by the access network and allows the access network to enjoy the benefits of NFV and SDN, while also enabling the network cloud architecture and service creation environment to extend from the core all the way to the end of the access network.

This architecture described in the previous section has not yet been fully implemented. As with any major new architecture, it will be introduced gradually and opportunistically, and it will evolve over time. Also, the underlying access network technology and standards themselves are continuing to evolve. Over the years, wireline access technologies have evolved from dial-up modems to asymmetric digital subscriber line (ADSL) to very-high-bit-rate digital subscriber line (VDSL) to GPON. At the same time, wireless technologies have evolved from 2G to 3G to 4G-LTE, and soon, 5G. Furthermore, with the growing evolution toward high-speed small cells requiring fiber access, and the introduction and evolution of fixed wireless access technologies as a substitute for wireline broadband access, the evolution of wireline and wireless access technologies is beginning to converge.

The impact of NFV and SDN on access network architecture results in a complex multidimensional evolution roadmap. The rest of this chapter will outline how the onset of NFV and SDN and

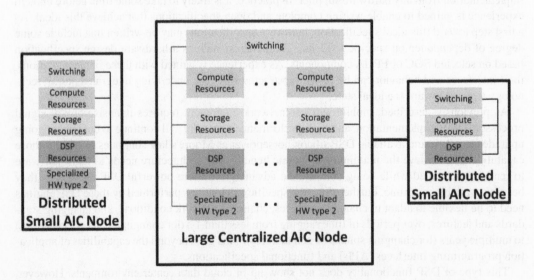

FIGURE 11.2 Extending AIC to access.

implementation of network cloud can converge the introduction of the ONAP access architecture with the ongoing technology evolution roadmap for advanced wireline access and advanced mobile wireless access.

11.4   WIRELINE ACCESS TECHNOLOGIES

The wireline access network is the part of a telecommunications network that connects subscribers to their immediate service provider using some form of media, such as optical fiber, unshielded twisted pair copper, coaxial cable, or a combination of these media types. Construction and maintenance of access networks make up a significant percentage of service providers' capital and expense budgets. The wireline access network is often referred to as "the last mile" although the actual distances can vary considerably. As shown in Figure 11.1, the components of the wireline access network include the access node in the service provider's network as one endpoint, a network terminating function (which could be part of a residential or business gateway) at the customer's location as the other endpoint, and the outside plant (OSP). The OSP consists of transmission media (fiber or copper) with various passive or active components connecting the two endpoints. Increasingly, the wireline access network is combined with wireless access points or cells at the customer's location for the final connection to the customer, which we will cover in a separate section of this chapter.

The ever-present challenge for network operators is to design and build higher bandwidth services to their potential customer base, while keeping costs constant or growing much more slowly than the amount of data delivered. As data usage by customers continues to increase dramatically year over year due to new applications and high-resolution video services, network operators must carefully manage the investment in their access networks to grow in line with revenue growth. The challenge of controlling costs to handle traffic growth calls for investing competitively in next-generation access infrastructure, which leverages SDN, NFV, cloud networking principles, a flow-oriented data plane, a decoupled control plane, and open hardware platforms with hardware abstraction. This approach promotes the use of commodity hardware to support flexible, rapid service creation, and delivery.

The new access architecture needs to be able to deliver existing services, and also new services created natively in the new architecture over legacy networks. This requires allowing legacy operations support systems (OSS) and business support systems (BSS) systems to interact with new access network components as if they were legacy components, as well as enabling legacy access nodes to add capabilities that permit them to be controlled by services and software components of the new architecture. Figure 11.3 depicts this in more detail.

Fiber-based point-to-multipoint architectures such as passive optical networks (PONs) are well suited for cost-effective, high-performance outdoor access networks spanning many kilometers. Copper-based technologies over coax or twisted pairs such as G.fast can distribute high speeds and capacities for shorter distances, such as within buildings or multi-dwelling units (MDUs). These technologies are also highly effective in combination. For example, a PON network can provide the data path to a building, and a G.fast distribution point unit (DPU) can provide distribution to different apartments or businesses within the building as shown in Figure 11.4

11.4.1   PON TECHNOLOGY

GPONs have been deployed for several years and will continue to be deployed for some time. GPON is a cost-effective technology that allows sharing a single fiber for most of the distance between the network access node and the customer, using passive optical splitters located near the customer and appropriate protocols to accomplish this sharing while allowing each customer to realize high capacity and peak speeds. The network access node for GPON is called the optical line terminal or OLT, and the customer node for GPON is called the optical network terminal or ONT (see Figure 11.1). The GPON fiber infrastructure utilizes 2.5 Gb/s downstream and 1.25 Gb/s upstream speeds, shared dynamically between individual subscribers on each of the OLT's fiber GPON ports.

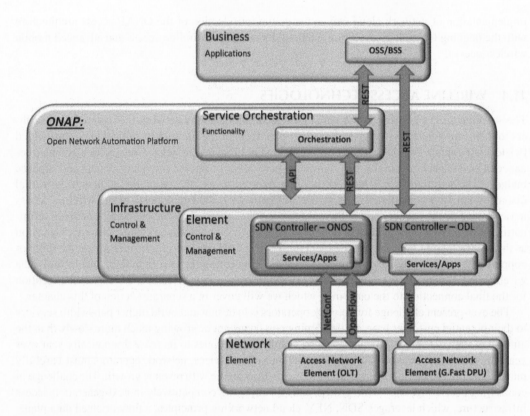

FIGURE 11.3 Block diagram of new architecture for wireline.

GPON technologies are continuing to evolve. NGPON2 is a new technology that operates at 10 Gb/s and can run over several different wavelengths using tunable wavelength components. This will permit much higher capacities by running multiple NGPON2 instances on a single fiber. However, at the current time, the costs of tunable components remain very high, making this technology uneconomical for many applications. Another next-generation PON technology that was recently standardized in the Telecommunication Standardization Sector of the International Telecommunications Union (ITU-T) is called XGS-PON. It is a 10 Gb/s symmetrical technology using fixed wavelengths, with four times the downstream speed and eight times the upstream speed of GPON. For most network operations, the overall approach for GPON evolution is multiphased

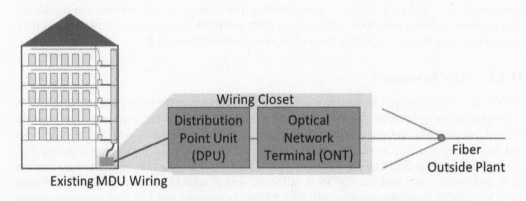

FIGURE 11.4 G.fast architecture for MDUs.

and will take multiple years to realize. A key part of this approach uses the new architecture to shift the economic value-added focus from hardware to software, allowing the software to be reused across multiple access technologies over time.

In the evolution from broadband passive optical network (BPON) to GPON, operators were able to reuse the OSP optical distribution network (ODN), which consists of the fiber optic cables, optical splitters, and other passive devices that make up the physical layer of the access network. Beyond this passive OSP, all of the active network elements, software, and management systems had to be replaced. In addition, BPON and GPON operated on the same upstream and downstream wavelengths, so they could not coexist on the same optical fiber. Each system required separate fibers on the ODN.

The passive OSP of a GPON network consists of the fiber optic cables, optical splitters, and other passive devices that make up the physical layer of the access network. To enable a graceful upgrade process, GPON and next-generation PON systems such as XGS-PON and NGPON2 have dedicated noninterfering wavelength assignments. In combination with the new software architecture, this will allow an elegant transition from GPON to XGS-PON and NGPON2. Network element hardware can be designed to support both XGS-PON and NGPON2. With open hardware specifications and hardware abstraction, the modular NFV and SDN software developed for GPON and XGS-PON can be modified and extended to support NGPON2 and future advanced PON access technologies.

The key to this elegant transition is the use of coexistence element (CE) devices or passive wavelength multiplexers in the central office where the OLTs are located, along with wavelength blocking filters or tunable optics at the customer location. This will allow GPON, XGS-PON and NGPON2 to operate simultaneously on the same fiber in the ODN, optimizing both the flexibility and extensibility of the new access network, while providing additional bandwidth management capabilities and common services across all of the technologies (see Figure 11.5).

The new access architecture coupled with XGS-PON provides a flexible access network suited to support:

- Best-effort consumer broadband services at 1 Gbps (and beyond) symmetric for both single family and MDUs
- Best-effort business broadband services at 1 Gbps (and beyond) symmetric

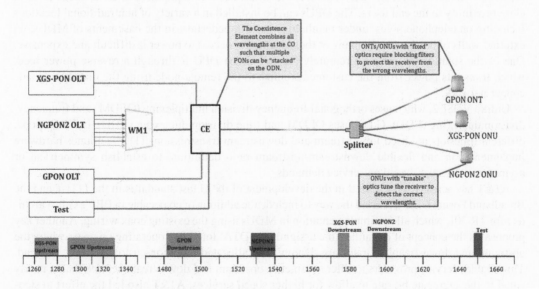

FIGURE 11.5 Fiber wavelength band plan with CE.

- Business broadband services with service-level agreement assurances for a wide variety of speed classes with network on demand (NoD) allowing customers to quickly increase or decrease bandwidth as needed
- Access transport infrastructure for mobility and cloud support

This architecture significantly reduces the complexity of incorporating current and future evolving access technologies, simplifies the development of new services, and reduces the ongoing operating cost of the network. It eliminates the need for totally new hardware and software for each access technology, and it allows new services to be developed once rather than having to be developed separately for each access technology. With the XGS-PON architecture in place, network operations are in position to implement NGPON2 and any other new innovative access solutions in the future, and to share these access technologies across business and consumer customers and services.

11.4.2 G.FAST TECHNOLOGY

In some cases, such as inside existing buildings, the installation of a new fiber optic infrastructure can be cost prohibitive and could delay the availability of next-generation broadband services. In these cases, the best way to distribute new high-speed broadband services is to use short-reach copper-based technologies that can support increased access speeds on existing indoor copper access infrastructure. G.fast is an important technology that can distribute high-speed broadband services over existing coaxial cable or twisted pair wiring within a building or MDU complex.

G.fast is the next-generation digital subscriber line (DSL) technology, with performance targets between 150 Mbps and 1 Gbps depending on loop length. G.fast can provide nearly 1 Gbps aggregate bandwidth, generally referred to as the sum of upstream and downstream bandwidths on very short copper loops (<100 m), which is more than five times the speed of existing VDSL2 technology. Currently deployed VDSL2 uses the frequency spectrum up to 17 MHz, whereas the first generation of G.fast technology will use 106 MHz frequency band and eventually move to 212 MHz in the future for higher bandwidth capability.

With G.fast operating on a higher frequency band and achieving higher speeds, it also means shorter transmission distances, greater power consumption, and a lower signal-to-noise ratio. The frequency band that is ultimately used is a compromise between performance, costs, and implementation. To achieve greater performance, the G.fast network equipment or DPU needs to be at close proximity to the end users. The DPUs can be installed in a variety of nontraditional locations including on telephone poles, under manhole covers, in pedestals, in the basements of MDUs, on external walls of homes, etc. In many of these locations, access to power is difficult and expensive. One of the solutions to power a remotely located G.fast DPU is through a reverse power feed, which transmits power from the customer premise to the remote node using the distribution side copper network.

Unlike VDSL2, which uses orthogonal frequency division multiplexing (OFDM) and frequency-division duplexing (FDD), G.fast uses OFDM and time division duplexing (TDD). In TDD mode, different timeslots are used for upstream and downstream transmission. TDD facilitates hardware implementation and flexible downstream/upstream ratio definition, to establish symmetrical or asymmetrical access, based on service demands.

AT&T has actively participated in the development of the G.fast standards in the ITU-T and the Broadband Forum (BBF), and led the way to include the addition of coax cable in BBF's G.fast architecture TR-301, which allows implementation in MDUs using the existing coax wiring. Another key proposal is the concept of dynamic time assignment (DTA) for G.fast operating on coax, where the upstream and downstream bit rates are dynamically allocated based on customer traffic demand. This enables service providers to offer symmetric upstream and downstream service that is nearly equal to the aggregate bit rate to allow for higher speed services. AT&T also led the effort to standardize the management of G.fast network terminating equipment (NTE) that is a standalone unit

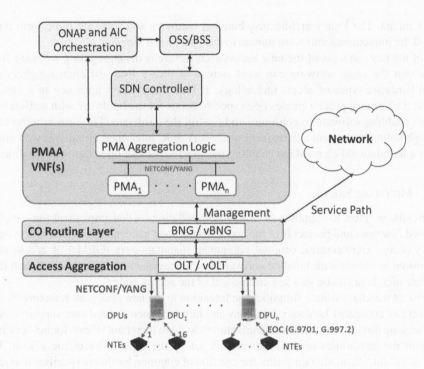

FIGURE 11.6   Management architecture of G.fast.

and not integrated with the customer premise equipment (CPE) or residential gateway. The NTE uses a Network Configuration Protocol (NETCONF) and Yet Another Next Generation (YANG) management model in combination with the embedded operations channel (EOC), which is the management communication channel between the G.fast DPU and NTE. This is critical for transitioning the management of G.fast devices (DPU and NTE) to an SDN-controlled management solution using NETCONF/YANG.

Figure 11.6 shows a high-level view of G.fast device management in the new access architecture. This is based on the BBF's TR301 management architecture, which introduced the persistent management agent (PMA) that provides persistence and modification of the DPU configuration when a DPU is not powered. A PMA Aggregator (PMAA) provides aggregation of PMAs, and the PMAA exists as a virtualized management layer in the AIC. ONAP and AIC orchestration control the instantiation and resiliency of PMAAs.

The OSS/BSS orchestration layer uses an SDN controller as an adapter to manage multiple PMAs, DPUs and NTEs through the PMAAs. The network solution includes passing in-band management traffic through the intermediate network layers (Broadband Network Gateways [BNGs] in Central Office routing, and OLTs/virtual OLTs [vOLTs] for access aggregation) in the network service path. The PMA manages its associated DPU and connected NTE modems using the NETCONF protocol and YANG data models. The DPU manages each NTE through an EOC protocol over each G.fast link, per ITU-T G.9701 (G.fast physical layer specification) and G.997.2 (G.fast physical layer management for G.fast transceivers). The DPU YANG model proxies the NTE inventory, status, faults, performance monitoring, and software download functions over the EOC to the NTE.

11.4.3   WIRELINE ACCESS HARDWARE

Wireline access technologies such as GPON, XGS-PON, NGPON2, and G.fast require specialized hardware functions to handle the protocols and functions necessary for shared multikilometer optical fiber access, and for high-speed transmission across existing short-distance copper twisted pair and coax media. The legacy architecture bundled hardware and software to perform these functions, and the implementations were unique to each particular supplier.

One of the key concepts of the new access architecture is disaggregating software from hardware, so that the same software can work across hardware from different suppliers and even different hardware types of access technology. This requires a new approach to access network hardware. This new approach creates open specifications for the hardware with well-defined open interfaces enabling software to configure and control the hardware. The current trend toward the use of high-volume merchant silicon (access network SOC integrated circuits) is well aligned and provides a foundational element for enabling the specification of open hardware platforms.

11.4.4 MERCHANT SILICON

Merchant silicon refers to suppliers of SOC-integrated circuits that implement the complex hardware-based functions and protocols of the access technology needed for network access equipment. In legacy access architectures, original equipment manufacturers (OEMs) of access equipment create custom hardware with bundled software and management systems, even though they often incorporate merchant silicon as a key component of the access node.

The rise of merchant silicon simplifies the transition to the new access architecture by making it easier to create consistent hardware functions and interfaces across hardware suppliers. AT&T has started working directly with silicon manufacturers to drive merchant silicon for network hardware, as many of the capabilities of the hardware are set by the capabilities of the silicon. The existence of merchant silicon also simplifies the creation of common hardware specifications, as will be explained in the following sections.

11.4.5 WIRELINE ACCESS HARDWARE STANDARDS AND OPEN SPECIFICATIONS

Access hardware relies on well-known and established standards processes that specify the interfaces and behaviors between the access node in the network (such as the OLT) and the equipment at the customer's premises (such as the ONT). The new access architecture will continue to use these standards for technologies such as XGS-PON and G.fast, but it will also require additional standardization efforts that have not previously been used. These new standardization efforts will create detailed hardware box specifications, so that multiple manufacturers can make hardware that will work with common software.

These new standards are being introduced through the Telecom Group of the Open Compute Project (OCP), where OLT hardware designs have been defined utilizing merchant silicon. With these open specifications, operators can require hardware suppliers to build access nodes that conform to these designs, and will have the flexibility to obtain interchangeable network element "white boxes" from multiple sources.

The OCP initially focused on creating open hardware specifications for server hardware based on the x86 microprocessor architecture for use in data centers. In one sense, an x86 processor embedded in a data center server product is analogous to merchant silicon embedded in an access node product. OCP's Telecom Group extends the scope of OCP specifications to encompass telecom hardware. The mission of the Open Compute Networking Project is to create a set of networking technologies that are disaggregated and fully open, enabling rapid innovation in the network space. The objective of the Open Networking Project is to facilitate and enable new and innovative open networking hardware and software standards, design creations and collaborations, project validation and testing, and OCP community contributions. Key drivers of OCP are common standard form factors and 100% standards-based white box hardware implementations, which enable the transition from existing proprietary solutions to solutions based on nonproprietary open specifications.

A full specification of an access node would be highly complex, requiring a detailed description of all of the capabilities and behaviors embedded in an access node SOC, as well as describing how that SOC would be designed into an access node. The specification of an access node becomes much more manageable by basing the design on a merchant silicon SOC for that access node.

11.4.6 Open vOLT Hardware Specifications

Network access equipment is often deployed in different environments. For example, deployments for a data center, a central office, and a remote outdoor plant enclosure would be of various sizes and would have different environmental requirements. The new access architecture would allow a wide range of designs to be configured and controlled by common software and to offer the same set of services. In order to jumpstart the process of creating open hardware specifications, AT&T worked toward the creation of a consistent set of hardware specifications for vOLTs that could operate in different environments. During the March 2016 OCP Summit, AT&T submitted three designs for 10 Gbps PON hardware to the OCP Telecom Group: the MicroOLT, the 16 port OLT pizza box, and the 4 port hardened "clamshell" OLT. Each design is optimized for a different deployment scenario.

MicroOLT—The MicroOLT (shown in Figure 11.7) is a PON optics module with an embedded Ethernet bridge with PON MAC and physical (PHY) layers in the enhanced small form-factor (SFP +) form factor (one of the enhanced versions of the small form-factor pluggable transceiver specification.) The MicroOLT is a component building block of a vOLT platform. It uses 10 Gbps XGS-PON optics to enable XGS-PON protocols and speeds at a fraction of the cost of using tunable dense wavelength division multiplexing (DWDM) optics.

This 10 Gb/s OLT plugs into low-cost white-box Ethernet switches or other Ethernet-enabled networking equipment with SFP+ sockets, instantaneously enabling them to become the network access node for a 10 Gb/s PON network. This SFP+ OLT design facilitates SDN and NFV in the new architecture, enabling it to become the hardware component of a vOLT. Figure 11.8 shows a high-level design of a MicroOLT used in the new access architecture.

The functionality of the OLT can be split into multiple parts to align with the generic network functions and isolate the functions specific to PON access technology. The MicroOLT contains the PON-specific functions including bridging, MAC Layer, and physical layer optics. The rest of the system can be generic functions of layer 2 switching for aggregation and layer 3 switching with traffic management and routing. These generic devices can be based on standalone data center switching components with SDN control and cloud orchestration.

Significant flexibility can be achieved by selecting PON or point-to-point Ethernet for each port on an Ethernet switch, on a port-by-port basis as needed. The MicroOLT eliminates the need to deploy a dedicated chassis, while using less power and requiring less space. Cost savings and added flexibility are achieved by removing the application-specific hardware and replacing it with best-in-class Ethernet switches.

16 Port OLT Pizza Box—The Open XGS-PON 1RU vOLT is a cost-optimized access design focused on NFV Infrastructure deployments, which support symmetric 10 Gbps PON access connectivity and provide up to 160 Gbps uplinks to the Top of Rack (ToR) or spine layer of the network. This design uses the Broadcom OLT PON MAC SOC (BCM68628), which supports SGPON1, XGS-PON, NGPON2, and 10G-EPON, and Qumran switch (BCM88470 QAX).

Figure 11.9 shows the main system block diagram for a 16-port OCP design consisting of eight BCM68628 devices connected to the BCM88470 QAX switch that supports up to 300 Gbps of traffic.

FIGURE 11.7   MicroOLT.

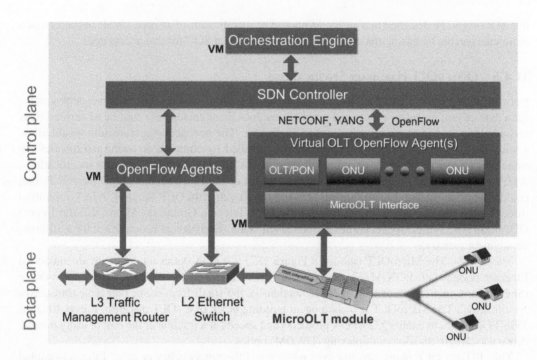

FIGURE 11.8 vOLT software architecture.

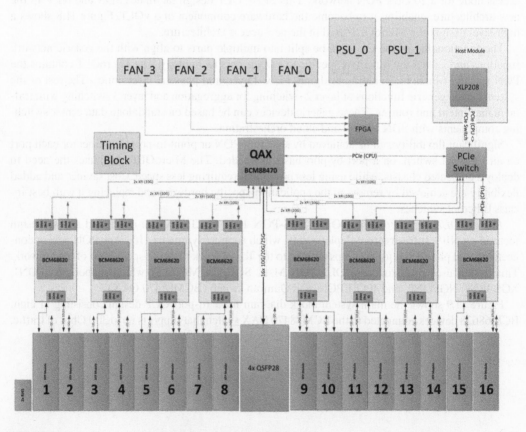

FIGURE 11.9 Main system block diagram of 16-port OCP design.

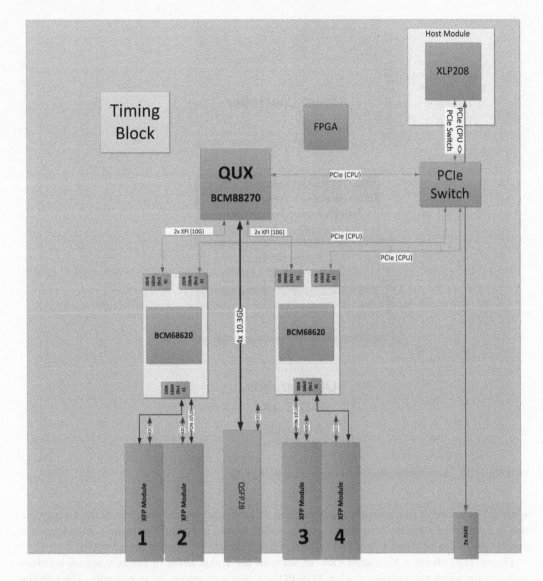

FIGURE 11.10 Main system block diagram of 4-port OCP design.

4 Port Hardened OLT—The Open XGS-PON 4-port remote vOLT is a cost-optimized access design focused on NFV infrastructure deployments, which support symmetric 10 Gbps PON access connectivity and provide 40 Gbps uplinks to the ToR or Spine layer of the network.

Figure 11.10 shows the main system block diagram 4-port OCP design consisting of two BCM68628 devices connected to the BCM88270 QUX switch that supports up to 120 Gbps of traffic.

AT&T plans to deploy vOLT hardware into the access network by using designs approved by OCP as new merchant silicon becomes available. These designs represent AT&T's first open specifications for virtual access node hardware, and can pave the way for open specifications for other virtual access technologies.

11.4.7 WIRELINE ACCESS SOFTWARE

The overall SDN, NFV, and ONAP architectures will extend to include support of wireline access nodes such as the vOLTs described in the preceding section. Some software functions are more

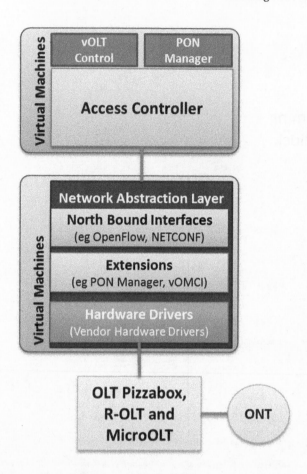

FIGURE 11.11 Software architecture of next-generation PON.

specific to wireline access components. These can include functions that come from the disaggregation of current software in conventional access nodes, as well as specific software components that enable services to be written once and applied across a variety of types of access nodes.

Figure 11.11 shows a high-level view of some of the components of the new access software architecture, using the next-generation XGS-PON access technology as an example. The specific types of hardware are depicted at the bottom. The network abstraction layer hides the details of the underlying hardware from higher software layers, enabling common software control and management for the different hardware configurations. Although it is not shown, this layer also provides abstraction for different types of access technology. The SDN access controller provides key control functions that integrate the access network components such as the vOLT control and PON management with the rest of the ONAP software architecture. It is the desire of AT&T that all updates to the controller source and applications are contributed back to the open source community to maintain a long-term viable product.

11.4.8 NETWORK ABSTRACTION LAYER

The network abstraction layer is a thin layer of software that provides security protection and a consistent interface for various underlying hardware components incorporated into the access network. The protocols and non real-time capabilities are disaggregated onto a micro service running in the AIC environment on standard x86 compute cloud in a virtual machine or container. Through

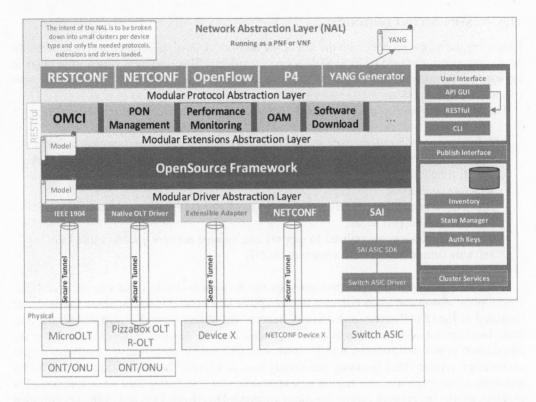

FIGURE 11.12 Block diagram of the network abstraction layer.

abstraction we can introduce new physical network devices more rapidly, reducing both development costs and time to market. The abstraction layer is intended to be open and to be developed collaboratively to gain industry adoption.

The overall network abstraction layer is composed of a common framework that ties together the three key abstraction layer components: protocol abstraction layer, extensions abstraction layer, and driver abstraction layer. Figure 11.12 shows the block diagram of the network abstraction layer.

The protocol abstraction layer provides the northbound interfaces to the various management and control systems through use of standard protocols. The framework should allow for the underlying protocols to be modular and extensible so that new or upgraded protocols can be introduced without modifications to the underlying network hardware. The extension abstraction layer provides an extension point to interface with disaggregated device features or scriptable device functions. The extension abstraction layer would utilize a standard RESTful interface and can run either as part of the network abstraction layer or in a separate virtual machine or container. Extensions will follow a model-driven architecture to simplify the integration of new capabilities and feature sets.

The driver abstraction layer uses a model-driven architecture to relate the device driver SDK[*] to a common framework and feature set. Drivers will follow a model-driven architecture to simplify the integration of future hardware. AT&T plans to pursue standardization of protocols that will ultimately simplify the driver model and create optimizations in the network abstraction layer.

As seen in the previous sections, the technology of the access network is evolving and will continue to evolve. The components of the network abstraction layer are designed to allow higher-layer SDN and application functions to be written independently of specific network access functions and hardware, enabling common software to support multiple SOC suppliers, multiple access protocols, and multiple access technologies.

---

[*] Software development kit.

11.4.9   SDN Access Controllers

Integrating the access network with the rest of the network cloud architecture requires access software components managed by local SDN access controllers. This integrates with a wide range of software components such as the ONAP subsystems (see Chapter 7) that also integrate with other components such as OSSs.

Three types of controllers are utilized in the network cloud architecture, as discussed in Chapter 7:

- *Infrastructure controllers* create and maintain underlying software environment by instantiating and managing virtual machines or containers used to support the access network. The distributed nature of the access network may require extended configurations of infrastructure controllers.
- *Application controllers* are utilized to support and manage required applications, including those that support access.
- *Network controllers* are utilized to support and manage network elements that form the software defined network infrastructure (SDNI).

There are two major functions performed by the network controllers that support access: the control plane management function, and the data plane flow control function. These functions are separated in the ONAP access architecture. Because the access network is distributed, the data plane functions for some access network components must be distributed, which can require the introduction of new components. AT&T's work with OCP and leading industry hardware suppliers for a new type of vOLT hardware has already been described. The distributed nature of access technology may potentially also require new approaches for distributing data plane functions. For example, distributed network control functions are needed to provide QoS and traffic prioritization to meet performance measures for shared media access and concentration points in the access network, or to dynamically reroute traffic in the event of failures.

Control plane functions of the access network controller can also be complex. One of the control plane functions is supplying event data into the ONAP DCAE system. In some cases, access event data are pushed up through the SDN controller to the data collection systems, while in other cases, access events may be pushed directly to the data collection systems by the network elements where low-latency real-time processing is critical.

11.4.10   Open Access Network Software

Just as open hardware specifications play a key role in achieving the benefits of the new architecture, open interfaces and open source software also plays a key role. Open interfaces are a critical component of the access network because they enable interoperability across multiple types of hardware, as well as between multiple hardware and software suppliers. As the new access architecture disaggregates software functions, open source software will be a key component. Open source software will provide multiple benefits, including the following:

- Community contributions for features and enhancements
- Elimination of duplicated development efforts, thus decreasing industry costs
- Common reference software for silicon and hardware developers
- Common software across a variety of suppliers to optimize interworking and integration functions

Some key open software components for the access network will include

- Industry standardization of YANG models for access network elements
- NETCONF for management communication of the network devices

- Protocols for control plane communication between the access device and SDN controllers
- REST for communication between access-related applications

It should be noted that the use of open interfaces and open source software do not necessarily imply specific commercial or support models. However, it will be critical to ensure that modifications and customizations of key software elements and models are incorporated back into source code to avoid the creation of splinter distributions that would hamper interoperability and adaptability as access network technologies continue to evolve.

NETCONF and YANG provide a good example of open software approaches. Consistent use of NETCONF and YANG for the management of network devices and the specification of data models plays a critical role in driving interoperability between access technologies and network elements from multiple suppliers.

YANG is a data modeling language used to model configuration and state data. These data are manipulated by the network configuration (NETCONF) protocol, using NETCONF remote procedure calls and NETCONF notifications. NETCONF and YANG, in and of themselves, provide a strong technology base for driving simpler, more effective, and more robust configuration management. Additional flexibility, adaptability, and cost savings will be achieved by the definition of open and standard YANG configuration data models for various types of access network elements.

The BBF has been actively working to standardize data models for access networks [6]. There are common YANG models (WT-383) that can be used across all technologies, and also specific YANG data models for standard access technologies such as G.fast (WT-355), G.hn (WT-374), and PON (WT-368). Use of these models by network access equipment, as well as continued extension and development of these models as new capabilities and access technologies are defined, will be essential for achieving the benefits of the new access architecture.

11.5   MOBILE WIRELESS ACCESS TECHNOLOGIES

Most of the approaches involved in applying the new SDN- and NFV-based access architecture to wireline access networks provide a foundation to apply the same architecture to mobile wireless access networks. The need for distributed compute resources and for specialized hardware with access SOC components carries over to mobile wireless networks. In addition, mobile wireless access needs other capabilities.

The needs of mobile wireless network nodes, such as today's 4G-LTE network and future 5G networks, extend beyond the needs of SOC-based access nodes. These extensions include the need for specialized acceleration hardware for some types of nodes, and the need to support a variety of configurations with different divisions of functionality between distributed local nodes and centralized nodes. These added requirements come from the support of a range of architectures needed to provide the most efficient deployment configurations based on the density of traffic and the variety of types of transport economically available in various parts of the wireless network.

Just as GPON access technology is evolving, wireless access technology and standards are also evolving. The 4G-LTE network continues to evolve and add new capabilities with every new release of specifications from the Third Generation Partnership Project (3GPP) [7]. Standards and development work is also ongoing for 5G, the next stage of wireless technology evolution. 5G includes a new radio interface technology called "Next Radio," generally designated as 5G-NR.

Early experiments and developments are already starting to apply SDN and NFV techniques to LTE radio access network (RAN) equipment. This early work generally takes a somewhat simplified approach by applying SDN and NFV only to the higher-layer switching and routing functions of the RAN nodes. The 5G network will be the target for a more comprehensive application of the SDN and NFV network cloud architecture. While it may seem obvious that the reason for this is that 5G represents a new technology point and therefore provide the opportunity for a new architecture, there is actually a deeper motivation for SDN and NFV in the 5G network architecture. The

5G network represents an evolution step that is different from previous evolutions from 2G to 3G to 4G-LTE. Because of this fundamental difference, achieving the promise of 5G technology will require the incorporation of SDN and NFV network cloud technology to achieve its full potential.

In the next section of this chapter, we describe three current LTE RAN configurations, and show how each configuration could use NFV/SDN technology.

11.5.1 LTE RAN CONFIGURATIONS

The basic structure of the LTE RAN for conventional early LTE deployments followed the distributed RAN (or dRAN) architecture. This architecture is similar to the architecture of earlier generations of RAN technologies, and is shown in Figure 11.13. In this architecture, all of the components of the RAN node are located at the cell site, including the radio frequency (RF) components (such as the antenna and the radio amplifier) and the baseband unit (BBU), which does the digital processing of the radio signals. The RF and antenna are typically located at the top of the cell tower, and the BBU is typically located at the base of the cell tower. The BBU and RF are interconnected by a short fiber interface that typically uses a protocol such as Common Public Radio Interface (CPRI) designed to operate for relatively short distances over a dark fiber [8]. The BBU then has a backhaul connection to the core network, typically using a high-speed Ethernet link.

Applying NFV and SDN techniques to the dRAN architecture is very challenging, because cell sites located at cell towers are distributed geographically. New deployments of cell sites are also shifting to small cells, which are even more distributed. In typical configurations, small cells may be self-contained units mounted on utility poles, light poles, or on the ceilings inside buildings.

In situations where the density of traffic and cell sites are very high, and where dark fiber transport is available to cell sites, network operators have been deploying a centralized RAN (cRAN) architecture, shown in Figure 11.14. In this architecture, only the RF equipment is located at the cell site. A collection of BBUs for a number of cell sites within a small distance can be located together at a central site, referred to as baseband pooling. The pool of BBUs at this site has a dark fiber carrying CPRI that runs from the centralized location to each cell site that is homed on that particular centralized site. There are several advantages to this configuration. The centralized pool of BBUs can provide reduced lease and operations costs. In some configurations BBU capacity can be configured to be shared across multiple cell sites, reducing capital costs. The BBUs can also be easily interconnected with high-speed and low latency links, which can facilitate some advanced features that help improve the performance of the RAN network by coordinating between the different cell sites. The

FIGURE 11.13   Block diagram of the dRAN architecture.

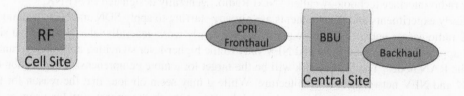

FIGURE 11.14   Block diagram of the cRAN architecture.

main disadvantage of this architecture is the need for dark fiber transport from the central site to each cell site, carrying the inefficient CPRI protocol. In many situations, this transport may not be available or cost effective, when compared with using Ethernet transport to connect with the cell sites.

The cRAN architecture provides more opportunities for applying SDN and NFV techniques. The central baseband pooling site consists of a collection of interconnected BBUs, with each BBU containing specialized DSP hardware and firmware, as well as general purpose processing and networking hardware and software. This provides an opportunity to implement the functions at the central baseband pooling site by disaggregating the functions of the BBU, and using the general architectural approach shown in Figure 11.2. In this architecture, the functions normally executed by the specialized DSP hardware are virtualized and executed on the specialized hardware component of the extended AIC architecture, and the other functions are executed in the general purpose servers and networking components.

Because of the distance and cost restrictions of the CPRI link, the cRAN architecture can only be effectively deployed in selected situations. Even in situations where the architecture is cost effective, the distance and cost restrictions limit the number of cell sites that can be homed on a single central baseband pooling site. This in turn limits the size of the gains that can be obtained from NFV and SDN and the network cloud architecture.

A more recent RAN architecture development is the split RAN (sRAN) architecture, as shown in Figure 11.15. This architecture is being developed to combine the advantages of the cRAN architecture with the cost-effective transport of the dRAN architecture. At the current time, this architecture is not yet in widespread deployment—it is at the prototype and development stage.

In the sRAN configuration, the hardware and functions of the BBU are decomposed into two separate components shown as BBU part 1 located at the cell site and BBU part 2 located at the central site. The lower layer protocol functions, which are the most sensitive to delay, are located in the part of the BBU located at the cell site, with a CPRI interface running up the cell tower to the RF components. These functions at the cell tower also greatly reduce the data transport requirements, so that the link between BBU part 1 at the cell site and BBU part 2 at the central site can be an efficient low latency Ethernet link instead of the CPRI link. These architectures are under development and evaluation. A key characteristic of these architectures is that the BBU functions must be split between the cell site and the central site. Each function of the BBU needs to be assigned to one of the locations. This leads to a wide variety of feasible split points in the RAN architecture. In general, when more functions are located at the cell site the speed and latency demands placed on the transport network are reduced, but the gains that can be provided by baseband pooling are also reduced. This means that the optimal split point in the architecture depends on the specific network and transport conditions in the network. Developing and deploying a sRAN architecture using conventional technologies is difficult because there is not necessarily a single optimal split point for all configurations, and because (unlike the cRAN architecture) a flexible sRAN architecture generally requires changes to the BBU hardware and software.

While the cRAN architecture can take advantage of SDN and NFV, the sRAN architecture *thrives* on it. The sRAN architecture generally will require a hardware redesign, so the move to the extended AIC architecture shown in Figure 11.2 can be a natural approach. The sRAN architecture requires disaggregating the software components of the BBU, which also aligns well with the transition to SDN and NFV. Most importantly, the optimal "split point" for the sRAN architecture varies

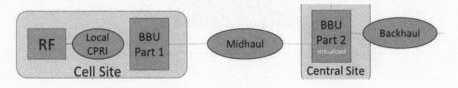

FIGURE 11.15   Block diagram of the sRAN architecture.

with specific traffic and network conditions, which makes legacy sRAN implementations difficult. In the new extended architecture, all of the BBU functions are disaggregated and virtualized, and the ONAP system will be able to instantiate and manage the virtualized functions as needed to optimize cost and performance. With the new extended AIC architecture, the split architecture is not limited to two types of locations as shown in Figure 11.15. As an oversimplified example, layer 1 functions could be implemented at the cell site in the figure, layer 2 functions at the central site, and layer 3 functions in a large data center site, potentially collocated or integrated with virtualized mobile core network functions (discussed in Chapter 14).

As the next section will show, 5G technologies and the 5G architecture will, in general, look more like the sRAN architecture, making them a strong fit for NFV and SDN implementation. But the advantages that this new architecture will provide to 5G go beyond the architecture. The key characteristic of 5G—the characteristic that will makes it much more than just another "G" in the succession of wireless technologies—will be directly enabled, empowered, and enhanced by the new NFV- and SDN-based architecture.

11.5.2 5G WIRELESS

As wireless technology moved from 2G to 3G to 4G, each generation was "faster and better" than the last, bringing significantly higher speeds and capacities, along with lower costs per bit. 5G will continue this trend and reach higher speeds and capacities, a feature of 5G known as enhanced mobile broadband (eMBB). But 5G will also break from this trend. The same 5G network that offers faster and better will simultaneously offer "slower and longer," by enabling a vastly larger number of inexpensive ubiquitous devices with low data rates and volumes and very long battery lives, as a part of the Internet of things (IoT). This is a feature of 5G known as massive machine type communication (massive MTC). The same 5G network will also simultaneously offer "surer and quicker" services with ultrahigh reliability and ultra-low delays, also known as critical machine type communication (critical MTC). This is what makes 5G more than just another G: it does not expand in one dimension; it expands simultaneously in multiple dimensions, offering a wider range of services on a single network. Figure 11.16 illustrates how 5G expands the dynamic range of wireless technology simultaneously in these three different directions.

As a result of these additional capabilities, 5G will enable many new classes of applications such as high-quality augmented reality, the IoT, autonomous vehicles, and industrial automation.

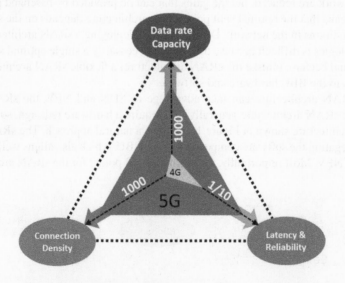

FIGURE 11.16   5G expands 4G capabilities in multiple dimensions.

All of these capabilities, and new combinations of these capabilities, will be supported on a single network. This not only brings about economies of scale, but also opens up new avenues for growth and supports increased revenue for network operators.

Different classes of applications are enabled by the three types of capabilities of 5G, based on their requirements:

- eMBB: These applications encompass all sorts of ultra-broadband services such 4 K video, augmented reality and tactile Internet. These applications typically require very high bandwidth and reasonably low latency. Throughput in Gb/s is targeted.
- Massive MTC: These applications represent a broad general category that includes a wide range of connected devices such as meters, environmental sensors, biological sensors, home security systems, appliances, and industrial monitors, to name just a few. Some envision that in the era of 5G almost everything will be connected to the Internet via an umbrella of networks. Some forecasts project that in the not-too-distant future the number of such devices worldwide could be in the tens of billions. While most of the devices for these applications are not bandwidth-intensive, they require coverage depth and the ability to support a battery life of up to 10 years.
- Critical MTC: This category of applications includes machine-to-machine communications that require ultra-low latency and extreme reliability. Examples of such applications include vehicle-to-vehicle communications for collision avoidance systems, industrial automation, and robotics.

In order to support this diversity of requirements, 5G will also support a range of new technologies, including technology to support a broad spectrum range. Much of the press regarding 5G emphasizes the use of centimeter and millimeter wave frequencies (largely above 24 GHz). These bands are new to mobile wireless networks, and will play an important role in 5G, especially for eMBB services. In the past, these very high-frequency bands have been off limits for mobile networks due to their poor propagation characteristics as well as the technical difficulty of developing low-cost transceivers that can be battery operated and fit in the form factor of a handset. Fortunately, recent developments in various technologies such as phased array antennas, advanced digital signal-processing capabilities, and mmWave radio frequency integrated circuits (RFICs) have made it possible to use these high-frequency bands as a part of the 5G network.

At the same time, 5G will also use licensed and unlicensed spectrum below 6 GHz. These bands will play an important role providing the coverage needed for massive MTC and the reliability needed for critical MTC. Hence, to realize the full capabilities of a 5G network will require the new mmWave bands in combination with more conventional spectrum bands. 5G will implement a range of new technologies to support these new capabilities and the new mmWave spectrum band. These are described below.

11.5.2.1 Advanced Multielement Antenna Structures

Figure 11.17 illustrates the concept of a phased array antenna. This is an essential technology needed in mmWave to overcome the propagation challenges. Conventional LTE systems typically use 2–4 antenna elements with independent signals in a multiple input multiple output (MIMO) configuration. Full dimension MIMO (FD-MIMO) extends this to tens of antenna elements, using techniques that are most effective in TDD spectrum bands. 5G will increase the effectiveness of FD-MIMO in FDD bands below 6 GHz, improving capacity and efficiency in sub 6 GHz spectrum. Due to the small size of antenna elements at mmWave frequencies, hundreds of antenna elements can fit in a reasonably sized antenna structure. This allows the system to compensate for propagation loss through beamforming, which can concentrate specific transmissions aimed at the location of each device. Because each beam can carry a different signal in the same frequency, this technology also enables a significant increase in capacity.

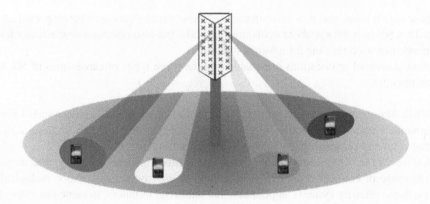

FIGURE 11.17 Phased array antenna with beam forming.

With the very high bandwidths used in the mmWave frequency band, and the very high number of antenna elements used in both phased array and FD-MIMO antenna structures, the use of CPRI in a cRAN structure becomes impractical because of the inefficiency of the protocol. As a result, the cRAN architecture in Figure 11.14 becomes impractical. To take full advantage of the large number of antenna elements in mmWave and FD-MIMO antennas, the advanced antenna arrays in 5G will need to execute some of the BBU-like functions in the antennas. This will make the overall configuration more analogous to the sRAN architecture in Figure 11.15, which is a strong match for implementation in the extended AIC architecture.

11.5.2.2 Dual Connectivity

Another characteristic of mmWave spectrum is that these signals are far more easily blocked by obstacles, so that turning a hand-held device from one position to another could rapidly block the mmWave signal. To overcome constant interruptions while using mmWave spectrum, 5G will use a technology called dual connectivity, shown in Figure 11.18. This allows a device to be simultaneously connected to multiple parts of a 5G network, or even to be simultaneously connected to a 5G network and an LTE network. For example, a connection using a traditional cellular frequency would provide a robust and reliable anchor for services and signaling. A second simultaneous connection to a 5G mmWave frequency can then be used opportunistically to provide high speed and high capacity. This ability of the network to simultaneously use multiple connections between a device and different components of the network is an essential aspect of the 5G network design.

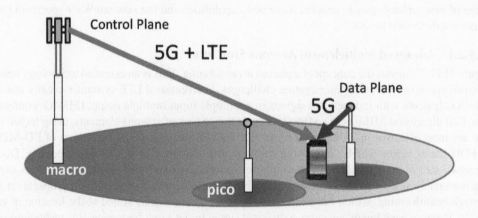

FIGURE 11.18 Dual connectivity.

Dual connectivity can benefit significantly from an implementation based on the new architecture. The separation of control plane and data plane functions aligns with the use of different frequency bands and even radio technology generations (LTE vs. 5G) for control and data. Also, SDN is ideally suited to handle the dynamic switching of traffic between different sites and parts of the network, while balancing traffic needs and priorities across multiple sites and technologies.

11.5.2.3   Ultrahigh Density and Self-Backhaul

Small cells are a very important aspect of the 5G network. The use of mmWave frequencies requires small cells because of their short range. Also, small cells are one of the most effective ways to improve the "capacity per area" of the network, an important 5G objective to handle the accelerating demand for more data. One of the biggest challenges in creating a dense network of small cells is the need for a dense transport network to support backhaul from the small cells. In 5G, each small cell will require backhaul speeds of tens of Gb/s. Self-backhaul is a technique that allows the 5G access radio to simultaneously provide access to users, and to make backhaul connections to access radios at other 5G cell sites, using the same spectrum. This can be a major factor in reducing the cost of deploying a very dense network of small cells, since fiber-based transport would not need to be deployed to each small cell location. Self-backhauling is a natural choice in the mmWave band. As shown in Figure 11.19, the beamforming phased array antenna panels can create separate beams dedicated to backhaul with minimal interference to the access beams used to reach customers devices.

The NFV and SDN capabilities of the new access architecture provide an excellent platform for implementing the self-backhaul capability. SDN can provide rapidly adaptable network configurations across a dynamic web of self-backhaul links that can adjust to changes in connectivity, and NFV can adjust resources between various access and backhaul links as needed to optimally handle changes in traffic patterns.

11.5.2.4   Flexible Carrier Configuration

The key characteristic of 5G is its ability to support multiple types of traffic on a single network. Traffic types as diverse as eMBB, massive MTC, and critical MTC create very different types of demands on the resources of the radio network. In the past, efficiently supporting these traffic types would require separate independent radio networks and carriers. In order to support these diverse traffic types on a single network, 5G requires a fundamental change in approach to the 5G radio carrier waveform. In earlier generations of radio technology, a single carrier used a single consistent subcarrier structure across the entire carrier. 5G breaks from that approach, and enables a single carrier to simultaneously support different types of separate self-contained subcarriers, as shown in Figure 11.20. As an example, eMBB traffic can use a subcarrier optimized for the highly efficient transfer of large amounts of data, using tight synchronization, automatic ranging, and coding optimized for maximum spectral efficiency. At the same time, Massive IOT traffic can use a

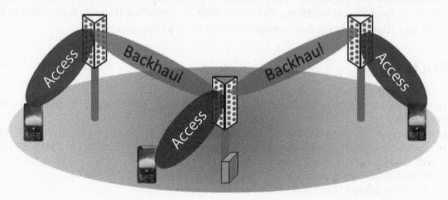

FIGURE 11.19   5G self backhaul.

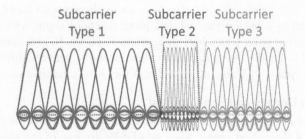

Subcarrier Type 1   Subcarrier Type 2   Subcarrier Type 3

FIGURE 11.20   5G example subcarrier structure.

subcarrier structure with narrow bandwidth carriers to enable efficient short information transfers and low peak power while maintaining high coverage, with asynchronous access techniques for low protocol complexity and signaling overhead to allow low-cost device implementations and long battery life. The amount of resources devoted to each type of subcarrier could be dynamically customized for each time and place to align with the traffic demands. Because the subcarriers are self-contained, 5G also has the potential to evolve to add new subcarrier types in the future optimized for new kinds of traffic and applications.

The new architecture with the extended AIC provides an ideal approach to realize the flexibility of the 5G carrier structure. The resources of a flexible specialized DSP processing pool can be allocated to follow dynamic adjustments to the different types of subcarriers. Network routing protocols and paths can be allocated and optimized for each type of traffic. Software updates to existing service and subcarriers can be isolated from changes to other services and subcarriers. New subcarrier types and new services can be added through software updates that include changes in the algorithms executed by the DSP resource pool.

Overall, the 5G access network design is revolutionary because of its flexibility and dynamic range. When 5G is first introduced it may not support the full range of capabilities described above, but 5G technology will evolve to incorporate all of these capabilities and more. The full 5G architecture is capable of supporting a wide range of services, dynamically allocating resources among services and spectrum bands, arranging and adapting transport configurations as conditions change, simultaneously connecting to devices across different generations of technology, and evolving to support entirely new classes of service in the future.

ACKNOWLEDGMENTS

The author would like to acknowledge the contributions to this chapter from colleagues in the Access Architecture and Analytics organization. In particular, many thanks to Arun Ghosh for contributions to the wireless sections, and Eddy Barker, Sumithra Bhojan, Aaron Byrd, Blaine McDonnell, and Tom Moore for their contributions to the wireline sections.

REFERENCES

1. https://www.fsan.org/task-groups/ngpon/
2. http://www.5gamericas.org/files/2214/7257/3276/Final_Mobile_Broadband_Transformation_Rsavy_whitepaper.pdf
3. http://www.5gamericas.org/files/3214/8833/1313/3GPP_Rel_13_15_Final_to_Upload_2.28.17_AB.pdf
4. http://www.semiconductors.org/clientuploads/Research_Technology/ITRS/2015/0_2015%20ITRS%202.0%20Executive%20Report%20(1).pdf
5. http://www.khronos.org/opencl/
6. https://www.broadband-forum.org/
7. http://www.3gpp.org/
8. http://www.cpri.info/

12 Network Edge

Ken Duell and Chris Chase

CONTENTS

12.1 Introduction 249
12.2 Edge Core Paradigm...249
12.3 Traditional Edge Platforms..250
    12.3.1 Vertically Integrated Edge Platforms 250
    12.3.2 Edge Application 252
12.4 Network Cloud Edge Platforms..253
    12.4.1 Disaggregated Edge Platforms 253
    12.4.2 Network Fabric 253
    12.4.3 Edge vPE VNF..256
12.5 EVPN: Flexible Access Grooming and Universal Cloud Overlay 258
    12.5.1 Access Scale and Resiliency..259
    12.5.2 Cloud Overlay Connectivity 260
    12.5.3 EVPN..260
12.6 Future Evolution of Network Edge 261
    12.6.1 Open Packet Processors...262
    12.6.2 Open Configuration and Programing of Packet Processors 262
    12.6.3 Open Control of Packet Processors 263
References...264

12.1 INTRODUCTION

The network edge provides layer 2 and layer 3 service functions reached by the wired and wireless access networks. In the canonical reference model, the network edge platform (aka provider edge [PE]) sits between the many types of customer access networks (Chapter 11) and core networks (Chapter 13) and provides service processing, operations, administration and management (OA&M), and traffic aggregation functions (Figure 12.1). In this chapter, we describe the drivers, both business and technical, for the recent rapid evolution of packet edge platforms. In many existing networks, Ethernet is the base link packet technology for service PE networks because Ethernet VLANs are the fundamental unit for grooming and aggregation from the access technologies. Current merchant switches and routers provide very dense and inexpensive Ethernet interfaces to build a high-capacity IP Ethernet fabric to flexibly join access and core through the virtual network function (VNF) service applications. This chapter covers this transition in detail and explores where the boundary between software and hardware implementation can be drawn, so as to optimize the performance, yet maintain maximum flexibility.

12.2 EDGE CORE PARADIGM

Packet edge platforms can be effectively framed in the context of the edge-core paradigm (Figure 12.1), which defines where packet processing function should reside in a network. In this paradigm, the core is purposely kept simple to provide high-speed reliable packet transport (e.g., removing the Internet route table and removing edge label switching functions). In addition to simplicity, the core network generally is not changed by an operations support system (OSS) on a per customer service

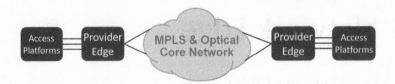

FIGURE 12.1   Edge, core networking model.

order. In contrast to the core, the edge is where the primary packet processing and policy needed for various packet services are implemented and edge platforms are often reconfigured by an OSS per customer service order.

In many ways, the overlay–underlay paradigm for data center network fabrics and servers match the principle of the edge-core paradigm in that the network fabric acts as the "core" and the router services or gateway functions running on servers act as the "edge."

The packet edge paradigm provides important solutions to efficiently manage the nature of service provider packet traffic. Service provider networks fundamentally differ from Web 2.0 data center networks in two major ways—the nature of the traffic and the geographic footprint.

The majority of traffic in a service provider's network is stochastic in that a network operator does not control when and how much bandwidth an end customer consumes. For example, the service provider cannot inform a customer that they have to wait until their neighbor finishes streaming a video before they can begin watching their video because of a capacity bottleneck. This necessitates that the network capacity be sized with enough overhead capacity to handle the peak load, not unlike the electrical grid.

Data center networks on the other hand have traffic patterns that are a mix of stochastic and deterministic traffic. The deterministic traffic can be calendared, or scheduled in time and with a specific number of bytes controlled at known and variable bytes per second, such as synching storage disk arrays. This allows, with proper closed-loop control of the deterministic traffic, very high utilizations in data center networks where the capacity headroom is monitored and used for deterministic traffic.

To maximize network efficiency, AT&T has evolved to have a single common core supporting multiple edge platforms. Because different services experience peak traffic at different times, having a shared common core provides a fungible capacity pool to handle traffic peaks for all services on a single network. In addition, a single pool of capacity is more future proof, from an investment standpoint. If there were two or more core networks, there is always the risk that core A capacity augment investment is misdirected when the demand growth later appears on core B instead of core A, thus stranding that augment investment.

The second way a service provider's network differs from data center networks is footprint. Service provider network buildings typically number in the several thousand across a large geographic area, whereas Web 2.0 data center networks number in the tens to hundreds. This is by design, so as to shorten the reach to the customer, whether by wireline or wireless access.

It is the nature of the traffic and the large edge footprint that drives the design and implementation of packet edge platforms today, and their evolution tomorrow.

12.3   TRADITIONAL EDGE PLATFORMS

12.3.1   Vertically Integrated Edge Platforms

For the past two decades, edge platforms have been a vertically integrated combination of a chassis, silicon (packet processing, fabric switch, CPUs), and software. The network operator had limited visibility into the inner workings of any particular platform (aka blackbox or opaque). Instead, equipment users and suppliers coalesced around an open standards model (IETF, IEEE, etc.) where the behavior of the platform was functionally described so that equipment could interoperate between service providers and between different equipment suppliers.

While the open standard model did work and enabled delivery of many sophisticated packet services to meet the ever-growing customer needs, it had three major issues. The first issue was time-to-market. It generally took a long time for multiple competing ecosystem players to first converge on a new standard and then it took even longer for multiple ecosystem suppliers to consistently implement the standard on multiple vertical platforms. The second issue was upgradeability. Often times when a new standard was created it was impossible to implement this new capability on existing already-deployed hardware and the new standard required hardware forklifts. The third issue was ambiguity. Despite best efforts by the standards committees, full interoperability has always been a difficult goal that is only achieved after multiple rounds of pairwise testing and bug fixes. This pairwise integration resulted in significant testing cost and lost time to network operators.

While the core network supports multiple services with a common pool of packet capacity, edge routers and switches generally supported one or two services per platform driven in large part by fundamental limits of vertically integrated packet edge platforms.

In practice, network edge platforms are limited or constrained in four dimensions: *port*—constrained is the number of physical ports facing access and the customer on the limited real estate of a chassis line card system; *logical*—constrained pertains to the size of tables that are constrained by memory size; *CPU*—constrained for processing signaling messages and basic router functions, and network pipeline; and *throughput*—constrained maximum Gigabits per second. For edge routers port, logical, and CPU constraints typically surface long in advance of pipeline throughput constraints. As a result, strategies such as using Ethernet multiplexing with low-cost Ethernet switches in front of the network edge platform to overcome port constraints and aggregation routers to combine traffic from multiple network edge routers to increase utilization prior to long-distance transport become necessary (Figure 12.2).

In addition, strategies of periodic fork-lifting CPU router cards and entire routers when memory limits are hit were a necessary evil that network operators had to manage. Frequently, routers had lowest common denominator issues where a new card may have new silicon, software, and features but if there was an older generation card in the router, the router only supported the lowest common denominator set of features. Managing many generations of hardware necessitated costly and time-consuming truck rolls over a large footprint to evolve the packet edge infrastructure. Failure to evolve had, at times, catastrophic headline events when the Internet route table exceeded the hardware limits of one popular early router (Lemos 2014).

For a period of time, there was an industry push to develop a single common packet edge platform for all services. However, this generally did not succeed in practice at scale because it led to service feature compromises and complex and difficult to predict interactions between how multiple services consumed logical resources.

As the speed of innovation increased, the flaws of blackbox edge platforms became acute and necessitated a fundamental rethinking of packet edge platforms, which we address in the next section. The next section gives examples of edge platform applications in typical service provider networks.

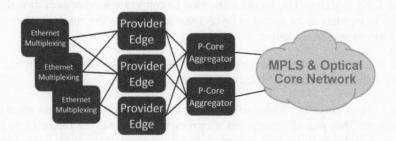

FIGURE 12.2 Overcoming constraints with Ethernet multiplexing and core aggregation routers.

12.3.2 EDGE APPLICATION

As the Internet grew in the late 1990s, large IP router backbone networks were built. Independently enterprise private data services were implemented on frame relay and ATM networks. These separate networks provided virtual circuit services as private line replacements and offered Ethernet transparent LAN bridging services.

About 1998, a new service-prover-based IP VPN service was introduced, implemented with the new Multi-Protocol Label Switching (MPLS) technology. MPLS also soon became an infrastructure technology for implementing Ethernet virtual private LAN and point-to-point psuedowire services. At the same time, the Internet backbones of most large ISPs were moved to MPLS implementations. In the early 2000s, the Internet and private data services were merged onto a common MPLS infrastructure. In the mid-2000s, Ethernet started to become the transport technology for both access and core trunking, supplanting the synchronous optical networking (SONET) technologies.

On this common IP MPLS technology built on Ethernet links, various services were implemented in the network edge where access and core met. These edges were the PE routers or service gateways that created public and private layer 2 and layer 3 services. A nonexhaustive list of edge network services on these service provider networks as are follows:

- Internet broadband service: Implemented on Broadband Network Gateways (BNGs, also called Broadband Remote Access Server). This provides the ubiquitous broadband consumer and small business Internet service. The BNG implements per subscriber IP address allocation and routing, security filters, class of service policy enforcement, and usage statistics. In the service provider space, these have been accessed by various types of DSL technologies and more recently passive optical network (PON) technologies.
- Enterprise dedicated Internet service: For large enterprises, PE routers provide similar functions to the BNG. They provide per customer IP address allocation and routing, security filters, denial of service (DOS) protection, and access connectivity monitoring.
- Internet peering: Implemented on large-edge IP routers that exchange Internet routes with other Internet transit providers. Internet peering is what creates the global Internet reachability across all the ISPs. These peering routers implement various security filters and route policies that protect the ISP infrastructure and the Internet at large.
- Service-provider-based IP VPN service: Implemented on IP VPN PE routers. This service provides private IP networks for interconnecting sites using IP routing. This is used both by enterprises to create private IP intranets and internally for service provider infrastructure purposes. While the service is implemented over a common network infrastructure, the isolation of VPNs is achieved via isolated virtual routing and forwarding instances in the PE routers and through MPLS in the core. In addition to providing a private IP routing exchange among VPN sites, the PE also enforces class of service traffic policies for each VPN site.
- Metro Ethernet service: Implemented on an Ethernet services PE router. This provides for point-to-point Ethernet virtual circuits (also called psuedowires) and multipoint virtual private LAN bridging. The services are used to connect customer sites directly to each other or to partners or as access to layer 3 services such as the enterprise Internet or IP VPN services mentioned above.

There are also a number of edge service functions that run between service networks rather than between access and core. Some examples of these are

- *Internet tunnel gateways*: These terminate IP tunnels such as IPSEC into other services. For example, they provide enterprises with remote access over the Internet into their private intranets. In the WiFi services space they are used both for implementing remote WiFi controllers and for securing unprotected WiFi radios. For mobile cellular services, they are used for connecting small metro cells or customer premise-based small cells (sometimes called femtocells or picocells) back into the mobile enhanced packet core (EPC) via the Internet.

- *Network-based proxies and load balancers*: These devices may front service provider data centers that provide hosting services or they may sit in front of network applications like DNS or network time protocol (NTP). Network load balancers provide a virtual IP interface that can balance traffic across a number of host endpoints. Proxies will implement reverse HTTP proxies for load balancing or security purposes into a data center or CDN.
- *Service provider-based network address translation (NAT)*: Provide NAT functions between network services. These are used to extend IPv4 address space and interconnect private IP address space to the public Internet. In mobility networks, they sit between the user bearer IP traffic and the Internet. They are used in broadband services to extend the IPv4 address space. Also, they are used between private enterprise VPNs and public application service providers.
- *Network bonding service*: This is implemented in a router that provides IP routing, route filters, and NAT between enterprise VPNs and public application service provider networks such as cloud and vertical application service providers (e.g., Amazon Web Services, PeopleSoft, or Microsoft Azure cloud and Office 365).
- *Session border controllers for multimedia services*: These provide a firewall security function and a proxy function for external video and video over IP services to reach IP multimedia cores within the service provider network.
- *Network-based firewalls*: Enterprise firewall and intrusion detection functions that reside in the service provider network between the Internet and enterprise private networks such as MPLS VPNs.
- *Internet security scrubbers*: These are specialized devices used for filtering DOS traffic. Targeted traffic from Internet services is diverted to these devices for filtering before being returned to its original path.
- *Various mobility gateway functions*: For example, serving gateway (SGW) and packet delivery network gateway (PGW) provide mobility edge networking functions discussed in the mobility chapter.

Prior to SDN, all the various edge service function examples given above have been implemented in dedicated hardware. SDN provides an architecture where these become VNFs that can be implemented across a uniform, shared hardware infrastructure.

12.4 NETWORK CLOUD EDGE PLATFORMS

The previously described issues with vertical or blackbox packet edge platforms necessitated a fundamental rethinking of packet edge architectures.

12.4.1 DISAGGREGATED EDGE PLATFORMS

The new packet edge architecture is shown in Figures 12.3 and 12.4 using the elements of AIC as described in Chapter 4. In this architecture what was a vertically integrated packet edge is now disaggregated into three major modular subsystems: (1) network fabric hardware, (2) edge VNF service software running on servers, and (3) control software. This will be covered in more detailed in the following three subsections.

12.4.2 NETWORK FABRIC

The first major component in the disaggregated packet edge platform is the network fabric. The most common network fabric topology is the Clos architecture (Clos 1953). The premise of this

Central Office / Data Center

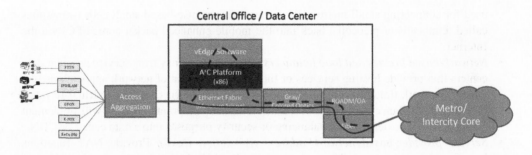

FIGURE 12.3 Disaggregated network function.

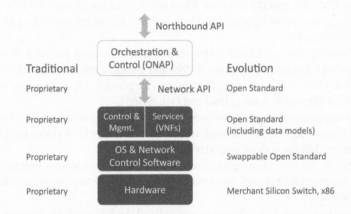

FIGURE 12.4 Moving to open standards for the network cloud.

architecture is to optimally synthesize very large switches by using multiple stages of smaller switch components. This architecture is named after Charles Clos, a Bell Labs researcher, who in 1952 established the mathematical theory behind optimal multistage switches. It has been used in traditional voice, private line, frame relay, ATM, Ethernet switches, and IP routers. Clos is also the primary architecture used in modern intra-data center switching to interconnect servers/storage and is central to the concept of the "scale out," or pay as you grow, networking strategy to scale existing network fabrics to achieve higher packet throughput requirements over time. While Clos is used for intra-platform and intra-data-center, it is not well adapted to inter-data-center or inter-central-office due to outside plant fiber topologies and layer 1 wavelength cost structures where partial mesh topologies are more cost efficient.

Evolving hardware ecosystem: Over a period of about 5 years, the cost of packet switching has dropped an order of magnitude, from $10,000 per 10GigE port in 2010 to $100 per 10GigE port in 2015 and these costs continue to drop. This is very material to a network operator because customer bandwidth continues to grow exponentially while the revenue associated with this bandwidth changes linearly.

In our observation, there are five main drivers for why there has been a three order of magnitude drop in cost. The first is *Moore's Law* and advances in silicon technology. The second is new *merchant silicon competitors* in the marketplace that aggressively compete against custom silicon offerings from traditional original equipment manufacturer (OEM) suppliers. The third is *simplification* in switching features provided by merchant silicon entrants that result in increased packet switching throughput. The fourth is *disaggregation* of vertical packet switch/router platforms from traditional OEMs into separable software and hardware modules. And, the fifth is *high volume commoditization* with merchant silicon by Web 2.0 companies leading to economies of scale.

The tremendous technological progress encapsulated in Moore's law is well documented elsewhere (Moore 2015). The other drivers deserve further comment as they shed light on architectural considerations discussed in this chapter.

Merchant silicon: Packet switching silicon can be classified into three broad categories based on a matrix of features versus throughput. *Custom silicon* is rich with features but has lower throughput per chip and has the highest cost per bit because of the number of chips required in a Clos architecture for equivalent throughput. Custom silicon for packet switching is only available by purchasing vertically integrated products from OEMs. At the opposite end of the spectrum is *data center merchant silicon* with very limited features or flexibility but with the highest throughput per chip and the lowest cost per bit. For many applications, a single system on a chip (SOC) suffices to attain target platform throughput. Broadcom Trident/Tomahawk is one family of SOC's that has been integrated into platforms for data center applications by dozens of OEM and original design manufacturer (ODM) suppliers.

In between there is *wide area network (WAN) merchant silicon*, which has a reduced feature set compared to custom silicon but still attains a cost per bit very close to data center merchant silicon. Broadcom Qumran/Jericho, Cavium Xpliant, and Barefoot Networks Tofino are examples of WAN merchant silicon.

WAN merchant silicon is important to Tier 1 network operators because it has critical features that are lacking in data center merchant silicon that are needed to (1) manage multiple services on the same platform (Ethernet and IP); (2) seamlessly connect legacy brownfield routers to new greenfield network fabrics; and (3) handle the bursty nature of customer traffic in the network.

Simplification: One of the reasons why Web 2.0 companies have so successfully driven down the cost per bit in packet switching infrastructure is because of a willingness to eliminate all but the most essential features required on switch silicon and this strategy has been very successful as evidenced by the success of data center merchant silicon in Web 2.0 data centers. We spent a year analyzing the ability of data center and WAN merchant silicon to meet our service requirements and determined that WAN merchant silicon in conjunction with VNF software can meet the requirements at a dramatically lower cost per bit compared to custom silicon. Current generation data center merchant silicon has too many requirement gaps to meet all of the varied packet service requirements, but this area is evolving rapidly with several merchant silicon entrants creating more flexible programmable pipelines with tremendous SOC throughputs. Given the rapid progress in technology, each new generation of merchant silicon will offer significant benefits, such as described in recent successful field trials (Making the Switch 2017).

Fabric disaggregation: Until recently, network fabric technology was a "blackbox." Blackbox is the industry term for a vertically integrated packet switching platform of hardware and software, with limited visibility to the inner workings often considered proprietary by OEM suppliers. Input/output behavior of the blackbox was defined by open standards (such as IETF and IEEE standards). This began to change with programs such as the Open Compute Project initiated by Facebook[*] and Google's Jupiter project (Singh et al. 2015) that have led to disaggregation of blackboxes into "whiteboxes" or "grayboxes" along three major dimensions—software from hardware, Clos from chassis, and third-party optics from branded optics.

While the disaggregation of software from hardware is covered later, the disaggregation of Clos from chassis is implemented by replacing the chassis switching backplane by optically cabling multiple identical small whiteboxes in a Clos topology. A chassis with a fixed backplane only allows for matching supplier line cards whereas, the elimination of a backplane has the benefit of avoiding vendor lock-in allowing pay as you grow scale out and homogeneous stock keeping unit's (SKU) at high volumes for economies of scale.

The third fabric disaggregation is facilitated by going to off-the-shelf optical components in place of branded versions from OEM suppliers. As the price per bit of merchant silicon has dropped

---

[*] www.opencompute.org.

dramatically, the price of optics, as a relative contribution to total cost, has become a more dominant factor. Optical plugs are generally developed by third parties. Historically, OEMs would only allow branded optics from third parties at added cost. Now with whiteboxes and grayboxes, optics can be procured separately from the box supplier and appropriate competitive bidding and new optics options are substantially driving down the cost.

High volume commoditization: Cloud data center standardization around x86 CPU hardware fostered a vibrant open software ecosystem. Similarly, for network edge, standardizing the hardware around switch merchant silicon could pave the way for a vibrant open software ecosystem. For a disaggregated ecosystem to operate efficiently, there needs to be a high volume of common underlying silicon or a robust open hardware abstraction layer that interfaces to any target silicon to maintain interoperability between many independent hardware and software players. To date, it has been the former model with Broadcom's Trident/Tomahawk providing the common silicon base for network operating system (NOS) software. But there are major efforts underway targeting the latter model with investments to create a robust open hardware abstraction layer (Section 12.7.2) so that software development investments can effectively compile and work on a large market of target hardware platforms. Hardware abstraction layers hold promise to enable multivendor silicon but this ecosystem is still immature.

"Whitebox" is an industry term for unbranded or unlabeled boxes that are provided by ODM or contract manufacturers that match an open specification and most commonly in a small form factor (1 or 2 rack unit—1RU or 2RU) using an SOC merchant silicon processor. By a chassis we mean closed and branded rack with a proprietary backplane to interconnect branded line cards that compose a router or switch.

The move from blackbox to whitebox by Web 2.0 companies has been tremendously successful in driving down the cost per bit for packet transport. However, adopting whiteboxes creates additional responsibilities. In a pure whitebox model, users are responsible for incremental lifecycle costs of hardware selection and development, ODM selection and development, integration, test, quality assurance, lifecycle management, and 7×24 platform support and maintenance.

A more refined view is emerging that falls between blackbox and whitebox known as "graybox." This optimizes the level of disaggregation based on platform volumes and total cost of ownership (TCO) analysis. There are two emerging graybox models. Model 1, which became commonplace in 2015, is where the platform is purchased from a third-party integrator that provides an ODM merchant silicon whitebox plus software plus support and maintenance for a fee above and beyond the ODM whitebox purchase price. Model 2, which gained some momentum in 2016, where an OEM sells the platform without a NOS but with a simple bootstrap loader such as open network install environment (ONIE) or preboot execution environment (PXE). In this model, the OEM still supports lifecycle management of the hardware but the NOS is disaggregated.

12.4.3 Edge vPE VNF

The virtual provider edge (vPE) VNF software that runs on standard cloud hardware is the second major component in the disaggregated packet edge platform. The purpose of this VNF is to provide the packet edge processing functions for all the various edge applications described in Section 12.3.2. These functions cannot typically be done by merchant silicon because of their reduced functionality support.

One starting requirement in new installations of disaggregated routers is that the packet service specification be at parity and fully interoperate with preexisting vertical routers in production. This is for the simple reason that for a new technology to be successful it must be easily adopted by the end customer and their existing networks. Once it is adopted at par, then it becomes easy to evolve the service with new value-added capabilities. As noted earlier, merchant silicon traded reduced functionality to significantly lower cost and to increase throughput when compared to custom silicon on traditional routers. These functional gaps mean that merchant silicon platforms cannot replace traditional routers. To fill these gaps, we turn to VNF software.

There are many evolving architectures for vPE software. vPE software, in general, can be disaggregated into five major modules (Figure 12.5) of input/output (IO) management, packet pipeline, data plane management, control plane management, and vPE OSS. The IO manager processes packets from the server network interface controller (NIC) card into the packet pipeline and includes MPLS/VLAN label management functions. One or more packet pipelines perform the match action processing of packets as well as scheduling functions such as class of services. The data plane manager determines how to direct packet flows to which pipeline. The control plane manager serves as a local autonomous control plane and responds in real time to BGP, bidirectional forwarding detection (BFD), etc. messages. And the OSS manages configuration of the vPE software as well as communicates telemetry upstream to ONAP (see Chapter 7).

For low throughput vPE applications (<1 Gbps), the five functional modules may run on a single CPU. For high throughput vPE applications (>10 Gbps), a common strategy is to split the functional modules across multiple server CPUs or even split across multiple servers to maximize performance using parallel processing. Interestingly, the first routers were in fact software running on general purpose CPUs. As routing started to take off, the ability of software and CPUs to handle increasing packet throughputs hit fundamental limits of CPU and software technology. The industry turned to custom application-specific integrated circuits to address the throughput requirements because of bandwidth growth and several companies emerged to create products that combined custom silicon and software into integrated vertical router solutions. This model dominated the networking industry for two decades. In a "Back to the Future" moment, what happened in the intervening years was Moore's law and high levels of integration of multiple CPU on a single chip resulted in increased performance to the point that software routers were able to displace vertically integrated router for multiple throughput speed tiers. In particular, the speed tiers of interest are tied to customer physical port speeds of 1 Gbps, 10 Gbps, 40 Gbps, 100 Gbps, and in the near future 400 Gbps.

Implementing a PE router as a vPE software running on x86-based servers cannot meet all use cases so physical PE routers are still required for some use cases. The tradeoff between PE VNFs and physical network functions (PNFs) is described in terms of *zone of advantage* as shown in Figure 12.6. The horizontal axis is the packet throughput and the vertical axis is the complexity

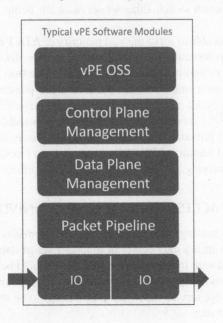

FIGURE 12.5 vPE software components.

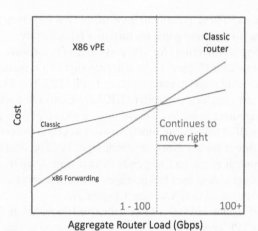

FIGURE 12.6 Zone of advantage.

of the packet processing function. Through testing for each use case the dividing line between the overall TCO of a VNF-based PE is compared to a PNF-based PE.

The zone of advantage dividing line is rapidly moving to the right. Two years ago, vPE software was not adequately optimized to achieve high throughput. However, once the software was rearchitected and reoptimized to overcome bottlenecks the throughput increased appreciably. At the time of this writing the zone of advantage for most use cases is able to support 10 Gbps user network interface (UNI) speeds with multiple 10 Gbps customers being served on a single server. This is an important milestone because, as of this writing, more than 95% of AT&T end customers attach to the network with UNI's less than 10 GE and hence more than 95% of AT&T packet service customers can be served with software-based routers over time as their deployment expands. With new servers and continuously improved and optimized software 100 Gbps port speeds are now being attained in test bed configurations and will move into production over the coming year.

Currently, enterprise Internet and VPN services utilize this new vPE technology, while residential BNG service and metro switch Ethernet services are being certified for production vPE operation.

The implications of being able to serve the vast majority of AT&T customers with software routers are many. The shift from vertically integrated routers to a new disaggregated router architecture finally manifests the vision of making packet edges multiservice fungible without compromises. In addition, new value added on-demand services are now made possible. With on-demand services truck rolls and long turn up cycles are replaced with software service chain of VNFs in conjunction with the vPE to provide greater service velocity than was ever possible before.

With the new capabilities provided by vPE VNFs it became necessary to rethink the architecture for interconnecting vPE and traditional PNF PEs using modern fabric technologies. Ethernet VPN (EVPN) is emerging as a unifying construct within cloud networking (Sajassi et al. 2016).

12.5 EVPN: FLEXIBLE ACCESS GROOMING AND UNIVERSAL CLOUD OVERLAY

The cloud architecture for hosting edge services in VNFs provides flexibility in the placement of VNFs (see Chapter 3) within a data center or among a set of data centers. The VM software orchestration and VNF hardware independence allow the VNFs to be moved as needed for capacity, maintenance or operational reasons, and for resiliency. This requires flexibility and scalability in how access reaches the edge service functions. It also needs to be able to move dynamically with the movement of a VNF instance.

12.5.1 ACCESS SCALE AND RESILIENCY

Current access wireline access technologies use an Ethernet link layer and aggregate or separate customers using VLAN tagging. For large networks such as AT&Ts, the scale can be enormous. For example, at AT&T, for enterprise and infrastructure, there are about 400,000 point-to-point and multipoint Ethernet access connections used for mobile cell site backhaul, enterprise inter-site connectivity and connectivity to IP-edge services. There are roughly 16 million broadband connections from DSL and PON aggregation nodes and these are groomed using VLANs. These access connections are spread across almost 5000 access offices and points of presence domestically and internationally. They need to be groomed and connected to the edge service VNFs residing in hundreds of AIC zones (see Chapter 4) that could potentially grow over time to thousands.

There is also a need to groom these connections into bundles rather than signal each one separately through the control plane. For example, carrying the 16 million broadband connections as individually signaled MPLS psuedowires would increase the existing MPLS signaled state in the network by about a factor of 50.

In addition to scale, the connectivity into the cloud needs to be resilient. Interoffice rerouting over a diverse transport topology is provided by distributed MPLS routing protocols (see Chapter 13). However, the new generation of merchant leaf edge switches are fixed form factor chasses. To overcome the time to repair these potential single points of failure, access nodes (e.g., optical line terminals [OLTs] and digital subscriber line access multiplexers [DSLAMs]; see Chapter 11) will be connected to two edge switches, with the pair of links acting a single logical link from the perspective of the access node.

EVPN with its point-to-point virtual psuedowire service, flexible cross connect bundling, and multihomed attachment support is ideal for meeting these access connectivity requirements across wide area core networks as well as metro and regional networks as shown in Figures 12.7 and 12.8. In Figure 12.8, A- denotes access connected to the leaf, C- denotes compute server connected to the leaf, P- denotes provider core connected to the leaf in the terms A-Leaf, C-Leaf, and P-Leaf.

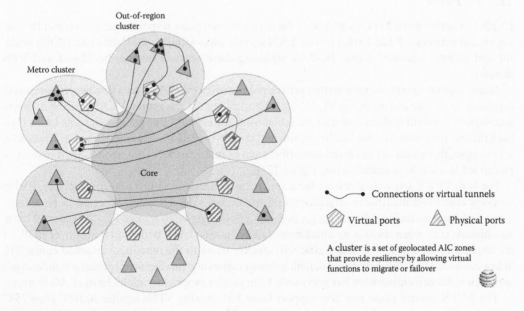

FIGURE 12.7 Metro clusters.

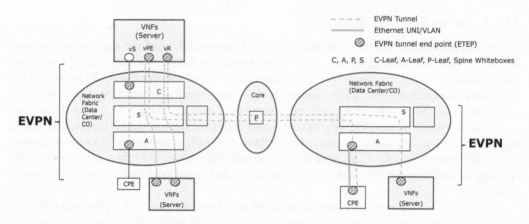

FIGURE 12.8 Interconnectivity provided by EVPN.

12.5.2 CLOUD OVERLAY CONNECTIVITY

Connectivity among edge service VNFs and cloud virtual routers requires a combination of layer 2 and layer 3 VNFs. VNFs on the same subnet require layer 2 Ethernet multipoint bridging. Or, they may only require a virtual IP network providing a default gateway and IP layer 3 routing. Then, many will require both, with a virtual network that does integrated intra-subnet bridging and default gateway IP routing (sometimes called integrated routing and bridging). And the resiliency requirement is that these VNFs see connectivity on their hosts via an NIC bond for redundancy, while the bond is actually a multichassis link aggregation group (LAG) across a pair of neighbor leaf switches.

EVPN meets the layer 2 and layer 3 virtual routing requirements and provides a standardized approach to support multichassis LAG redundancy.

12.5.3 EVPN

EVPN is a multiservice MPLS technology for both Ethernet point-to-point, multipoint, and IP routing virtual networks. It has virtual private LAN service supporting protocol data unit (PDU) bridging and address learning. It uses BGP for signaling these services among the PEs of an EVPN domain.

It also supports point-to-point virtual private psuedowires (VPWS) with a flexible cross-connect capability to rewrite and push/pop VLAN tags. This last feature can groom VLANs across different access ports combining them into a single psuedowire. Effectively, this psuedowire acts like a logical Ethernet port with its own locally significant, double-tagged VLAN address space. Therefore, for example, thousands of broadband subscribers across multiple PON OLT and DLSAM nodes can be carried in a single psuedowire (see Figure 12.9).

For both VPWS and virtual private local area network service (VPLS) Ethernet services, EVPN supports redundant multihome attachments on the access side. These can operate in an active–active fashion with traffic splitting across both. EVPN will signal to remote PEs the status of these attachments (i.e., when there is an attachment failure, it can send a bulk withdrawal over BGP for the impacted psuedowires) so that traffic will quickly move to the remaining attached egress PE. When combined with the distributed default gateway capability, this feature provides a standardized approach to implementing what has previously been vendor proprietary multichassis LAG features.

The EVPN control plane can also support layer 3 IP routing VPNs similar to RFC4364/2547 VPNs in wide use for service provider-based VPNs. More importantly for cloud, EVPN supports IP routing with a distributed default gateway capability for integrated routing and bridging in a

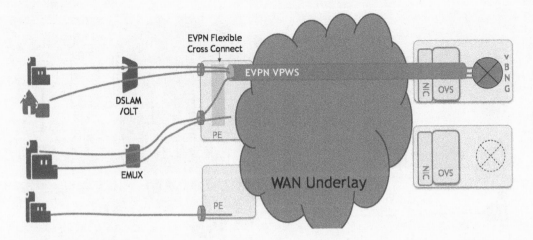

FIGURE 12.9   EVPN in nominal operating state.

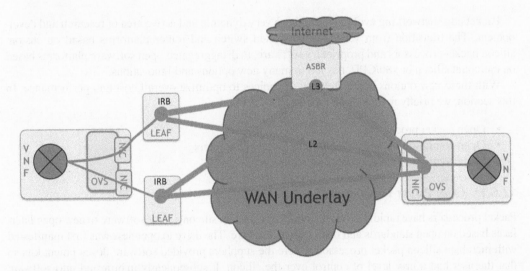

FIGURE 12.10   Integrated routing and bridging in EVPN.

single virtual network. This avoids having to take an extra hop to tandem route host VNFs through a default gateway router (see Figure 12.10).

EVPN BGP signaling uses route target communities to auto-discover PEs with endpoints participating in a particular virtual network service. This dynamic discovery is event based allowing very fast rehoming of services and rerouting for state changes (e.g., failure of one side of a dual attachment or injection or withdrawal of a layer 3 route). This is ideal for cloud movement of a VNF instance, allowing the associated customer access VLANs to move with the VNF (see Figure 12.11).

12.6   FUTURE EVOLUTION OF NETWORK EDGE

In this chapter, we reviewed the drivers, both business and technical, for the recent rapid evolution of network edge platforms. Disaggregation of hardware and software for packet processing has fundamentally changed the architecture and cost structures for the network edge. This leads to a new interplay and intersection between software and hardware to optimize the performance, yet maintain maximum flexibility. In this section, we examine the future evolution of the network edge.

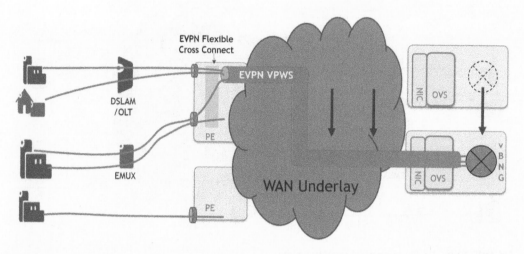

FIGURE 12.11   EVPN rerouting psuedowires.

Packet edge networking evolution remains a very dynamic and active area of research and development. The transition from vertically integrated switch and router platforms based on custom silicon packet processors and proprietary software, to disaggregated open software platforms based on merchant silicon or x86/CPU, has led to many new options and innovations.

With these new options come an array of choices to optimize overall cost and performance. In this section, we briefly review three broad evolution themes.

- Open packet processors
- Open configuration and programming packet processors
- Open control of packet processors

12.6.1   OPEN PACKET PROCESSORS

Packet processors have rapidly evolved from closed custom silicon and/or software to new open interfaces based on open standards and open source software. The drive to openness was first manifested with merchant silicon packet processors where the suppliers provided software development kits so that the user had a new level of control over the silicon. It subsequently manifested into software packet processors running on CPUs with maximum flexibility and openness such as Open vSwitch (Open vSwitch) and Vector Packet Processor (The Fast Data Project). And in the near future, suppliers will be developing highly programmable ASICs for high throughput packet processing.

In the idealized future, packet processors should be open, abstracted, and highly programmable with only a few differentiating variables such as throughput (ASIC > Server throughput), buffers for class of service (Server > ASIC buffers), and logical table sizes (Server > ASIC tables) to allow for simplified selection, application, and control. These packet processors should be accompanied by robust open source distributed control plane stacks that include multiservice unifying constructs such as EVPN.

12.6.2   OPEN CONFIGURATION AND PROGRAMING OF PACKET PROCESSORS

The most active area of research and development is in the area of configuring and programming packet processor data planes.

Configuring a vertically integrated switch or router is a painstaking exercise involving writing many lines of commands in a unique language that varies across suppliers as well as across platforms from the same supplier. This is loosely equivalent to programming a CPU in assembly language, requiring lots of fine tuning in multisupplier networks to tune each platform to meet a common carrier specification for services such as layer 2/Ethernet VPNs or layer 3/IP VPNs. Out of necessity, the academic community began to propose high-level abstractions for configuring packet pipelines.

A major step forward in overcoming the painstaking process of configuring packet platforms arrived with OpenFlow, which provides a standard open interface to directly access and manipulate the forwarding plane of network processors in switches and routers (McKeown et al. 2008).

The next major leap occurred with the arrival of the top-down approach of P4 (programming protocol-independent packet processors). OpenFlow is a "bottoms-up" approach that focused on packet pipelines that were well defined and with a relatively fixed set of standard protocols. In contrast, P4 is a domain-specific language that is compiled to the specific target packet pipeline, whether software or an ASIC, and controls switches "top-down" by first specifying their forwarding behavior using the match action abstraction and then populating the forwarding tables (Bosshart et al. 2014). This led to a new level of abstraction that allowed network engineers to rapidly develop and validate new packet processing concepts.

Another step forward is the Switch Abstraction Interface (SAI), which defines an abstraction interface for switching ASICs. The interface provides a means for the same software to control multiple suppliers' switching pipelines while keeping the programming interface consistent. This specification also allows exposing supplier-specific functionality and extensions to existing features (Subramaniam 2015).

In a similar vein, Domino is an imperative language to program the data plane of high-speed programmable routers and help network engineers design such programmable routers in the first place. It also was a step forward from P4 in providing a construct for algorithms that require changing switch state, or stateful processing, as it modifies or inspects the packet (header or data) while it transits the switch (Sivaraman et al. 2016).

While there are many other packet programming advancements, we highlight these four, (OpenFlow, P4, SAI, and Domino) to illustrate an important principle in the evolution of network edge platforms. A key business principle in networking is backward compatibility so that each advance should be able to replicate and maintain the previous state of the art. This does not mean each new applications should be the sum of the requirements of all previous applications but, if desired, backward compatibility can be maintained so as to greatly simplify a user's transition from one technology to the next.

Specifically, OpenFlow demonstrated that it could configure a simpler fixed-form ASIC that previously required low-level assembly-language-like configuration. A P4 program has been created that can fully replicate OpenFlow (McKeown and Rexford 2016). SAI can be used as an interface for a P4 program (Kodeboyina 2015) and a Domino language has a complier backend that can auto-generate P4 (Sivaraman et al. 2016). So in each successive wave of advancement with OpenFlow, P4, SAI, and Domino, the principle of backward compatibility is maintained.

These continuing advancements not only make the network engineer's job easier but also we are convinced they will free up time so network engineers can focus on developing new innovations in services and performance of packet edge platforms. In addition, these abstractions are superior to open standards in networking in that being a program they provide unambiguous specification of system behavior across multiple target application technologies.

12.6.3 OPEN CONTROL OF PACKET PROCESSORS

In this section, we examine the newly emergent open control of packet processors. We will not be reviewing the state of SDN controllers and standard control applications such as orchestration, load balancing, network measurement, security algorithms, and congestion control, which are well covered elsewhere (Kreutz et al. 2015; Stallings 2016; and Chapter 7). Rather we will briefly examine use cases and propose new research areas.

With the choice of multiple packet processor technologies available in a Network Cloud environment, there comes a need to decide where to most efficiently process packet flows across the many available options. This notion of choice also extends across multiple network layers such as optical and packet layers, as well as across access, edge, and core portions of the network.

In a tier one provider Network Cloud environment with multiple services such as broadband, video distribution, voice, mobility, virtual private networks, and software defined-WAN (SD-WAN) networks, there will be many millions of packet processors available that are ideally part of a broad fungible pool of resources. This leads to the notion of hybrid architectures where a combination of distributed controllers with centralized assist, can continually optimize packet flows across the pool of packet processors that are parameterized by a small set of differentiating features such as throughput, buffers, and logical table sizes.

The most powerful application of such control will be for resiliency and restoration where lower reliability components (two or three 9s unit availability) are combined into a pool of resources to enable restorations both locally and across geographies to provide end-to-end services with high reliability (five or six 9s of availability).

For the same reason that domain-specific languages have been created for programming packet pipelines, we believe that creation of open domain-specific languages with high level of abstraction for closed-loop control of packet processors will be a fruitful and essential evolution of packet edge technologies to address these use cases.

REFERENCES

Bosshart, P. et al.; P4: Programming protocol-independent packet processors; *ACM SIGCOMM Computer Communication Review*; Vol. 44, No. 3, 88–95, July 2014.

Clos, C. A study of non-blocking switching networks; *Bell System Technical Journal*; 32, No. 2, 406–424, March 1953.

Kodeboyina, C.; An open-source P4 switch with SAI support; 2015 (http://p4.org/p4/an-open-source-p4 -switch-with-sai-support/).

Kreutz, D. et al.; Software-defined networking: A comprehensive survey; *Proceedings of the IEEE*; Vol. 103, No. 1, 14–76, January 2015.

Lemos, R.; Internet routers hitting 512K limit, some become unreliable; Ars Techica, August 13, 2014 (https:// arstechnica.com/security/2014/08/internet-routers-hitting-512k-limit-some-become-unreliable/).

Making the Switch; Disruptive Telecom White Box Collaboration Accelerates and Opens the Platform, Powering Unprecedented Network Performance and Insights; AT&T, April 4, 2017 (http://about.att. com/story/white_box_collaboration.html)

McKeown, N. and Rexford, J.; Clarifying the differences between P4 and Openflow; 2016 (http://p4.org/p4/ clarifying-the-differences-between-p4-and-openflow/).

McKeown, N. et al.; OpenFlow: Enabling innovation in campus networks; *ACM SIGCOMM Computer Communication Review*; Vol. 38, No. 2, 69–74, April 2008.

Moore, G.; Gordon Moore: The Man Whose Name Means Progress, The visionary engineer reflects on 50 years of Moore's Law; *IEEE Spectrum*; March 30, 2015 (http://spectrum.ieee.org/computing/hardware/ gordon-moore-the-man-whose-name-means-progress).

Open vSwitch; Linux Foundation Collaborative Project (http://www.openvswitch.org/).

Sajassi, A. et al.; BGP MPLS Based Ethernet VPN; draft-ietf-l2vpn-evpn-11; April 2015; and EVPN VPWS Flexible Cross-Connect Service; draft-sajassi-bess-evpn-vpws-fxc-00.txt; January 2016 (ieft.org).

Singh, A. et al.; Jupiter rising: A decade of Clos topologies and centralized control in Google's datacenter network; *Proceedings of the 2015 ACM Conference on Special Interest Group on Data Communication*, pp. 183–197, London, United Kingdom, August 17–21, 2015.

Sivaraman, A. et al.; Packet transactions: High-level programming for line-rate switches; *SIGCOMM* 2016 (http://web.mit.edu/domino/).

Stallings, W.; *Foundations of Modern Networking: SDN, NFV, QoE, IoT, and Cloud*; Pearson Education Inc., Indianapolis, Indiana, 2016.

Subramaniam, K.; Switch Abstraction Interface (SAI): A Reference Switch Abstraction Interface for OCP; Open Compute Project; 2015 (http://www.opencompute.org/wiki/Networking/SpecsAndDesigns#Switch_ Abstraction_Interface).

The Fast Data Project (FD.io) and Vector Packet Processing (VPP); Linux Foundation (https://fd.io/).

13 Network Core

John Paggi

CONTENTS

13.1 Optical Layer 266
    13.1.1 Optical Technologies 268
    13.1.2 Flexible Software-Controlled Optical Networks...272
    13.1.3 Open ROADMs 273
    13.1.4 Future Work in the Optical Layer...275
13.2 MPLS Packet Layer 276
    13.2.1 IP Common Backbone..276
    13.2.2 Evolution of MPLS 277
        13.2.2.1 Basic MPLS Transport..277
        13.2.2.2 Fast Reroute and Hitless Rearrangements 280
        13.2.2.3 Distributed Traffic Engineering: Bandwidth Aware Routing...................282
    13.2.3 Segment Routing...283
        13.2.3.1 Packet Forwarding under Normal Conditions 283
        13.2.3.2 Explicit Routing 284
        13.2.3.3 Faster Restoration Using Segment Routing 285
        13.2.3.4 Segment Routing Benefits...285
        13.2.3.5 Segment Routing versus RSVP-TE...285
    13.2.4 Core Router Technology Evolution...286
    13.2.5 Route Reflection...286
13.3 SDN Control of the Packet/Optical Core..287
    13.3.1 Centralized TE..287
    13.3.2 Multilayer Control 288
    13.3.3 Optimization Algorithm Implementation...291
Acknowledgments..291
References..292

The network core is the layer of the network that provides high-speed connectivity between the edge platforms described in Chapter 12. The data plane of the network core comprises an optical layer and an Multi-Protocol Label Switching (MPLS)-based packet layer.

Historically, networks were developed and deployed for specific applications (e.g., Internet services, private VPN services, mobility services, etc.). As the industry evolved to all-IP networks, service providers were motivated by economic and operational considerations to migrate their application-specific silo networks to a model that consisted of application-specific edges interconnected by a common, converged core (Figure 13.1).

The current focus on NFV/SDN is causing further evolution of the network core to include a software-controlled, reconfigurable open optical transport layer, the introduction of high-density/low-cost merchant silicon-based packet technologies, the virtualization of the route reflector (RR) function, and a shift from distributed protocols to a hybrid control architecture that marries the best of distributed and centralized SDN control. The following sections will expand on these themes.

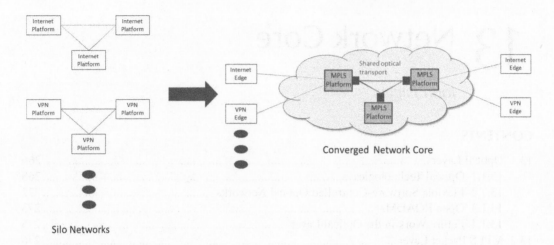

FIGURE 13.1   Migration to converged network core.

This chapter is organized into the following major sections:

- Optical transport layer
- MPLS-based packet layer
- SDN control layer

Core requirements include

1. Ubiquitous connectivity.
2. Scale—carries many petabytes of traffic in the average business day. The core has no restrictions on the physical number of edges it can anchor. All service-specific routing is removed from the core to eliminate any scale limitations.
3. Availability—the network core is designed for resiliency. This is achieved by designing layers of defense that include component redundancy within the network elements (e.g., redundant route processors and switch fabrics), redundant network elements within a central office, dual connectivity between the edge platforms and the core, and network-based restoration schemes capable of rapidly responding to failure conditions. The core network is equipped with advanced capacity management and engineering processes and tools, and sophisticated failure-detection and response systems.
4. Service-agnostic—the addition of new services does not drive the need for any changes to the core.
5. Security—the core infrastructure must be cloaked from attacks from external users.
6. Performance—the core is designed, engineered, monitored, and operated to ensure minimum latency, jitter, and packet loss.
7. 100G—across both optical and routing platforms. Large demands by both service providers and Web 2.0 companies have driven high volumes and competitive pricing.

13.1   OPTICAL LAYER

The optical layer provides high speed (100G today, growing to 400G in the near future), reliable, cost-efficient, and flexible communications between central offices. The communications happen over optical fibers and is enhanced by the use of optical transport platforms in the central offices. Section 13.1.1 will expand on the functions provided by the optical transport platform but let us start with a very short introduction to the fiber optic cables themselves.

TABLE 13.1

Most Common Optical Fiber Types Deployed in Terrestrial Networks

Fiber Type	ITU Specification	At 1550 nm	
		Effective Area (um²)	Chromatic Dispersion (ps/nm/km)
Standard single-mode fiber (SSMF)	G.652	~85	17
Dispersion shifted fiber (DSF)	G.653	~45	0
Non-zero dispersion shifted fiber (NZDSF)	G.652	~55–75	~4.5

Fiber optic cables are made from thin strands of very pure glass roughly the diameter of a human hair and capable of acting as a waveguide carrying optical signals around corners long distances with minimal distortion and loss [1]. Each fiber has a core that provides the medium in which light is transmitted, a cladding that reflects light back into the core when there are bends, and a buffer layer that provides mechanical protections and strength.

Table 13.1 lists the three most common types of optical fiber deployed in terrestrial optical networks, along with their approximate chromatic dispersion and effective area. The attenuation of all three fiber types is typically between 0.2 and 0.25 dB/km. It is now established that, of the three fiber types in Table 13.1, SSMF is optimum for state-of-the-art, 100 Gb/s-per wavelength optical systems utilizing coherent detection and digital signal processing in the receiver for impairment mitigation [2–4].

The core network can have wide geographic reach. In AT&T's case, the fiber-optic network includes more than one million fiber route miles globally, and consists of three main segments: dense metro networks serving customers within a city or region, long-haul networks that spans the United States to provide intercity connectivity, and an international fiber network connecting multinational customers.

AT&T's domestic intercity optical backbone network is shown in Figure 13.2. Denser regional and metro fiber deployments also exist but make the chart too busy to be meaningful. High fiber-count cables that bundle typically several hundred fibers are deployed in a mesh network architecture.

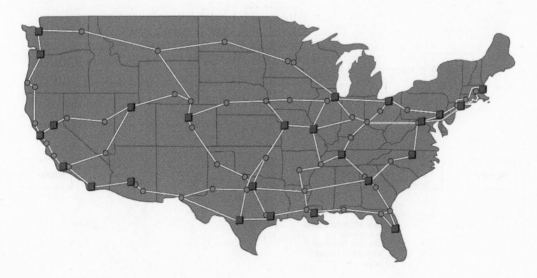

FIGURE 13.2   AT&T domestic 100G long haul optical backbone.

13.1.1 OPTICAL TECHNOLOGIES

There are exciting advances in photonic technology. The first theme is around technologies that make it possible to build a flexible optical network where optical connections can be created between arbitrary endpoints through centralized software control. This level of flexibility is a huge enabler for process automation, improved reliability, and capacity on-demand services. The second major theme is open systems. Today's optical systems are closed proprietary systems that must be sourced from a single supplier. Industry efforts are now moving toward a model where a flexible optical network can be built by mixing and matching components from multiple suppliers—reducing cost and increasing the velocity of introduction of innovative technologies.

This section starts with a short primer on the basics of optical networking and then expands on the flexible technologies and open optical systems.

On the top of Figure 13.3, we show a simple triangle connection between three routers. There would not typically be a dedicated pair of fiber supporting each of the connections (A–B, B–C, and A–C). Instead, the interconnections between the routers in each of the three offices are supported by an optical transport system that provides the following functions:

1. Multiplexing—since the deployed fiber is a precious resource its utilization needs to be maximized. To do so, technology is deployed in central offices that multiplex many optical signals onto a single fiber pair.
2. Reach—optical amplifiers (OAs) are installed inline along the fiber to boost the power of the optical signal to reverse the effects of propagation loss.
3. Switching—consider what is happening in office B in the figure. The signals from A–B and B–C are terminated in this office. However, the signal from A–C passes through office B. The optical transport system should be able to selectively allow some signals to be added/dropped while others get expressed.

Today's typical commercially available optical transport systems can multiplex 96 signals, each operating at 100 Gb/s onto a single fiber pair and transport the composite signal several thousands of kilometers purely in the photonic domain. Higher speeds and longer distances that tradeoff one parameter for another are possible (e.g., higher speed wavelengths can be supported at shorter distances or with lower channel counts).

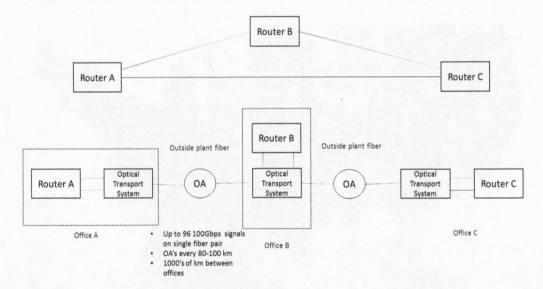

FIGURE 13.3 Functions of an optical transport system.

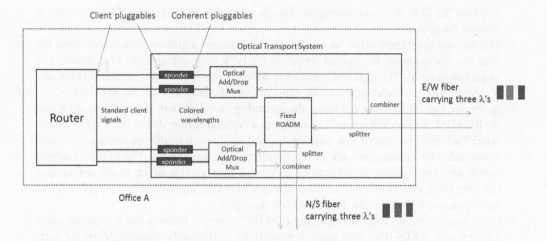

FIGURE 13.4 Block diagram of a fixed optical transport system.

Figure 13.4 drills down the next level and shows a high-level block diagram of the subsystems within today's fixed optical transport systems.

- The transponder (xponder) maps a client interface* that uses a standard wavelength (also referred to as "gray" optics) to a unique "colored" wavelength.† This is done to condition the optical signal so that it can be combined with other wavelengths for transport over a single fiber. The wavelengths are combined using a technique known as dense wave division multiplexing—which defines standards for spacing and packing the wavelengths. For example, the router might transmit a signal with a standard wavelength of 1310 nms on each of its interfaces toward each transponder. The output of the blue transponder might be a wavelength of 1529.55 nms while the green wavelength might be 1530.33 nms. Transponders perform optical–electrical–optical (OEO) conversions as the wavelength shifting function must be performed in the electrical domain. Each O–E conversion requires lasers and photodetectors, which are costly components; hence, the use of unnecessary transponders is to be avoided.
- The add/drop mux combines multiple wavelengths coming from separate transponders into a single photonic signal. In the figure, the top add/drop mux is combining the blue and green wavelengths for transport over the E/W fiber. The add/drop mux on the bottom combines the blue and red wavelengths for transport over the N/S fiber.
- If an optical spectrum analyzer was used to look at the optical signal on an outgoing fiber, what would be observed is a composite signal with multiple wavelengths all propagating the length of the fiber. Because they operate at different wavelengths, they do not interfere with each other and filters can be used at the far end to separate the constituent signals.
- A fixed reconfigurable optical add/drop multiplexer (ROADM)‡ provides the ability to selectively switch specific wavelengths coming in on one fiber to exit via another fiber. Note the violet wavelength in the picture that comes in the N/S fiber and gets switched

---

* A router was chosen to illustrate the role of the client. The optical network also supports dedicated services in which case the client would be a customer CPE, which could be a router, Ethernet switch, OTN switch, etc.

† The term "gray" is used to indicate that if a human could observe the light signal it would have the same color on all client interfaces. "Colored" means each would be seen as a unique color. In reality, the wavelengths used for optical communication are slightly outside those visible to humans.

‡ The term ROADM is sometimes used to refer to the entire optical transport system and other times more narrowly to refer to the photonic switching function. Here, we use the latter definition.

to go out the E/W fiber and vice versa. This is referred to as "expressing" the wavelength through the office.

- The splitters/combiners allow us to combine express and add/drop wavelengths onto the fiber. So, for example, the optical signal on the E/W fiber contains the blue, green, and violet wavelengths. The blue and green were added in this office via the add/drop mux and combined with the violet signal which came in from the N/S fiber via the ROADM.
- At the far end, the splitter replicates the incoming signal to both the add/drop mux and the ROADM. The add/drop mux separates the wavelengths intended to be dropped in this office back into the constituent wavelengths and sends each to the appropriately colored transponder for conversion back to the standard gray wavelength. The ROADM would switch any wavelengths intended to be expressed through this office to the appropriate outgoing fiber. For simplicity, we have shown a ROADM supporting two outgoing fibers but in practice higher degree* ROADM's are deployed.
- Early dense wave division multiplexing (DWDM) systems literally had a separate stock keeping unit (SKU—the code used to identify an individually orderable part) for each color of transponder. The second generation (early 2000s) brought the advent of tunable lasers, which allowed for the deployment of a common transponder that could then be configured through the management interface to the appropriate color.

As explained, ROADMs allow us to add, drop, or express wavelengths in a central office. To illustrate the value of expressing a wavelength, consider a case where a 100 Gb/s optical signal originates in office A, traverses through office B, and terminates in office C. Fixed DWDM optical systems that only have multiplexors must terminate all optical signals in office B, bring the 100 Gb/s optical signal through a transponder back to a standard client signal, and then connect to a second transponder in that office that is part of the line system that connects office B to office C. As shown in Figure 13.5, these back-to-back transponder configurations add extra equipment and cost to the circuit. A ROADM allows us to express the optical signal through office B in the photonic domain without having to purchase any transponders in office B. Transponders are only needed at the "tails" of the connection to shift the wavelength back to the standard client interface.

A transceiver converts the electrical inputs into optical outputs and vice versa. Today's transceivers are packaged as hot pluggable modules that are accessible from the front of the equipment and interchangeable by the user. They are available at many speeds, reaches, fiber type (single mode and multimode) and can be sourced from a variety of suppliers due to multisource agreements that create electromechanical commonalities.

Pluggable optics have several advantages over integrated optics:

1. Modularity—the failure rate of optics is higher than integrated circuits. When the optics fail, a pluggable can be replaced in the field as opposed to replacing an entire line card.
2. Supplier independence—in theory, a service provider can purchase the optics independently of the line cards. In practice, many telecom original equipment manufacturers (OEMs) require that you purchase the optics from them, usually at a premium to open market prices. The OEMs argue that they are responsible for the entire system from a maintenance perspective and must control the optics. They typically implement firmware-based validation checks that will reject optics they have not sourced. As part of the new architecture, AT&T has taken a firm stance with its suppliers that they must support AT&T's use of third-party optics.
3. Pay as you grow—line cards on a packet fabric box support tens of interfaces. The cost of optics is now dominating the overall cost of a packet/optical reference connection. As

* The term "degree" refers to the number of distinct fiber pairs a ROADM can support. If a ROADM terminated separate N, E, S, and W fiber pairs, we would refer to it as having four degrees.

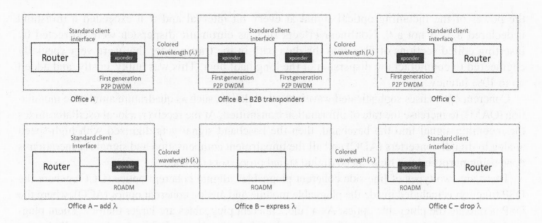

FIGURE 13.5   Expressing wavelengths.

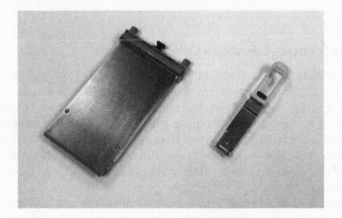

FIGURE 13.6   100G Pluggable Form factors (CFP on left, QSFP28 on right).

a result, a common practice is to purchase only the quantity of optics needed to support immediate capacity needs and to add more as the demand grows.

4. Right tool for the job—pluggables are available that support differing reaches (distances). Longer reach equates to better quality optics, higher power consumption (e.g., temperature stabilized lasers or higher output power lasers for longer reach), and higher cost.

The pluggables supporting a new data rate start in a large footprint and get smaller as the technology matures, enabling increased density with time. For 100G, it started with a "CFP"* and the smallest footprint commercially available now is a "QSFP28" (Figure 13.6).

A CFP is shown on the left. It has 24 Watt power dissipation and the physical size of roughly a man's wallet. A QSFP 28 is shown on the right. It has 3.5 Watt power dissipation and is the size of a pack of gum. Note the optical connectors on the front panel. It is hard to see but there are electrical connections on the back of each unit that would mate with a receptacle on the router.

Recently, pluggable optics is being used on the line side of the transponder as well. Today, all the high-speed connections (100 Gbit/s or more) above 80 km reach use coherent optics [5]. Earlier generation optical systems operated using simple on/off keying. The receiver sampled

---

* CFP stands for C Form-factor Pluggables. The use of the C is reference to the Roman numeral for 100, as the initial purpose of this form factor was for 100 Gbps optics.

the power of the incoming optical signal at every bit interval and if it exceeded a threshold it declared a 1, if not a 0. Nonlinear effects such as chromatic dispersion were corrected by inserting coiled sections of fiber into the data path in the transponder that were very specially engineered to counteract the dispersion of the long haul fiber. This was bulky, costly, and added up to 10% latency.

Coherent optics uses sophisticated waveform modulation, such as quadrature amplitude modulation (QAM), to increase the rate of information transmitted. At the receiver, a local oscillator mixes the incoming signal into the baseband, then the baseband signal gets digitized with high-speed analog to digital converters (ADC) and all the impairment compensation and signal regeneration is done using purpose-built high-speed digital signal processors (DSPs).

There are two types of line side coherent pluggables: digital coherent optics (DCO), where the DSP function is included inside the pluggable module and analog coherent optics (ACO), where the DSP is outside the pluggable optics. As a rule, coherent pluggables are larger than the client pluggables, because the output power of the transmitter is hotter. While DCO modules usually have a larger footprint than ACO due to the additional power-hungry DSP function inside, ACO modules have an external analog high-frequency interface that requires special engineering and sacrifices performance.

13.1.2 FLEXIBLE SOFTWARE-CONTROLLED OPTICAL NETWORKS

Currently deployed ROADMs are referred to as being *fixed*. With today's technologies, each add/drop port has a fixed color and direction. Once a transponder is cabled to an add/drop port, the color and direction of that wavelength are determined. Referring to Figure 11.4, to reconfigure the ROADM to allow a wavelength currently going out the E/W interface to instead egress the N/S fiber, cables would need to be physically moved from the upper transponders to connect to the lower add/drop mux ports. Note that the current design also creates process and system complexity as manual records of the connections between transponders and add/drop ports needs to be maintained and updated as changes are made.

Advances in photonics, most notably the availability of high-density, cost-effective wavelength selective switches, have made it commercially viable to build a layer of photonic switching that allow the output of a transponder to be connected to any outgoing fiber on the ROADM. By doing this, the limitations that exist in today's ROADMs are removed. The industry uses the terms colorless, directionless, and contentionless to describe this new level of flexibility that has been incorporated into ROADMs [6].

Figure 13.7 illustrates how wavelengths can be created and rerouted completely through software control. The green wavelength was initially built. The transponders supporting the green wavelength could later be repurposed to use a different color (purple) and path, without manual dispatch. Alternatively, the green wavelength could be removed and each transponder could be repurposed to create a new wavelength to any other transponder in the optical network.

This flexibility can be exploited for many uses:

Restoration: Consider a case where a fiber is cut or an OA fails. Today's networks rely purely on robust L3 network designs and over-provisioned capacity to restore around the failure until the cable can be respliced or the failed equipment replaced. This creates a compromised position during the failure and exposure to service outages if subsequent failures occur. With a flexible ROADM, the optical wavelengths can be rerouted around the failure point and service quickly restored. The failure can then be repaired on a scheduled basis.

Network optimization: Networks grow over time. At one point in time, the best path to route a connection between offices A and Z may be to traverse intermediate offices B and C. If over time, sufficient demand existed directly between A and Z it might be more efficient to build an express fiber between those two offices. With a flexible ROADM, it is now

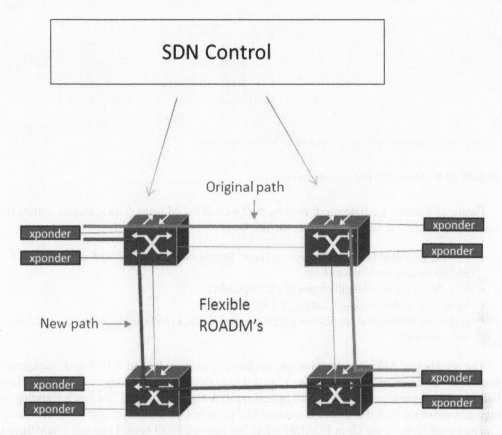

FIGURE 13.7 Flexible ROADM's enable software reconfiguration of wavelengths.

possible, through software, to reconfigure the original connections to use the express path. This would provide a lower latency path to the customer and because there is less intermediate equipment it would have higher reliability at a lower unit cost. This process of cleaning up the point-in-time accretive decisions of the past to optimize utilization of the resources is often referred to as optical "defrag."

On demand services: Over time, as we predeploy transponders, wavelength services can be turned up, down, and moved via software without the need to dispatch technicians or move cables, thus enabling a whole new paradigm of capacity on demand services.

Currently, AT&T has deployed a rich footprint of flexible ROADMs in the intercity network managed by an SDN controller capable of creating, deleting, and rearranging optical wavelengths via software control. These systems support 100G wavelengths today and have been designed to be extensible to 400G wavelengths as the technology becomes technically and economically viable.

13.1.3 OPEN ROADMs

Today's optical networks are closed systems with all components (transponders, ROADM, amplifiers, etc.) sourced from a single supplier. Several industry efforts have been launched recently to remove the supplier lock-in and enable an optical network to be constructed by mixing and matching components from different suppliers. This section will specifically describe the Open ROADM activity. Visit the OpenROADM.org website to learn more about this effort, view the latest participants, and download the specifications [7].

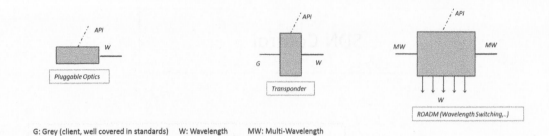

G: Grey (client, well covered in standards)   W: Wavelength   MW: Multi-Wavelength

FIGURE 13.8   Open ROADM reference architecture.

Figure 13.8 shows the framework used by the Open ROADM team to disaggregate optical networking functions and define four reference interfaces:

- G—the gray interface coming from a client. Standards already existed for this interface and no further work was needed.
- W—the single wavelength output of a transponder.
- MW—the multiwavelength output of a ROADM.
- API—the northbound application programming interface (API) to a management and control subsystem.

The northbound APIs from the photonic devices will connect into an SDN-based management and control architecture as shown in Figure 13.9. The SDN control will be layered—the lower layer will focus on the management of the L0 optical layer. A higher layer will focus on integrating the management of the packet and optical. Section 13.3 describes the multilayer control in more detail.

The initial focus of the Open ROADM effort has been on 100G optical systems where there is broad industry adoption that has driven high volumes and competitive price points. One of the next steps is to investigate whether and how to extend open ROADM to both lower and higher speeds.

One of the key design objectives was to remove the distributed proprietary functions that result in today's closed systems. In some cases (e.g., forward error correction [FEC]), this was accomplished by selecting and specifying a single implementation in the specification. In other cases, functions implemented on the hardware elements today were moved into the controller layer by making the management of the function accessible through the API. For example, the programming of optical parameters such as amplifier output levels is done today through proprietary feedback loops built into the equipment. The new architecture provides hooks through the APIs to enable this to

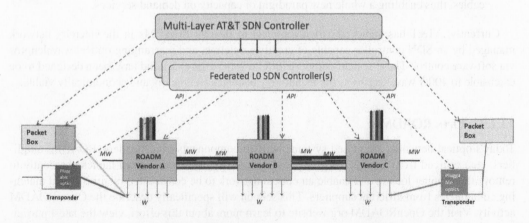

FIGURE 13.9   SDN control of packet/optical layer.

happen from the SDN controller. Today's closed systems are disaggregated into separate hardware and software components; the network equipment is simplified by removing functionality from it and moving those functions to a central controller, and this simplification is used as an enabler to creating open systems.

The initial open specifications targeted metro applications. The distances in metro are shorter relative to the nationwide intercity network, enabling system performance to be traded for openness.

The open ROADM program will overcome the vendor lock-in inherent in the deployment of today's closed optical systems. It creates a new sourcing model and intense competition among suppliers who will invest in improving the price/performance and functionality of their products. This model has obvious benefits for service providers but also creates opportunities in the supplier space as service providers will be more amenable to using products from new market entrants as the need to pick an established supplier with strong staying power is removed.

There are other efforts to define open optical systems using "alien wavelengths." In this approach, the ROADMs must be sourced from a single supplier but the transponders can come from a third party. But there is one big catch—the transponders on either end of a wavelength must be bookended and need to come from the same supplier. In contrast, the open ROADM effort allows complete flexibility in mixing and matching ROADMs and transponders from different suppliers. A requirement to bookend transponders would translate to a need to segment transponders into supplier pools, hence restricting the value of having flexible rearrangements.

The OpenROADM.org specifications include the common data models (device-level, network-level, and service-level templates) that enable uniform control and management of a multivendor optical network. The models include the "hooks" that allow for supplier-specific extensions across the multiple vendor implementations. The data models are specified using the YANG [8] data modeling structure and the prescribed API is Netconf [9].

There is an additional transformative dimension to the open project. Today, all management and control of metro optical networks is done using Telcordia-developed operations support systems (OSS). The origins of these systems were put in place in the mid-1980s as part of divestiture to ensure continued OSS support for the Regional Bell Operating Companies. In order to harness the power that comes with flexibility and to enable new services, the open ROADM program is departing from this approach. It will use a new management and control architecture that leverages SDN control, the use of model-driven designs, and the ONAP platform discussed in Chapter 7 (or an equivalent).

The key functions of the controller and management layer are as follows:

- ROADM controller (equipment and link discovery, equipment inventory, topology, wavelength connection computation and activation)
- Alarm surveillance and performance monitoring
- Work flow management
- Capacity management and planning tools

13.1.4 FUTURE WORK IN THE OPTICAL LAYER

The open ROADM specs were initially developed for metro but can be extended for use in the intercity domain. The diameter of a metro network is several hundreds of kilometers while in intercity the requirement is for several thousands of kilometers of reach. The longer reaches will require more sophisticated modulation and coding schemes and require more complex, and costly, hardware.

Initial implementations of the ROADM controller shown in Figure 13.8 will be based on custom development. The industry would benefit from the development of an open ROADM controller platform functions. This will help lower the barrier to entry into the open ROADM for service providers and incent investment and development of value-added applications by third parties.

A hot topic of work in the data center market is to explore the technical feasibility and economic value of moving the optics to being fixed devices installed during manufacturing on the router/ switch boards. The thinking is that tightly integrating the optics with the electronics removes the cost of the packaging associated with pluggability and the need for separate testing. An industry effort called Consortium for Onboard Optics (COBO) [10] has been launched and has strong representation by the optics and switch/router community.

13.2 MPLS PACKET LAYER

The purpose of the core packet layer is to provide global interconnectivity for the edge services platforms described in Chapter 12. This section covers the edge/core design paradigm, the evolution of the MPLS implementation, the transformation for core routing technologies, and the evolution of RR design.

13.2.1 IP COMMON BACKBONE

An important design decision early in the development of the all-IP network is the adoption of an edge/core paradigm. There are two key drivers:

1. Efficiency—by aggregating the flow of traffic from many edges for transport across the core, the utilization of the costly long haul optical resources is improved.
2. Separation of functions—the edge is focused on the agile delivery of new and innovative solutions meeting enterprise and consumer needs. Edge platforms need to cost-effectively anchor many customer access connections, both high- and low-speed, and historically TDM, synchronous optical networking (SONET), and Ethernet. The core is all about scalable and reliable bulk transport. The edge and core will each evolve at their own pace. A new services offering should never create any requirements to evolve the core.

AT&T uses the term IP Common Backbone (CBB) to describe the core. Today's IP CBB carries over 100 petabytes of data a day. It is a seamless global network providing services in hundreds of countries. There are common design paradigms used globally—giving AT&T scale, efficiency, and reliability.

The CBB supports all of AT&T's enterprise and consumer services including

- Consumer broadband Internet access
- IPv4/IPv6 internet connectivity
- IPv4/IPv6 VPN service
- Ethernet services
- VOIP services
- Cloud services
- Mobility infrastructure and internet access

MPLS technology [11] is used to implement the packet core. Figure 13.10 uses the standard industry framework in which edge functions are referred to as the provider edge (PE) and core functions as the provider (P); the PE's encapsulate packets into MPLS overhead for transport across the P core.

MPLS is used to build a set of tunnels that deliver traffic from an ingress P router across the core to the edge-facing port on an egress P router. Each P router advertises to its subtending PEs the tunnels the PE should use to get information to distant egress PEs. The ingress PE packages its service-specific data into an MPLS container and pushes it into the tunnel advertised for the

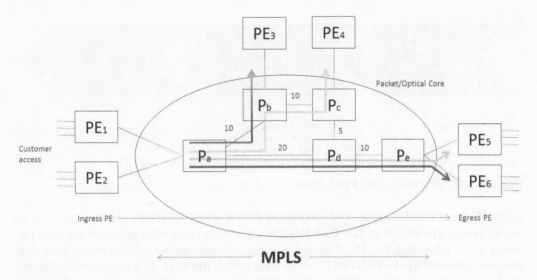

FIGURE 13.10 MPLS packet core.

egress PE. The core delivers the MPLS packet to the egress PE choosing the best route available through the core. It does so without needing to know anything about the contents of the container or the eventual end customer the contents will be sent to. Its job is simply to efficiently and reliably transport the MPLS container to the egress PE. The ingress PE only needs to specify which egress PE the container should be delivered to; it does not need to worry itself with the messy details of how to best get it there. The next sections will discuss how the tunnels are routed and built, their resiliency, and their adaptability.

For simplicity, we have only shown the tunnels originating from P_a in Figure 13.10. Tunnels do originate from all other P routers as well.

The nodes in the network use an Interior Gateway Protocol (IGP)[*] to exchange routing information including topology (which nodes have adjacency to others), link metric, and state (link up/link down) information that enable each node to make informed decisions regarding the best path to route tunnels toward a destination. Typically, mileage is used as the metric causing nodes to route along the lowest latency path but there can be other considerations. Note, for example, that the path of the tunnel from P_a to P_c was via P_b. There is a path via P_d that was not chosen because it has a higher cumulative cost (sum of metrics of the links in the path).

13.2.2 Evolution of MPLS

Figure 13.11 shows the major evolutionary steps of the MPLS-based packet core.

13.2.2.1 Basic MPLS Transport

The first phase of the MPLS journey focused on building a single MPLS-based core to provide common transport for all services. This was the first realization of the edge/core paradigm and produced a service-agnostic core.

In building the MPLS core, tunnels need to be created and routed through the core. The technical term for a tunnel is a label switched path (LSP). Edge routers insert a small amount of additional overhead to each packet (beyond the IP overhead) to enable MPLS forwarding across the core. The overhead contains a label, which is the identifier for the tunnel, that is used to indicate the desired

---

[*] The two most popular routing protocols are Open Shortest Path First (OSPF) and IS–IS. There are many resources readily available for the interested reader.

- Fast Re-Route
- Hitless Re-arrangements
- Bandwidth Aware Routing

FIGURE 13.11   Evolution of MPLS packet core.

egress PE. Each P router has a set of subordinate PEs. It learns about their existence through the routing protocol mentioned earlier. Each P router advertises to each of its neighbors the label they should use to reach each PE. The P routers exchange this information with their neighbors, both their subordinate PEs as well as other P routers. Each node is free to assign a new label for a destination PE; there is no requirement that it reuse the labels that it received from its adjacency. But a given router must advertise the same label for each PE on all of its interfaces and the label assigned for each PE must be locally unique. The protocol used for label distribution is called Label Distribution Protocol (LDP) [12].

Assuming all of the nodes in the network have labels informing them how to reach each PE over each of their outgoing interfaces, each node now uses its routing database to determine the lowest cost path for each destination PE. This path identifies the next hop—the next adjacent router in the lowest cost path toward that destination. It finds the label that the next hop has advertised to use for that destination PE and creates a binding in its forwarding table that looks like—"If a packet comes to me with label X, swap the label to Y (label next hop wants me to use for the same PE associated with X) and push the packet out the interface connected to the next hop on the shortest path to the destination."

Figure 13.12 shows a simple network topology including the metric (the term weight is used interchangeably) for each link. For simplicity, we have only shown the label advertisements for PE3.

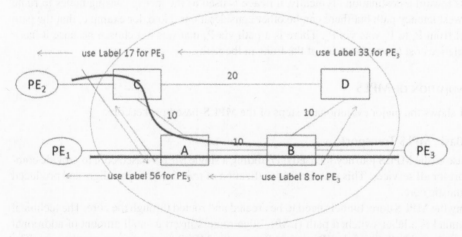

- A would load rule: if you receive a packet with label 56, swap the label to 8 and send packet on interface to B
- B would load rule: if you receive a packet with label 8, remove (pop) the label and send packet on interface to PE_3
- C would load rule: if you receive a packet with label 17, swap the label to 56 and send packet on interface to A
- D would load rule: if you receive a packet with label 33, swap the label to 8 and send packets on interface to B

FIGURE 13.12   Basic LDP operations.

As we can see, this process creates LSPs or tunnels from PE1 to PE3 and PE2 to PE3. The tunnel from PE2 to PE3 goes through C–A–B. When PE2 wants to get a packet to PE3 it:

- Pushes label 17 into the MPLS header and sends it C
- C then swaps the label to 56 and sends the MPLS packet to A
- A then swaps the label to 8 and sends to B
- B removes the MPLS overhead completely (the expression used is it that "pops" the label) and sends the packet to PE3

Prior to deploying MPLS, the core had to know how to route IP packets. It maintained entries in its forwarding table for each unique IP address. The size of the Internet routing table today is in the high hundreds of thousands. Once AT&T completed the migration of the core to MPLS, the Internet routing table was removed from the core routers. The forwarding tables in the core routers was reduced from one with many hundreds of thousands of entries to one that had the size on the order of the number of PEs in the network (thousands). This dramatically reduced the amount of memory and processing that a core router needed to perform and replaced large and complex forwarding tables with much simpler tables that swapped labels.

Consider what happens if the link between A and B from the network model shown in Figure 13.12 is lost. The nodes will communicate the link state change to each other using their routing protocol and dynamically shift the flow of traffic bypassing the failed link. The LSP connecting PE1 to PE3 now gets routed along the path A–C–D–B and the LSP connecting PE2 to PE3 gets rerouted to go from C–D–B. The new LSP is realized via changes in the forwarding tables of A and C. In short, the network is adaptive and resilient (Figure 13.13).

However, there is a catch; it takes time for all of the nodes to receive routing updates indicating the link failure and to recompute their forwarding tables to include the new paths. For example, A is close to the failure and is the first to detect it and update its forwarding tables. It will now forward packets destined for PE3 to C. But until C is informed of the failure and has a chance to update its forwarding table, it will continue to forward packets destined for PE3 to A! This creates what is called a micro-loop—a spiral death dance for packets. There is a field in the header known as time-to-live (TTL) that gets decremented by 1 each time the packet traverses a node. If the field is decremented

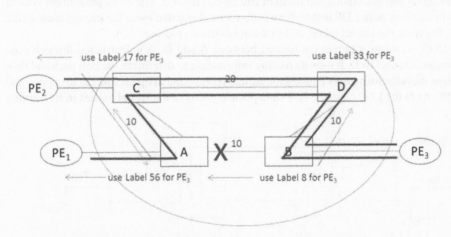

After Convergence, the new rules would be -
- A would load rule: if you receive a packet with label 56, swap the label to 17 and send packet on interface to C
- B would load rule: if you receive a packet with label 8, remove (pop) the label and send packet on interface to PE3
- C would load rule: if you receive a packet with label 17, swap the label to 33 and send packet on interface to D
- D would load rule: if you receive a packet with label 33, swap the label to 8 and send packets on interface to B

FIGURE 13.13 Network state after a link failure with basic LDP.

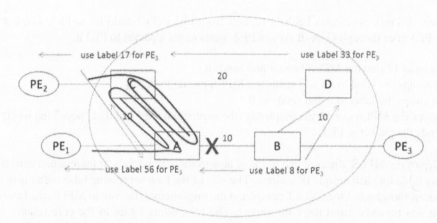

After link A-B fails but before the network converges, a routing loop exists between A and C -
- A would load rule: if you receive a packet with label 56, swap the label to 17 and send packet on interface to C (new rule)
- B would load rule: if you receive a packet with label 8, remove (pop) the label and send packet on interface to PE₃
- C would load rule: if you receive a packet with label 17, swap the label to 56 and send packet on interface to A (old rule)
- D would load rule: if you receive a packet with label 33, swap the label to 8 and send packets on interface to B

FIGURE 13.14 Micro-loop exists after failure but before network fully converges.

to zero, the packet is dropped. The process of updating all nodes in the network with new routing and forwarding information in response to a state change is known as "convergence" (Figure 13.14).

The time to convergence follows a stochastic distribution. There are many parameters to consider—what is the diameter of the network, how big are the routing and forwarding tables, and what other activities are happening in the network that are competing for CPU resources on the route processors. Most flows converge in a matter of seconds but the tail of the distribution could extend out tens of seconds for the outliers.

13.2.2.2 Fast Reroute and Hitless Rearrangements

These micro-loops can be avoided by the use of a protocol called Resource Reservation Protocol (RSVP) [13] to signal and maintain a full mesh of end-to-end tunnels. The conceptual improvement this approach brings over pure LDP is that there is now a head end that owns the management of the tunnel, knows the route the tunnel takes, and that can control its routing.

In Figure 13.15, we show a tunnel configured between A and D. A is instructed through configuration to create a tunnel to D. It uses its routing information to determine the best path and then sends a signaling message using RSVP along the path to reserve and assign resources to support the end-to-end LSP. As in the LDP case, the LSP data plane consists of a series of entries in forwarding

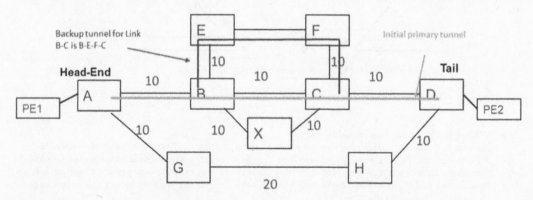

FIGURE 13.15 RSVP signaled tunnels.

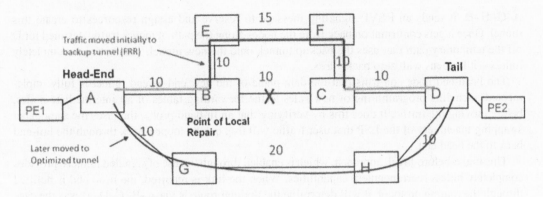

FIGURE 13.16   FRR and hitless rearrangement.

tables instructing each node to perform a label push, swap, or pop function. The specific label bindings to be used by each node are communicated in the RSVP signaling messages.

Each node in the core is instructed to create backup tunnels for each link. A link is an adjacency between a pair of nodes. In Figure 13.15, we show a backup tunnel created for the link between B and C. It is not shown, but there would be other tunnels also using the link between B and C—for example, the tunnel from E to D would also use this link. The backup tunnel protects all of the tunnels that are using a given link.

The backup tunnel is created automatically by the node using its routing protocol and RSVP. Each node finds the lowest cost alternate path (that obviously excludes the link being protected) with the constraint that none of the fiber supporting the links in the backup tunnel share common resources[*] with the fiber supporting the link being protected. Through an offline process that draws on detailed fiber routing databases, we define shared risk link groups (SRLGs) to identify the pools of links that have diversity violations and we download that information into each core router. In the example, the link B–X–C would have been lower cost but was not chosen to avoid the possibility of coincident failures.

With node B having established the backup tunnel using RSVP signaling messages, nodes E, F, and C create label bindings in their forwarding tables that implement the backup LSP. The rules are active, but as long as B is forwarding traffic to C they sit quietly unused. Node B creates the binding to direct traffic to E but for now while the link between B and C is healthy, it sits idle.

Figure 13.16 illustrates the responses of the network to a link failure. When the link between nodes B and C is lost, node B detects[†] the failure and immediately activates the binding to direct traffic to E. Nodes E, F, and C already have their forwarding rules in place and need not be touched to implement the restoration. All of this happens in 50 ms or less. We call this process fast reroute (FRR). It is fast because it is localized, preconfigured, and only requires that we detect the failure and flip a single bit in B to use the backup tunnel.

The term "point of local repair" is used to describe the function performed by B. It is responding to repair the loss of link BC by replacing it with a backup tunnel. But note that the route the end-to-end tunnel from A to D is taking is not optimal.

Knowledge of the failure of the link BC was flooded through the network via the routing protocol. All head-end nodes with tunnels traversing this link began to compute new best paths excluding link BC from their routing database. In our case, node A found the new tunnel path along

---

[*] Common resources could mean the different wavelengths have been multiplexed onto the same fiber, or that they ride different fiber but in the same bundle, or that they are in different bundles but share the same conduit.

[†] The optical interfaces on the router can detect when the incoming optical signal becomes so faint that we cannot decode the information on it or when the bit error rate exceeds a configured threshold. This condition gets reported to the controller on the router and would be the trigger to initiate FRR. It would also be reported to management systems to take action to troubleshoot and repair the underlying problem.

A–G–H–D. It sends an RSVP signaling message to reserve and assign resources to create this tunnel. Once it gets confirmation back from the nodes along the path, it moves traffic destined for D off the temporary path that uses the backup tunnel, onto the new tunnel. This action is completely hitless—it happens with zero packet loss.

The head-end node confirms that the data plane of the new end-to-end tunnel is fully implemented, including programming of new rules to the forwarding tables of all intermediate nodes, before moving the traffic. It does this by verifying that an in-band probe, that uses the same label swapping machinery of the LSP that user traffic will use, can be looped back through the tail-end back to the head end.

This make-before-break approach, which is enabled through the use of signaled tunnels, provides completely hitless rearrangement capabilities. When the link is repaired, the head end is notified through the routing protocol, it will determine the optimal route is via A–B–C–D (as was the case originally before the failure), signal it, verify it, and hitlessly move traffic back to the new tunnel.

The hitless rearrangement is not only used for reoptimization after a failure, it is also used during planned events. AT&T deploys two core routers in each office. When a maintenance event is performed on one router, it is first "costed out." This is accomplished by increasing the metric assigned to each of its links causing all head ends with tunnels traversing this node to recompute new paths for their tunnels that avoid this node. Hitless rearrangement makes this a completely nonintrusive process.

13.2.2.3   Distributed Traffic Engineering: Bandwidth Aware Routing
The implicit assumption is that there is sufficient bandwidth available along the reoptimized route to support not only the traffic that was there but also the rerouted traffic. It is common policy to deploy additional capacity in the core above and beyond what is needed to support user traffic demands during sunny day scenarios to be available when capacity is temporarily lost due to fiber cuts or equipment failures. We loosely refer to this as restoration capacity. In AT&T's case, the network is engineered to protect 100% of the critical traffic during all possible single link (the loss of all bandwidth between a pair of adjacent nodes), span (the loss of all facilities that share common fiber routes), and complete node failures. However, if multiple failures occur simultaneously, there may not be sufficient remaining capacity to support the offered traffic load. This results in congestion and possibly* packet drops.

To improve upon this, traffic engineering (TE) was implemented to make the routing of tunnels sensitive to the offered traffic demand and the available capacity in the network. The routing protocol and RSVP-TE are expanded to advertise not only topology, metric, and link state but also the capacity of a link and the actual measured user traffic on the link. This information is distributed among all core nodes and they use this knowledge to route their tunnels on the lowest-latency path that has sufficient available bandwidth to support the needs of the tunnel. This is referred to as Constrained Shortest Path First (CSPF) routing in contrast to the open shortest path (open meaning simply that all links on the path have their link state up).

Because all head-end nodes are acting independently, it is possible that a head end could signal to set up a tunnel only to find that the resources have been grabbed by another head end. The RSVP protocol reserves the resources on each node as the setup message propagates along the path. The signaling message indicates the required bandwidth for this tunnel and each node keeps a record of the resources it has committed and reserved on each of its links. If the signaling message reaches the tail-end successfully meaning it has been able to reserve resources at each hop, the response messages back to the head end commit the resources to this tunnel. If at any point a node is unable to support the request because it has exhausted the capacity of a required link, it rejects the

---

* Routers have buffers that are able to temporarily hold data when the short-term traffic demand exceeds the bandwidth of an outgoing interface. This results in queuing delays but not loss. If the load is sustained, eventually buffers will be overrun and packets will be dropped. This can be partially mitigated by higher layer protocols, such as Transmission Control Protocol (TCP), which are able to detect packet congestion and loss and respond by throttling the rate at which data are sent across the network.

reservation. The response messages back toward the head end then releases any reserved resources in upstream nodes. The head end then updates its routing database based on what it has learnt and tries to find another path.

Not only do these mechanisms support failure scenarios but also they work in response to changing traffic demands; for example, a new app going viral causing a temporary shift in traffic demands.

In the figures, the links between core nodes have been shown as a single line. In practical applications, a link is composed of many parallel optical interfaces that are logically bundled together into something called a link aggregation group (LAG). Even with the deployment of 100G interfaces between the core nodes, LAGs are still required as the capacity demand on the largest links is rapidly approaching 1 terabit/s. The core routers on either side of the LAG distribute traffic across the member links using a hashing algorithm that ensures the same flow* always hashes to the same member link but provides good distribution of flows between member links. A control protocol ensures that the insertion/removal of member links is coordinated from both sides of the interface.

Before bandwidth-aware routing was available, an entire LAG was brought down even when only one or a small number of member links within it failed. This was necessary because the link would not have sufficient capacity in its diminished state to carry offered traffic resulting in congestion and packet loss. With bandwidth-aware routing the remaining member links are kept in service and the remaining capacity is effectively used to carry traffic.

TE has had an extremely positive impact on AT&T's network. It has dramatically improved overall customer experience, improving the utilization of the network while maintaining and enhancing the survivability. The network is more intelligent; it seeks out and finds available bandwidth. Higher utilization means the deployment of proportionally less restoration capacity translating to lower unit cost.

13.2.3   SEGMENT ROUTING

With RSVP-TE, a network with N nodes will need N*(N-1) tunnels. All nodes need to keep state information about *all* TE tunnels for which they are the head end, tail end, or mid-point. This drives up resource needs on the router. One emerging and promising technology that might help reduce the state that needs to be retained by nodes while supporting similar functions as RSVP-TE, is segment routing.

Segment routing provides a mechanism for source-based routing without requiring intermediate routers to maintain state information. The source router determines a path and encodes it in the packet header as an ordered list of segments [14]. The intermediate routers follow the instructions provided in the segments and forward the packet based on the segment. For MPLS forwarding, the segment will be represented by MPLS label and list of segments as MPLS label stack. The basic principles of segment routing are explained through a few examples below.

13.2.3.1   Packet Forwarding under Normal Conditions

Consider a simple network shown in Figure 13.17. Each router assigns a prefix-segment for its loopback address (e.g., 16006 for R6 router, etc.), which will be globally unique (similar to loopback address) and floods it via the core routing protocol. All routers in the network will learn the prefix-segment and will use this label to send the traffic to the destination router. There is no need for another protocol like LDP to advertise the labels. In the example below, if router R1 wants to send the packet to R6, it determines the shortest path is R1–R2–R3–R6 and will add label 16006 and send it R2. R2 will swap 16006 to 16006 and send it to R3, which will then forward the packet to R6.

---

* A flow is the exchange of data between a source and destination, It is important to keep the packets from the same flow on the same link to avoid getting the packets out of order. Higher layer protocols like TCP will drop packets if they are received out of order causing retransmission and poor customer experience.

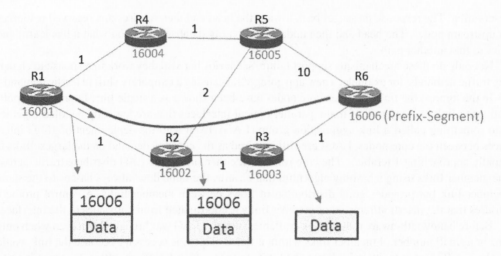

FIGURE 13.17   Segment routing with "shortest path first."

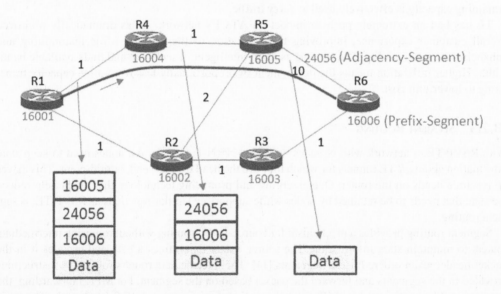

FIGURE 13.18   Explicit routing using segment routing.

13.2.3.2   Explicit Routing

One of the main attractions of segment routing is the ability to explicitly route packets. Figure 13.18 shows how a packet from R1 to R6 can be forced to take a higher cost path—such a case may be to avoid congestion in shortest path.

At the source router, one somehow determines (e.g., via an SDN controller) the desired path (R1–R4–R5–R6) and then pushes the appropriate labels on the stack to get to the destination. To force traffic on a specific link (such as R5–R6) even when its cost is high, segment routing defines an "adjacency segment" that can be used for this purpose. In the above example, R5 assigns label 24056 for adjacency segment between R5 and R6.* The adjacency segment is local in nature, the label value is *not* globally unique (i.e., it can be used by other routers for its adjacency segments). In the above example, R1–R4–R5–R6 path can be reached by first sending the packet to R5 (shortest

---

* The adjacency segments were also distributed via the routing protocols for each node.

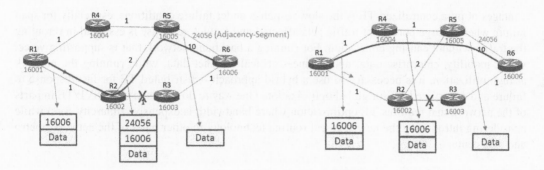

FIGURE 13.19   Fast restoration immediately after failure and after convergence.

path from R1 to R5 is R1–R4–R5) and then use the R5–R6 link to reach the destination. This can be represented as label stack {16005, 24056, 16006} and R1 router pushes this on the data packet and forward it to R4. When packet reaches R4, it will forward it to R5 after popping label 16005. R5 will see 24056, which is the adjacency segment R5–R6 and it will pop the label and send the packet to R6 using R5–R6 link. In this entire process, only the source router will need to construct the label stack that determines the explicit path (which can be done using SDN controller) and the rest of the routers follows the label swapping/pop operations. There is *no* state information about the path in intermediate routers R4, R5, and R6.

13.2.3.3   Faster Restoration Using Segment Routing

The ability to push additional labels (segments) at any intermediate node provides the capability to do fast restoration without RSVP/TE. This is illustrated in the following example.

For protecting R2–R3 link failure, router R2 will precompute paths to the various destinations that use R2–R3 in their path. In the left side of Figure 13.19, R2 determines that packets to destination R6 can be restored by sending the traffic first to R5 and then using link R5–R6. When R2–R3 link fails, R2 will detect the failure, push label 24056 (16005 is not necessary as R2 is adjacent to R5) and forward the packet to R5. R5 will recognize label 24056 is its adjacency segment and it will pop it and send the packet on R5–R6 link. Thus, the traffic can be restored under 50 ms. Once the network recognizes via the core routing protocol that R2–R3 link has failed, router R1 will find alternate path via R4 and forward the traffic to R4 directly and the packets from R1 to R6 no longer use the longer R1–R2–R5–R6 path (FRR path). Note—the backup path state is NOT present in any other router except router R2.

13.2.3.4   Segment Routing Benefits

In addition to removing LDP for label distribution and supporting FRR (without the need to configure TE tunnels), segment routing has other use cases such as service chaining (using service segments), controlling egress peering by the source rather than egress border nodes, and class of service-based routing.

13.2.3.5   Segment Routing versus RSVP-TE

TE using segment routing would require a centralized controller.* The controller will have full visibility of the network and can compute explicit paths (if necessary) and push these paths to the router using Path Computation Element Communication Protocol (PCEP) [15]. One of the disad-

---

* Distributed control using segment routing is still emerging but it is a hard problem due to lack of getting accurate traffic matrix (trying to create an accurate traffic matrix is nontrivial and would end up losing the benefit of SR by introducing too many counters to track flows) and reservation mechanism to resolve contention when two independent entities trying to grab the same resource.

vantages of pure centralized TE is the slow response under failure conditions, especially for span failure where a large portion of traffic gets affected and a quick response is essential in rerouting the traffic without causing congestion. For running a backbone network that is supporting voice, video, mobility, enterprise data, and business critical Internet data, while running the network at high utilization, it is necessary to use a hybrid approach—distributed TE for fast response to failure and centralized TE for global optimization. One way to do this is to use RSVP/TE in parts of the network that requires TE optimization where bandwidth is expensive (intercity core) while considering introducing the new segment routing technology in other parts of the network (metro and data centers).

13.2.4   Core Router Technology Evolution

Traditionally, the core routing function has been fulfilled by large-scale routers that contain custom ASICs capable of supporting over 50 terabits/s of throughput. Recent advances in merchant silicon promise to have sufficient scale and functionality to support core routing requirements. Systems based on these chip sets are beginning to become commercially available. Time will tell if the reality lives up to the promise and if these systems can fully displace custom routers.

The common fabric introduced in Chapter 12 supports many functions. The WAN functions previously supported by a core router might now be fulfilled by what is called a P leaf. It is not necessarily a dedicated box; instead one should think of the P leaf as ports on a shared multitenant fabric.

Figure 13.20 offers a compelling argument for merchant-silicon-based routing. Nine bays of equipment can be replaced by one-half of a bay—nearly a 20X improvement. The footprint, power consumption, and unit cost all scale roughly the same.

13.2.5   Route Reflection

In the introduction to MPLS, it was described that the ingress services edge would request the transport of packets to a distant egress edge. Expanding further, each service edge, either through static configuration or a dynamic protocol called Border Gateway Protocol (BGP), knows the set of addresses that are accessible behind it. Each edge needs to then advertise this information to all other edges in the network. Using a simple IPv4 Internet service as an example, one PE may know that the address 200.201.202.203 is accessible through it. It needs to shout to the rest of

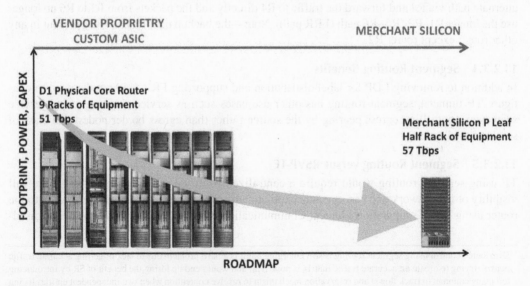

FIGURE 13.20   Packet core technology evolution.

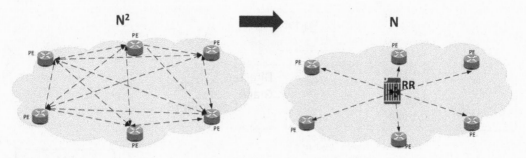

FIGURE 13.21 Why use a RR?

the PEs in the network "hey, if you want to send packets to 200.201.202.203 send them to me." It could do this by sending individual messages to each other IPv4 PE in the network. This creates N^2 sessions to exchange routing updates. The industry developed a RR as a meet-me point for exchanging routing state. Each PE establishes one session to the RR to send and receive routing state (see Figure 13.21).

Because they can become a single point of failure, RRs are always deployed with at least $1 + 1$ redundancy. There are very sophisticated schemes for load balancing, scaling, and creating resiliency in the RR infrastructure.

Historically, RRs were implemented using traditional routers. Since there are no customer interfaces on this equipment, there is no data plane traffic at all. There was typically only a pair of redundant, low-speed interfaces to connect the RRs to core routers for the purposes of exchanging the routing state. Today, the RR function can be deployed as a VNF running on the AIC cloud. In addition to a tremendous reduction in unit cost, there is also dramatic improvement in performance since the servers in the AIC infrastructure are much more state-of-the-art than the control processor that was embedded in RRs based on routers.

13.3 SDN CONTROL OF THE PACKET/OPTICAL CORE

As discussed, the performance of the packet core has improved considerably with the implementation of TE using the RSVP protocol. The bandwidth-aware nature of this protocol allows traffic flows to be moved around the network to find available capacity and avoid congestion.

However, eventually there are limitations because these protocols are distributed. The head-end node for each tunnel is independently calculating routes. Often these requests collide, and several tunnels will need to retry to find a path meeting the bandwidth needs. As the network gets more congested, these retries occur more frequently and it takes a longer time to converge.

The way to improve performance (i.e., faster convergence) and to eliminate congestion is to move to a centralized approach using SDN control. An SDN system has visibility to data from the entire network to form a global view of demand and capacity, and with efficient algorithms, it can better optimize latency and congestion. An added advantage is that this can lead to higher utilization of the network, since extra capacity to handle these congestion cases will not be needed.

13.3.1 CENTRALIZED TE

The core packet network uses distributed RSVP-TE signaling to respond to traffic changes, failures, etc. Distributed protocols have been hardened over many years of operation, and perform extremely reliably in the network. But as discussed above, they do not always find an optimal solution, and may leave the network in a state of congestion, even though capacity might be available in other places.

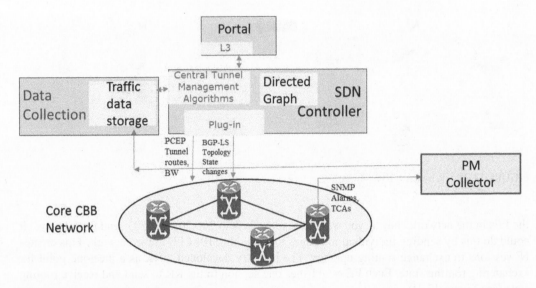

FIGURE 13.22 Centralized TE architecture.

However, centralized solutions are not necessarily robust, because the centralized SDN controller needs to be able to communicate to the network to understand the failure, come up with a solution, and distribute that solution to the network. The same failure that needs to be handled by the SDN controller may impact the communication from the controller to the network.

The right solution is a combination of both—distributed and centralized control.

The distributed RSVP-TE protocols will continue to respond immediately after a failure or traffic spike and reroute tunnels to minimize congestion. Capacity management processes will insure that there is enough capacity in the network to protect critical traffic during single node, link, or span failures. The communication from all of the network nodes to the SDN controller will be part of the protected traffic so this communication should be reliable. However, it is prudent to take a fail-safe approach and design the system so that in the event that communication to the central controller is lost, the distributed protocols will still be able to route traffic to minimize congestion.

Figure 13.22 illustrates the high-level design for centralized SDN control of traffic engineering. Two new protocols are deployed in the network to communicate to the SDN controller. The BGP-link state (BGP-LS) protocol [16] will update the controller on which links are up and how much capacity they have. The path computation element protocol (PCEP) will update the controller on the status of all the MPLS tunnels—their end points, routes, and signaled bandwidth. The existing SNMP protocol will also be used to further improve the granularity and timeliness of data the controller can access.

The SDN controller will be invoked by network changes communicated via the BGP-LS and PCEP protocols. The SDN controller will analyze the routing performed by the distributed protocols and decide whether there is a better way to route the tunnels to make use of the available network capacity as indicated by the BGP-LS protocol. Heuristic algorithms can perform such calculations in a subsecond timeframe and yet come very close to the theoretical optimal solution.

13.3.2 MULTILAYER CONTROL

Centralizing TE will have some positive impact to network performance, but it is even more exciting to look at the possibility of converging the management and control of the L3 and flexible optical network.

Flexible ROADMs allow the optical segment of new trunks between routers to be rapidly provisioned. The need for new trunks could be due to traffic spikes or steady organic growth across all packet services. In order to take advantage of this automation, "tails"—which consists of a 100G router port cabled to a transponder, must be prebuilt. The SDN control uses these pools of tails to be able to automatically create new trunks without manual intervention. The tail can be used to create a trunk in any direction out of the router office.

Today's capacity management process attempts to forecast the traffic demands between specific city-pairs several months into the future. The forecast is then implemented by engineering, ordering, and implementing capacity between these city-pairs. Once in place, the capacity is static. The actual traffic patterns inevitably differ from the forecast.

Flexible ROADMs allow us to vastly simplify the capacity management process. Today's process in which we forecast and build capacity between city-pairs (a N^2 problem) can be replaced with a simple consumption model in which we manage an inventory of tails in each of the N core offices. SDN-controllable ROADMs allow end-to-end L3 trunks to be built when and where the demand materializes, eliminating forecasting error and ultimately improving utilization and customer experience.

The process of creating a new L3 trunk requires convergence of the management and control of the L0 and L3 networks. The high-level steps are as follows:

1. Build and inventory the tail (router port cabled to transponder).
2. Build the end-to-end optical wavelength between the two desired routers. This will be done via the ROADM Node Controller (RNC).
3. Add L3 configurations including activating routing protocols, assigning metrics, activating link failure detection mechanisms, etc.
4. Add the trunk to an existing LAG or create a new LAG.

Figure 13.23 shows the high-level design for the multi-layer controller. It builds on the capabilities put in place for centralized traffic engineering and adds control of the L0 network via an interface to the ROADM Node Controller and configuration management of the L3 packet routers via a Netconf/YANG interface.

This fast provisioning process under SDN control can be applied to failure situations also. Trunks that fail due to a fiber cut or an OA failure can be rerouted over a fiber path that is intact. Or, it may be necessary to configure additional trunks between other nodes to provide an alternate path for the traffic on the failed link. The overall result would be to reduce the number of active trunks that are needed to carry traffic for both normal and failure events. Even when you include the additional tails needed to create these restoration trunks, because of the sharing between all possible directions out of an office, the network has less equipment and less cost overall.

SDN can also be used for even more complex control involving customer trunks that are accessing packet services. With virtual PE functions distributed across all of the AIC nodes in the network, and customer packet service homing onto a particular PE in a particular office, if there is a fiber failure on the link connecting the office, the traffic can be rerouted via SDN control through the fiber network to get back to the office. If the office has a failure and the PE function needs to be relocated in a different AIC node, the SDN control can first recreate the PE function in the different AIC office, remove the existing trunk/wavelength to the first AIC office, and create a new trunk from the customer to the new AIC office. In that way, customer traffic can be restored after failures that today cannot be addressed.

SDN control can also be used to offer new services. The packet network tends to have utilization that varies during the day by more than a factor of two or more. Since the network sizing is designed to the busy hour, there is a lot of idle capacity during the non-busy hours. Customers could request "bandwidth calendared" services to run at some point in the future. These are services that are not time critical, such as backing-up data from one data center to another. The SDN controller will manage the acceptance of these services into the network at a specific point of time where capacity is assured. Assuming the customer agrees to the time, duration, bandwidth, etc., the SDN controller will then create MPLS tunnels dedicated to the customer at the appropriate time, and do whatever

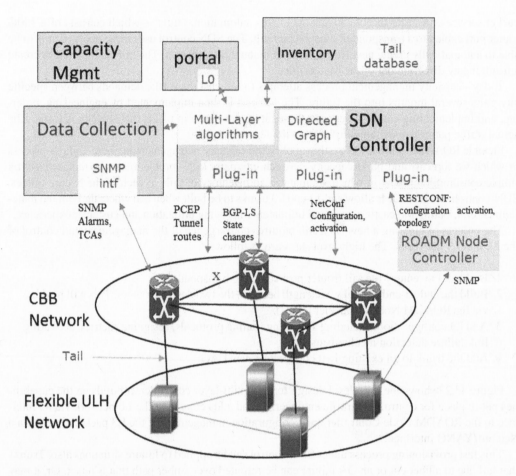

FIGURE 13.23 Multilayer control.

centralized tunnel management and trunk/wavelength creation might be necessary to support the customer demand.

Figure 13.24 depicts the historical and projected core utilization trend. It is important to note that not only has utilization improved but survivability has also improved and the amount of packet loss while the network converged after a failure or while costing a core node for maintenance has been dramatically reduced.

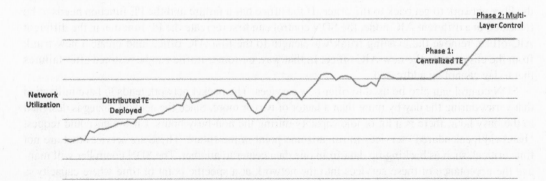

FIGURE 13.24 Historical and projected core utilization.

TABLE 13.2

Comparison of Algorithms

Multicommodity Flow Algorithm for Bin Packing	Fast Heuristic Algorithm for Bin Packing
• Efficiency: 100%	• Efficiency: >99%
• Requires every TE tunnel to be split into arbitrarily small parcels, each of which can be routed independently (implementation not practical)	• Only requires some of the largest TE tunnels to be split into a small number of equal-sized subtunnels (implementation practical)
• Requires solving a linear program (LP)	• Requires iterated application of CSPF (constrained shortest path) routing with rearrangement of order of tunnel routing, uses the fact that usually a vast majority of tunnels can go on shortest path and finer adjustments are needed only for a small fraction of tunnels, partial routing (X% of actual bandwidth of a tunnel with a starting value of X = 100 and geometrically decreasing X until feasible), a combination of random reordering of tunnel routing and giving preference to tunnels that were not fully routed previously, keeping track of best solution obtained so far and using that after a prespecified stopping time
• Run-time ~10 s of minutes for a typical Internet service provider (ISP) backbone with 5–10K tunnels. For a larger backbone or for edge-to-edge TE tunnels, run-time can be hours	
• If other constraints such as latency upper limits on individual flows or the requirement of higher priority traffic to be routed ahead of lower priority ones are needed, then either the LP formulation fails or run-time increases greatly	
• Due to long run-times, not suitable to react to traffic pattern changes or topology changes in orders of seconds as is typical in an ISP network	• Run-time in subseconds to a few seconds for a typical ISP backbone and ~10 s for edge-to-edge TE tunnels
• It is not feasible to run the algorithm several times (can take many hours) to identify the best candidates for adding new IP/optical links (during traffic surges/failure events) or removing IP/optical links (after the event is over)	• Easy to add latency upper limit constraints or routing of higher priority traffic ahead of lower priority ones without significantly increasing the run time
• A network capacity planning exercise requires running the algorithm hundreds of times simulating many failure scenarios and time-of-day traffic variations. This will take several days with the multicommodity flow approach and so is not feasible	• Can react to fast traffic pattern changes and topology changes (~seconds)
	• It is feasible to run the algorithm several times (takes several seconds) to identify the best candidates for adding new IP/optical links (during traffic surges/failure events) or removing IP/optical links (after the event is over)
	• It is feasible to do network capacity planning exercise since even running the algorithm hundreds of times would take a few minutes

13.3.3 Optimization Algorithm Implementation

The optimal bin packing can be done by solving a multicommodity flow problem using linear programming. However, this approach is slow and is not suitable for repeated application under changing traffic patterns and topology changes under unpredictable failures. Furthermore, it is also not suitable for short-term capacity engineering (optimal choice of adding/deleting resources among an available pool) or long-term capacity planning (deciding where to keep a pool of resources so that it can be used optimally over many different failure scenarios).

AT&T has developed a fast heuristic algorithm that runs in subseconds to seconds and so is suitable to respond to fast traffic/topology changes and also for short- and long-term capacity engineering. Furthermore, its efficiency is 99% or better compared to the true optimal algorithm (using multicommodity flow). Table 13.2 explains the key benefits of the fast heuristic algorithm compared to the multicommodity flow approach.

ACKNOWLEDGMENTS

The author would like to acknowledge the contributions of Raghu Aswatnarayan, Martin Birk, Gagan Choudhury, Bruce Cortez, Lynn Nelson, and Kathy Tse.

REFERENCES

1. The fiber optics association guide to fiber optics, [Online]. Available: http://www.thefoa.org/tech/ref/basic/fiber.html
2. ITU-T G.652 Characteristics of a single mode optical fibre and cable, November 2009. [Online]. Available: http://www.itu.int/rec/T-REC-G.652-201611-I/en
3. ITU-T G.653 Characteristics of a dispersion-shifted single-mode optical fibre and cable, [Online]. Available: https://www.itu.int/rec/T-REC-G.653
4. ITU-T G.655 Characteristics of a non-zero dispersion-shifted single-mode optical fibre and cable, November 2009. [Online]. Available: https://www.itu.int/rec/T-REC-G.655
5. K. Roberts, Electronic dispersion compensation beyond 10 Gb/s, Digest of IEEE/LEOS Summer Topical Meetings, pp. 9–10, 2007. [Online]. Available: http://ieeexplore.ieee.org/document/4288305/
6. M. Birk and K. Tse, Challenges for long haul and ultra-long haul optical networks, in *Optically Amplifed WDM Networks*, Academic Press, Chapter 10, p. 277.
7. Open ROADM home page, March 2016. [Online]. Available: http://openroadm.org/home.html
8. M. Bjorklund, RFC 6020, YANG—A Data Modeling Language for the Network Configuration Protocol (NETCONF), October 2010. [Online]. Available: https://tools.ietf.org/html/rfc6020
9. R. Enns, RFC 4741, NETCONF Configuration Protocol, December 2006. [Online]. Available: https://tools.ietf.org/html/rfc4741#section-1.1
10. Consortium for on Board Optics, [Online]. Available: http://cobo.azurewebsites.net/
11. E. Rosen, A. Viswanathan and R. Callon, RFC 3031, Multiprotocol Label Switching Architecture, January 2001. [Online]. Available: https://tools.ietf.org/html/rfc3031
12. L. Andersson, I. Minei and B. Thomas, RFC 5036, LDP Specifications, October 2007. [Online]. Available: https://tools.ietf.org/html/rfc5036
13. R. Braden, L. Zhang, S. Berson, S. Herzog and S. Jamin, RFC 2205, Resource ReSerVation Protocol, September 1997. [Online]. Available: https://tools.ietf.org/html/rfc2205
14. C. Filsfils and K. Michielsen, Introduction to Segment Routing and Its Concepts, [Online]. Available: http://www.segment-routing.net/tutorials/
15. J. Vasseur and J. Le Roux, RFC 5440, Path Computation Element (PCE) Communications Protocol, March 2009. [Online]. Available: https://tools.ietf.org/html/rfc5440
16. H. Gredler, J. Medved, S. Previdi, A. Farrel and S. Ray, RFC 7752, North-Bound Distribution of Link-State and Traffic Engineering (TE) Information Using BGP, March 2016. [Online]. Available: https://tools.ietf.org/html/rfc7752

14 Service Platforms

Paul Greendyk, Anisa Parikh, and Satyendra Tripathi

CONTENTS

14.1 Introduction 294
14.2 New Service Design Solutions with SDN/NFV 298
    14.2.1 Virtual Service Framework 298
        14.2.1.1 Software-Defined Service Framework 299
        14.2.1.2 Virtual Service Control Loop Framework..304
        14.2.1.3 Deployment Considerations 305
14.3 Pivot to SDN/NFV: Methodology, Process, Skills...306
    14.3.1 Service Creation 306
    14.3.2 Service Design Methodology 307
    14.3.3 Testing Methodology 307
14.4 Current and Virtualized Service Platform Use Cases 310
    14.4.1 IMS Service Platform 310
        14.4.1.1 Current Implementation 310
        14.4.1.2 SDN/NFV Implementation 312
        14.4.1.3 Key Design Principles Applied...314
        14.4.1.4 Georesiliency, Topology, and Scalability..316
        14.4.1.5 Performance 317
        14.4.1.6 Hybrid Architecture (VNFs and PNFs) 318
    14.4.2 Evolved Packet Core 318
        14.4.2.1 Advantages of Virtualizing EPC 318
        14.4.2.2 Virtual Logical Platform Architecture and Design 319
        14.4.2.3 Network Design 322
        14.4.2.4 Capacity and Scaling Design 323
        14.4.2.5 Integration with ONAP..324
    14.4.3 BVoIP Services 324
14.5 Topics for Future Study 326
References..329

This chapter discusses service platforms that support end-to-end service implementations in an IP network. Examples of these platforms include the IMS—IP Multimedia Subsystem-based platforms [1] that support Voice over LTE (VoLTE), Rich Communication Services (RCS) including video and messaging, consumer Voice over IP (VoIP) services, and Business VoIP (BVoIP) service platforms, which support business customers, and the EPC—Evolved Packet Core [2], which supports mobility voice and data services. The current implementations of service platforms are mostly based on proprietary equipment provided by vendors. SDN and NFV are the key technologies that enable addressing many of the problems with the current implementations. A design framework that uses software defined networking (SDN)/network function virtualization (NFV) to enable new solutions and virtualized implementations of service platforms under SDN control, is described in this chapter. In addition, the chapter also discusses the skills and methodologies pivot needed to implement service designs that use SDN/NFV.

14.1   INTRODUCTION

The evolution of the wireless mobility networks and service platforms that support a variety of voice and data services can be characterized as follows. Third-generation (3G) networks had a goal of providing differentiation between service and platform. 3G provided standardization and open interfaces with backward compatibility with Global System for Mobile communication (GSM) and Integrated Services Digital Network (ISDN), to utilize the significant prior investments. Support of multimedia, wideband radio access, and utilization of the Internet protocol (IP)-based network for data and control transport were important goals of 3G. The higher data rate made it possible for telecommunication service providers to offer multimedia services (such as video streams/conferencing and VoIP) with variable quality-of-service (QoS) needs. It also enabled new service architectures standardized by the Open Mobile Alliance and provided intelligent services to GSM and Universal Mobile Telecommunications Service (UMTS). Telecom operators adopted the service architectures enabled by the various 3GPP public land mobile network (PLMN) circuit-switched (CS) and packet-switched (PS) releases and implemented corresponding service platforms, striving for compliance to the 3GPP standards.

With the industry definition of the long-term evolution (LTE) radio access network (RAN) architecture, telecom operators adopted LTE as the most viable access option for maximizing system coverage, throughput, and capacity. The LTE/EPC infrastructures were developed and launched around 2010. LTE/EPC supports data rates beyond 3G high speed packet access (HSPA), provides higher efficiency for the spectrum, and makes all IP services possible. The LTE implementation required a new LTE radio network supported by the EPC. The voice and short messaging services (SMS) were mapped to an IP-based application to effectively use the all-IP LTE network for all voice and data services. IMS data and VoIP solutions were integrated for service continuity with existing networks.

On the VoIP service platforms side, in the early 2000s, many telecom operators started looking into migrating their entire infrastructure to IP. At that time, there was no generally known architecture available that would support all possible applications, specifically VoIP and multimedia. VoIP was primarily viewed as a technology to either provide an enterprise with the ability to use the existing data network to also carry voice, or provide carriers with a way of bypassing access or egress charges. There was no known comprehensive architecture in the industry that a large service provider could potentially use to support all of its services across an IP network. Telecom operators started to define several major characteristics of the desired architecture and contained within it the key architectural principles that IMS is based on. Separation of access, session management, and service logic layers were the most critical requirements. The ability to support access technologies of the future without impacting the rest of the architecture was also deemed critical. The ability for application logic to be totally independent of the access technology being used at the time of a service call, the ability for an application to be applied to a session over various access technologies, and the ability for several independent applications to be applied to a given call leg were among other important required characteristics.

There was no single network architecture supporting voice, video, and multimedia services. Telecom operators had several separate, standalone, unconnected network islands supporting voice, video, and multimedia services for its various entities. Any traffic between these networks was done via the public switched telephone network (PSTN). The vision for a common architecture for real-time services combined these various networks into a single IP-based network to provide end-to-end IP connectivity, thus improving both network and capital efficiency as well as enabling interoperability and next-generation services. With IMS standards and the associated architecture specifications reaching a level of acceptance in the telecommunications industry and becoming the de facto standard for real-time, multimedia services over wired and wireless networks, telecom operators adopted IMS as the basis for their network architecture, and started to deploy IMS-compliant functions supplemented with additional functions in the network as appropriate and as business needs required.

The current architecture for real-time services over IP is based on a layered framework with the following basic concepts:

1. Infrastructure core for session control and management on top of a converged IP/MPLS (Multiprotocol Label Switching) network
2. Each access technology communicates via a border controller, which provides a uniform internal view
3. Packet core infrastructure that supports packet processing gateway functions
4. To provide a service, the corresponding application server (AS) is plugged in
5. Once a new access technology is supported, it can be used by all existing and future AS
6. Once a new AS is deployed, it can support all existing and future access technologies
7. Application programming interfaces (APIs) expose key network capabilities for application developers

Figure 14.1 shows these basic concepts. The current implementations of this architecture are multivendor environments deployed in telecom service providers' data centers where the physical implementation of most network elements typically consists of a collection of vendor-specific hardware and chassis configuration that is integrated with other infrastructure components including routers, switches, firewalls, and load-balancers. Vendor-specific load-balancing solutions are used for most network elements. The most prevalent database technology used for network databases is relational databases. With the advent of cloud technologies and the industry-wide push to move toward network function virtualization and software-defined networking, the evolution strategy for service platforms moves the physical network functions (PNFs) to virtualized network functions (VNFs) implemented as software that runs in a cloud environment, controlled by a set of policy-based controllers with the ability to perform automated recovery and dynamic scaling of the VNFs.

In the present mode of operation (PMO), services are designed and instantiated on a network platform for real-time services, which consists of physical network elements with tightly coupled hardware and software. Vendors provide proprietary hardware for network functions and each

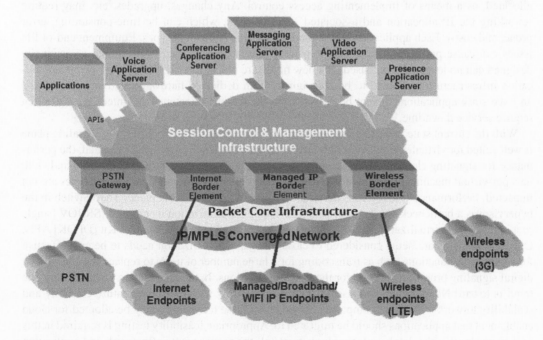

FIGURE 14.1 Conceptual view of service platform architecture.

vendor also provides its own element management system (EMS). This makes it very difficult to select a best in-breed vendor for a network function.

The concept of having one platform for all services presents several problems. The composition of services on the platform is predefined and consists of required as well as nonessential capabilities. New service rollouts and software upgrades typically require changes to existing network elements. This requires extensive regression testing of the platform and existing services, thereby extending the roll-out period of the new service. Session state is tightly coupled with network elements and is lost if network elements at a site fail. Also, upon element failures, stable sessions are lost and there is a possibility of registration storms, which requires overprovisioning network resources to handle storm scenarios. Network platform rollouts (capacity augmentation or upgrades) also have a very lengthy cycle time as a result of a fixed deployment approach based on geographic regional boundaries.

In the current model of introducing new services, especially those that require the deployment of PNFs, it typically takes 12–18 months or more to realize a new service, depending on the complexity of the service and the number of network functions involved. Much of that time is spent in building the laboratory, performing laboratory testing, field deployment, and production testing, prior to service turn-up. Many services are deployed as silos requiring dedicated infrastructure resources. Investment in network infrastructure is made at least 6 months before it starts getting utilized for many services. To meet the business continuity needs in cases of network outages, infrastructure is deployed with $1 + 1$ redundancy per component at a location and across geographical locations in most cases. In addition, capacity is calculated to accommodate the busy hour of all services and applications, including spikes in traffic for mass event applications (like voting for American Idol), resulting in excess network infrastructure that does not get used at all times.

The current architecture is expensive to grow and maintain. Deployment of new network elements and applications requires additional hardware, which must be individually architected and planned prior to deployment. This process can take months. Capacity expansion requires deploying additional dedicated compute resources. This takes careful planning and execution, and additional dedicated capital specifically for each application (not sharable) well in advance of demand.

The current architecture requires every interface to be defined, with every IP address statically allocated, as a means of implementing access control. Any changes, upgrades, etc. may require reworking the IP allocation and associated security rules, which can be time-consuming, error prone, and costly. Each application is designed for specific equipment types. Equipment end-of-life issues can cause problems for capacity expansion, since the same hardware model as originally designed can no longer be purchased. If a new hardware model is used, the network element/application infrastructure may need to be redesigned. With dedicated hardware, maintenance can be an issue since applications cannot be moved from the hardware instance. Maintenance will often require service downtime.

With the current state of virtualization technology and cloud infrastructures, the signaling plane is well suited for virtualization. Although virtualization adds some processing overhead, the performance for signaling elements is acceptable. The ability to provide guaranteed network bandwidth on a per-virtual machine (VM) basis is critical to ensure that end-to-end signaling latencies are not impacted. Performance is a key consideration for virtualizing the media plane. The vSwitch in the hypervisor is a bottleneck for i/o processing. Various I/O optimization techniques (SR-IOV [single root input/output virtualization], faster vSwitches, Intel Data Plane Development Kit [DPDK] APIs, CPU pinning, etc.) are being considered to close this gap. Virtualization needs to be cost effective for some media functions such as transcoding for a large number of users to replace the specialized digital signaling processors (DSPs) for these media functions. In the area of data virtualization, the trend is toward NoSql cloud databases, which are massively scalable. They feature elasticity and scalability to work with cloud computing. A target database technology may be adopted for cloud enablement that applications should be migrated to. Appropriate feasibility testing is required in this area to ensure that the database technology meets all the requirements of network and application databases before they are considered for migration to a NoSql database.

At a high level, there are some key paradigm shifts (Figure 14.2) that should be considered when architecting network functions for SDN/NFV:

1. A shift from a model with integrated software and purpose-built hardware vendor deliveries to a model that focuses on software-only vendor deliveries that can be run on a cloud platform.
2. A shift from dedicated hardware resources that are tightly managed and optimized for peak traffic to using shared commodity-based infrastructure where hardware resources are plentiful and managed as a commodity.
3. A shift from assuming highly reliable hardware to software-based recovery from single component failures where software is distributed on multiple VMs that are load-balanced. QoS policies (anti-affinity rules, availability region, etc.) are used to ensure reliability.
4. A shift from designing solutions to handle millions of subscribers with large failure groups to having flexibility in real-time sizing with decomposition of the large solution into many smaller solutions with smaller failure groups and the ability to scale to different sizes at different locations.
5. A shift from monolithic components that include many functions to functional decomposition of network functions into multiple VNFs.
6. A shift from always instantiating every function to instantiating functions on an as-needed basis.

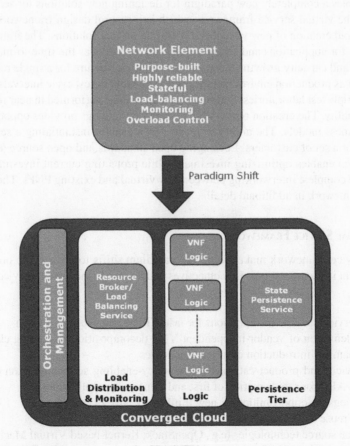

FIGURE 14.2 Paradigm shift: architecting for SDN/NFV.

7. A shift from stateful components with a 1+1 active/failover redundancy model to stateless/ transaction-stateful front-ends with long-lived state decoupled from logic for greater scalability and reliability.

8. A shift from per-network component congestion control (e.g., Session Initiation Protocol [SIP] [3] and diameter [4] overload controls) to minimizing the need for per-hop overload control by dynamically instantiating resources ahead of time using predictive traffic forecasting. Continuous monitoring of resources identifies resource constraints before they are realized. Overload control is achieved using scale-out by dynamically instantiating components as needed.

9. A shift from a model where each element monitors the downstream elements (e.g., via SIP OPTIONS) that it sends messages to, to a model where common cloud capabilities keep track of elements and route traffic to the appropriate set of network elements to achieve uniform monitoring, routing, and management. Cloud monitoring and routing capabilities keep track of failed components and route requests to in-service components. Standard (e.g., transmission control protocol [TCP] monitoring) or VNF-specific monitoring mechanisms may be supported.

10. A shift to a model where network functions run ubiquitously across multiple cloud availability regions for ease of geo-redundancy.

14.2 NEW SERVICE DESIGN SOLUTIONS WITH SDN/NFV

SDN/NFV technologies open, simplify, improve the scaling of, and increase the value of networks. SDN/NFV enables a completely new paradigm for designing new solutions for services. This section describes the virtual service framework, which is a general design framework that uses SDN and NFV for rapid creation of new virtualized real-time service solutions. The framework provides business agility for applications and services launched and reduces the time-to-market (TTM) for service roll-out and capacity growth. A common cloud infrastructure for a single reusable test environment as well as production environment significantly reduces test cycle intervals and costs associated with multiple test laboratories. Software upgrades can be performed in near real-time for new feature functionality. The creation of the VNF and service catalogs provides opportunities for new third-party business models. The ability to reuse VNFs enables instantiating a service as needed for a customer or a set of customers. Leveraging open standards and open source (e.g., OpenStack, restful APIs, etc.) enables optimizing investment while protecting current investment by enabling coexistence and complete interworking between new virtual and existing PNFs. The sections below describe the framework in additional detail.

14.2.1 VIRTUAL SERVICE FRAMEWORK

The virtual service framework makes some key paradigm shifts to address the problems encountered with current implementations. Key objectives and drivers of the virtual service framework are as follows:

1. Virtual service-specific instantiations for faster TTM and capacity growth
2. Optimal definition of vendor-independent VNF decomposition leveraging cloud technology for seamless introduction of new technologies
3. VNF, service, and product catalogs and a policy-enabling service creation environment with open APIs for rapid creation of first and third-party applications
4. VNF catalogs for opportunities for new third-party business models
5. Ability to reuse VNFs across services reducing TTM and cost
6. Use of open source technologies (e.g., OpenStack, Kernel-based Virtual Machine [KVM], open source databases) to lower cost and leverage industry innovation

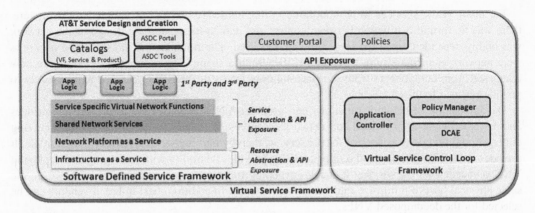

FIGURE 14.3   Virtual service framework.

7. Policy-driven auto self-healing and scaling
8. Enable a virtual cloud-based test model to speed certification and reduce capital cost
9. Enable scaling of services at a lower cost

The virtual service framework defines the software-defined service framework (SDSF). It pivots from the concept of one universal platform for all real-time services to one where services are instantiated on demand on a common commodity-based infrastructure. It leverages the basic principles of NFV that decouples hardware from software and transforms dedicated network functions into software-based VNFs that can run on a common cloud infrastructure while maintaining service requirements for performance, availability, reliability, and security in a multitenant cloud environment. It facilitates mechanisms to customize and instantiate only the essential network functions that are required for a specific service by decomposing current network functions into granular smaller VNFs. Session state is decoupled from the VNFs. This minimizes registration storms and enables a better user experience where a user's session is maintained if the VNFs at a site fail. Services are deployed in virtual zones with smaller failure domains, impacting fewer customers under failure scenarios. N+K georesiliency is used for efficient use of resources.

The virtual service framework defines an ecosystem for service definition and creation. The set of decomposed VNFs forms the basis of a VNF catalog that can be used by service designers for rapid and on-demand creation of a service. Beyond service definition and creation, the virtual service framework defines a virtual service control loop framework for rapid and dynamic orchestration and management of a virtual service. It leverages cloud elasticity and automation to enable automated recovery and auto-scaling. Automated recovery rapidly brings the service back to a complete redundant mode in near real-time and improves the overall service reliability. The ability to auto-scale virtual services using a common pool of cloud resources enables adapting to changes in traffic in near real-time.

Figure 14.3 shows the virtual service framework and the functions that it comprises. The sections below describe the virtual service framework in more detail and highlight specific use cases for virtual services.

14.2.1.1   Software-Defined Service Framework

The SDSF is a virtual network-function-centric architecture. It uses a layered approach where the layers are defined in terms of a cloud service model. It is based on design principles of decomposition of network functions and standardization of technology used for real-time services. Cloud-based paradigms are used to achieve both these characteristics. Each layer of the SDSF is exposed to first- and third-party developers for service design and creation and enables new business opportunities.

For many years, services were monolithic in that hardware was tightly coupled with software, there was no formal distinction between control and data systems, and the design of the network was highly dependent upon the service being delivered. The fundamental units of these networks were network elements strung together like beads on a string. The monolithic services are now composed from components from four decomposed layers of the SDSF, chained appropriately to provide services to the end user.

Decomposition of network functions into granular VNFs enables time to market for services and drives efficient use of cloud and network resources. It enables instantiating and customizing only essential functions as needed for the service, thereby making service delivery more nimble. It provides flexibility of sizing and scaling and also provides flexibility with packaging and deploying VNFs as needed for the service. It enables grouping functions in a common data center or on the same physical host to minimize intercomponent latency. In addition, best-in-breed vendors can be selected for the decomposed VNFs.

The approach for decomposing network functions is to use the following guiding principles:

1. Decompose if the functions have significantly different scaling characteristics (e.g., signaling vs. media functions, control vs. data plane functions, etc.)
2. Decomposition should enable customizing a specific aspect of the network function on instantiations (e.g., the interworking function may need to be customized specific to each carrier interconnect instantiation)
3. Decomposition should enable instantiating only the functionality that is needed for the service (e.g., if transcoding is not needed, it should not be instantiated)

Today, the registration and presence states of the user are tightly coupled with the network functions and are not replicated across disaster recovery sites. In case of a site failure, the registration/presence state of users is lost. This requires all user devices, which were registered at the failed site, to reregister and republish their presence information. This causes a massive registration/presence storm at the disaster recovery site and increases the risk of failures at the disaster recovery site due to the additional load. This also requires overdimensioning to handle capacity needed during registration/presence storms. Also, today, the state of stable sessions is tightly coupled with the network functions and is not replicated across disaster recovery sites. In case of a site failure, the state of stable sessions being handled at that site is lost. This impacts users and has a less-than desirable user experience. Decoupling long-lived state including the registration/presence state of users and session state of stable calls from the virtualized network function in a cloud enables storing the state in a separate database tier. The database tier can be implemented as a cloud-based state persistence service using a cloud database and a caching solution. The long-lived state is replicated across a set of disaster recovery sites by the database tier and is available at those sites. In case of a site failure, user devices can continue being registered with the network since the registration state is available at the disaster recovery sites. Also, devices do not need to republish their presence information as it will be available at the disaster recovery sites. In case of a site failure, stable sessions of a user will stay up since the state of the stable sessions is available at the disaster recovery sites. It improves the overall reliability of the network. Decoupling state from the network function also improves scalability as the two tiers can scale independently of one another. This improves cost efficiencies since additional capacity no longer needs to be deployed to handle registration/presence storms.

Vendor-proprietary load-balancing mechanisms that are built into many current network functions are decoupled to enable use of standardized cloud-based load-balancer solutions where possible. Databases tied into network and application functions are also decoupled to enable use of open-source scalable cloud databases where possible. Standardization of technologies provides operational benefits where common instantiation, configuration, scalability, and availability paradigms can be used. This significantly simplifies the manageability of service functions.

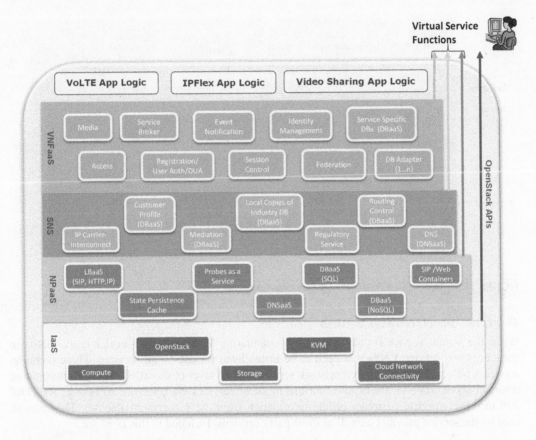

FIGURE 14.4 Software-defined service framework (SDSF).

Figure 14.4 depicts the principles of SDSF by illustrating a high-level view of decomposition of functions and showing the different layers needed to support virtualized services. These layers are described in more detail in what follows.

14.2.1.1.1 VNFs as a Service (VNFaaS)

This layer includes decomposed VNFs that are instantiated and customized on a per-service basis. They form the basis of the decomposed VNFs that are part of the VNF catalog available to service designers. At a high level, they may include the following:

1. Access functions that include session border functions for SIP, web, etc. communications
2. Media functions including audio/video transcoding, playback, recording, etc.
3. Session control
4. Registration and user authentication
5. Event notification functions that enable application logic to subscribe to and receive events (e.g., changes to a customer profile)
6. Identity management
7. Federation for federated services
8. Service broker

Figure 14.5 shows an example of functional decomposition of network functions.

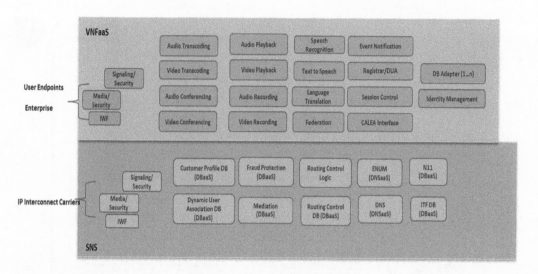

FIGURE 14.5 Decomposition of network functions (example).

14.2.1.1.2 Shared Network Services

While the virtual service framework drives instantiating VNFs on a per-service basis, there are some common service VNFs that need to be instantiated for use by all services. These common service VNFs make up the shared network services (SNS) layer of the SDSF. The pre-instantiated common services/VNFs are exposed for use by other services via reference APIs (e.g., RESTful API to retrieve/update customer profile information from the subscriber database). They can be used by the service provider as well as third-party services. Included in this layer are

1. IP carrier interconnect, which a telecom operator would set up with other carriers independently of the services.
2. Routing logic.
3. Routing databases and local copies of industry databases or interfaces to those databases.
4. Customer profile database that can be used by all services to retrieve and update customer-profile information. This is a scalable cloud-based database that can be used to mine customer-specific information to provide value-added services.
5. Mediation function and database that are used by services to send call detail record (CDR) information. Besides providing data for billing purposes, this database serves as the basis for information used for correlation and analysis to detect certain faults.
6. Regulatory service that interfaces with the government regulatory systems.
7. Common Domain Name Server (DNS) instantiated and used for inter-VNF communication and for communication from devices external to the network.

14.2.1.1.3 Network Platform as a Service

The network platform as a service (NPaaS) layer exposes a set of common cloud services that can be used by real-time services. It enables use of a common set of technologies that can be instantiated and managed in a common way. These services can be exposed to third parties and can provide a new revenue stream for the telecom operator.

Some NPaaS services that may be provided are described below:

1. *Database as a service (DBaaS)*: Database services provide a common pool of high-performance database resources, data replication, and a distributed data cache. The database resources may be based on object stores or relational databases (e.g., SQL [Structured

Query Language], NoSQL (Non SQL) key/value pair, NoSql document, etc.). The service should provide throughput and predictability for all workloads including reads, writes, inserts, and deletes. It will support a mechanism for logically presenting a single view of distributed data for large databases such as the home subscriber server (HSS) and will support a distributed in-memory data cache for low latency access (e.g., 2 ms response time). It will support mechanisms for replication across all or a subset of availability regions.

2. *Load balancer as a service (LBaaS)*: Load balancer/resource broker services support the capability of distributing requests across VMs in a VNF cluster and across multiple VNF clusters. They monitor the VNFs using standard (e.g., TCP) or application-specific (e.g., SIP OPTIONS) monitoring capabilities. They support distributing requests to local as well as cross-site VNFs using local and global load balancing. If an entire local pool of VNFs fails, the global load-balancing capability can distribute requests to a pool at another site. Session-aware (e.g., SIP, diameter, HTTP, etc.) load-balancing capabilities and client/server persistence (stickiness) are required for some VNFs (e.g., transaction-stateful network elements). IP source-based load-balancing mechanisms are required for some VNFs. Distribution of requests can be using either the round-robin or least utilized algorithm. The load-balancer service maintains a view of the utilization of each VNF so that it can distribute requests to the least-utilized VNF as appropriate. It keeps track of failed VNFs and distributes requests only to in-service VNFs. It marks an entire VNF cluster to be out of service when a preconfigured number of VMs in the VNF cluster fail. It supports dynamic binding for new dynamically instantiated VNFs that it needs to add to its distribution list. There can be multiple instantiations of this service if needed that may be customized to the needs of the network functions that it sends requests to. Each instantiation will need to maintain the availability state of the network functions that it monitors.

3. *DNS as a service (DNSaaS)*: The DNS service supports address resolution and performs translation of domain names to numeric IP addresses. It must be ubiquitously deployed within the cloud infrastructure. Name resolution requests are directed to the nearest location based on the source IP to minimize latency for certain scenarios. The DNS service must support a localized DNS view, that is, a request from a network function should resolve to the local network function. It provides APIs to create and manage name resolution (DNS) records. For example, a DNS record can be created using the APIs when a new VM is created. It supports split-horizon DNS entries by providing a domain name for every instance that is launched. This domain name should resolve to the private IP address of the VM instance if queried from within the same availability region, and to the public IP address if queried from outside the cloud infrastructure or from another cloud availability region.

4. *State persistence service*: The state persistence service provides persistence of state in the cloud for components that are stateful. It supports the architectural trend of stateless applications in a cloud and enables network functions to focus on logic. It supports replication of state across all or a subset of cloud availability regions. It supports a highly scalable cache for quick access to state information.

5. *vTaps as a service*: The vTaps service will provide a common way to collect data from the cloud infrastructure for session tracing. Examples of protocols required for session tracing include signaling protocols such as SIP, HTTP, H.248, and diameter and media/messaging protocols such as real-time transport protocol (RTP), message session relay protocol (MSRP).

14.2.1.1.4 Infrastructure as a Service

The infrastructure as a service (IaaS) layer provides the cloud infrastructure or network functions virtualization infrastructure (NFVI) (referred to as AIC in Chapter 4), which comprises the physical hardware and operating environment. Separate physical storage servers are also provided along with the compute nodes. The computing resources, for example, CPU and memory are abstracted by the hypervisor layer. The hypervisor layer makes virtual resources available for deploying virtualized applications. The hypervisor that will be used for the SDSF is an OpenStack supported KVM hypervisor. The hypervisor abstracts the physical resources in the form of VMs. VMs are independent of each other and offer dedicated virtual resources to the virtualized applications. Applications along with tools can be loaded onto the VMs and executed in support of a virtualization project. Refer to Chapter 4 for the IaaS, which is provided by the NFVI.

14.2.1.2 Virtual Service Control Loop Framework

The virtual services control loop framework enables on-demand rapid and reliable instantiation of real-time services on a cloud infrastructure. It supports near real-time correlation of events for policy-based advanced elasticity and auto recovery. The framework includes instantiating the service components and the connectivity required to all virtual and physical components of the service. The framework enables pivoting to an operational paradigm where services can be automatically scaled as well as recovered from software and hardware failures.

The virtual service control loop framework aligns with the Open Network Automation Platform (ONAP) framework (discussed in Chapter 7) for closed-loop automation as shown in Figure 14.6. It provides dynamic auto-recovery and auto-scaling capabilities for virtual services. The flow for closed-loop automation is as follows:

1. Virtual functions send fault and performance data to Data Collection Analytics and Events (DCAE).
2. DCAE performs correlation and analytics on data received at both the virtual function level as well as the service level. For example, at the virtual function level, it may detect that a VM needs to be rebooted. At the service level, it may detect that the rate of incoming registrations is much higher and is exceeding a threshold for a service.
3. Policy is consulted for events generated by DCAE as needed and includes recommended actions for specific events.
4. The controller may be requested to perform auto-recovery actions, for example, reboot a virtual function.
5. The orchestrator is requested to perform an action if an instantiation needs to be performed in real-time for auto-recovery (e.g., a new virtual function needs to be instantiated) or for auto-scaling (e.g., to expand a service footprint).

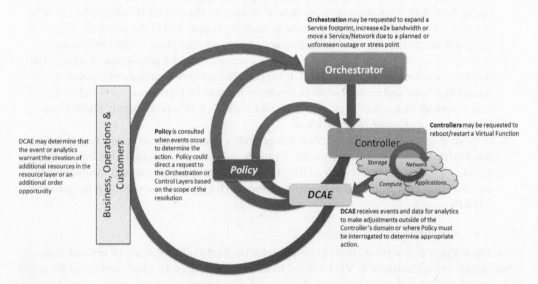

FIGURE 14.6 Closed-Loop Automation.

When a control loop action is taken, an event is sent to the ONAP portal for operations visibility into any auto-scaling and auto-recovery action. During service designs, the set of events and performance data required for control loop automation is identified. Correlation and analytics at the virtual function level and at the service level (across virtual functions that are part of the service) is designed, and the appropriate policies at the virtual function and service levels are created for the recommended actions.

14.2.1.3 Deployment Considerations

Real-time services such as VoLTE have an operational requirement to constrain failures to a localized region in the spirit of avoiding a disaster that impacts customers nationwide.

The virtual service framework defines the concept of a virtual zone (see Figure 14.7) to enable using cloud capacity where it exists and moving from a regional to a national model. A virtual zone is a cloud availability zone that contains a grouping of virtualized registration and session control network functions that supports a set of subscribers from a localized region. For subscriber expansion at a localized region, the subscribers are typically homed to virtual zones within the localized region. If there are capacity constraints at cloud availability zones within the localized region, the new subscribers may be homed to virtual zones that are located outside the localized region temporarily, until the time when capacity is available at cloud availability zones within the localized region.

This approach enables rapid expansion of subscribers while still meeting the operational requirement of avoiding a nationwide impact to subscribers if a data center fails. It enables utilizing cloud capacity where it exists, avoids stranding available capacity and also enables a flexible design for georesiliency.

With the implementation of virtual zones, sizing of the equipment at each zone will be small so as to enable deployment of just the capacity that is needed. Also, with the concept of smaller virtual zones, we achieve smaller failure domains. The impact of a site failure is limited and failure of any one site in this model will not result in loss of service or degradation of service to an entire region.

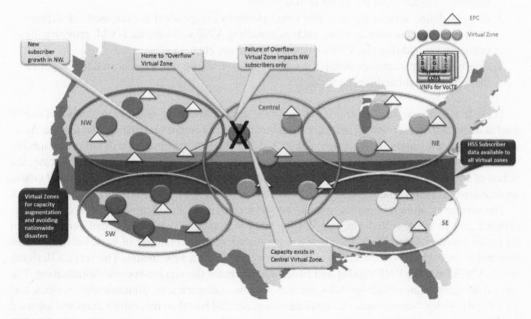

FIGURE 14.7 Virtual zones.

14.3 PIVOT TO SDN/NFV: METHODOLOGY, PROCESS, SKILLS

Delivering services using SDN/NFV requires a change in mindset, methodology, processes, a pivot of skills as well as cross-organizational alignment to deliver on the promise of virtualization for service platforms. A significant emphasis on training and educational sessions for SDN/NFV, ONAP, and change in methodology and processes is required to enable a scalable model for designing and implementing virtualized services.

14.3.1 SERVICE CREATION

SDN/NFV enables creating and rolling out services quickly. The ONAP service design and creation environment (covered in Chapter 7) supports the specification of real-time services in near real-time. It provides the ability to specify custom parameters and policies for service instantiation within the cloud. The service designer can make specific selections from the VNF, SNS, and NPaaS services. These are all represented in catalogs that are available to the service designer to create services that form the basis of a "service catalog." End-to-end services are created using resources from the various catalogs and are exposed as "product catalogs" to external customers/service designers based on policies that determine combinations of services and resources that have been certified and can be used to create a product. The service templates contain details of a service including its chaining logic, business policies, engineering rules, installation recipes, service policies, dimensioning of virtual resources, and service configuration parameters.

The definition of a service includes four components:

1. Service data model files define "what" the service is comprised of. They contain pointers to specific VNFs that needed to be instantiated for the service being defined.
2. Business policies and engineering rules define "how" the service should be instantiated. They contain the constraints of the service such as QoS, affinity rules, and security zones. For example, if there is a voice component to the service, the appropriate QoS will be included so that when the service template is executed, appropriate resources are allocated for the service. Dimensioning defines the sizing of the service with respect to virtual resources needed from the cloud infrastructure.
3. Recipes define service policies and cloud platform independent scripts used for deployment and lifecycle management, such as installing VNF software on a VM, provisioning connectivity, auditing the VNFs when they come up, etc.
4. Configuration parameters define the service-specific configurations required to customize the VNFs instantiated for the service.

These four components of a service definition are used to define the service templates that can be used as an input to the service orchestrator to automatically instantiate and manage the service. As an example of customization for a service such as VoLTE, the QoS parameters will indicate a specific "Class of Infrastructure" that the service provider needs to offer this service on. In this case, the "Class of Infrastructure" selected will indicate that support for media is required. This will enable appropriate cloud infrastructure resources to be allocated so that the voice quality will be preserved.

The service creation process for virtualized services changes significantly. With every new virtualized service being rolled out, the VNF catalog is populated with VNFs that can be reused for future services. Once the VNF catalog is populated with the majority of basic reusable VNFs required to create a service, creating a new service pivots to a new model. The service designer selects VNFs from the VNF catalog and customizes them for the service-specific instantiation. The service designer creates chaining rules, service-specific configurations, dimensioning, recipes, and associated policies. Service-specific templates are generated based on this information and are used for the service-specific instantiation.

14.3.2 SERVICE DESIGN METHODOLOGY

The traditional waterfall approach for designing services pivots to using the Agile approach with a DevOps model. Design teams partner upfront with all the other organizations, vendors delivering the VNFs are required to deliver software releases in Agile fashion with shorter timeframes for integration and testing in an iterative mode. The high-level design of a virtualized service and the VNFs that comprises the service is documented and the Agile process is used to kick off small scrum teams that then work through the various aspects of the design, network connectivity, interworking with PNFs, instantiation, control loop scope, etc. This approach enables rolling out service VNFs and capabilities iteratively for testing in the laboratory as well as production environments. A common service design process, common templates to document the design, common templates to provide requirements to partner organizations, and common solutions that apply across all service designs are expected to reduce the overall service design life cycle. As the VNF catalog gets populated and service designers pick their VNFs from the catalog, there should be a further reduction in the service design life cycle.

14.3.3 TESTING METHODOLOGY

Figure 14.8 shows the key paradigm shifts required in an SDN/NFV-based architecture. In today's environment, approximately 50% of the testing effort is consumed certifying new hardware platforms. By instantiating our network functions on a virtualized cloud platform, the hardware platform no longer needs to be tested, and the testing effort can be appropriately reduced. Essentially, virtualization isolates the test environment from hardware obsolescence. By using virtualization, orchestration, and SDN control on a common cloud platform, virtual compute resources or VMs can be spun-up as needed, and virtual storage allocated instantaneously to create a laboratory environment.

The service testing effort can be viewed and validated in layers, where each layer builds on the layer(s) below. Functionality is validated in layers and functions already validated in lower layers are reused for new services. This is pictorially represented in Figure 14.9.

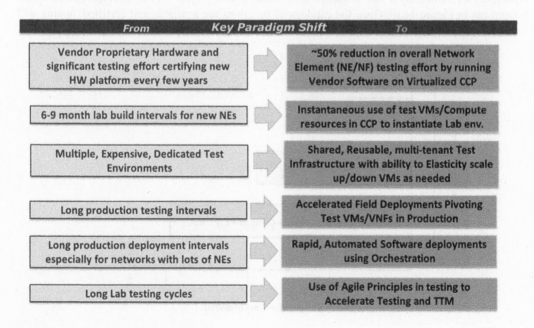

FIGURE 14.8 Paradigm shifts in testing.

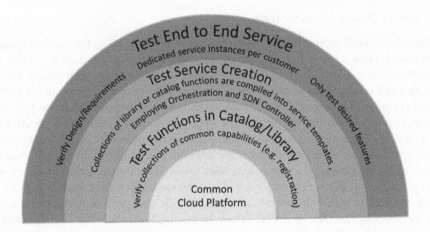

FIGURE 14.9 Testing vision (2020).

Certification testing of all virtual functions is accomplished on an existing production cloud infrastructure. First, testing is performed on common capabilities that are provided by a catalog of network functions. Once the individual VNFs have been validated, they do not need to be revalidated for use in new service offerings and the testing effort shifts to validating groupings of VNFs that make up a service. Full testing and certification of the service will allow the test team to focus on the next higher layer in the model. Once the service and service templates have been validated, testing efforts will focus on the highest layer in the model: end-to-end service (or product) testing. This is where multiple services are combined into a product to provide the desired end-user functions. With the lower layer VNFs, services and service templates fully tested, they can be used as building blocks to quickly assemble a new end-to-end service.

With the deployment of a cloud infrastructure and the development needed to support real-time communication services, all testing (laboratory and field) can be achieved in 6–9 months. This represents approximately a 50% reduction from today's testing cycle time and is represented by the bottom bar in Figure 14.10. Once the catalogs of functions are available and certified, testing will focus on service creation and E2E Service testing, and services can be realized in approximately

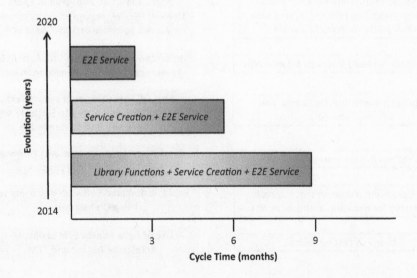

FIGURE 14.10 Testing cycle time reduction.

3–6 months as represented by the middle bar in Figure 14.10. Looking further ahead, once all the catalogs needed to realize a service have been certified and the service creation environment has been realized, testing will focus on end-to-end service flows and capabilities and services can be realized in a matter of days or weeks. This interval is represented by the top bar in Figure 14.10. Furthermore, customers will be able to build services themselves using exposed APIs, which could eliminate the need for further testing altogether. Of course, the testing intervals will vary based on the number of VNFs required, complexity of features/service, and quality of VNF software.

Using a production cloud location to instantiate the VNFs is a key enabler that allows us to shift from an environment where field testing follows laboratory testing resulting in a long test interval, to a model where we can accelerate field deployments by pivoting test VMs and VNFs into production VMs and VNFs once laboratory testing on those VMs has completed. Using production VMs also enables parallel field testing. Using service orchestration, we will also be able to significantly reduce field deployment intervals. Upgrades and configuration changes are automated and can be fully deployed in a matter of hours or days once the code has been proven to be stable in production.

In the PMO, testing progresses in a serial or "water-fall" fashion and field testing (network validation testing, operational readiness testing, Service First Field Application [FFA]) follows the completion of laboratory testing. In the future mode of operation (FMO), field testing substantively overlaps laboratory approval for use (AFU) testing on the same set of VNFs, resulting in a dramatic reduction in cycle time. Once laboratory AFU testing on VMs and VNFs completes, those VMs and VNFs are "pivoted" into production and a much smaller amount of additional field testing is required. Figure 14.11 depicts this.

Transitioning to a fully virtualized network will take time to realize. During the transition to an all virtualized network, PNFs that have not yet been virtualized will need to interconnect with newer VNFs to provide an end-to-end test environment. The goal is to leverage existing PNF laboratory equipment, and not to replicate those PNFs, in order to minimize laboratory costs. Since testing is done in a production cloud environment, laboratory certified VM/VNF templates or Snapshots can be directly applied to instantiate production VMs/VNFs. OpenStack supports sharing Glance

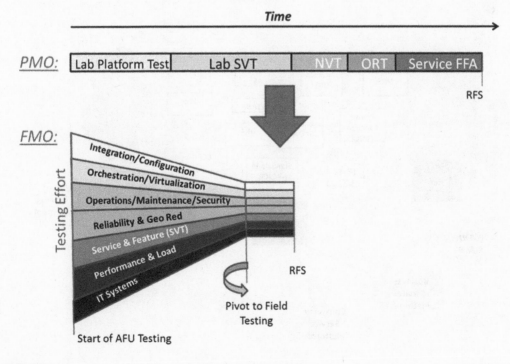

FIGURE 14.11   Pivot from lab to production to reduce cycle time.

images across cloud sites. This minimizes human error associated with network-wide deployments of new VMs/VNFs. Even initial field deployments will look more like capacity expansion efforts than new technology deployments.

14.4 CURRENT AND VIRTUALIZED SERVICE PLATFORM USE CASES

This section provides a view of some of the key service platforms and services, their current implementations, and virtualized implementations.

14.4.1 IMS SERVICE PLATFORM

IMS is an architectural framework that provides a standardized way to deliver multimedia services over an IP network. IMS further enables fixed-mobile convergence by providing the ability to access services over wireless and wireline access. The session control layer isolates the access layer from the applications layer to enable access to agnostic IMS-based services.

14.4.1.1 Current Implementation

Current IMS service platform implementations are multivendor platforms that provide voice call processing and messaging support to multiple VoIP services such as VoLTE. IMS service platforms maintain the layered architecture with the access, core, and AS layers. Access nodes contain border elements (such as session border controllers [SBCs]) that provide access from a geographic region and interface with the core network (CN) functions. The CN functions interface with various ASs that provide services.

Figure 14.12 shows the various components of an IMS-based service platform and the interfaces to other networks. In addition, the platform also contains surround support components for operations and maintenance functions including vendor-provided EMSs, software upgrade servers, auditing tools, and physical probe elements with interfaces to OSS, BSS, and billing systems.

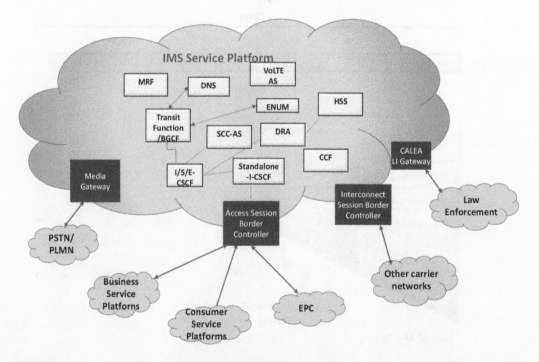

FIGURE 14.12 IMS service platform components and interfaces.

IMS service platform implementations are based on the following design goals:

1. The design should be scalable in the long term and should not be constrained by transitional designs.
2. The platform is a shared infrastructure supporting multiple services on a shared core.
3. There are well-defined and limited interfaces to client service infrastructures, such as SBCs, ASs, OSSs, and other shared elements and services.
4. The design must meet carrier-grade requirements of five 9 s (99.999%) availability.
5. Geo-redundancy is required.
6. The design needs to allow operations to easily maintain and troubleshoot failures and must include design for audit tools, call trace probes, backup/restore, console server, jumpstart/kickstart, and software upgrade servers.
7. Minimal configuration changes should be required for growth of components.

As shown in Figure 14.13, the IMS service platforms follow a layered architecture.

The AS layer contains various ASs that support voice and multimedia services. Below are some of the ASs supported:

- *CPM*—converged IP messaging server; provides SMS-over-IP service.
- *VoLTE AS*—telephony AS; provides calling feature and specialized service logic in call processing for VoLTE.
- *UCE AS*—user capabilities exchange AS; enables users to discover service capabilities of the devices of their contacts, thereby enhancing communications.
- *PCM AS*—personal communication manager AS; provides access to call logs and other portal services.

The core contains the session management and media network elements including

- *CSCF*—call session control function; the serving CSCF (S-CSCF) is the principal call processing element. When a user endpoint (UE) registers with the network, the UE is assigned to a particular S-CSCF. The interrogating CSCF (I-CSCF) is used for making queries

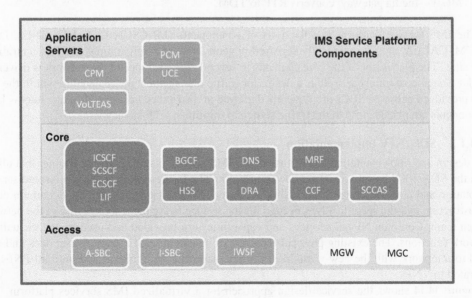

FIGURE 14.13 IMS service platform layered architecture.

to the HSS, and for assigning the S-CSCF to the UE. The emergency CSCF (E-CSCF) handles emergency services, such as 911 calls.

- *BGCF/TF*—breakout gateway control function/transit function; provides optimal routing to the TDM network. The transit function is also included in this element, which supports IP interoperation with other carriers.
- *HSS*—home subscriber server; master database with subscriber profiles, performs authentication.
- *DNS*—domain name system; allows network elements to address one another using symbolic names.
- *ENUM*—E.164 Number to Uniform Resource Identifier (URI) mapping; translates telephone numbers into Internet addresses.
- *DRA*—diameter routing agent; finds the HSS that corresponds to a particular subscriber since the subscriber's data are not common to all HSSs.
- *CCF*—charging collection function; receives accounting request (ACR) information from the network elements and assembles them into billing records, or CDRs.
- *MRF*—Media resource function; provides tones and announcements.
- *SCC-AS*—service centralization and continuity AS; accommodates call continuity when a UE leaves LTE coverage, such as moving into a 3G-only area.
- *LIG*—lawful intercept gateway; interface for communications assistance for law enforcement act (CALEA) intercept requests.

The access layer contains the border controllers and gateways:

- *A-SBC*—access session border controller; serves as a firewall, handles IPv4-to-v6 interworking and other services. Implements the proxy CSCF.
- *I-SBC*—interworking session border controller; similar to the above but used between untrusted systems.
- *DBE*—data border element; provides a stateful firewall, deep packet inspection, and other services.
- *IWSF*—IMS web security function; provides a secure interface for a variety of endpoints.
- *MGCF*—media gateway control function; handles signaling for the corresponding MGW.
- *MGW*—media gateway; converts RTP to TDM.

The IMS service platform is a fixed set of components (I/S-CSCF, CCF, HSS, DRA, DNS, ENUM, CALEA, etc.). It is typically deployed in georedundant configurations at a few centralized core sites. The placement of the core elements in several core locations in the network is driven by vendor design constraints as well as a desire for traffic management and balancing within the service provider's network. SBCs are typically deployed at distributed access node sites. Access layer components are placed close to the traffic source to minimize RTP latency issues.

14.4.1.2   SDN/NFV Implementation

The design and implementation of the virtualized IMS service platforms is in a manner that aligns with the SDN/NFV-based virtual service framework. The framework introduces a layered service adaptation and abstraction approach with lower layers such as IaaS provided by a common cloud infrastructure and the specific VNFs needed for the service being the responsibility of the network design team. Common NPaaS services and capabilities are provided and used by the virtualized network functions. The existing physical IMS service platforms will coexist for services and will need to interwork with the new virtualized service platforms for a time period until all PNFs are migrated to VNFs.

Figure 14.14 shows the service design approach for a virtualized IMS services platform. The virtualized IMS services platform design is based on the virtual service framework and aligns with

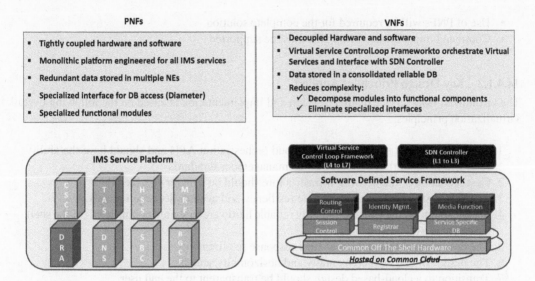

PNFs

- Tightly coupled hardware and software
- Monolithic platform engineered for all IMS services
- Redundant data stored in multiple NEs
- Specialized interface for DB access (Diameter)
- Specialized functional modules

VNFs

- Decoupled Hardware and software
- Virtual Service ControlLoop Frameworkto orchestrate Virtual Services and interface with SDN Controller
- Data stored in a consolidated reliable DB
- Reduces complexity:
  ✓ Decompose modules into functional components
  ✓ Eliminate specialized interfaces

FIGURE 14.14   Virtualized IMS service design approach.

the new architecture and design principles for SDN/NFV. Reusable decomposed VNFs are defined, designed, and added to the VNF catalog. The reusable VNFs can be used for service-specific instantiation on an as-needed basis for services. For example, the reusable VNFs may be instantiated for the VoLTE service and can be instantiated separately for a consumer VoIP service. Some VNFs that are part of the SNS (e.g., for IP-carrier interconnect) will be shared across services.

Figure 14.15 illustrates the SDSF that is one aspect of the complete virtual service framework for the virtualized IMS service platform. Some points to note about the design include

- The infrastructure layer is provided by the common cloud infrastructure
- The NPaaS layer exposes a set of common cloud platform services required that can be used
- The virtualized solution is a combination of VNFs and PNFs

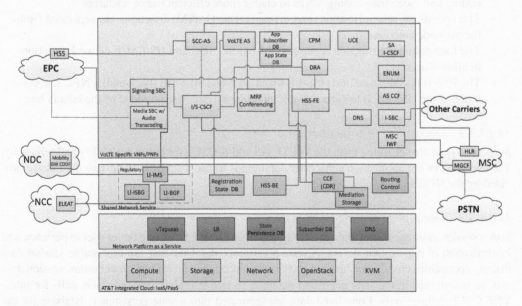

FIGURE 14.15   Logical view of virtualized IMS platform.

- Use of PNFs will be required for the complete solution
- Continued interworking with PNFs must be supported

14.4.1.3 Key Design Principles Applied

The virtualized IMS service platform design and implementation is based on the following overall virtualization principles:

1. The design should support standard cloud orchestration APIs and should have the ability to evolve in support of emerging cloud management standards
2. Cost reduction of the network infrastructure should be a major consideration
3. The design should meet carrier-grade resiliency and availability requirements
4. It should address the shift from highly reliable hardware to shared commercial off the shelf hardware of sufficient quality
5. The design should meet the real-time response requirements
6. The design should be incrementally and horizontally scalable
7. Transition to a cloud-based design should be transparent to the end user
8. Support for runtime BSS/OSS capabilities should be integrated
9. Relevant security compliance requirements must continue to be supported

The following principles are applied to the overall implementation.

14.4.1.3.1 Decomposition of Functions

As stated earlier, decomposition of PNFs into granular VNFs is a key principle applied to virtualize IMS service platforms. The decomposition of the IMS network functions is illustrated with a few examples:

- The SBC is decomposed into signaling access, media access, interworking (IWF), and audio transcoding VNFs to enable more efficient use of resources.
- The media server is decomposed into the media signaling, media processing, audio transcoding and video transcoding VNFs to enable more efficient use of resources.
- The operations, administration, and management (OA&M) functions are separated from the network functions.
- The load-balancing function is separated from the SBC and I/S-CSCF network functions to allow elasticity.
- The HSS is decomposed into HSS-FE (front-end) and HSS-BE (back-end) VNFs. The evolution of the HSS-BE is to use a scalable cloud-based database provided by the NPaaS layer.

14.4.1.3.2 Decoupling of Subscriber Data

Subscriber data are separated from the VoLTE AS and S-CSCF for better scalability and reliability. The evolution of the application subscriber database is to use a scalable cloud-based database provided by the NPaaS layer.

14.4.1.3.3 Decoupling of Long-Lived State

Stable session and registration data are separated from the S-CSCF for a better user experience and minimization of registration storms. A state persistence database that can host stable session data that are accessible across georedundant sites is introduced. For initial implementations, session data may be maintained in the state persistent database to enable maintenance of stable calls for intra-VNF CSCF failures only. Long-lived data are separated into a state persistence database for the VoLTE AS for better reliability.

14.4.1.3.4   Scalability at VNF Level

An emphasis on N+K local redundancy is applied to enable horizontal scalability of VNFs, reduce the capital expenditure, and increase site-based reliability for VNFs where possible, including the CSCF and VoLTE AS. This also enables services to be massively scalable where a specific VNF can be scaled as needed, independent of other VNFs.

14.4.1.3.5   Resiliency

An emphasis on N+K georesiliency is applied to reduce the capital expenditure and overprovisioning of network functions. There is a heavy emphasis on software-based recovery of service VNFs from cloud failures. All VNFs support either $1 + 1$ or $N + K$ local redundancy. Locally at a site, the VNF VMs are distributed across multiple availability zones using anti-affinity rules for additional resiliency. In addition, the number of sites that a service is distributed at accounts for cloud hardware and software upgrade downtimes to ensure meeting the overall reliability needs of a service. As additional cloud sites become available, virtualized IMS service VNFs can be deployed in a much more distributed fashion at a greater number of sites.

14.4.1.3.6   Smaller Failure Domains

Virtualized IMS service platforms will take advantage of broad cloud deployments to deploy smaller VNF pools or clusters at each site, so that each site will support a smaller number of subscribers, thus limiting the impact of a site failure. It also enables the services to be massively scalable.

14.4.1.3.7   National Subscriber Provisioning

Subscriber databases will implement a "National Subscriber Provisioning Distribution" model to minimize stranding of capacity and allow more efficient use of database resources.

14.4.1.3.8   Maintaining Quality of Service

Differentiated services code point (DSCP) markings by VNFs are used to maintain quality of service for specific traffic types (media, signaling, OA&M) routed in the cloud environment.

14.4.1.3.9   Licensing

The mechanism of a universal license key or a pool of licenses provided by VNF vendors is used to enable automated instantiation, automated recovery, and scaling of traffic in a cloud environment.

14.4.1.3.10   Common NPaaS Services

Common platform services are provided by the NPaaS layer of SDSF and include the following:

- DNS services that can be used to distribute traffic
- Cloud databases that can be used to store application and network subscriber data
- Cloud databases that can be used to store state data
- Load balancers that can be used to distribute load to the VNFs
- vTaps services that can be used for probes

14.4.1.3.11   VNF Catalog

A list of reusable VNFs that are added to the VNF catalog for IMS service implementations is shown below:

- CSCF
- Application subscriber database
- State persistence database
- Access signaling SBC
- Access media SBC

- Interconnect signaling SBC
- Interconnect media SBC
- Transcoding
- Audio conferencing MRF
- Video conferencing MRF
- HSS-FE
- HSS-BE
- CCF
- Virtualized probes
- DBE

14.4.1.3.12   Integration with ONAP

With the virtual services control loop framework and integration with ONAP capabilities, the following critical automation capabilities are designed and engineered for virtualized IMS service platforms:

1. Rapid service creation and orchestration
2. All VNFs will send fault and performance data directly to the DCAE component of ONAP
3. DCAE will support performance reports and data collection from VNFs, PNFs, and vTaps for performance monitoring and fault management
4. Universal license keys or pools of licenses will be supported to enable auto-scaling
5. Auto-recovery will be supported for all VNFs via closed-loop automation
6. Elasticity will be provided initially by triggers from operations and with control loop automation for scaling actions based on predictive analysis and preprovisioned business policies and rules over time

14.4.1.4   Georesiliency, Topology, and Scalability

In order to maximize the use of resources, from a georesiliency standpoint, the following principles should be applied to virtualized IMS service platforms:

1. Failure of a site should not result in a nationwide impact to subscribers.
2. Localized onsite software or hardware failures should be contained.
3. Localized onsite failures due to software defects or misconfiguration should not propagate to result in a nationwide impact.
4. VNFs should be deployed in N+K redundancy at local sites where possible.
5. Individual VNFs should have the ability to fail over to one or more georedundant sites.
6. It should be possible to deploy multiple VNFs for the same network function at a site.
7. Registration state and stable session state should be replicated across all core sites that are part of a georedundant site pool.
8. Each VNF must have high availability software architecture that supports geographic resiliency.
9. Latency requirements must be met for postdial delay. This drives the geographic placement of some of the virtualized IMS service platform network functions.
10. Latency requirements must be met for mid-call features under failure scenarios using the session state from the state persistence database.

A virtual zone contains a grouping of registration and session control virtualized network functions that supports a set of subscribers. A target view would call for deployment of many virtual zones of smaller capacity in each zone. A greater number of cloud sites will enable smaller failure domains and increased efficiency for resource usage in the future. Although each site is mapped to a zone, upon failure of a site, failover can be executed to wherever the infrastructure (VMs, network)

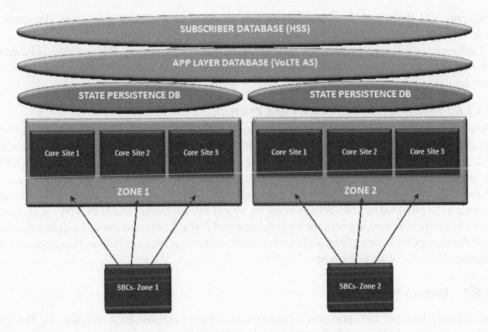

FIGURE 14.16   Georedundancy for virtualized IMS service platforms.

capacity exists. As an example, a deployment may include two virtual zones containing three core sites each and five or six access sites. With this, the geographic deployment of the core sites in each zone follows N + K geo-resiliency where $n = 2$ and $k = 1$. This configuration can potentially result in a 30% capital expenditure savings over a $1 + 1$ georesiliency strategy. Implementation of a nation-wide subscriber database system eliminates the need for regional deployments. Figure 14.16 shows the change to this configuration as an example. Based on an analysis of performance and latency needs for the VoLTE service, a minimum of five sites are needed to meet latency requirements but this does not meet reliability requirements. A minimum of six sites must be deployed to meet reliability as well as latency requirements. Placement or location of the sites is key to meeting the latency requirements. Each site with access network functions distributes new call requests (registration messages) to all three core sites in its georesiliency zone. Under failure of one core site, the other two core sites can handle the load. Each core site operates at 66% capacity under sunny day scenario. Under failure of one site, the remaining two core sites operate at 80% capacity. Virtualized SBCs round-robin requests to the three core sites in its georesiliency zone. With the hybrid implementation of VNFs and PNFs, each core site would use the geographically closest sites for access to the PNFs.

14.4.1.5   Performance
The IMS signaling and media VNFs have special requirements for real-time response times and low latencies. A combination of service-level agreement (SLA) constraints defined in service templates and cloud orchestration mechanisms to enable them are used to meet these performance requirements for virtualized network functions. The cloud is expected to support guaranteed minimal latency of handling signaling messages (e.g., 20 ms), low latency and jitter, and guaranteed high throughput for media processing. The virtualized media handling VNFs, media SBCs and media servers use acceleration techniques of Intel's DPDK APIs, SR-IOV, or CPU pinning and nonuniform memory access (NUMA) (see Chapter 4) as supported by the cloud, in order to meet the required media processing performance requirements.

Virtualized infrastructures typically use oversubscription mechanisms to optimize processor utilization. This could negatively impact performance of network functions since some virtualization functions running on the same physical host may be deprived of resources. The cloud needs to support VM flavors where oversubscription is disabled to guarantee consistent allocation of resources to VMs for network functions with stringent performance needs.

Lastly, site selection and placement of VNFs at cloud sites supporting IMS services is based on ensuring that the end-to-end latency objectives for services meet the required minimum postdial delay for calls under sunny-day scenarios.

14.4.1.6 Hybrid Architecture (VNFs and PNFs)

While the target end-state vision is to have all services use only VNFs, in reality, there will be a period of time where there will be virtual as well as PNFs. Once VNFs of a service are instantiated, connectivity is provisioned to existing PNFs of the service during orchestration. The VNFs are provisioned with parameters (e.g., fully qualified domain names of existing PNFs) to enable VNFs to discover and interface with the existing PNFs. End-to-end cross-domain operations will continue to be provided by OSS/BSS systems for hybrid environments. The IMS VNFs will coexist and will require interworking with the already deployed PNFs. Continued use of a single-subscriber home domain for both virtualized and existing implementations requires interworking between the virtualized and physical components.

14.4.2 EVOLVED PACKET CORE

The currently deployed EPC platform is based on a number of purpose-built components. The EPC platform, as shown in Figure 14.17, consists of a number of elements including the serving gateway (S-GW), packet data network (PDN) gateway (P-GW), policy and charging rules function (PCRF), mobility management entity (MME), HSS, HTTP-proxy, firewalls, load balancers, DNS, and supporting tools. We shall focus on a subset of this architecture, shown in the green square box below.

14.4.2.1 Advantages of Virtualizing EPC

Virtualization affords many opportunities to introduce new features and capabilities that were previously unavailable in a physical compute solution. By deploying a virtualized packet core, many benefits of cloud computing can be realized:

- The ability to achieve capacity and performance increases through rapid advances associated with common off-the-shelf (COTS) hardware.
- The capability to migrate fully developed environments to new hardware without starting from scratch is achieved as the software is abstracted from the hardware layer once virtualized.

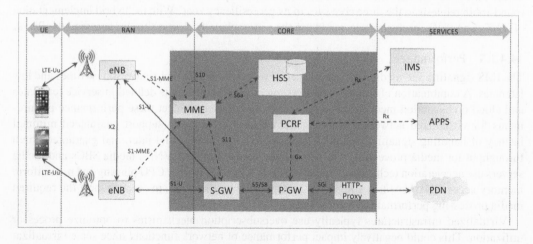

FIGURE 14.17 EPC architecture.

- The potential to pursue best-in-class fault, configuration, accounting, and performance (FCAP) solutions, which are universally applicable across virtualized EPC nodes is realized.
- A reduction in the timeframe and learning curve associated with placing packet core nodes into service through automated deployment methods is achieved.
- A reduction in the installation time for capacity augmentation by using prepositioned COTS hardware is achieved; what used to take 6 months to rack, stack, and power in a data center can be reduced to days if not hours once virtualized systems are fully developed.
- The ability to independently grow VNFs within the EPC to adapt to changes in the call model to avoid congestion or risks to the business is achieved.

With virtualization, EPC nodes will need to interface with other infrastructure components to include

- KVM as the hypervisor OS (Operating System).
- SDN overlays for managed communications between VNFs and systems deployed within a cloud.
- Use of OpenStack tools to support the deployment of VNFs and their associated VMs.
- Integration with the ONAP platform to
  - Automate the instantiation and configuration of EPC VNFs
  - Monitor the VNF components
  - Automate the recovery of VNF components

thereby, significantly improving the in-service time.

All VMs of virtual EPC VNFs are designed with either N + 1 or N + K redundancy to ensure the realization of the above benefits. Standalone single VM instances are very rare in the virtual EPC design.

14.4.2.2 Virtual Logical Platform Architecture and Design

The logical architecture of the virtual EPC platform is very similar to its current physical counterpart, except that all the elements are now software systems deployed on common hardware. The elements in the virtual EPC platform logical architecture are shown in Figure 14.18. In this section, we shall discuss the virtualization of the elements (MME, HSS, PCRF, SAEGW, Proxy) shown in the figure.

14.4.2.2.1 S-GW and P-GW VNF Components

SAEGW or System Architecture Evolution [2] Gateway elements combine Serving Gateway (S-GW) or PDN (Packet Data Network) Gateway (P-GW) functions in a logical network function. AT&T's current SAEGW has the potential to function as an S-GW-Only, P-GW-Only, or both. S-GW or P-GW elements are typically deployed in AT&T's LTE mobility networks.

- **S-GW (Serving Gateway)**: The S-GW performs many functions while connecting to the e-NodeBs on the RAN network, and the Mobility Management Entities (MMEs) and P-GW elements on the mobility network. Details of these functions can be found in various 3GPP documents [8]. Below is a short summary of these functions:
  - Anchor for inter e-NodeB handover
  - Anchor for 3GPP-defined mobile core elements
  - Lawful intercept
  - Packet routing/forwarding
  - Packet marking for transport (both uplink and downlink)
  - User accounting and QCI (Quality of Service Class Identifier) granularity for inter-operator charging

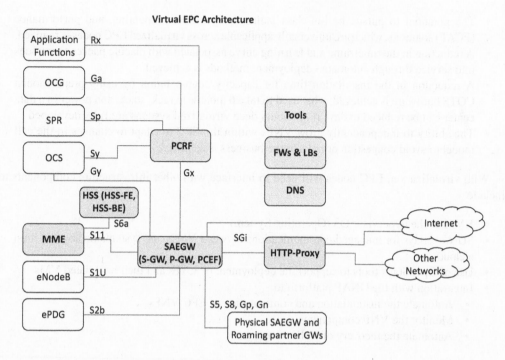

FIGURE 14.18   Virtual EPC design.

- **P-GW (PDN Gateway)**: The PDN Gateway acts as an interface between the 3GPP LTE network elements and other PDNs such as the Internet, and the SIP-based IMS networks. Details of P-GW functions are well described in various 3GPP documents [8]. A short summary of the functions is as given below:
  - PDN gateway
  - Per-user packet filtering
  - Lawful intercept
  - User Endpoint (UE) IP address allocation
  - Packet marking for downlink transport
  - Service level charging support for uplink/downlink

14.4.2.2.2   Policy Control VNF Components

Policy and charging rules function (PCRF) is the network element responsible for policy and charging control in the mobility network. It connects to a subscriber repository and applies charging rules based on the policies defined for the services enabled for a subscriber. These rules are sent to the P-GW for enforcement. PCRF also connects to IMS system VNFs (like P-CSCF) and other policy-enabled applications to provide the policies.

The PCRF consists of load balancers, policy directors, policy engines, session managers, and administrative and control function components. A single PCRF VNF is designed to support multiple instances of SAEGW VNFs. The components of PCRF are designed to scale independently.

14.4.2.2.3   Mobility Management Entity VNF Components

Mobility management entity (MME) is a key element in the control plane of the LTE mobility network. It supports subscriber and session management. Some of the functions that the MME provides include the following:

- Subscriber authentication by interacting with the Home Subscriber Server (HSS)
- Negotiation of security with the user endpoint (UE)
- Reachability of an idle UE
- Activation/deactivation of bearer sessions
- SAEGW selection for initial attach and inter-eNodeB handover
- Termination of Non-Access Stratum signaling
- Enforcement of roaming constraints
- Lawful intercept of signaling
- MME selection for handoffs with MME change

The virtualized MME is designed to work with nonvirtualized MME instances also.

14.4.2.2.4 Home Subscriber Server VNF Components

The Home Subscriber Server (HSS) is a master repository for a given user/subscriber, and contains data elements including user identification, user addressing, user profile information, authentication vectors, and user identity keys. Some of the functions that HSS provides are

- Functions to store and update the following types of information:
  - User identification, numbering, and addressing information
  - Security data, including user identity keys
  - User registration and user profile
  - Location information
- Generation of security keys for authentication, ciphering, etc.
- Support for session control and management by providing authentication

In the virtualized implementation, the HSS is decomposed into HSS-FE (front-end) and HSS-BE (back-end) VNFs. The evolution of the HSS-BE is to use a cloud database that enables scalability and the ability to provide data where needed.

14.4.2.2.5 User Data Repository VNF Components

The virtualization effort has enabled the opportunity to work toward 3GPP defined user data convergence (UDC). The goal of virtualization efforts is to build a user data repository (UDR) that provides all types of user/subscriber information to various network functions involved in providing mobility services to customers. By deploying a UDR, many network functions separate state into a select set of virtual functions, thereby allowing other network functions to reduce state-based processing. This enables these functions to scale independently, which improves efficiency, speed of recovery, and scalability.

14.4.2.2.6 HTTP-Proxy VNF Components

The HTTP-proxy provides the following features:

- HTTP inline optimization for transparent traffic
- Addressable traffic (MMS)
- Subscriber-based opt-in/out in HTTP-proxy
- Information forwarding/header enrichment
- TCP proxy traffic (TCP tunnel) for transparent traffic
- HTTP tunnel for transparent traffic
- Simple and secure client-based service entitlement
- Security enhancement with HTTP-proxy trusted and nontrusted IP addresses

- IPv4 and IPv6 user plane support
- Centralized Multi Service Administrator (MSA) management
- Sponsored data

The HTTP-proxy consists of management nodes, administrative and monitoring nodes, traffic servers, distribution servers, data cache servers, and access servers. The design allows for a small number of management nodes to handle multiple groups of traffic, distribution, access, data cache, administrative, and monitoring servers. All these HTTP-proxy components scale independently to achieve optimal use of cloud resources.

14.4.2.3 Network Design

Cloud infrastructures use an OpenStack Neutron plugin to enable L3 overlay networks via a distributed virtual router deployed in the hypervisor OS. A corresponding local SDN controller is used to create network service instances and implement service chaining features to support scaling (scale-out and scale-in) of network functions such as virtual firewalls, load balancers, DNS, IPs, etc. This is discussed in Chapter 4.

14.4.2.3.1 Overlay Network Design

Virtual networks are logical constructs implemented on top of the physical networks. Virtual networks replace VLAN (virtual local area network)-based isolation and provide multitenancy. Each tenant or an application can have one or more virtual networks. Each virtual network is isolated from all the other virtual networks unless explicitly allowed by security policy. By using virtual networks configured by the SDN local controller, networking of VMs and virtual services is no longer a manual process.

Cloud infrastructure virtual networks can be connected to (and extended across) external physical networks via MPLS/BGP layer-3 VPNs using gateway routers. Virtual networks are also used to implement service chaining of VNFs. VNFs are modeled in the local SDN controller system as "service-instance" and may be implemented by one or more VMs.

An example of NFV insertion is shown in Figure 14.19. The traffic is forwarded from network A to 20.0.0.1 via one or more VNFs. Service instances present routing information in network A such that subnet 20.0.0.0/24 is routed through the VNFs.

14.4.2.3.2 Connectivity to Mobility VPN and Mobility Legacy Networks

The cloud infrastructure data centers must interconnect with the existing data centers and the outside world. The approach to realizing this integration is shown in Figure 14.20. This figure lays the foundation for building the mobility network in a virtual EPC. Connectivity to networks outside of the cloud is accomplished via a pair of the IP-edge routers performing SDN gateway functions for the local SDN controller. When an overlay network that needs to communicate with networks in a different data center is created, a route target gets assigned to that network in the local SDN controller (route_targets parameter in OS::Contrail::Virtual Networks Heat resource).

In this manner, a predefined set of route targets is established between local SDN control nodes and the IP-Edge routers. The IP-edge routers are preconfigured to import any prefixes tagged with route targets into the corresponding routing instances—which in turn are mapped into the network backbone VPNs. Connectivity to existing networks is shown in Figure 14.20 as a pair of

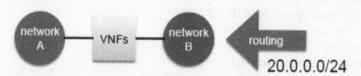

FIGURE 14.19 NFV insertion.

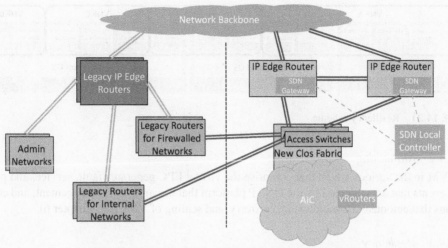

Network Integration Between the Physical and Virtual Data Centers

FIGURE 14.20 Connectivity to legacy networks.

interconnects between the legacy routers and the Clos fabric in the cloud. Routing policies are setup between the legacy and new routers in such a way that interconnects are only used for local legacy ←→ new traffic.

14.4.2.4 Capacity and Scaling Design

The virtual EPC platform is deployed in multiple geographically diverse cloud locations. Each cloud location may consist of at least three OpenStack tenants:

- EPC tenant (VNFs: SAEGW, MSP traffic service, DNS, test tools, firewalls, load balancers, and NAT)
- Support tenant (VNFs: PCRF and MSP management)
- Tools tenant (VNFs: tools server and test tools administration)

There are typically one support and one tools tenant per cloud location. There can be multiple EPC tenants per cloud location. This stratification allows for better capacity management and scaling of the virtual EPC platform. Once an EPC instance reaches its maximum capacity, a new EPC instance is automatically instantiated, and activated.

14.4.2.4.1 Scalability

The EPC platform is designed to scale VNFs independent of each other. The following VNFs all scale independently of other VNFs:

1. SAEGW (S-GW, P-GW and PCEF)
2. HTTP-proxy
3. PCRF
4. DNS
5. Firewalls and load balancers
6. Supporting tools

In addition, components within each VNF can scale independently of other components in that VNF. This capability allows an instance of the EPC to react to business needs dynamically, with optimal utilization of data center resources.

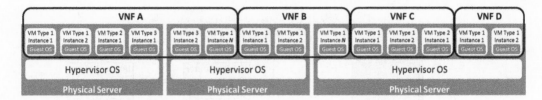

FIGURE 14.21 Resiliency scheme.

14.4.2.4.2   Monitoring

Every VM in the various VNFs that comprise the virtual EPC generates fault, service, and performance events that are collected by the ONAP platform that has analytics, management, and control functions that automate the monitoring, recovery, and scaling of VMs (see Chapter 6).

14.4.2.4.3   Resiliency

Each VNF in a virtual EPC instance consists of several types of VMs (traffic servers, control nodes, database engines, traffic directors, load balancers, etc.) (Figure 14.21).

In such an environment, resiliency is achieved via the following methods:

1. Anti-affinity rules applied to VMs that should not share a host
2. Distribution of VMs across availability zones
3. Maintaining availability zone affinity across failures and scaling functions
4. Geo-diverse distribution of virtual EPC instances

14.4.2.5   Integration with ONAP

The following capabilities of ONAP are used in lifecycle management of the various EPC VNFs and their components:

1. Instantiation of VM groups using OpenStack heat templates.
2. Design of VM groups in a manner that allows similarly scalable VMs to scale-out and scale-in.
3. Automatic configuration of the various VMs in each VNF instance using multiple configuration tools (such as NETCONF/YANG, Ansible, etc.); this configuration function is invoked during different portions of the VM life cycle (instantiation, recovery, rebuild, scaling, etc.)
4. Monitoring of both the infrastructure and the VM components; analytics engines are driven by rules that allow real-time correlation of different types of fault, performance, and service events.
5. Recovery of failed VMs, networks, and other virtual resources using automation (based on correlations performed during monitoring).

In addition, the operations teams have the ability to intervene in the automation functions performed by ONAP during a VNF instance's lifecycle, via a real-time portal.

14.4.3   BVoIP Services

Most network functions for service platforms that support BVoIP services are also currently supported on vendor-proprietary hardware. The physical implementation of the current AS complexes typically consists of a collection of commercially available hardware, including server blades in a chassis configuration, integrated with other supporting components such as layer 2 and layer 3 switches, external disk arrays, and load balancers. Most ASs run on a standard Linux operating

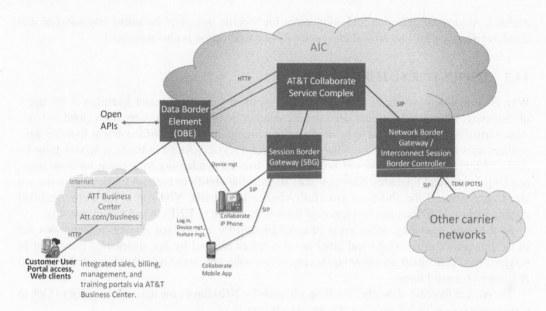

FIGURE 14.22 AT&T Collaborate implementation.

system, with the application software executing in either a commercially available AS container (e.g., IBM WebSphere, Tomcat/Apache), or in a vendor-provided package.

AS complexes also typically include one or more database servers, again using commercially available database technology (e.g., Oracle) or a vendor-provided database. An AS complex configuration may consist of one or more instances of a variety of ASs and database types, each assigned to a specific blade server populated in a chassis.

The strategy for virtualization of the network functions that support BVoIP services is based on needs for capacity enhancements, end-of-life hardware and software triggers, and new product needs. Design and implementation of the VNFs is in a manner that aligns with the new architecture and uses the virtual service framework. At the time of writing this book, AT&T has deployed initial implementations of some BVoIP services including mobile call recording and AT&T Collaborate, which is described here as a BVoIP use case.

AT&T Collaborate is a hosted BVoIP service that is intended to serve all business segments. The customers run the range of small businesses with 1–5 lines or extensions, all the way to large enterprises with thousands of lines and extensions. As shown in Figure 14.22, AT&T Collaborate service elements are deployed in AIC. The DBE is used to provide secure data access to the platform. The session border gateways are used to provide secure SIP access to the platform for voice and video calling. The Collaborate service complex is a grouping of servers that perform service logic, routing, media functions, device management, voice messaging, desktop sharing, and supporting services. The complex connects to other carrier networks via SIP interconnect where available and TDM trunking otherwise.

The virtual service framework is used to deliver the Collaborate service. Most of the design principles applied to the virtualized IMS service platforms (refer to Section 14.4.1) apply to this service also. The SBC is decomposed into signaling SBC and media SBC. The signaling and media SBC VNFs are reused and instantiated for access and network interconnect purposes. Subscriber data are decoupled from the AS logic. VNF VMs are deployed across multiple cloud availability zones at a site using anti-affinity rules. Many reusable VNFs are added to the VNF catalog with this service, including the signaling SBC, media SBC, DBE, and the media server. The decomposed SBC, media server VNFs, and the ASs included in the Collaborate service complex can be reused for other services. AT&T Collaborate is one of the first virtualized services for which control loop automation using the virtual service control loop framework for auto-recovery of failed VMs is realized. Automated detection and notification for specific use cases including international call fraud, registrations failing, no dial tone, and call storm detection is also supported.

14.5   TOPICS FOR FUTURE STUDY

With the initial deployment of virtualized services, there were some major learnings in the areas of resiliency. Virtualized services need to have a heavy emphasis on recovery from cloud failures. Also, virtualized services need to be deployed in a highly distributed fashion across multiple geo-resilient sites in order to meet the reliability needs of a service. As an example, a service must be deployed minimally at four sites in order to meet five 9's availability to account for downtimes resulting from cloud upgrades. Services must support the ability to redirect traffic from a site for cloud upgrades and the ability to gracefully shut down specific VNFs at a site. Fully qualified domain names (FQDNs) must be supported for routing traffic to VNFs in order to enable this.

At the time of writing, some areas of consideration for virtualized service instantiations are in the Internet of things (IoT) and other novel services enabled by 5G, distributed placement of services, next generation access agnostic core, cloud solutions, standards, database evolution, and efficiency of control loops.

The current forecast of nearly 20 billion IoT units by 2020 forces the network operator to look at service enablement of IoT services at scale, which entails

- Enabling agile creation of vertical solutions via service-specific overlays.
- Leveraging SDN/NFV to enable designing reusable common network/platform capabilities.
- Using SDN/NFV and decomposition/composition of VNFs for a modular architecture.
- Using network slicing and service-specific instantiation to minimize the overall cost per service.
- Using API exposure via a mature common service enablement platform to create and deliver broad IoT enablement services.
- Leveraging common service enablers (e.g., voice, video, etc.) and light-weight technologies (e.g., WebRTC).
- Enabling horizontals across third-party verticals via a federated approach.
- Using common core smart/predictive analytics and policy-based closed loop response capabilities.
- Leveraging broad cloud deployment locations to place latency-sensitive capabilities and applications close to the edge as needed.
- Leveraging partners and hosted platforms for vertical-specific IoT businesses where needed.
- Supporting flexible deployment options and support structures:
  - Hosted on a cloud and managed by the service provider
  - Hosted and managed by the vendor on the service provider's cloud
  - Hosted and managed by the vendor

Horizontal/cross-vertical capabilities across vertical platforms will be enabled using cloud-driven and API exposure principles and API/platform federation. Each vertical platform is supported on a cloud and exposes its APIs to an API development studio, which is part of a service enablement platform. This studio is used to develop horizontal/cross-vertical applications that connect to the vertical platforms via APIs. The APIs are used by the application to trigger cross-vertical capabilities as needed. An example is enabling quick passage of an ambulance by adjusting traffic lights via API federation across the "Connected Car" and "Smart Cities" IoT verticals. An application can be built to connect the connected car and smart cities verticals via APIs exposed by the API platform. Another example is delivering emergency medicine in case of accidents or disasters via API federation across the drones and mHealth verticals. The services would leverage reusable service enabler capabilities that are currently supported by the network to expand the breadth of IoT services offered including WebRTC voice and video, messaging, notification, speech recognition, VoLTE, etc. These capabilities can be combined with any service provider's network for IoT service offerings. Examples of this are use of speech recognition for connected car services, WebRTC real-time video and telematics data feed to drones, WebRTC video calling capabilities for healthcare services, WebRTC real-time video streaming from drones for aerial view capture, etc.

Also, these services would use reusable core capabilities hosted on the cloud. The reusable modules would be in the following areas: device management, id management, authentication, data ingestion, data transformation, data storage, smart analytics, policy engine, policy management, portal capabilities, and API exposure. These reusable modules can be instantiated for an IoT service with the appropriate configuration on an as-needed basis. This is depicted in Figure 14.23.

A key component of the 5G roadmap involves virtualizing the mobile core and distributing that core closer to the network's edge to support both 5G and the IoT in a cost-effective way. Broad cloud deployments will enable access, mobile core, and service enabler virtualized components to be deployed in a highly distributed manner with latency-sensitive functionality deployed closer to the end consumer. The increased bandwidth and speed of 5G and billions of connections expected for IoT are drivers to push forward with these broader cloud deployments. The distributed approach uses a virtualized mobile core deployed on cloud locations as close to the edge of the network as is economic, to support millions and billions of things connected to the network for IoT and other applications. Similarly, the distributed approach can be used to deploy applications and other components appropriately to serve the needs of an end-to-end service. Distributing the user plane to edge locations is especially important to better serve many of the novel IoT services. Certain capabilities such as analytics and policy enforcement functions can be distributed with latency-sensitive functions deployed at the edge while there will be centralized deployment for analytics and policy enforcement across edge locations with long-term trend analysis and forecasts. Additional considerations for placement include dynamic reallocation in real time based on traffic load thresholds and based on locations (e.g., policies that are triggered when there is heavy traffic on certain highways during certain time periods). Examples of a latency-sensitive IoT service are connected driverless cars where a decision needs to be made in a split second for turning at a signal light and analysis and decision making based on real-time video and telematics feed from drones. Examples of services

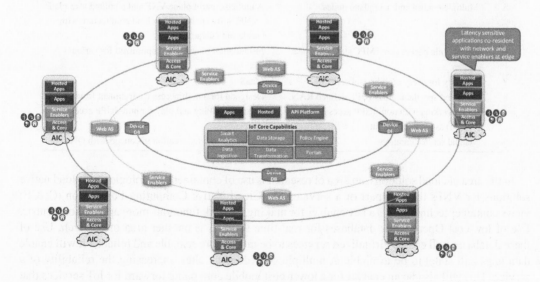

FIGURE 14.23 Distributed placement of IoT and other novel services.

that are not latency sensitive are asset tracking and asset management where sensor data can be sent to a centralized location. In summary, broad deployments of cloud locations enable a service-oriented CN architecture and provide the flexibility to support new service models.

A Next Generation "x" Core (NGxC) is AT&T's vision of the upcoming next-generation multiaccess packet core architecture and its functions. The "x" in NGxC indicates a multiaccess core. It means exploring the virtualization and integration of wireless and wireline core functions including layer 3 and above such as policy, authentication, signaling, etc. It is envisioned that the NGxC will make full utilization of the SDN/NFV technologies. The NGxC will have a modular architecture for composition and decomposition of NGxC VNFs. This will help in creation of an optimized NGxC instance for the specific need of a use case. It will perform most of the functions of the existing 4G mobile core (e.g., authentication, charging, QoS, policy, mobility, handover, etc.), and will support emerging 5G use cases, while achieving synergies with wireline access core functions. These functions, however, will be performed in the NGxC environment and may not necessarily follow the 4G architectural concepts. In addition, NGxC is envisioned to perform newer functions to meet the requirements of the use cases outlined in the 4G America's 5G White Paper [6]. Some of these new functions are as follows:

- Access agnostic
- Mobility as a service (MaaS)
- Separation of control and user plane

The architectural concepts for the NGxC are influenced by the NGMN white paper [7]. The table below gives a summary of key evolutionary changes from today's (4G LTE) architecture to NGxC.

No.	From	To
1	PNFs	VNFs
2	Nodal boundaries	Composition and decomposition of VNFs
3	Hierarchical network deployment	Flexible network deployment
4	Nodes with combined control and user plane functions	Complete separation of control and user planes for independent scaling and distribution. Standard interface between control and user plane. Decomposition of bundled user plane functions
5	Multiple control and user plane nodes	A unified control plane VNF and a unified user plane VNF. Software based on cloud architecture with scale-out characteristics
6	One mobile packet core (MPC) to suit all use cases	Instantiation of an NGxC optimized for various use cases
7	Mobility for every mobility subscriber	MaaS
8	Designed specifically for 3GPP defined RAN. Interworking with non-3GPP access (e.g., Wi-Fi) as an afterthought	Designed for a multiaccess environment from the beginning. Plug and play for non-3GPP access types
9	Manual service creation	Automatic service creation in concert with ONAP

In the area of cloud solutions, an area of research is use of container technologies and cloud native solutions for VNFs that are part of a service. The Cloud Native Computing Foundation (CNCF) views container technology as a key enabler for making network functions more application-centric. Use of low-cost OpenSource databases for real-time services is another area of research. Use of these databases will enable virtualized services to be much more scalable and reliable. It will enable data (e.g., call state) to be available at multiple georedundant sites, increasing the reliability of a service. This will also be an enabler for a lower-cost mobile core going forward for IoT services that will require massive scale. Another area of research is the efficiency of control loops and dynamic instantiation. Currently, virtualized services that require high availability of five 9's are deploying highly redundant configurations. A highly efficient ecosystem for fault detection and rapid dynamic instantiation of VNFs could enable minimal redundant configurations in the future, thus further lowering the overall cost of virtualized services using NFV and SDN. Lastly, comprehensive API exposure of network functions across all service domains to enable third parties to build services using VNFs deployed in the cloud, will be an enabler of supporting additional business models to provide the network service.

REFERENCES

1. 3GPP TS 23.228: IP Multimedia Subsystem (IMS).
2. 3GPP TS 23.402: 3GPP System Architecture Evolution.
3. IETF RFC 3261: SIP: Session Initiation Protocol.
4. IETF RFC 6733: Diameter Base Protocol.
5. 3GPP TS 23.060 V13.4.0 (2015-09): General Packet Radio Service (GPRS).
6. 4G Americas' Recommendations 5G Requirements and Solutions, October 2015.
7. NGMN White Paper, 2-17-2015.
8. 3GPP TS 23.002: Network Architecture.

15 Network Operations

Irene Shannon and Jennifer Yates

CONTENTS

15.1 Introduction 331
    15.1.1 Operations and Engineering Teams 333
15.2 Impact of NFV and SDN on Network Operations 335
    15.2.1 NFV and Its Impact on Operations 335
    15.2.2 NFV-Related Challenges 337
    15.2.3 Role of ONAP in Network Operations 338
        15.2.3.1 Deploying and Deleting VNFs 339
        15.2.3.2 Network and Service Monitoring 340
        15.2.3.3 Control Loops: Repairing Network Impairments and Resolving
                Customer Impact 340
        15.2.3.4 VNF Change Management 342
        15.2.3.5 Configuring ONAP to Enable Operations Automation 343
    15.2.4 ONAP-Related Challenges 343
        15.2.4.1 Safe Policies 343
        15.2.4.2 Identifying Operations Policies 346
        15.2.4.3 Increasing Software Complexity 346
15.3 Transforming the Operations Team 346
15.4 Migrating to the Network Cloud 347
    15.4.1 Rolling out Network Cloud Technologies 348
    15.4.2 Bootstrapping SDN/NFV Deployments Leveraging Current Knowledge
        and Experience 348
    15.4.3 Legacy and SDN/NFV Networks Coexisting in Harmony 348
15.5 Topics for Further Study 349
Acknowledgments 350
References 350

This chapter describes the traditional role of network-facing operations and how moving to a network cloud will impact that role. We intend to motivate the need for innovation in the delivery of network-based services by demonstrating the cost drivers in operations. The move to NFV/SDN and ONAP enables not only faster time to market for new services but also significant reduction in operational cost, a main component of a service provider's cost of doing business.

15.1 INTRODUCTION

Traditional network operations cover three large categories of work: providing the quantity of network elements configured appropriately to deliver economical and quality service (network provisioning), ensuring that existing and planned elements in the network efficiently provide necessary service levels (network administration), and preventing faults or correcting existing faults (network maintenance).

Network operations is complex. A key contributor to complexity is the highly distributed nature of the network itself. There is physical equipment in tens of thousands of locations including business premises and residential neighborhoods. Much of the equipment is subject to harsh environmental conditions, which can lead to failures and performance impairments. Many of the services that run on the network infrastructure have strict performance requirements, which leads to complex network designs to ensure reliability and resiliency. Over the last several decades, the application of computer technology to network operations has achieved productivity gains that have supported the explosion of growth in wireline and wireless data network traffic.

Network provisioning includes planning, engineering, and implementation. Over the last 20 years, the planning horizon has become smaller due to the faster pace of technology and service innovation. In the 1970s and 1980s, long-range network planning (also called fundamental network planning) would look at the next 20 years to determine appropriate plans for the evolution of the network. Today, the fundamental planning cycle is much shorter, focused on 6–8 years out. Current network planning provides estimates of the quantities of network elements in the next 1–3 years and, even with this shorter planning horizon, it is difficult to predict how user demand and technologies of the future might impact the growth in demand on network elements. A recent example is the introduction of the first smart phone in 2007, which led to a 100,000% increase in data traffic on AT&T's wireless network over a 7-year period.

Network administration ensures that network elements are used efficiently and deliver on predefined service objectives. In order to administer a network effectively, it is critical to have data on network traffic, equipment, software, and the performance of the services that end users consume, for example, voice, data, and messaging services. Most of the data required are obtained through automated systems, although service performance data are the least mature. Performance management can be either proactive or corrective. Performance metrics are defined for services and network elements and are monitored continuously. Time series data are evaluated to identify existing performance degradations and to determine whether there are trends that could lead to performance degradation. There are both manual and automated processes to handle a performance degradation. If the root cause is capacity related, then capacity management is triggered, which may require a long lead time to resolve in traditional networks. The administration function is also focused on the utilization of network elements to ensure that they are being used in the best way possible. A goal of network administration is the configuration of network elements so that they meet a targeted utilization. In order to meet the reliability requirements of critical services such as voice applications, redundant equipment and capacity has been deployed in the network and this can lead to inefficiencies.

Network administration includes the network management function that ensures that the network remains well behaved (carrying its engineered traffic load) in the face of traffic anomalies or equipment issues. Many network management controls have been automated in the last several decades so that the network itself can autonomously respond to potential or realized anomalous conditions. For example, when a natural disaster strikes in a region, network traffic management controls are used to ensure that congestion does not spread to other areas of the network. In addition to automated controls, there are also manual network management controls that are used in unusual conditions that require human judgment. For example, controls are typically put in place to ensure that calls originating from a disaster area get preferential treatment to calls originating outside the area but seeking to terminate there.

The final broad category of network operations is maintenance of the network. Preventative maintenance involves routine servicing of equipment and procedures like generator runs to detect and mitigate potential troubles before they impact service levels. You can think of this like proactively changing the oil in your automobile so that you do not risk issues that can arise when there is not enough oil to keep engine parts lubricated. In addition to preventative maintenance, there is corrective maintenance, also referred to as fault management. There has been significant work to automatically detect network events (failures and performance impairments). Where it is economically feasible to do so, automated response to a detected event is executed to achieve continuous service levels. An example of this is the facility restoration work done in the 1990s to detect failures such

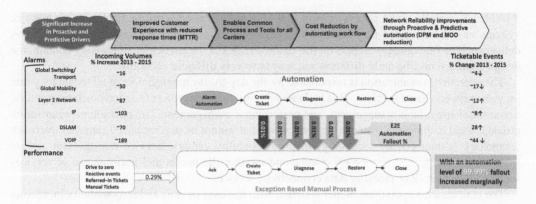

FIGURE 15.1 Role of automation.

as cable cuts in the core network and to determine routes around the failure using spare restoration and protection capacity. The system automatically issues commands to network cross-connect systems to implement the new routes. Most facility faults on the core network are handled seamlessly and quickly, restoring service on the order of minutes. Modern IP networks can also automatically change how they direct packets when the network detects that a router is no longer responding or that a link between two routers has failed. This is accomplished within the routing elements themselves and does not require any manual intervention.

In 2000, AT&T launched a significant effort to transform network operations and move toward an adaptive management model. There was a recognition that most of the network management capabilities were developed to react to specific types of network issues. Over the last 15 years, AT&T has moved much of its operations to the predictive or proactive stance by leveraging many sources of data to identify potential issues before they are visible to the end customers. Work on adaptive operations is underway and one of the fundamental principles of the SDN/NFV transformation is to use data in a virtuous cycle to automate, learn, and adapt.

Operations models and processes are dynamic and are influenced by a variety of factors such as new technologies, new network-based services, operating economics, the market landscape, regulatory policies, and corporate business objectives. Significant changes to modern operations began to take place in the 1970s when engineers realized how computer technology could be applied to network operations to make those operations more efficient. Computer-based operations support systems (OSSs) enabled many operational functions to be centralized and performed remotely. As the number of network elements increased by orders of magnitude, the number of operations personnel required to perform planning, administration, and maintenance increased at a much slower rate due to the increasing functionality of operations systems including the ability to correlate related alarms in a fault condition, auto-test to assist in fault isolation, and perform automated actions to clear fault conditions. Figure 15.1 illustrates related statistics over a 3-year period. This figure illustrates that even though the number of network alarms has been growing, in some cases significantly, there has been automation delivered to ensure that ticketable events (events that require human operator engagement) do not scale linearly with the number of alarms.

15.1.1 OPERATIONS AND ENGINEERING TEAMS

In a typical large network operator, like AT&T, the operating structure consists of an engineering team that performs fundamental and current network planning and an operations team. The engineering team is responsible for introducing new technology into the network, designing the network, implementing the network, and adding additional capacity to the network. Engineers are specialists in particular network technology domains such as layer 0/layer 1 transport, layer 2 Ethernet, layer

3 IP/MPLS, mobility radio access networks (RAN), and IP-based services. Each of these network systems must be designed to predetermined performance and capacity requirements as well as deliver specific functionality. Based on the characteristics of the particular technology domain, the design approach may be quite different and may have very different economics.

The operations organization is responsible for the day-to-day management of all technology that provides network-based services. This includes all lifecycle management functions such as upgrades, decommissioning, fault management, and performance management. The operations organization provides local technicians who can perform work that cannot be accomplished remotely. Network technology is managed by technology-focused reliability centers. Service management centers focus on the end-to-end service experience and monitor the health and performance of services such as Voice over IP or mobility data connectivity service.

The maintenance operations model is tiered, with the level of skill and responsibility increasing in higher tiers. For many strategic, going forward technologies and services, work that was traditionally classified as Tier 1 has been fully automated. Examples of these Tier 1 work tasks include reviewing and storing log files, verifying port configurations, and verifying circuit connectivity. The Tier 2 organization performs most of the maintenance work in the network and also does most of the administration work that is not already automated. Tier 2 technicians are more skilled than Tier 1 and can handle most tasks without significant oversight. They are permitted to make changes to network systems such as changing an access control list, a routing configuration or upgrading software on network elements. They are the primary team involved in routine trouble resolution. Tier 2 monitors the performance of the network as well as determines whether action is needed to address performance degradations.

The Tier 3 organization can move between design work and operations support seamlessly. It is called in to assist on complex network problem resolution and restoration. This team also works on chronic or systemic problems and typically works with the network element vendor to do so. The Tier 3 organization has a close partnership with the engineers who design network services and there is collaboration on the network management plan for network elements and services. Tier 3 designs and certifies the complex procedures that are required for replacing network elements, inserting new network technology, or upgrading network software. It does the technical forensics required to determine root causes of complex network outages and creates the action plans to address any gaps that are uncovered during root cause analysis.

The AT&T operations organization structure also includes a Global Technology Operations Center (GTOC), a single, site-redundant network operations center shown in Figure 15.2. The GTOC manages outage escalations and provides operational oversight and coordination across the various technology-focused reliability centers, customer care, and external bodies such as the FCC. It also serves as the single communications channel to internal senior leadership so that consistent and accurate information on network events reaches all stakeholders. In addition, the GTOC develops and maintains critical processes used across the entire operations organization such as the appropriate way to prepare and execute changes to the production network.

The GTOC has a formidable task since AT&T's global network connects more than one billion devices in over 180 countries. There are typically over 400 network incidents each day; events that are sufficiently big to warrant GTOC engagement. There is an established command and control process that provides guidelines on how AT&T will respond to network service degradation or faults, traffic anomalies, and congestion/capacity issues. Telecommunications providers such as local exchange carriers, interexchange companies, satellite companies, and cable companies all have FCC reporting requirements for outages that exceed certain thresholds. The incident managers in the GTOC perform the critical role of outage reporting and FCC filing preparation.

Although there has been significant investment in the development of automation in the last several decades, the rate of growth and change in the network is starting to outpace our ability to deliver automation as a separate set of software applications running on top of the network ecosystem. In fact, we have added complex features and new elements solely for the purpose of managing

FIGURE 15.2 Global Technology Operations Center.

the network at tremendous scale (e.g., traffic engineering, RFC3107, fast reroute, label switching, etc.). Another significant driver of operations work is the many different elements that are being used in the network infrastructure, each one being a unique platform. For example, there are in excess of 50 specific network elements performing unique functions that are involved in providing end-to-end mobility service! This proliferation of unique platforms has been an obstacle to commoditizing many operational functions and applying automation to reduce the operations workload.

While the preceding section captures the current state of the affairs, the emergence of new technologies, especially NFV and SDN, presents a unique opportunity to streamline and automate network operations functions. This is discussed in the sections that follow.

15.2 IMPACT OF NFV AND SDN ON NETWORK OPERATIONS

As discussed in earlier chapters, NFV and SDN are driving a fundamental transformation in how networks and network-based services are designed and managed. From an operations point of view, it is changing how everyday functions are performed, such as turning up new network elements and lifecycle management. This in turn enables corresponding improvements in operational efficiency and, over time, improved customer experience. It impacts the operations teams' roles and responsibilities and even the skillsets of the personnel involved. The following sections first discuss NFV and then focus on SDN in the context of AT&T's software-defined networking platform known as ONAP (see Chapter 7). We examine some of the impact on network operations, challenges associated with operating legacy networks in parallel, and the implications this has on operations' organizational structure and skillsets.

15.2.1 NFV AND ITS IMPACT ON OPERATIONS

As is typically the case with the introduction of new technologies, NFV presents a double-edged sword. It provides opportunities to reduce operational expenditure (OPEX) and improve customer experience through the rapid turn up of new services and the ability to spin up new network capacity in response to changing network conditions. However, NFV also introduces additional layers of complexity from an operational point of view and puts more onus on the operator to integrate technologies that were traditionally integrated by a vendor.

As discussed in Chapter 2, traditional network elements were implemented using customized hardware and ASICs with tightly integrated software and a vendor-specific element management system (EMS). The integration of all of these was thus provided by the vendor, and the ecosystem appeared as a "black box" to the network operator. The interfaces through which operator OSSs interacted with the equipment to deploy, configure, and manage the network elements were vendor specific. The alarms, logs, and the majority of the performance measurements reported by the network elements and EMSs were different for each vendor and even type of element. Even common metrics—for example, accessibility and retainability [3GPP_KPIs, CELLULAR_QOS] in mobility environments—have very different meanings as measured and reported by equipment from different vendors. Operating a network with such diversity necessitates tremendous investment in custom interfaces from OSSs to each of the different elements. But it also meant that the operations organization had to build significant vendor and element-specific expertise within their teams, resulting in operational silos with dedicated teams focused on specific types of network elements.

With traditional physical network elements, deploying new elements involves sending technicians to each and every site where a new network element is to be deployed. When hardware fails, technicians again have to be dispatched to the physical location of the hardware to repair or replace components (e.g., line cards) or entire network elements. When hardware is out of service, depending on the resiliency options deployed, customers may also be out of service. Alternatively, the network may be operating in a simplex mode where a second similar failure could result in customer impact. Thus, rapid repair is imperative; technicians are either deployed on site 24×7, or are rushed to the location on demand. Spare hardware for each and every component has to be readily available for rapid deployment, thus requiring significant inventory lying idle across the network.

The most fundamental operational transformation enabled by NFV may be the opportunity to reduce operational overhead by, using the industry phraseology, turning "pets into cattle." With physical network elements, each network element is akin to the family pet; it has to be lovingly cared for and highly managed. When it is "sick," it is tenderly nursed back to health. In contrast, efficiently scaling networks with the plethora of network functions that need to be supported means that different network functions need to look nearly identical and be treated accordingly, as with cattle. When a network function is "sick," it is simply (and rapidly) replaced with another one.

Moving to a "cattle" model using NFV requires the following: (1) standardizing management interfaces to maximize commonality across VNFs from different vendors and (2) being able to rapidly spin up new network functions on demand and migrate customer traffic to these with minimal customer impact.

In a world built on custom hardware and ASICs, it was typically deemed infeasible to standardize how network elements were deployed and configured, and standardize the data generated by them, as functions were quite different. However, the move to commodity hardware and software implementations of network functions opens up the opportunity to fundamentally transform network operators' operating models by standardizing much of the functionality across VNF interfaces.

As VNF management interfaces become increasingly standardized, enabling common mechanisms for deploying, configuring, and managing VNFs, the diversity of domain knowledge and operational know-how required to manage the plethora of different types of network functions deployed in carrier networks will be reduced. OSSs and SDN platforms such as ONAP will be considerably simpler and cheaper to build as we reduce or even eliminate the need to implement vendor-specific interfaces to manage VNFs and to process vendor-specific logs, alarms, and measurements. Having standardized interfaces should also simplify outage management. For example, in a traditional network, localizing and resolving the more esoteric and complex service issues often necessitates involving teams representing a wide range of network elements. When there is a complex outage, many individuals join outage conference bridges (audio calls organized to do rapid triage), with the intent of identifying which specific network function (network element) is impaired and thus identify the corresponding operations team responsible for resolving the issue. This is an inefficient use of resources, driven by the complexity of some of the issues that arise and the diversity of the network elements and domain knowledge required to support these elements. As proprietary hardware is eliminated and the management interfaces to the VNFs become increasingly standardized, the goal is that operations teams' engagement will be increasingly simplified, with corresponding OPEX improvements. Simplified outage management should also result in reduced outage durations, thereby improving customer experience.

With NFV we eliminate the need for specialized hardware, instead leveraging commodity servers and storage in the cloud. Server and storage capacity is predeployed, with cloud capacity augmentations and hardware replacements performed on a periodic basis—without urgency or technician dispatches. Deploying a new instance of a network function or repairing a failed VNF now becomes a software exercise. Similarly, there is no need to store spare hardware across the network for each of the different network element types deployed. With NFV and ONAP, deploying a new network function can be managed entirely remotely, using predeployed server capacity and filling out HEAT templates. And when a VNF's hardware or software fails, it can be replaced simply by rapidly spinning up and configuring a new VNF on available cloud capacity in the same or even a different physical location. This simplifies the operations tasks involved in deploying new network capacity, and responding to traffic surges, hardware or connection failures and performance degradations. It also promises to improve customer experience, through faster service provisioning and faster resolution of service-impacting network conditions (failures, overload conditions, and performance degradations).

The ability to dynamically spin up resources also introduces new opportunities for other operations functions. To further illustrate the operational impact of NFV, let us consider the example of VNF change management—and specifically the situation in which the network operators are responsible for upgrading the software on a VNF. Changes are a frequent occurrence in service provider networks and consume significant operational resources. Network designers and operators go to significant length to minimize customer impact. In legacy, hardware-based environments, network engineers have permanently deployed spare network element capacity to carry traffic during software upgrades, worked with vendors to implement hitless software upgrades, or in extreme cases where neither may be available or practical, they simply have to accept that maintenance activities may result in customer service disruptions. In contrast, with NFV we can spin up new versions of the software on demand—this can be used to streamline change management activities by spinning up a new VNF with the updated software version; traffic can then be moved over to the new VNF with minimal service impact. Thus, we can cost effectively minimize customer impact without having to deploy costly spare network hardware. This does require some spare capacity in the cloud that can be used to spin up the new VNF but it is taken from a pool of capacity that supports multiple VNFs. This approach impacts how change management is performed, and requires corresponding ONAP functionality and analytics to enable automation.

15.2.2 NFV-RELATED CHALLENGES

Although NFV promises exciting benefits, it comes with associated challenges. Capacity planning for a shared cloud infrastructure supporting large numbers of VNFs that can be dynamically spun up and down as needed has its challenges. Accurate forecasts are vital; as is timely deployment of new server capacity. However, achieving this in a very dynamic world and with variable VNF deployment timelines is challenging, especially as large numbers of new VNF technologies are being rolled out. With so much riding on this shared infrastructure, if capacity is not available when required, it could have tremendous impact both on VNF deployment timelines, and on service reliability and performance.

Standardizing VNF interfaces to provide common mechanisms for deploying, configuring, and managing network functions requires an industry-wide commitment. Making this transformation a reality is a tough task, and will require significant partnerships and push from standards bodies, the open source communities, network operators, and VNF vendors. Standards bodies are the traditional forum used by the telecommunications industry to converge on agreements, but increasingly service providers are leveraging engagements and contributions in open source communities to drive change.

Another challenge results from the introduction of new network interfaces that must be managed by network operators. After years of eliminating network layers (e.g., ATM), virtualization is introducing additional complexity by introducing separate layers—the physical layer (compute and storage platforms, also referred to as the hardware layer) and logical (VNF/software) layer(s). Historically, these two layers were combined within a single, integrated hardware/software platform (network element) and a given type of network element was managed by a proprietary EMS and a single operations team. With the new division introduced, different operations teams will typically manage the hardware and the software layers. An issue at the hardware layer (cloud servers) that impacts the software layer (VNF) will alarm at both layers. Without appropriate intelligence and automation, the alarming on the hardware and software layers could result in both the hardware operations team and the team managing the VNF to simultaneously respond to the issues reported and investigate root cause in parallel. This would be a tremendous waste of valuable operational resources, as in reality only the hardware team needs to investigate; the VNF team may have to mitigate the issue by re-instantiating the failed VNF on new hardware (if not an automated function), but it would be a waste of resources looking for potential root causes of the impairment at the software layer. This effect would be compounded if multiple VNFs are supported on common hardware.

Thus, the separation of the hardware and software layers necessitates additional intelligence to capture and dynamically track the relationship between the layers so that hardware faults and change management activities can be correlated with impacts on the corresponding VNFs. Of course, tracking cross-layer dependencies, correlating events across network layers, and coordinating change activities across operations teams managing different network layers is something that is far from new. Technologies and processes exist and are widely used today for achieving this; however, leveraging these in the network cloud will require new alarming and correlation rules/policies.

Similarly, after years of driving convergence (taking functions that once resided in separate devices and combining them into a single system), there is an industry movement now reversing this trend to pursue disaggregation. Disaggregation refers to the breaking up of existing systems, where applicable, into their basic modules. Whenever possible, the disaggregated modules are implemented using standard commercial off-the-shelf servers or white box switches, and are based on open-source software. For example, a reconfigurable add-drop multiplexer (ROADM) was traditionally a monolithic system provided by a single vendor with a corresponding EMS. Through disaggregation, a ROADM (also referred to as an Open ROADM in Chapter 13) may be broken up into separate transponders, wavelength selective switches and a backplane—all potentially provided by different vendors. Disaggregation promises greater flexibility, innovation, and the ability to use "best in class" for each technology. However, disaggregation also results in there being significantly more components to manage in a network, which complicates localizing and resolving issues. Just like with the separation of hardware and software, disaggregation results in more interfaces to manage, new alarming (as new interfaces are exposed) and corresponding additional correlations required to localize issues. This requires an intimate understanding of the different modules and their roles, interfaces and failure modes.

The ONAP platform and associated analytics and policies is responsible for this tracking and correlation between hardware and software layers, between different network layers, and between disaggregated network functions.

15.2.3 Role of ONAP in Network Operations

As discussed in Chapter 7, ONAP is the "brain" providing the lifecycle management and control of the software-centric network resources, infrastructure, and services. ONAP lets us rapidly onboard new services and from an operations perspective, it provides a framework for real-time, policy-driven software automation of network management functions.

The ONAP platform together with supporting intelligent analytics, recipes and policies enables service designers and operators to automate network provisioning, network administration, and network maintenance functions in a consistent and comprehensive manner across all VNFs. Customers benefit through the rapid introduction of new and innovative services, reduced service provisioning times (potentially minutes instead of months) and improved service performance as automation typically responds to impaired conditions faster than human operators. The automation enabled by ONAP also allows the operations group(s) to deal with massive customer demand growth—absent corresponding equivalent revenue growth—without corresponding expansion in the size of the team(s).

The automation of network management functions is not new; there is a tremendous amount in place today. However, network management is all too often introduced as an afterthought in how technologies and services are designed, built, and deployed. In contrast, network management automation is at the very heart of the network design with ONAP, the focus being on bringing common automation and reusable analytics and tooling across diverse VNFs and services.

In the next section, we will look at how ONAP facilitates different operational tasks, including service and VNF provisioning, capacity management, traffic management, fault, performance and service quality management (SQM), and change management.

15.2.3.1   Deploying and Deleting VNFs

New service deployment, capacity augmentation, traffic management, and even fault and performance management all rely on ONAP's ability to automatically spin up and configure VNFs.

VNF provisioning, also known as instantiation, describes deploying and configuring a new VNF. Parameters for configuring the VNF, such as IP addresses, need to be either automatically selected or input to the ONAP system—ideally from automated, external interfaces. ONAP is then responsible for determining where to spin up the new VNF, spinning up the VNFs on spare cloud capacity, configuring the VNF, and then validating that the new VNF has been successfully instantiated.

Historically, network planners have expended significant time and effort designing long-term capacity augmentations; they had to plan well ahead due to the long lead times related to deploying physical network elements and (where required) provisioning high capacity links between offices. However, it is impossible to accurately predict needs too far in advance, and thus capacity ends up being suboptimally deployed. This leads to additional complexity in operating the network, with operators needing to adjust traffic routing (particularly under failure scenarios) according to available capacity.

ONAP and NFV together enable VNF capacity to be made available where and when it is needed. However, we still need to interconnect data centers. For very-high-speed links such as those used between IP routers, connecting two offices has historically required provisioning times on the order of months as planners first design the circuits and then field technicians manually connect circuits together at intermediate offices to establish connectivity. However, ONAP in combination with ROADMs (see Chapter 13), enable connectivity between data centers to be automatically established in a matter of minutes—a vast improvement compared with months! Going a step further, the control plane can now be integrated between the IP and optical layers, enabling intelligent cross-layer decision making and real-time provisioning of capacity in response to changing network conditions (e.g., failures, changing traffic patterns).

Historically, considerable time has been spent by network operators in managing traffic in response to failure conditions, changes in traffic patterns, and planned maintenance activities. Operators had to plan how they would respond to conditions as they arose, typically leveraging offline tools that aid the decision making. Once decisions were made regarding what actions to take, operators executed these actions. Depending on the routing technologies used, the actions could involve DNS updates or routing protocol configuration changes (e.g., OSPF weight changes), for example. In a virtualized environment, this would also include spinning up or down VNFs. ONAP, combined with intelligent algorithms operating on data collected and stored by ONAP, will automate this process enabling the removal of human operator.

Thus, ONAP, in combination with smart analytics and appropriate data plane technologies (NFV, ROADMs), is creating a more intelligent, dynamic network that can more effectively adapt to changing conditions. However, as noted in Section 15.2.2, there is additional complexity introduced in the shared infrastructure to enable this simplicity and flexibility in managing the VNFs.

15.2.3.2 Network and Service Monitoring

As discussed in Chapter 16, individual network functions (virtual and physical) generate a tremendous volume and range of fault, performance, and log data. These measurements capture information that allows network engineers and operations teams to design and capacity plan the network cloud, and to answer questions such as, Are the network elements up and operational? Are they performing well? Is the CPU too high or the element running out of memory? Are network links dropping packets?

However, it is not sufficient to simply provide a network function (or network element) view for managing networks and services. What ultimately matters is the customer experience regarding the services that are carried over the network. Are voice calls clear? Are videos being successfully received with high quality, without stalls and pixilation? Are customers able to rapidly access web content? End-to-end service or application monitoring provides measurements that attempt to capture that end-to-end user experience—from source to destination, across the wide range of diverse end points and across different services and applications.

Service performance monitoring is challenging to scale, courtesy of the vast number of end points and the wide range of services and applications. There are two primary techniques that exist—active and passive monitoring. Active probes send test traffic across networks, emulating customer services; passive probes "listen in" to customer traffic to provide aggregated visibility into customer experience. Both are typically leveraged in large-scale networks in a bid to construct a good view of how the services are performing, aggregated across customers.

Data collectors operating on ONAP are responsible for collecting the logs, traps, counters, and service monitoring measurements required for operation tasks (and beyond), and for passing the measurements to long-term storage for historical analyses. ONAP can also leverage the measurements for a range of real-time operational tasks, including real-time capacity management, identifying network and service performance conditions, and validating change management activities.

15.2.3.3 Control Loops: Repairing Network Impairments and Resolving Customer Impact

ONAP *control loops*, as defined in Chapter 7, are integral to fault and performance management, as well as real-time capacity management and even traffic management. The basic function of a control loop is to detect, isolate, and mitigate issues, whether these are "hard failures" (e.g., network and/or services are "down"), "soft failures" (performance degradations), or capacity crunches. Ideally, issues can be mitigated via automated responses, such as by rebooting a virtual machine (VM), re-initiating failed VNFs, rerouting traffic, or spinning up new VNFs. However, not all issues can be automatically resolved—either because the issue cannot be automatically localized or the mitigating response is either not currently automated or fundamentally cannot be automated. In such situations, the control loop will notify a network operator (human being) via ticketing or similar mechanisms so that the operator can intervene to resolve the issue.

The biggest challenge in resolving faults, performance, and service degradations is in localizing and troubleshooting issues—identifying *where* the problem is, and *what* is causing it. Without knowing where the issue stems from, and in many cases the specific root cause of an issue, the ability to take effective actions is seriously inhibited.

A single underlying issue will typically result in a plethora of alerts and alarms. To illustrate this, let us consider an example of a server that is dropping packets due to high server load. This issue should be alarmed on in the physical layer, where the packet loss, high load, and corresponding high CPU should be detected and alerted on. Any VNFs that reside on the server that are also impaired will also alarm. Let us assume that one of these VNFs is a network router, which is experiencing high packet loss and thus generates a corresponding alarm. This router is in turn leveraged to support a mobility network service, connecting two mobility network VNFs. These mobility VNFs will also detect the packet loss experienced between the two VNFs and alarm, and corresponding mobility service alarms will also be generated. Thus, we have alarms being generated at multiple network layers, on individual components and representing the overall service experience. With all of these different alarms being simultaneously generated, how can ONAP know what automated actions to take? Even if these alarms are simply passed on to network operators to resolve, without automation to combine these alarms, multiple independent operations teams will all be notified of the issues being experienced at the different network layers, and they will all independently start their investigation. However, in reality, the only team that can actively repair the underlying issue is the team responsible for the server that is dropping the packets; the operations teams or automation at the higher layers may be able to take actions to mitigate the impact of the server issues (e.g., by moving the corresponding VMs to a different physical server), if they can isolate the issue to the offending server.

We use *correlation* [CORRELATION] to associate the alarms across physical and different network software layers, and between neighboring network elements and at the service level. Correlation and other alarm dampening techniques reduce the number of automated responses or tickets opened with operators by many orders of magnitude compared with the number of alarms. In the example above, the plethora of alarms generated at the hardware, IP, and mobility element and service layers can all be associated together and isolated to the offending server. By generating a single, correlated alarm, the impaired server will be automatically identified, as will be the impact of that impairment on corresponding VNFs and network services. With this resulting signature of the event identified, ONAP can then look up policies (rules) to determine what action(s) to take—whether this be automated actions such as moving the impacted VNFs to a new server, or ticketing for further human investigation and repair.

The control loop events that are detected, the correlations that are performed and the resulting actions to be taken are all specified by service designers and network operators via ONAP *policies*. In our above example, policies are defined to specify the alarms that are to be generated for the server, VNFs and service layer monitoring, and the specific correlation rules that inform ONAP's correlation engine on how to associate these alarms to conclude that they are related and that the server is the underlying offender. Policies also govern ONAP's responses to the detected issue; in this example specifying that the impacted VNFs should be migrated to a new server, taking the failed server out of service, and logging the underlying server issue so that the failed hardware can be scheduled for repair. Policies can become rather sophisticated; for example, allowing for different actions to be attempted depending on the success of previously attempted actions, and dynamically invoking automated, active tests on VNFs and service performance. However, although the move to this policy-based mode increases flexibility, it also introduces additional risk, requires the lifecycle management of policies and also introduces the need to carefully understand potential policy interactions. We thus need to be careful to provide a balance between the flexibility and the risk, complexity, and manageability of policy-based control. We discuss this more in Section 15.2.4.1.

As service designers and network operators design control loops, they should keep customer experience in the forefront of their policy designs. Service quality management (SQM) complements traditional fault and performance management, by detecting issues that may otherwise be flying under operations' radar (i.e., were not detected within the network) and by quantifying the customer impact of known network impairments.

Service performance measurements, as discussed in Section 15.2.3.1, can be used to detect degraded customer experience. By relating degraded customer experience with known network faults and performance impairments, we can identify and consequently mitigate issues that may not have been detected within the network (i.e., silent failures). And by relating the customer impact to the known network events, we can prioritize resolution activities based on customer impact, or schedule customer impacting mitigation actions to minimize overall customer impact. In addition to needing sophisticated, service-specific active and/or passive measurement infrastructures, SQM requires sophisticated analytics that can be executed on ONAP as DCAE microservices to quantify the customer impact of network events [TONA, MERCURY] and to detect abnormally poor customer experience [ARGUS]. SQM also requires detailed knowledge of how customer traffic is carried across the network(s), including across network layers. We refer to this as the *end-to-end service path*.

The automation enabled by ONAP and associated event detection, correlation, SQM, and other analytics offers the ability to reduce OPEX by automating operational troubleshooting, mitigation, and repair actions. However, we can also improve customer experience, as automated troubleshooting and repair will typically reduce the time that it takes to investigate and resolve an issue—thereby reducing customer impact. However, it is important to note that the effectiveness of ONAP will be governed by the effectiveness of the policies and analytics (microservices) that define the control loops.

15.2.3.4   VNF Change Management

Networks are continually undergoing change—as new technologies are introduced, new features are deployed, and software bugs are fixed. These updates typically involve deploying new software updates or patches, or changing configurations—typically network wide. Before changes are rolled out, they are extensively tested, and then introduced in a small field deployment (first field application [FFA]). Once successfully soaked in the field on a small scale, the software update, patch, or configuration change will typically be rolled out network wide over a period of time.

The process by which a change is deployed on a given network function (virtual or physical) is defined via a *method of procedure (MOP)* or *workflow*. The heart of the workflow is the changes that are to be executed on a network function—these may be software updates or patches, or configuration changes. But any change that is executed must be validated—to verify that it was successfully deployed and that there were no negative side effects. Thus, checks are defined both before and after the change is executed; the results of these pre- and post-tests are compared to determine whether the change was successful or not (e.g., [MERCURY]). If the change was not successful, then the change may be rolled back, or a different mitigation action may be executed.

Historically, network changes were performed using a combination of ad hoc scripts and manual intervention, with each operations team typically creating their own scripts. Systems would track change management schedules for different technologies and attempt to identify conflicts across different activities both within the same technologies and across network layers. Complex, manual "deconflicting" would then be involved in trying to deal with conflicts identified.

In contrast, ONAP provides a consistent and systematic approach for change management across all network functions, thereby eliminating the need for ad hoc scripts and providing consistent process across network function technologies. Change management workflows are specified via ONAP through simple drag and drop capabilities, leveraging predefined, reusable recipes that describe the mechanisms for executing the change (software and/or configuration update) and the health checks. Changes are scheduled using sophisticated scheduling algorithms that account for activities across different network functions of the same type, across network layers and along the service path to avoid conflicts. As you can imagine, with hundreds of thousands or more network functions, complex service paths, multiple network layers, and a dynamic network, it can be extremely complicated to achieve such scheduling, necessitating sophisticated optimization algorithms and detailed and accurate cross-layer topologies. However, if achieved, this avoids the need to do manual deconflicting.

Finally, ONAP also automates the execution of the workflow across network functions, and roll back procedures should issues occur. As the changes are progressively rolled out across the network, sophisticated analytics such as those described in [LT_MERCURY_1, LT_MERCURY_2] can be used to verify that the changes are not resulting in unexpected negative performance impacts, and are realizing desired positive performance impacts. Should the desired goals not be realized, it may be necessary to halt the rollout until a solution is identified and can be deployed.

Automating the change management process at scale reduces OPEX, and also reduces the risk of human operational errors ("Plant Operating Error," or POE).

15.2.3.5 Configuring ONAP to Enable Operations Automation

One of the primary goals of ONAP is to put the definition of new automation in the service designers' and operators' hands, and to minimize any software development required to introduce new functionality. As an example, ONAP enables users (service designers and operators) to define control loops without requiring software development—specifying events to be detected, correlation rules that localize the underlying location and root cause of an issue, and defining actions to be taken to automatically mitigate an issue, or to alert a network operator for further investigation. As another example, ONAP enables workflows to be specified from predefined building blocks to define how change management activities should be executed.

This dynamic configurability of the ONAP environment ensures that the operations teams are in control of the policies and workflow that manage networks and services, and do not require that development teams be engaged to make changes unless new platform or analytics capabilities or network function interfaces are required. This improves operations' ability to introduce new capabilities, enabling faster deployment of new automation for new conditions observed in the network, and correspondingly improving customer experience.

15.2.4 ONAP-Related Challenges

Although ONAP promises exciting benefits, as with NFV, it comes with associated challenges.

15.2.4.1 Safe Policies

With the tremendous power of automation comes risk. A "bad" automation, such as a "bad" control loop, introduced either accidentally or maliciously, could have significant negative impact on network and service. In theory, a control loop could do something like shut down the entire network—clearly not a typically desired outcome! We thus need to introduce processes and technologies to minimize the risk of introducing policies that could result in negative outcomes. Security mechanisms are of course vital here—we must ensure that only those approved to author policies have access to do so. We also need to consider how we validate policies—both to ensure that a policy achieves its desired intent, and that it does not have negative side effects or negative interactions with other policies in the system. ONAP's policy framework thus uses a range of techniques to ensure *"safe policies"*—minimizing the risk associated with introducing policies that have an inadvertent effect on the outcome of automation.

Figure 15.3 illustrates the life cycle of a policy—from its design and authorship through to when it has reached its "end of life." For illustrative purposes here, we consider the example of control loop policies, which specify the events (issues) that are to be detected, and the correlations and actions to be performed in response to the detected events.

Policies are designed, authored, and entered into the ONAP platform by service designers and network operators. As a new policy is introduced, it should pass through a series of validations—which may include laboratory testing, simulation, and offline analysis. Once it has passed all of these tests and the policy has been approved for deployment, it is then activated in the field where it runs until it is no longer needed, when it can be deactivated (either removed, or put in a "sleep" mode monitoring the events without taking automated actions). In addition, a new policy may result

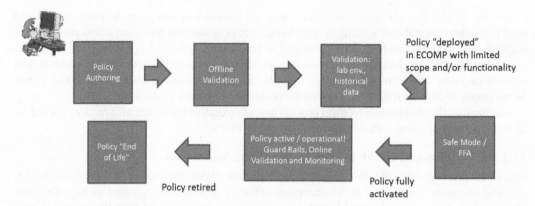

FIGURE 15.3 Policy life cycle.

in modifications to a preexisting policy or new operational data or experiences may indicate that a policy needs to be modified.

To mitigate risk, even a well-tested new policy will typically be deployed into the network incrementally. This may involve turning up the policy in a small portion of the network, and/or for a period of time having network operators validate the event detection in a control loop policy before an action is performed. Consider the example in Figure 15.4 of the simple control loop policy, which automatically restarts a VM if it is determined to be unresponsive. There is a particular *signature* that defines that a VM is unresponsive, and this signature could be relatively complex in its definition. Thus, as the policy is first introduced into the live network, it may be executed in a "safe mode" where only the condition ("if VM unresponsive") is executed, without the automated action (reboot VM) being performed. Instead, ONAP either waits for a human operator to approve the action recommended by the system before it is performed, or simply opens a ticket for manual investigation and repair when the condition is detected. An operator monitoring this new policy can validate that the conditions are being successfully detected; once the operator is satisfied with the policy operating in safe mode, it can be fully operationalized so that the actions are also automatically executed. Such a process can also be performed with control loop policies that involve multiple steps, such as that illustrated in Figure 15.5. With such a policy, each of the different conditions can be validated using the above approach before the next condition in the sequence is activated.

Strict security mechanisms and the testing and validation executed on policies before policies are introduced into the ONAP ecosystem will reduce the risk of "bad" policies reaching the network. However, given the tremendous impact that policies will have, we want to ensure that we introduce additional safety mechanisms for active policies—just in case.

> If VM unresponsive then
> Reboot VM

FIGURE 15.4 Simple control loop policy, rebooting a VM.

> If VM unresponsive then
> Reboot VM
> If VM unresponsive and restart_fail
> Rebuild VM
> If VM unresponsive and rebuild_fail
> Ticket (human intervention)

FIGURE 15.5 Multistep control loop policy.

During runtime, guard rails limit the potential impact of inappropriate policies. As an analogy, guard rails on mountain highways are placed at the edge of the highway to provide protection against cars catapulting off the edge of a cliff. Under normal conditions, cars should not hit the guard rails. However, should a car spin out of control and hit the guard rail, the guard rail is designed to prevent a catastrophic outcome. Policy guard rails are similar; they are essentially overriding policies, which when they detect a violation of an invariant condition (a condition that should always be true), prevent further relevant actions from being taken.

As a concrete example, consider an invariant that states that at least 80% of all network interfaces must be up and operational at all times. If this invariant is violated (too many network interfaces have been shut down), then all control loop actions that involve shutting down interfaces should be overridden, and the interface shut downs prevented. A guard rail such as this would prevent a "bad" policy from shutting down the network—something which a policy as simple as that in Figure 15.6 would achieve if the appropriate guard rails were not in place.

Designing effective guard rails requires appropriate domain knowledge and operational experience regarding the type of conditions that can arise. Thus, service designers and operators should be ultimately responsible for designing and managing the guard rails; similarly, if a guard rail is ever "hit," the operations team should be notified and be responsible for investigating and resolving the conditions that triggered the event.

Policy testing and automated validation techniques need to be complemented with an overarching governance process designed to manage how policies progress through the policy life cycle. The governance process provides a gating process for determining when policies should be moved to field deployment, into and out of safe mode, and to decide when to remove policies from the ecosystem. This governance process would typically be owned by service designers and network operators.

In addition to validating individual policies, we must also consider how multiple policies interrelate. As the number of policies in the ONAP infrastructure scales up, it will become increasingly challenging to understand how different policies interact and to ensure that these policies are consistent. As an example, we need to be able to identify when two different policies may create potential conflict—for example, two different policies with the same priority may under the same conditions specify different actions. As a simple example, consider the two policies captured in Figure 15.7. The first policy says that ONAP should reboot a VM if its available memory drops below 20 GB; the second policy says to migrate the VM if the available memory drops below 15 GB. The two policies have overlapping conditions (when available memory <15 GB) where there are two different actions specified—reboot the VM in the first policy versus migrate the VM in the second. Which action should the system take? These policies *conflict*, and should not coexist in the ONAP system unless one is defined to be of higher priority than the other.

ONAP uses advanced analytics to validate individual and combinations of policies before field deployment, including checking for conflicts such as that illustrated above. When issues are detected, feedback is provided to policy designers so that policies can be redesigned to avoid these types of issues.

If interface load <= 100% then
Shut down interface

FIGURE 15.6   Example "bad" policy.

Control loop policy 1:
If VM available memory < 20GB then
Reboot VM

Control loop policy 2:
If VM available memory < 15GB then
Migrate VM

FIGURE 15.7   Conflicting control loop policies.

15.2.4.2   Identifying Operations Policies

Another significant challenge is identifying the operational policies, particularly related to control loops. ONAP's policy framework empowers operators and designers to control ONAP's behavior. However, the domain knowledge that is to be captured in ONAP policies is often distributed across many network operators and engineers, making it a challenge to synthesize. As new VNF software is introduced into field environments, there are likely software bugs that are not even known to network operators and even the VNF vendors; they are only revealed over time as events are investigated in depth and common patterns are revealed. Machine learning offers opportunities to come to the rescue here by helping service providers automatically learn policies. In the case of control loops, machine learning enables signatures and actions to be automatically captured by mining vast amounts of historical alarm, performance, and log data. Machine learning can identify which events are statistically correlated, and are thus candidates for alarm correlation. Machine learning can also be used in conjunction with logs capturing human operator actions and other network data to identify which signatures (correlated events) typically trigger operators to take specific actions. For example, using historical or laboratory logs that capture when operators manually reboot VMs along with a range of other network data sources (alarms, logs, performance measurements), machine learning techniques can be used to identify what combination of alarms, performance, and logs typically trigger a human to perform a VM reboot. This information, combined with expert domain knowledge, can be transformed into policies that automate future reboot actions.

15.2.4.3   Increasing Software Complexity

ONAP and virtualization together provide a meteoric shift to an increased reliance on software, with a corresponding reduction in our reliance on hardware. As the industry gains more experience with these technologies, we will learn whether this comes with a corresponding increase in the number and complexity of software bugs, and whether this creates a greater challenge in troubleshooting. As we gain further experience, we expect that this will drive new understanding of the tooling required to support troubleshooting of these issues.

15.3   TRANSFORMING THE OPERATIONS TEAM

Despite the fact that traffic volumes continue to grow exponentially, revenue growth for service providers does not follow the same curves. Thus, operations teams typically cannot be significantly expanded, even in the face of a vast number of new, virtual network functions and control capabilities (ONAP) being deployed. It is also unlikely that the economics will make it attractive to immediately replace all of the legacy physical network elements—even as each given network function is being virtualized. Thus, service providers will likely take a "cap and grow" approach to initial VNF deployment; leaving physical network functions (PNFs) in place and adding new growth traffic onto virtualized functions. This is more capital efficient, but leaves operations teams having to support both physical and virtual versions of the same network function, thereby significantly increasing operations responsibilities. Thus, achieving this without expanding the operations team size will necessitate significant enhancements to legacy technologies (PNFs) automation, as well as taking advantage of the improved efficiencies promised through ONAP.

The virtualization of network functions also changes the roles that operations teams perform. The operations role becomes more closely aligned to that of a traditional software company, such as Google or Facebook. However, there still remains a very strong role for networking expertise as the applications running on the "cloud" are network functions that provide networking capabilities and leverage networking protocols.

As network functions are virtualized and use a common hardware platform, there is a single operations team that is responsible for the hardware across different network functions. This contrasts with physical network elements where, given the complexity of typical hardware-based network functions, operations teams were often dedicated to specific network element types from specific vendors. If, for example, a carrier had a dual vendor policy, supporting two different vendors for a given network function would often result in having two distinct and dedicated teams supporting each given type of network element from each vendor. Similarly, as VNFs become more like "cattle" rather than "pets," such that different network functions from different vendors leverage common interfaces, alarm types and generate similar logs, then operations personnel and teams will be more readily able to work across different types of network functions from different vendors. This will reduce the number of operations teams required, with a corresponding reduction in the number of teams engaged in complex, cross-element troubleshooting calls. Fewer teams should also translate to faster resolution of these complex issues. SDN also promises to allow us to collapse network layers (e.g., IP and optical), with corresponding integration of the operations teams.

As automation continues to advance through ONAP and beyond, lower operational tiers (tier 1, tier 2) will continue to be reduced and/or eliminated entirely. Tasks previously performed by humans will be handled by ONAP and other automation. The introduction of NFV is also reducing the need for onsite work forces and rapid, on-demand visits to central offices. However, the increased complexity and reliance on software drives an increase in tier 3 skills, where personnel are required to perform sophisticated debugging of complex software-based issues. Depending on organizational structures, roles, and responsibilities, the operations teams may also have a greater need for advanced software skills, and for policy developers who can use ONAP's policy interface to enable new automation.

One of the bigger challenges for an operations team that spans the complexity of AT&T's networks is to prioritize automation efforts. Effectively achieving this necessitates having a detailed understanding of the operations teams' work drivers, and where operations personnel time is being spent. This is often not easy to identify—however, it is vital to being able to effectively make decisions on where expenditure should be focused.

Aggressively rolling out NFV/SDN technologies requires rapid development, certification, and deployment. Adoption of agile software development techniques [AGILE_1, AGILE2] and DevOps principles [DEVOPS, DEVOPS_PHOENIX] for ONAP development and through vendor engagements are key to achieving this goal. Tight collaboration and engagement between architects, developers, and operations teams (both VNF and ONAP) is vital to achieving these goals; all teams must view themselves as being in the same boat and invested in taking a shared risk, calculated with due diligence. DevOps is also vital to achieving "continuous integration/continuous delivery" (CI/CD). ONAP enables the rapid deployment of VNFs; but rapid deployment of ONAP is also key to achieving the desired speed of innovation and VNF integration. Automation is thus also key to managing ONAP itself, including enabling rapid software instantiation and changes. Automation is also vital to minimizing human errors, providing a repeatable process by which changes are made and providing intelligence that enables ONAP faults and performance issues to be rapidly detected, localized, and resolved. The ONAP operations team is responsible for deploying ONAP instances, updating ONAP software, and detecting, troubleshooting, and resolving ONAP issues. ONAP-like functionality and automation also applies to operating ONAP, enabling many of the ONAP operations functions to also be fully automated.

15.4   MIGRATING TO THE NETWORK CLOUD

Rolling out NFV and ONAP technologies is a journey that must be executed with minimal customer impact—even in the face of the rapid change at both the control plane and data plane. Given the tremendous scale and diversity of networks like AT&T's, the reality is that not all traffic can be suddenly switched to being carried on the new SDN/NFV infrastructure. A typical customer connection will likely traverse both virtual and physical network elements for the foreseeable future, and thus virtual network functions will need to operate in harmony with physical network elements; similarly, ONAP will need to seamlessly operate with legacy OSSs and business support systems (BSSs).

15.4.1 ROLLING OUT NETWORK CLOUD TECHNOLOGIES

Unlike in the past where automation has essentially been an afterthought, automation is at the very heart of the Network Cloud vision. SDN and NFV together offer a "clean slate" opportunity, rather than continuing to build on legacy systems that have evolved along a complex journey over time. However, achieving this vision at scale while also ensuring that we continue to deliver carrier grade network performance and superior customer experience is akin to changing out the engine of the car while driving down the highway at 100 miles an hour. And the operations team is at the heart of performing this engine change.

AT&T is aggressively deploying VNFs and ONAP across different networks and services. Such an aggressive rollout equates to tremendous change. Given the tight service requirements of carrier-grade services, this change must be executed with acceptable risk and minimal customer impact. So, how do we manage this risk while balancing the speed of new technology introduction?

Scalable testing of new technologies before they are deployed, and controlled introduction of technologies in the field mitigates risk. However, keeping pace with the speed of change necessitates that as much of this testing and field validation be as automated as possible.

As both the data and control planes are being simultaneously updated, and with new policies being introduced to manage alarming and event correlation, we can expect that operations teams may not always have perfect visibility into network impairments. However, as discussed in Section 15.2.3.3, end-to-end service monitoring can be used to detect customer-impacting issues, which may otherwise fly under operations' radar. Such monitoring tests the integrity of the end-to-end service paths; inherently covering both the physical and virtual network technologies. Note that to be able to relate end-to-end service measurements with VNF and PNFs, we must have detailed end-to-end service paths, which fully describe how traffic is carried across virtual *and* physical elements (VNFs and PNFs) and across network layers. These can be nontrivial to obtain, but are vital to being able to relate end-to-end measurements with relevant network observations.

15.4.2 BOOTSTRAPPING SDN/NFV DEPLOYMENTS LEVERAGING CURRENT KNOWLEDGE AND EXPERIENCE

There is tremendous domain knowledge and expertise that has been gained by carriers in operating legacy PNF technologies. Much of this experience will be relevant in the new, software-centric VNF environment. How can we prime the new ecosystem to leverage the benefits and experience of the past? Today, policies exist in the legacy OSSs that manage today's networks—for example, correlation rules for alarm dampening and event localization/root cause analysis. As physical network elements are replaced by VNFs, we may be able to leverage such policies in the new world. The complexity of transforming legacy rules to apply them to VNFs depends on factors such as the similarity between physical elements and their corresponding VNFs.

15.4.3 LEGACY AND SDN/NFV NETWORKS COEXISTING IN HARMONY

There is a tremendous investment in the legacy, hardware-based network, and large numbers of different network elements that are to be virtualized. Thus, even with the most aggressive of virtualization deployment plans, it is unavoidable that we must live in a hybrid physical and virtual world for the foreseeable future. In addition, there are also functions within some network elements that simply cannot be fully virtualized or run on off-the-shelf compute hardware—optical transponders and the mechanics of the optical switching in ROADMs, cell site (eNodeB) remote radio units, and the physical infrastructure interconnecting servers, for example.

ONAP was originally conceived to manage VNFs, but it can also manage PNFs. Many of the ONAP benefits—including being policy driven, automated change management, and enabling automated responses to network conditions through control loops—can also apply to PNFs. ONAP can scalably enable new functionality that does not widely exist in legacy OSSs today, including automating operations responses to network impairments. And of course, via ONAP, operators can specify this through policies, putting this definition into the operators' hands enabling faster introduction of new functionality than could be achieved through software updates to legacy OSSs. If ONAP is leveraged to manage physical network elements, it can enable legacy OSSs and BSSs to be retired, thus eliminating the cost of supporting them.

There are numerous examples of where ONAP can be used to automate simple operator responsibilities on PNFs—from automating actions as simple as rebooting an impaired mobility cell site, to sophisticated traffic management in response to localized high network load, outages and planned maintenance activities, and to automated change management. In combination with intelligent algorithms, ONAP can also provide self-optimizing network [SON] functions—dynamically tuning cell site parameters, for example, to improve customer experience as network conditions change [e.g., OUTAGE_MITIGATION] (see Chapter 8 for more details regarding SON).

As previously discussed, customer traffic will typically traverse both physical and virtual network elements, managed by legacy OSSs and ONAP. Thus, operators will need to be able to support both the physical and virtual worlds in parallel, using legacy OSSs and ONAP-enabled intelligence. These domains cannot be considered in isolation; traffic will be routed back and forth between the physical and virtual elements as network conditions change, change management activities will need to be coordinated across domains, and troubleshooting complex network and service issues will regularly necessitate debugging complex interactions between the two domains.

15.5   TOPICS FOR FURTHER STUDY

In this section, we identify and describe some topics for further discussion.

1. *VNF decomposition to enhance operability*: Early virtualization activities across the industry have often focused on taking software from physical network elements and essentially migrating that software to the cloud with minimal changes for it to run in the cloud. Virtualization provides us with an opportunity to rethink how we structure network software, and decompose functionality. However, these VNFs are often not being designed from the ground up to operate in the cloud. How can we decompose VNFs while keeping operational complexity in check? How can we go a step further and even simplify the operability of VNFs by redesigning them as we virtualize? An example of this is the Edgeplex architecture, which decomposes edge router functionality to maximize the flexibility with which customers can be managed in a bid to improve reliability and manageability [EDGEPLEX]. There are significant research opportunities to explore how to most effectively design VNFs to improve operability.

2. *Network automation intelligence*: SDN and NFV together introduce new opportunities to systematically automate network functions and to introduce new flexibility and rapid, dynamic responses to changing network and customer conditions. However, achieving this requires sophisticated intelligence—knowing what to automate, when, and how across a large number of VNFs and PNFs. Automatically resolving network issues, for example, will require analytics and policies to detect anomalous network conditions, to accurately correlate events to effectively localize issues and sophisticated logic to reason as to what actions should be taken. If the analytics fail to accurately detect and localize an event, or if the responses are inappropriate, then erroneous actions could be taken, which could potentially introduce more harm than good. Thus, there are significant challenges associated with how to identify effective intelligence (analytics, workflows, policies, and available platform actions) that should be introduced into the ONAP ecosystem to enable operations automation. How can we scalably learn the policy rules required here? Can we leverage historical (PNF) rules and operational experiences? How can we create policies that are

reusable across diverse VNFs? How can machine learning be used to automatically identify policies, such as correlation rules and actions to be taken in response to detected conditions? And how can we ensure the integrity of the ecosystem in the face of bad or missing data, which is inherent in large-scale systems?

3. *Safe policies*: As discussed in Section 15.2.4.1, policies introduce great flexibility through the ability to rapidly enable new functionality via ONAP. However, with this flexibility comes risk. How can we most effectively balance the benefits versus risk here, and create and use technology to minimize the risk? How can we most effectively police and manage policies—ensuring that (bad) policies that can cause harm are not introduced to the system, and carefully managing interactions between different policies?

4. *Interworking between legacy and SDN/NFV technologies*: As highlighted in Section 15.4, there are significant challenges involved in introducing new technologies into large-scale, operational legacy networks. Such technology integration must be done seamlessly, without customer impact. How should legacy and SDN/NFV technologies interwork? How can we most effectively and seamlessly migrate from legacy technologies to SDN/NFV? To what extent can legacy technologies be retired?

ACKNOWLEDGMENTS

The authors wish to acknowledge Taso Devetzis, Juan Flores, Mark Francis, Mohammad Islam, and Mike Paradise for the insight and wisdom contributed to this chapter.

REFERENCES

[3GPP_KPIS] 3GPP TS 32.450 V9.1.0 (2010-06)—Key Performance Indicators (KPI) for Evolved Universal Terrestrial Radio Access Network (E-UTRAN): Definitions (Release 9), ETSI, 2010.
[CELLULAR_QOS] G. Gomez and R. Sanchez, *End-to-End Quality of Service over Cellular Networks*, John Wiley & Sons, 2005.
[CORRELATION] https://en.wikipedia.org/wiki/Event_correlation.
[MERCURY] A. Mahimkar, Z. Ge, J. Wang, J. Yates, Y. Zhang, J. Emmons, B. Huntley and M. Stockert, Rapid detection of maintenance induced changes in service performance, *Proceedings of the Seventh Conference on emerging Networking Experiments and Technologies*, Article No. 13, Tokyo, Japan, December 6–9, 2011.
[LT_MERCURY_1] A. Mahimkar, H. H. Song, Z. Ge, A. Shaikh, J. Wang, J. Yates, Y. Zhang and J. Emmons, Detecting the performance impact of upgrades in large operational networks, *Proceedings of the ACM SIGCOMM 2010 conference*, pp. 303–314, New Delhi, India, August 30–September 3, 2010.
[LT_MERCURY_2] A. Mahimkar, Z. Ge, J. Yates, C. Hristov, V. Cordaro, S. Smith, J. Xu and M. Stockert, Robust assessment of changes in cellular networks, *Proceedings of the ninth ACM conference on Emerging networking experiments and technologies*, pp. 175–186, Santa Barbara, California, December 9–12, 2013.
[ARGUS] H. Yan, A. Flavel, Z. Ge, A. Gerber, D. Massey, C. Papadopoulos, H. Shah and J. Yates, Argus: End-to-end service anomaly detection and localization from an ISP's point of view, *Proceedings of the 31st annual IEEE International Conference on Computer Communications (INFOCOM)*, Orlando, Florida, March 25–30, 2012.
[TONA] Z. Ge, M. Kosseifi, M. Osinski, H. Yan and J. Yates, Method and apparatus for quantifying the customer impact of cell tower outages, U.S. patent 9426665.
[SON] J. Ramiro and K. Hamied (Editor), Self-Organizing Networks (SON): Self-Planning, Self-Optimization and Self-Healing for GSM, UMTS and LTE, Wiley, 2011.
[OUTAGE_MITIGATION] X. Xu, I. Broustis, Z. Ge, R. Govindan, A. Mahimkar, N.K. Shankaranarayanan and J. Wang, Magus: Minimizing cell service disruption during planned upgrades, *Proceedings of the 11th ACM Conference on Emerging Networking Experiments and Technologies*. Article No. 21, Heidelberg, Germany, December 01–04, 2015.
[DEVOPS] M. Loukides, What is DevOps? *O'Reilly*, 2012.

[DEVOPS_PHOENIX] G. Kim, K. Behr, G. Spafford, The Phoenix Project: A Novel about IT, DevOps, and Helping Your Business Win, *IT Revolution Process*, 2014.

[AGILE_1] https://en.wikipedia.org/wiki/Agile_software_development.

[AGILE_2] C. Larman, *Agile and Iterative Development: A Manager's Guide*, Addison-Wesley, 2004.

[EDGEPLEX] A. Chiu, V. Gopalakrishnan, B. Han, M. Kablan, O. Spatscheck, C. Wang, and Y. Xu, EdgePlex: Decomposing the provider edge for flexibilty and reliabity, *Proceedings of the 1st ACM SIGCOMM Symposium on Software Defined Networking Research*. Article No. 15, Santa Clara, California, June 17–18, 2015.

16 Network Measurements

Raj Savoor and Kathleen Meier-Hellstern

CONTENTS

16.1 SDN Data and Measurements 354
16.2 Use of Real-Time Network Data with SDN 356
    16.2.1 Example of an IP/Optical Network with SDN Controller 356
    16.2.2 Where and How to Measure Real-Time Network Data 357
    16.2.3 Centralized TE Using SDN Controller to More Efficiently Manage TE Tunnels 358
    16.2.4 Dynamically Manage and Reconfigure the Mapping between IP and Optical
        Layers Using ROADMs 359
    16.2.5 Use of Available Spare Capacity to Offer Bandwidth Calendaring Service 359
16.3 Network Capacity Planning 360
    16.3.1 Current Network Capacity Planning Process 360
    16.3.2 Highlights of Benefits from SDN 361
    16.3.3 Traffic Matrix Data 362
    16.3.4 Traffic Forecast 362
    16.3.5 Wavelength Circuits 363
    16.3.6 Layer 0 and Layer 3 Resources 363
16.4 SDN Controller Measurement Framework 364
    16.4.1 Objectives for a Measurement Framework 365
    16.4.2 Measurement Framework Overview 366
    16.4.3 Component-Level Measurements 367
    16.4.4 Service and Network Measurements 368
    16.4.5 The Case for Decoupling Service and Network Path Measurements
        from Network Elements 368
16.5 AT&T's SDN-Mon Framework 370
    16.5.1 Network Deployment Example for the SDN-Mon Framework 372
16.6 Telemetry Measurements 372
16.7 NFV Data and Measurements 373
    16.7.1 NFV Data Model 374
    16.7.2 NFV Infrastructure Telemetry Data Model 374
        16.7.2.1 OpenStack Telemetry Service: Ceilometer 374
        16.7.2.2 OpenStack Telemetry Service: Ceilometer Examples 375
16.8 NFV Data Measurement Framework 377
    16.8.1 Organic NFV Data Measurement Model 377
    16.8.2 Passive vProbe Measurement Data Model 378
    16.8.3 Active vProbe Measurements Data Model 379
16.9 VNF Reporting Metrics 379
    16.9.1 VNF Resource Consumption and Operational Metrics 379
    16.9.2 VNF SLAs and KPI 380
    16.9.3 VNF Resiliency Reporting 380
16.10 VNF Scaling Measurements 381
    16.10.1 NFV Service Auto-Scaling 382
    16.10.2 Scaling Triggers 382
    16.10.3 Scale Out: Illustrative Example 383

16.11 VNF Efficiency Measurements and KCI Reporting..383
16.12 VNF Measurements for Optimal Placement and Sizing 384
16.13 Areas for Further Study 385
Acknowledgments...386
References...386

Software-defined networking (SDN) and NFV present many new challenges to the traditional operator model of operating services. VNFs are instantiated based on resource demands and the SDN-controlled logical connections used to connect services move dynamically based on utilization, policy, and resiliency designs. SDN and NFV data models have to enable measurements that ensure parity with the legacy service data models while adding new entities and relationships that cover design artifacts of virtualized and SDN-controlled environments.

As VNFs are added, moved, and decommissioned, the measurement points must also be defined, moved, and retired at the same rate. In a flat network architecture with multiple VNFs coresident on hosts in each zone supported by an SDN local overlay network, the traditional static approaches to measure resource utilization, collect flows used for troubleshooting, and advanced service analytics are no longer clearly marked. Virtualization of the network infrastructure and dynamic movement of VMs also renders traditional physical probes, fiber splitters, and traditional port mirroring functions obsolete, requiring new virtual probe (vProbe) measurement methods for SDN and NFV environments.

These NFV and SDN real-time streaming measurements require the ability to process and analyze massive amounts of data streams originating from different sources in the network infrastructure such as the VNFs, cloud infrastructure telemetry sources, SDN controllers, and vProbes. Real-time analytics processing offers network operators new opportunities to determine where cloud capacity must be added, and allows them to quickly respond to service assurance issues. Applying real-time analytics in a closed-loop mechanism enables operators to automate workloads and optimize outcomes.

This chapter focuses on the network data and measurements required by algorithms, SDN controllers, and other value-added applications. Also, the focus here is on the needs of data and measurements for traffic and network capacity management. Chapter 15 looks at the data and measurements needed for network operation and alarm detection and handling.

16.1   SDN DATA AND MEASUREMENTS

To illustrate how data can be measured and used to optimize network utilization, consider Figure 16.1, which depicts the overall architecture of an SDN-enabled backbone network (see Chapter 13). Much of the material presented in this subsection on the SDN network architecture is drawn from the *IEEE Communications Magazine* article [1] authored by some of the contributors to this chapter. The data plane is an IP/MPLS over a wave division multiplexing (WDM) optical network. The optical layer uses flexible ROADMs that provide open interfaces to allow wavelength reconfiguration. Such interfaces enable centralized control of the optical network to dynamically and remotely establish/release wavelength circuits, thus adjusting the layer 3 logical topology and capacity. Extending SDN control from layer 3 to the optical layer is important to automating and optimizing network provisioning and operation tasks, such as capacity augmentation and failure restoration. Many vendors support such control interfaces in their new products. This figure highlights the explicit use of real-time data coupled with efficient and intelligent optimization algorithms to dynamically manage a large IP network, both logical and physical layers.

The IP/MPLS routers support traditional distributed routing. They also provide open interfaces enabling centralized SDN control. MPLS still plays a critical role to ISPs in many operations, including routing, traffic engineering (TE), priority control, service provisioning, and failure restoration.

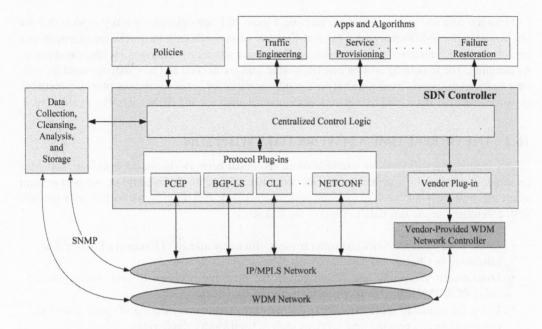

FIGURE 16.1 SDN network architecture.

Accordingly, the SDN control in this architecture works at the level of label-switched paths (LSPs). Some prior works perform fine-grain traffic control based on 5-tuple flows identified by source/destination IP addresses, protocol, and source/destination ports. While such an approach is beneficial to certain systems (such as data-center networks), it is unnecessary and technically challenging for large-scale backbone networks. Instead, with the aid of the optimized routing algorithm, the SDN controller will monitor and, if necessary, step in to manage these LSPs to react to traffic and network conditions while maintaining high reliability.

Separate from the data plane, a global SDN controller communicates with the routers and ROADMs through various interfaces to perform monitoring and control tasks. For the optical layer, the key control capability is to establish/release wavelength circuits to adjust the layer 3 logical topology. The controller performs such operations through a vendor-provided control module or an open RESTful interface to an open ROADM controller, which then proceeds with light path computation and ROADM configuration. For the IP/MPLS layer, the key control function is to monitor traffic load and link state to perform globally optimized LSP operations, such as establishing/releasing LSPs, adjusting the paths/bandwidth of LSPs, and multipath routing. The controller provides interfaces to support customized policies and various applications, such as centralized TE, dynamic service provisioning, and optimized failure restoration. The logically centralized controller could be implemented in a physically distributed way to achieve scalability and resilience. The specific controller design and deployment are beyond the scope of this chapter. Interested readers could refer to related materials, such as References 2–4.

The interfaces between the control and data plane are extensible to support multiple protocols. While OpenFlow is frequently discussed in the SDN community [5], so far it is not the most critical protocol in the carrier network architecture. Figure 16.1 illustrates a few indispensable protocols and data: Border Gateway Protocol-Link State (BGP-LS) for link status, capacity and per-class bandwidth usage; Path Computation Element Communication Protocol (PCEP) supports interactions between the controller and routers in terms of reporting and readjusting route and bandwidth of LSPs; Simple Network Management Protocol (SNMP) on traffic measurement on each link and LSP at minute-level frequency. The measurement data can complement BGP-LS and PCEP to achieve global optimal TE. Subsequent sections of this chapter will provide detail uses of these data.

From the data and measurement perspective, Figure 16.1 demonstrates two key aspects that are not available in pre-SDN networks: (1) real-time use of network data for traffic management in a coordinated and centralized control manner and (2) much shorter time horizon for integrated capacity planning. The remaining section describes what data are needed and how they are used for real-time traffic management (real-time network condition data); for iterative capacity planning (network resource inventory data), and for network performance analyses and SDN-controller monitoring.

16.2   USE OF REAL-TIME NETWORK DATA WITH SDN

In this section, we will show an example of an IP/optical network (including backbone and edge locations) controlled by a centralized SDN controller (Figure 16.2). In addition, we will explain where and how we would need to measure real-time network data. Finally, we will identify how the SDN controller can use this data to support the following:

- Centralized TE using SDN controller to more efficiently manage TE tunnels, LSPs or flows (discussed in Chapter 11)
- Dynamically managing and reconfiguring the mapping between IP and optical layers using ROADMs to enable faster provisioning of resources
- Using the available spare capacity in the network (particularly during off-peak hours) to opportunistically provide new services such as bandwidth calendaring

16.2.1   EXAMPLE OF AN IP/OPTICAL NETWORK WITH SDN CONTROLLER

Figure 16.2 expands upon what was depicted in Figure 16.1 to illustrate an example of an integrated IP/optical network and its interaction with the SDN controller. E[i] represents the IP edge routers; B[i], the IP core or backbone locations; and O[i], the optical nodes (ROADMs). A subset of the optical nodes is collocated with an IP core location. There are two core routers, A and B, per core location. All unicast traffic originates/terminates at the edge routers and each such router is connected

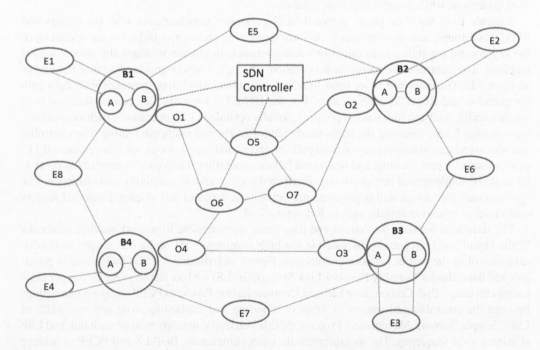

FIGURE 16.2   Example IP/optical network with backbone (B) and edge (E) locations.

to at least two core routers (same or different locations) using physically diverse paths. In addition, there may also be point-to-multipoint multicast traffic and all the endpoints of such traffic are also edge routers (it is of course possible for the same router to perform both edge and core functions).

A subset of all possible pairs of IP routers is connected via IP links to form an IP network. An IP link between two different core locations needs to be routed over the optical network. As an example, the IP router A in location B2 may be connected to the IP link B in location B4 over the sequence of optical nodes O2-O7-O6-O4. The SDN controller can control the edge routers, the core routers, and the optical nodes. The SDN controller is logically shown as a single centralized entity but it may be functionally separated into one controlling the IP network and one controlling the optical network as shown in Figure 16.1. Furthermore, for the purpose of reliability and disaster recovery, it makes sense to have one active SDN controller and one or more standby SDN controllers located geographically in different places.

16.2.2 WHERE AND HOW TO MEASURE REAL-TIME NETWORK DATA

The key measurement types and locations are the following:

- *Edge-to-edge measurements*: It is possible to measure and report on all edge-to-edge unicast and multicast flows in a small network. However, this is not scalable for large ISP networks since there may be hundreds of millions of flows in such a network. Instead, one can aggregate all traffic flows by type between a specific source-destination pair of edge routers to a small number of TE tunnels (point-to-multipoint tunnels in case of multicast flows). Studies have shown that typically we do not have to sacrifice efficiency when doing such aggregations. One mechanism for measuring and reporting on edge-to-edge TE tunnels is to use PCEP. It can report the up/down status, the detailed route, the signaled bandwidth, and the desired bandwidth of each TE tunnel to the SDN controller. The reporting from the routers can be periodic and/or following significant changes to any of the TE tunnel attributes.
- *Core-to-core, edge-to-core, and core-to-edge measurements*: Even though aggregating individual edge-to-edge traffic flows to TE tunnels cuts down on the number of measurements and managed entities, it is often necessary from a scaling perspective to aggregate further by using a small number of core-to-core TE tunnels for every pair of core routers. In such a scenario, it is also desirable to measure and report the edge-to-core and core-to-edge traffic flows using edge-to-core and core-to-edge TE tunnels. In a large ISP network with many more edge routers than core routers, the total number of core-to-core plus edge-to-core plus core-to-edge TE tunnels is usually significantly lower than the total number of edge-to-edge TE tunnels. The PCEP mechanism described above for measuring and reporting edge-to-edge TE tunnels may also be used for measuring and reporting core-to-core, edge-to-core, and core-to-edge TE tunnels.
- *Individual IP link measurements*: In order for the SDN controller to efficiently route edge-to-edge and core-to-core traffic, it is essential to obtain the up-to-date status of each IP link that can be on the path of the traffic flow. A mechanism to do this is to use the BGP-LS protocol that allows routers to report to the controller the up/down status, total capacity, and total reservable bandwidth per traffic class for each link in its OSPF area.
- *Individual optical node and link measurements*: Since IP links are routed over optical links, it is essential to monitor the up/down status of optical links and, if necessary, change the routing of the IP links based on that information.
- *Measurements on the SDN controller*: For the purpose of reliability and disaster recovery, there should be one active SDN controller and one or more standby SDN controllers. The status of each of the SDN controllers should be continuously monitored and if it is determined that the active controller is unable to control the network, one of the standby controllers should become the active controller.

It should be noted that while some of the above data are generally available today, in the pre-SDN environment, it is not used proactively for real-time traffic management in a coordinated and centralized manner. In the SDN environment and through the controller managing the routers, ROADMs and other network elements along with efficient optimization algorithms, one can use these real-time data to optimally manage the transport network maintaining the highest level of reliability and to allow ISPs to provide new services. While some of these techniques were referred to in Chapter 13, a few examples of the use of real-time traffic measurements in an SDN network are given below.

16.2.3   Centralized TE Using SDN Controller to More Efficiently Manage TE Tunnels

The traditional routing of TE tunnels uses a distributed mechanism in which each TE tunnel is routed by the head-end router. Each head-end router knows the spare bandwidth available on each IP link and the path and bandwidth used by tunnels controlled by the router but it does not have a global view of the path, and the bandwidth, of all tunnels. Furthermore, two different head-end routers may try to access the same spare bandwidth at an IP link to route their tunnels. This may result in inefficient routing. In order to eliminate this inefficiency, we use the end-to-end measurements (edge-to-edge or core-to-core) and the individual IP link measurements mentioned above, and feed that to the SDN controller. The routing inefficiency is eliminated since the SDN controller has the global view of the path and the bandwidth of every tunnel in real time and it is a single logical entity making routing decisions (as opposed to independent routing decisions by many head-end routers). The packing efficiency of distributed versus centralized routing, is shown with the help of an example in Figure 16.3 [1,6]. The best-possible packing is achieved by the multicommodity flow (MCF) mechanism where each TE tunnel may be split into infinitesimal subcomponents and each such component may be routed in the most efficient manner. We designate the efficiency of MCF as 100% and show the efficiency of the other algorithms in comparison to that. Since distributed routing is done by many different entities, the efficiency is not deterministic and can typically range from 90% to 98%. However, the efficiency of the centralized algorithm used by the SDN controller can usually exceed 99%. Even though the MCF mechanism has the best efficiency, it is not practically realizable since it requires arbitrary splitting of TE tunnels and typically needs a run time in the tens of minutes. In contrast, the centralized algorithm used by the SDN controller is practically realizable and can run in seconds.

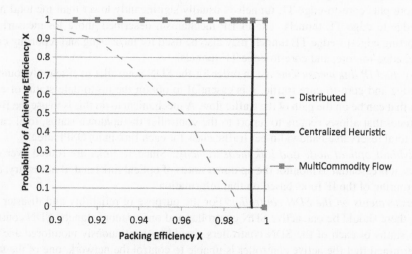

FIGURE 16.3   Packing efficiency of distributed versus centralized routing.

16.2.4 DYNAMICALLY MANAGE AND RECONFIGURE THE MAPPING BETWEEN IP AND OPTICAL LAYERS USING ROADMs

Traditional IP networks typically cannot be run at an average utilization greater than 50% (under no-failure condition) for two primary reasons:

- There is a significant lead time for procuring new IP/optical equipment (typically 3–6 months) and so essentially we need to build a network that can accommodate almost 6 months of capacity growth.
- Spare capacity needs to be designed in the network to account for traffic surges and unscheduled failures of single routers and single optical links (fiber spans).

Dynamically reconfigurable ROADMs allow us faster provisioning with lead-time cut to just a few weeks. Furthermore, instead of keeping all the IP links connected all the time (including spares for failure restoration), it is only necessary to keep the endpoint tails (a combination of a router port and a transponder) and some spare optical regenerators that are ready for rapid connection in the event of a failure or traffic surge. At any given instant only those tails are connected that are needed to support the traffic flow at that instant. As the traffic condition and network condition changes, the tails can be dynamically reconfigured (including optical regenerators, if needed) to match the new needs. This approach heavily relies on real-time traffic measurements that were mentioned before. First, end-to-end (edge-to-edge or core-to-core) and individual link measurements as well as quick failure detection at the packet IP and optical layers provide the trigger for capacity addition or deletion. Second, we also need a real-time view of available tails and regenerators to identify the optimal choice of link addition or deletion.

Use of the above approach allows the average network utilization (under no-failure condition) to be increased from about 50% to about 70%. This approach does require some preconfigured spare endpoint tails and spare optical regenerators. However, overall savings compared to the traditional approach can be substantial.

16.2.5 USE OF AVAILABLE SPARE CAPACITY TO OFFER BANDWIDTH CALENDARING SERVICE

Even after achieving a highly efficient network using techniques described above, there will still be some spare capacity available in the network. In Figure 16.4, we consider the intercity IP links of a typical ISP network. The horizontal axis shows the individual IP links arranged from largest to smallest capacities and the vertical axis shows the capacity and the maximum traffic at those links using a normalized scale. Each link has two directions with symmetric capacity but typically asymmetric traffic loads. For this reason, we show the traffic separately in the dominant direction and the other direction. Due to significant concentration of traffic in certain regions of the network, the link capacities are usually not the same and have significant variation. We note that even though the maximum traffic in the dominant direction is close to the capacity, there is significant bandwidth left in the other direction. Furthermore, even the dominant direction gets close to the capacity only during peak hour and there may be significant spare capacity during off-peak hours.

Based on accurate and up-to-date traffic measurements available at the SDN controller and forecasting, it is possible to offer bandwidth calendaring service that can use the spare capacity opportunistically. For the same example, Figure 16.5 shows the spare sellable traffic among a set of data centers as a function of time of day and using a normalized scale. The "typical" case assumes that the demand for bandwidth calendaring is symmetric among the data centers and independent of where capacity is available while the "maximum" case assumes that the bandwidth calendaring demand matches closely with the available spare capacity. It is observed that there is some bandwidth calendaring opportunity during the network peak period and significantly higher opportunity during off-peak hours.

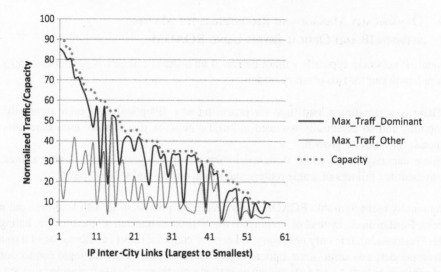

FIGURE 16.4 Maximum traffic versus capacity of IP link bundles.

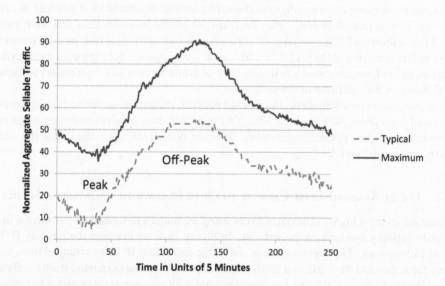

FIGURE 16.5 Spare sellable traffic among a set of data centers.

16.3 NETWORK CAPACITY PLANNING

The previous section demonstrates how the real-time network data can help drastically improve traffic management of the transport network with minimal network resources and without sacrificing high level of network reliability. In this section, we will discuss how SDN can also transform the way we address various capacity planning activities. Again, SDN transformation allows timely information, additional intelligence, process integration, and automation. Before we appreciate these benefits, let us briefly review the current network capacity planning process.

16.3.1 CURRENT NETWORK CAPACITY PLANNING PROCESS

As network traffic increases network capacity needs to be augmented to accommodate the increase in traffic. Network elements typically utilize modular designs; planners have to decide both when new elements are required and when new modules must be added to existing elements. Planners must also decide when link capacity must be augmented; augmenting link capacity may sometimes require augmenting element capacity. At other times link capacity can be augmented by using existing element capacity.

Planners start with data on the current network configuration and current network traffic. The data on the current network configuration includes routers, ROADMs, and the links used to connect the various elements as well as element and link capacity that are not currently operational but that are in the process of being added. These data are derived from inventory and configuration management systems.

Planners must also forecast how much traffic a network will be expected to carry. Planners will start with how much traffic a network is currently carrying; this data are derived from performance management systems. Planners then estimate how much growth is expected to occur, this is based on trend analysis of historical data as well as input on new service offers, market conditions and customers. For network capacity planning, a traffic matrix is required, which indicates how much traffic must be carried between different origin–destination pairs. A traffic matrix may be estimated from flow data like NetFlow or JFlow or from an inference method like tomo-gravity [7] that utilizes link-interface measurements.

Planners then use modeling tools to simulate the impact of the estimated traffic on network capacity and decide where new capacity is needed. An important aspect of simulating the impact of the expected traffic is simulating the effects of failures or taking capacity temporarily out of service for maintenance. A single incident like a fiber cut could result in the failure of multiple network elements.

Prior to the use of SDN, IP service planners would decide between which pairs of routers link capacity would be needed, which would then be used as an input to the capacity planning of the layer 1 network. Turning up the new capacity could often take months. With SDN, links between routers can be turned up dynamically in response to network failures or changing traffic conditions. As a result, interrouter links can be run at significantly higher utilizations but spare capacity must be planned in the layer 1 network to accommodate interrouter links being turned up on demand.

16.3.2 Highlights of Benefits from SDN

Some of the key benefits of SDN in the capacity planning process are described below:

- Pre-SDN
  - Need to build end-to-end IP links that are statically routed over optical links.
  - Long lead-time (several months) for ordering IP/optical equipment. Decisions about where capacity will be augmented are made several months before the capacity will actually be utilized.
  - Forecast accuracy is very important. If the forecast is off, then capacity may be in the wrong place even after several months. This is particularly true since the entire end-to-end IP link is static.
- Post-SDN
  - Only need to build endpoints (tails consisting of router port plus colorless/directionless transponder) and spare regenerators at strategic places and make the connection dynamically between any two available tails (including regenerators if needed) depending on the traffic and network failures.
  - Faster provisioning with SDN cuts down equipment ordering lead-time from several months to a few weeks. Decisions about where capacity will be augmented are made only several weeks before the capacity will actually be utilized.

- Uncertainty about forecast accuracy is less important since capacity planning operates on a time scale of a few weeks. Even if the forecast is off, there are no stranded static end-to-end IP links but rather with endpoints (tails) and possibly spare regenerators. The penalty is low since each tail can be repointed to a different direction as needed and the spare regenerators are pooled resources and can be used elsewhere.

The remainder of this section defines some of the key measurements needed for network capacity planning in the SDN environment.

16.3.3 TRAFFIC MATRIX DATA

Deriving the traffic matrix data required for network planning can be challenging. One option is to use a flow measurement product such as NetFlow [8] or JFlow. These products offer multiple options for collecting data on network traffic with different levels of granularity. Traffic is often collected for each 5-tuple, that is, source IP address, destination IP address, protocol, source port, and destination port. Because of the volume, data sampling is often used. Flow measurements may only be collected on selected router interfaces either because some elements do not support the collection of flow measurements, or to reduce the volume of the data generated. For network capacity planning the origin and destination routers where the traffic enters and exits the network is required. It may be necessary to infer the origin and destination routers from the origin and destination IP addresses, which requires extracting routing tables from the routers and doing routing table lookups. An alternative approach to derive a traffic matrix, for example, if flow measurements are not available or incomplete, is to use an inference method like tomo-gravity. Inference methods do not generate the actual traffic matrix but the results are often adequate for modeling. Such methods use mathematical techniques to derive traffic matrices from link interface measurements, which are easier to collect; if incomplete flow measurements are available they could be used to improve the quality of the resulting traffic matrices. The new SDN capabilities coupled with new opportunities for virtual probes will enable faster and more accurate collection of measurements in more places.

16.3.4 TRAFFIC FORECAST

It is common practice for network operators to model their traffic and derive traffic forecasts for the future. The traffic forecast is then used as input to capacity planning to determine how to augment the network, for example, installing additional regenerators, transponders, router ports, and establishing new wavelength channels among specific routers. A fundamental approach for traffic forecast is to study historical traffic measurement to quantify the trend of growth. Based on the quantified growth rate and the current traffic load, an estimated future traffic load is then derived.

A critical question is how to improve forecast accuracy. A conservative forecast could lead to insufficient capacity augmentation, which in turn results in network congestion and degraded quality of service (QoS). An aggressive forecast would cause subsequent capacity planning to add excessive resources to the network and thus incur unnecessary cost. In practice, it is a difficult challenge to obtain an accurate traffic forecast because the network carries heterogeneous services that could have very different usage patterns and growth rate. Some traffic may peak during the day and some may peak during the night. Some traffic may cross the entire backbone network and some, being largely localized, may span just a couple of hops. Some services may have stable volume and some may grow rapidly.

In a pre-SDN environment, the only viable approach to improving forecast accuracy was to attempt fine-grained modeling. Given the service heterogeneity, it helps to take a fine-grained approach to classify traffic into different types and model and forecast each type of traffic separately.

For example, media streaming services could be a significant percentage of the total traffic and they often demonstrate large variation and high growth rate. Such services could be highly localized in that most of the traffic does not traverse the entire core backbone network. By understanding the characteristics and routing of such traffic, one could develop a model specifically for media streaming and thus increase the forecast accuracy. At the same time, incorporating the localized routing characteristics of such traffic could also improve the forecast accuracy of the overall traffic load.

SDN takes a different approach to traffic forecasting. The need to have an accurate long-term forecast is replaced by cycle-time reduction. Given the uncertainty of traffic growth, it is virtually impossible to achieve good forecast accuracy over a long period of time. The traditional traffic forecast and corresponding capacity augmentation has a long cycle time, mainly because operational processes are manual, discrete (nonintegrated), and inherently lengthy. In particular, it is virtually impossible to substantially speed up the process of labor-intensive wavelength provisioning with traditional ROADMs where many constraints such as wavelength color, direction, and contention, complicate installation and configuration. An SDN-enabled optical network helps to reduce the cycle time for service provisioning. Although manual installation is still necessary, it is less complex with colorless, directionless, and contentionless ROADMs with open interfaces. In particular, the analysis and wavelength provisioning can be automated by using the optical network control interfaces. With this reduced cycle time, traffic forecasting is made easier and is more accurate. With dynamic reconfiguration at the packet layer, coupled with the flexibility of an SDN-enabled optical network, the network is better able to adapt to traffic uncertainties.

16.3.5 WAVELENGTH CIRCUITS

In an IP-over-WDM architecture, routers are interconnected using wavelength circuits. A wavelength circuit may span multiple hops traversing fibers, ROADMs, amplifiers, regenerators, etc. In practice, two routers could be interconnected by a bundle of wavelength circuits where not all the wavelengths in the same bundle take the same physical routing path. Also wavelength circuits interconnecting different pairs of routers may traverse the same set of physical elements, for example, the same fiber segment. When the element shared by two wavelength circuits fails, both wavelengths will be interrupted; these two wavelengths are said to belong to the same shared risk link group (SRLG).

The complete information of each wavelength circuit needs to be maintained, and the status of each wavelength needs to be monitored to facilitate trouble shooting during operation. For example, if multiple wavelength circuits fail simultaneously, it is very likely that a shared device among these wavelengths is having a problem. By checking the SRLG data, the operator could quickly narrow down the root cause of the failure and thus speed up troubleshooting. The wavelength circuit information is also important for capacity planning. The routing and SRLG data are used to determine additional resources needed under each potential failure scenario.

16.3.6 LAYER 0 AND LAYER 3 RESOURCES

In traditional networks, both layer 0 and layer 3 are static in that adding and removing connections are usually performed manually. An SDN-enabled optical network facilitates dynamic service provisioning in that a wavelength circuit can be programmatically established or released on demand, thus changing the layer 3 network in a matter of minutes. Such dynamic provisioning can be used to establish new wavelengths to add capacity to the network on the fly. It can also be used to perform failure restoration. Figure 16.6 shows three cases that leverage optical layer restoration. In case (a), a line card failure on router1 disconnects router1 and router2. The standby connection between router1 and ROADM1 is used to reconnect the two routers. In case (b), a failed fiber disconnects router1 and router2. At both ends, the line card-transponder connections become idle. The SDN controller can reuse such resources to make a new wavelength along the alternative path, where an

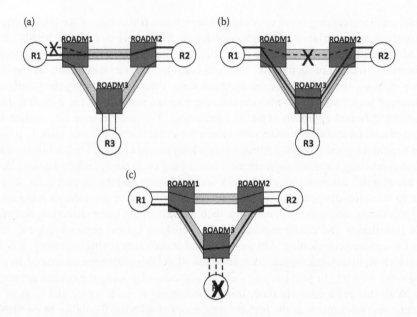

FIGURE 16.6 Optical layer dynamic restoration (dotted lines are broken lightpaths, and thick lines are restored lightpaths). (a) Port failure: Switch to standby port, (b) Fiber failure: Switch to alternative lightpath, (c) Router failure: Set up new lightpath between R1–R2.

idle wavelength is available. In case (c), router3 fails completely. The SDN controller configures ROADM3 to reuse the released wavelengths and transponders to add new capacity between router1 and router2, which helps to deliver traffic previously traversing through router3.

An essential requirement to perform dynamic provisioning is to have complete and accurate description of layer 0 and layer 3 resources, including router ports, transponders, tails, regenerators, amplifiers, WDM reach table, etc. The layer 0 resource monitoring and maintenance could be very complicated and tightly coupled with the specific equipment. In practice, the specific control of layer 0 is performed by a vendor-provided controller, which provides an open interface to interact with the layer 3 or global SDN controller. This design greatly simplifies the design and operation of resource monitoring and maintenance.

16.4 SDN CONTROLLER MEASUREMENT FRAMEWORK

There is a continued need to evolve measurement technologies as the network continues to transform. Just as with prior measurement implementations, the service provider needs to monitor compliance with service-Level agreements (SLAs), quickly identify network impairments that impact customer service or reduce network capacity, and isolate such impairments for repair. With the transformation, there is the added emphasis on the automation of detection of network impairments, requiring the SDN-related measurement solution to scale to enable such automation.

The network transformation to incorporate NFV and SDN, coupled with the advances made in measurement technologies, offers us an opportunity to refresh and optimize the current measurement framework for use with SDN-controlled network elements, whether virtual or physical. We are now in a position to leverage these transformations to redesign SDN-related measurement solutions, reaping cost, and efficiency benefits.

However, as we move forward, we need to be aware that many of the functional principles implemented in the past measurement designs still apply, with this transformation offering us the opportunity to optimize the measurement tool footprint that implements them. For example, measurements were traditionally organized into the following categories (and they continue to be relevant to an SDN measurement solution design):

- Service measurements
- Network measurements
- Component measurements

Service measurements typically focus on the customer's experience across the network. Network measurements focus on the health of the underlying network and interconnections of network elements that carry service traffic. Component measurements focus on the health of individual network elements. Therefore, service measurements help identify whether a network problem impacts the customer, whereas network measurements and component measurements help zoom into possible root causes for customer-impacting troubles and further identified events that reduce provisioned network capacity.

Each of these functions has traditionally been implemented in separate tool sets, developed with little correlation to each other, often requiring several specialized measurement components. In addition, these tools were built with a heavy reliance on network element vendors delivering measurement agents bundled into their products. This not only resulted in vendor-proprietary solutions that complicated the network element implementations, but also increased the cost of their development, and delayed the delivery of network features.

With the advent of NFV and SDN, we are offered the opportunity to decouple and virtualize many of these measurement features, with the overall impact of simplifying the network elements. Such decoupled measurement functions in turn become candidates for implementation as atomic components. Decoupling offers us the opportunity to optimize the measurement tool set, utilizing the same tool to measure multiple functions. Although the desire is to decouple measurement features from network elements, it is not always a straightforward task to identify whether a measurement feature is a candidate for decoupling. We will provide guidance on what measurement features make sense to decouple further in this chapter, as well as present the overall framework for SDN-related measurements.

16.4.1 OBJECTIVES FOR A MEASUREMENT FRAMEWORK

The following objectives are kept in mind while refreshing the prior measurement framework to work with the NFV/SDN network:

1. Improving the customer service experience, including providing customer-specific reporting.
2. Meeting the service obligations, including
   a. SLA compliance monitoring for customers
   b. Meeting legal obligations
   c. Supporting day-to-day business operations
3. Helping reduce network capital expenditure (CAPEX), including optimizing network capacity and utilization.
4. Helping reduce network operating expense (OPEX), including
   a. Service impairment detection
   b. Diagnosing and isolation of network issues
5. Abstraction of anonymized data close to where it is collected to reduce network load associated with transmitting raw measurement data.
6. Correlation between service and network measurement data, so isolation can be done automatically. This enables the building of a targeted feed of anonymized data to subsequently generate key performance indicators (KPIs) and other metrics to automatically point to a network impairment.

7. Keeping measurement network impact to a minimum (e.g., <0.5% of network bandwidth).
8. Minimizing the measurement instrumentation footprint needed to accomplish the objective. For example, we could prefer using a single measurement stream to cover both performance and reachability measurements.
9. Focus on monitoring close to network operator's service boundary with the customer (i.e., network edge), versus within the core network.
10. Enabling flexibility to craft new measurement signatures to quickly validate customer experience for specific application traffic.
11. Rapidly off-boarding measurement data. Such data would need to be forwarded to ONAP utilizing one or more standard, scalable, and secure methods.
12. Complying fully with new architectural specifications including configuration and reporting, using NETCONF/YANG [9], RESTCONF/YANG [10], APACHE AVRO [11], or a functionality equivalent protocol such as Google remote procedure call (RPC) (gRPC) [12], highly leveraging open source software.

16.4.2 Measurement Framework Overview

Traditionally, the network elements deployed are optimized for a core set of functions such as routing or switching packets. However, because of the various user communities—such as customers, operations groups, and network designers, and their unique requirements, additional measurement requirements are imposed on the specific network element. As illustrated in Figure 16.7, the network element shown in the inner box is focused on routing packets, for example, as part of providing a VPN service, and is optimized to do that task.

To ensure that the network is conforming to SLAs and that there are no connectivity impairments, we start to add service measurement requirements on top. This could include building an instance of the Two-Way Active Measurement Protocol (TWAMP) [13] sender and responder function instrumentation into the network element to measure SLA compliance.

However, this is still not sufficient to monitor the proper functioning of the network, and to ensure that the network operations group has all the appropriate information to diagnose problems within the network, we would be required to overlay additional instrumentation. Some examples would include

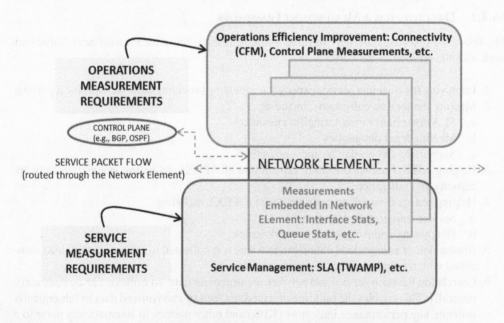

FIGURE 16.7 Network elements embedded with supporting measurement instrumentation.

instrumentation to measure connectivity across the network utilizing a protocol such as the IEEE Connectivity Fault Monitoring (CFM) [14]. Another example would include instrumentation to measure the proper functioning of the control plane, such as monitoring OSPF protocol announcements.

As can be seen from the above examples, the network element will very quickly get complex, with the addition of measurement instrumentation that is not at all directly related to its core function. Building such additional instrumentation has been costly in terms of delays in the time to deploy new features and has typically taken vendors several release cycles to fully conform to the additional measurement requirements.

We could certainly look into adjunct physical devices colocated with a network element, to specifically address the additional measurement functions; however, as the network scales to include several hundred thousand of such network elements, deploying adjunct monitoring devices quickly becomes infeasible.

The network transformation offers a great opportunity to optimize such a measurement tool footprint in a couple of ways. First, by decoupling these measurement functions from network elements, we can adopt a build once and apply everywhere paradigm with applications across the myriad of products that comprise the network. With the introduction of virtualization technologies across the network, we now have the ability to decouple such measurement functions in a cost-effective manner, and distribute them across the network as needed. Second, we can aggregate multiple types of such decoupled measurements to derive multiple observations from a single measurement stream. For example, if we take a "service measurement" view, we could reuse a single measurement deployed within a VNF to measure both performance and reachability. Adding analytic capabilities to the overall solution, such as within ONAP, allows us to reuse the measurement data collected to derive further measurement-related information across multiple layers of the network.

With this measurement framework transformation, component-level monitoring will continue to provide an in-depth network element measurement function, however, its scope is reduced to only collection of native data that has close affinity to the function of the network element and is not easily separable. For example, when a router receives or sends a packet on its interface, it needs to inherently keep a count of that packet for various reasons. We recommend that such natively derived data continue to be collected on the router. Another example would be error counts kept by the router related to internal packet handling. However, any synthetic measurement functions that may have been previously integrated into the network element should be decoupled and implemented separately.

The overall measurement framework consists of the following:

1. Component-level measurements, focusing on measuring the health of individual network components
2. Service and network measurements between edges of the network, and includes the use of active, passive, and hybrid techniques
3. Measurement instrumentation orchestration
4. Telemetry, focused on efficient transport of measurement data between remote and central collectors

The measurement framework outlined herein is extensible to include other measurement disciplines within the common framework. We continuously evaluate the need to include such new mechanisms, as and when the need arises.

The next few sections cover the details of each of these areas.

16.4.3 COMPONENT-LEVEL MEASUREMENTS

Component-level measurements cover network-element-specific measurements and are described in detail in Chapter 3. This functionality typically includes monitoring the utilization of CPU, memory, storage, links, and queues, monitoring of component-level errors, etc.

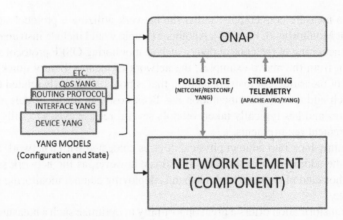

FIGURE 16.8 Component measurement instrumentation framework.

The overall approach is to implement component measurements for network elements via a YANG model-based approach, as shown in Figure 16.8. Utilizing YANG models that cover not only the configuration but also the operational state of various feature components supported by each network element, one can transform a fairly rigid component measurement solution, utilizing legacy technologies such as command line interfaces (CLIs) and SNMP, to a flexible measurement solution, centered on data models.

Essentially, with ONAP, the solution can utilize the YANG models supported by a managed network element to quickly understand the context of measurement data and therefore craft the appropriate analytics instrumentation. With the ability to receive such model-centric data both synchronously (i.e., polled/response via NETCONF [10] or RESTCONF [15]) or asynchronously via a granular publish and subscribe streaming telemetry mechanism, the solution presents a powerful enabler for automation of functions that rely on such component-based measurements.

As you will see further in this chapter, the same model-driven measurement framework will be applied to other parts of the overall solution, thus presenting a common framework for ONAP to interact with all network-based measurement artifacts.

16.4.4 Service and Network Measurements

Service and network path measurements are very important to the overall measurement solution. As the diagram below illustrates, these measurements allow the network operator to monitor SLA/ service level objective (SLO) compliance when placed at the service edge of the network. On the other hand, when placed at the infrastructure edge, such as within AIC (see Chapter 4), they help measure network path performance. Network path performance measurements help identify service-impacting network impairments that might affect traffic related to multiple services and customers that traverse the common network core.

16.4.5 The Case for Decoupling Service and Network Path Measurements from Network Elements

Traditionally, network operators have relied on network element vendors to provide much of the measurement instrumentation for services and infrastructure monitoring. This has often led to vendor-proprietary solutions and has complicated network element implementations.

A common challenge that is consistently faced has been the inability to decouple such measurement functionality from the network elements. However, we do see several merits to decoupling these measurement functions from network elements, including the ability to build once and apply everywhere across the myriad of products that comprise the network. With the introduction of virtualization technologies across the network, we now have the ability to decouple such measurement functions in a cost-effective manner, and distribute them across the network as needed.

However, the transition to a more decoupled measurement design ought to be done carefully. The following criteria tests whether it makes sense to decouple a measurement function:

1. Is measurement functionality enhanced?
2. Is cycle time for network feature development improved?
3. Is cost of the network element and measurement function development reduced?
4. Does it offer an opportunity to optimize the measurement functions?
5. Does it provide an opportunity to normalize measurement data to construct a common overview of the network?
6. Does it have a minimal operational impact?

In general, measurement functions that perform the service or network monitoring role more often meet the criteria for decoupling while component measurements targeted for decoupling, need to be evaluated more carefully.

We refer to native or embedded measurement functions (e.g., error counts, collection of component level statistics such as CPU and memory utilization) as *organic* functions to indicate the inherent coupling to the network element. The functions that do not belong to this class, termed *inorganic* or *synthetic*, can be decoupled from the network element and bundled into a separate VM, which we will call *SDN-Mon*. Examples would include the service and operational measurement instrumentation such as TWAMP [13], CFM [14], and control plane measurements, as shown in Figure 16.9.

A sampling of candidates for decoupling, which can be off-boarded to the SDN-Mon VM, is summarized below:

• Service performance monitoring using active measurement techniques
• Service reachability monitoring using active measurements
• Control plane monitoring instrumentation utilizing passive observations
• Analytics applied to raw information generated by network elements (e.g., packet stats, flow stats), with resulting KPI generation
• Bump-in-the-wire type monitoring for "sniffing" packets passing between network edges
• Deep packet inspection techniques applied to captured packets

To summarize, decoupling these measurement functions from the network elements allows for consolidation of these functions into a common SDN-Mon VM. Consolidation will facilitate the

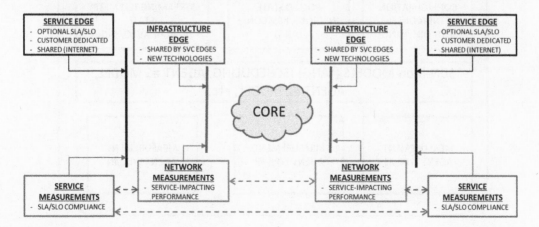

FIGURE 16.9 Service and network measurement instrumentation.

optimization of measurement instrumentation by eliminating overlapping functionality, resulting in simplifying the overall measurement solution. In addition, this approach also allows us to procure the measurement technologies from entities that specialize in such, rather than relying on network element developers, whose priority is to ensure that the network element meets its core functional definition first.

16.5   AT&T's SDN-MON FRAMEWORK

AT&T's SDN-Mon defines a common framework that allows for decoupling of all inorganic (synthetic) measurement functions from network elements, and encapsulates them within a common framework that appears to ONAP just like any other managed network element would.

The encapsulated diverse set of measurement functions in the SDN-Mon framework is shown as measurement agents in Figure 16.10. Examples of such measurement agents would include the ones listed above for service monitoring, control plane monitoring, or packet capture.

Measurement agents would in turn interface to ONAP as follows:

1. Configuration via the use of a model-based NETCONF or RESTCONF north-bound API. This would include configuration of the measurements within the context of a specific measurement agent, as well as configuration of the reporting of the measurement results via the API.
2. Retrieval of measurement results via polling, utilizing a model-based NETCONF, or RESTCONF north-bound API. Such an interface is limited to supporting low volume retrieval from measurement agents, and would be utilized in case of a diagnostic situation.
3. Retrieval of measurement results via a streaming telemetry interface, utilizing a model-based APACHE AVRO [11] API. This interface is designed to support retrieval of measurement results from measurement agents at a high volume and is based on a publish/subscribe paradigm.
4. SDN models that provide the following functions:
   a. The ability to schedule measurements for each of the measurement agents via the configuration interface, as well as schedule reporting of the measurement results via the polled or streaming telemetry interfaces.
   b. Measurement-specific models that encapsulate the measurement agent function within a YANG model. An example of such a model would be the TWAMP YANG model [13].

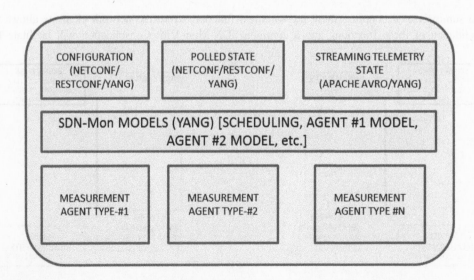

FIGURE 16.10   AT&T's SDN-Mon framework.

The framework allows for a flexible deployment of the various SDN-Mon functions. For example, in one instance, supporting a bundled deployment model, wherein the entire framework and measurement agents would be deployed as one single integrated component. In another instance, it could be deployed in a distributed model, wherein the common framework components would be deployed centrally, whereas the measurement agents would be distributed remotely. In either deployment scenario, the overall experience related to measuring the network would be identical.

In order to support a large-scale deployment of measurement solutions, and to accommodate the diversity of the measurement technologies needed for an overall solution that provides the flexibility in configuring and collecting measurement data, it is important that we adopt a common, standards-based approach in the building of SDN-Mon.

One example of such a standard is the Internet engineering task force (IETF) large-scale measurement of broadband performance (LMAP) framework and its corresponding YANG model [16,17]. Using LMAP, we will be able to bring commonality to configuration across the diverse measurement functions, as well as implement a common approach to scheduling measurement results reporting. Further, the requirement to orchestrate such measurements using the NETCONF or RESTCONF protocol and YANG model-based API will provide a common method for ONAP to configure the measurements.

For collecting measurement results, it is recommended that the following technologies be implemented for transmitting the measurement results instantiated in compliance with the YANG models:

1. For transactional-based results collection: NETCONF and RESTCONF
2. For streaming telemetry-based results collection: Apache Foundation's AVRO or functionally equivalent specification such as gRPC [12], with JSON object payload

One example of a measurement agent would be the active performance measurement agent implementing the TWAMP. Another example would be a packet capture measurement agent. In either case, using LMAP will provide a common interface and data models to configure the measurements and subsequently aggregate the results for forwarding to ONAP.

The use of a common SDN-Mon framework, with the ability to expose a common set of interfaces to a diverse set of measurement agents, enables AT&T to select from the best of breed vendor solutions for measurement agent instrumentation. This in-turn allows for minimizing, and eventually eliminating the need to build measurement vendor proprietary configuration and collection solutions.

As mentioned earlier, the SDN-Mon VM will look very similar to any other managed network element as far as ONAP is concerned, as shown in Figure 16.11.

Comparing the SDN-Mon solution in Figure 16.11, with the component measurement framework outlined in Figure 16.8, one can easily observe that the interfaces to ONAP are identical in both

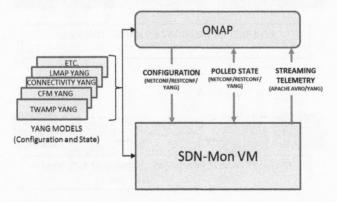

FIGURE 16.11   Interfacing the SDN-Mon framework to ONAP.

cases. The only difference in the interface is the YANG models that are used to structure the data for interfacing to ONAP, since the list of YANG models will be dependent on the set of managed functions.

16.5.1 NETWORK DEPLOYMENT EXAMPLE FOR THE SDN-MON FRAMEWORK

There are several deployment scenarios possible with SDN-Mon, but we have chosen only two of them to illustrate the utility value of this framework.

Figure 16.12 illustrates two methods used to implement an SDN-Mon VM for measuring path performance, which is in-turn utilized for SLA compliance measurements:

1. SDN-Mon is service function chained through adjunct network elements. In this scenario, the SDN-Mon VMs generate a representative measurement packet stream between themselves, traversing the corresponding network elements. This deployment scenario could be used to validate SLA compliance for a pair of communicating network elements. In past implementations, one would expect the network elements to embed interoperable protocols to generate such measurements, complicating network element development.

2. Generating a representative measurement packet stream between SDN-Mon VMs located in proximity to groups of network elements. This deployment scenario could be used to validate SLA compliance between locations such as between two cities, but not specific to any communicating pair of network elements. In past implementations, one would expect at least one of the network elements at each location to embed interoperable protocols that generate such measurements, complicating network element development.

As observed from the above examples, SDN-Mon enables the decoupling of such measurement functions from the network element. The SDN-Mon VM would encapsulate such measurement functions, while unburdening the network element from implementing such functions. This allows the network element to focus on its primary function, which in this case is forwarding packets.

16.6 TELEMETRY MEASUREMENTS

Transporting and handling measurement messages generated by VNFs and Operations, Administration and Maintenance (OAM)-VMs to collection systems, poses challenges. Given the large-scale, geographically distributed set of message producers that are connected to the high-speed

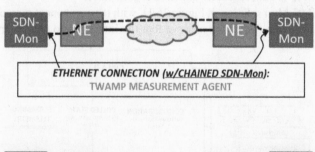

FIGURE 16.12 Sample of SDN-Mon deployment scenarios.

core network via a diverse set of access technologies, it is important that we look at this transport design to ensure that it can handle the volume of messages that will surely only increase over time.

It is to be expected that any implementation will see quite a bit of diversity in message volumes, based on the source and type of the messages generated. For the purposes of simplifying the telemetry technology selection, this can broadly be classified as follows:

1. *Low volume message producers.* Such producers would include traditional alarm streams, such as SNMP traps and SYSLOGs that are converted to functionally equivalent messages for a SDN implementation. The volume of this category of messages is expected to be on the average, mostly in low kilobytes/s range for any given producer. The other characteristic of such messages is that they will be typically identified within the integrated network YANG model's defining configuration and operational state. This set of messages is also considered historical in most cases, often generated as a result of onboard processing; or they are transactional in nature, and at best might only represent close to real-time status.

2. *High volume message producers.* Such producers would include nontraditional streaming data such as anonymized flow records and other types of "raw" data that might be used for closed-loop automation. The volume of this category of messages is expected to be on the average, in the megabytes/second range for any given producer. These messages are also expected to be identified within the integrated YANG model's defining configuration and operational state, with their format conforming to some defined YANG model. This set of messages is considered real-time, or at worst, near-real-time, and their timely delivery to the target is important.

   Keeping in mind the above considerations, the following choices are specified for telemetry:

   a. For traditional low-volume message producers, the use of NETCONF/YANG or RESTCONF/YANG is recommended

   b. For diagnostic related low-volume, one-time message producers, the use of NETCONF/YANG or RESTCONF/YANG is recommended

   c. For the high-volume message producers, Apache AVRO or functionally equivalent protocols such as gRPC [12] with JSON object payload is recommended

It is expected that over time, a single solution might be picked to meet all telemetry requirements, but given the newness of these technologies, the above two selections will give us the flexibility to grow each message domain into maturity. While a preference is indicated for telemetry solutions, much work needs to be done to qualify such solutions and one can expect further refinements in this area as more operational experience is gained.

16.7 NFV DATA AND MEASUREMENTS

As background before we discuss NFV measurements, let us review the network service measurement life cycle in the traditional legacy networks as defined by the telecommunications management network (TMN) model defined by ITU-T for managing open systems in a communications network. TMN is part of the ITU-T Recommendation Series M.3000.

Service network management was grouped into five functional areas, designated FCAPS—fault, configuration, accounting, performance, and security:

- Fault management—detect and notify faults encountered by the network element
- Configuration management—expose configuration, inventory, and software attributes
- Accounting management—collect usage information of all network element resources
- Performance management—monitor and report performance measurements at service and network level
- Security management—secure and authorize access to network devices, resources, and services

Each functional area is then integrated in a layered architecture, and measurements are exposed at the network element and element management system layer (EML):

- Network element layer (NEL) defines interfaces for the network elements, instantiating functions for device instrumentation.
- EML provides management functions for network elements on an individual or group basis. It also supports an abstraction of the functions provided by the NEL. Examples include collecting statistical data for accounting purposes, and logging event notifications and performance statistics.

16.7.1 NFV Data Model

The NFV data model has flattened and abstracted the original legacy network service data model. Further, it has made the NFV data model independent from EMS as there is a uniform management system (ONAP) now responsible for the NFV management.

The NFV model maintains parity with the legacy network service data model and also enhances it where needed. It relies on telemetry data from the native cloud infrastructure and organic VNF specific measurements. The NFV measurements are further enhanced where needed by vProbes (both passive and active), which are covered in Sections 16.8.2 and 16.8.33.

16.7.2 NFV Infrastructure Telemetry Data Model

The NFV infrastructure telemetry data represents event notification of the baseline resource consumption at the infrastructure level, and will typically include the following:

- Compute, memory, and storage telemetry data monitors (CPU utilization rate, memory utilization rate, Disk I/O operations rate, etc.)
- Network interface monitor statistics (bytes, packets, rate, errors, drops, latency, jitter)—some measurements may be available via native telemetry service (OpenStack Ceilometer) and others may have to be collected via SDN controllers or underlay network switches
- Native and hypervisor logs, events, and alarm data

Each native cloud environment will define its own telemetry service that exposes the native telemetry data. Since AIC is based on OpenStack and KVM (see Chapter 4), the telemetry service is supported by Ceilometer.

16.7.2.1 OpenStack Telemetry Service: Ceilometer

Ceilometer[*] is an open source component of the OpenStack telemetry data collection service that provides the ability to normalize and transform data across all current OpenStack core components. Its data can be used to provide resource tracking, capacity planning/management, and alarming capabilities across all OpenStack core components. This section is compiled using reference information for Ceilometer 6.1.1 (2016) project of the OpenStack Foundation.

The Ceilometer project was started in 2012 with the initial objective of enabling analytic engines to use this single source to transform events into billable usage records, which was labeled as "metering." As the project gained steam and began collecting an increasing number of meters across multiple projects, the OpenStack community started to realize that a secondary goal could be added to Ceilometer: become a standard way to collect measurements, regardless of the purpose of the collection (Figure 16.13).

---

[*] Ceilometer 6.1.1 Documentation, OpenStack Foundation, Update Version July 4, 2016.

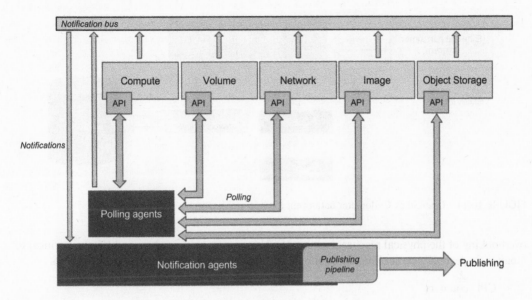

FIGURE 16.13 OpenStack Ceilometer data/event collection architecture.

This diagram is a representation of how the collectors and agents gather data from multiple sources. The Ceilometer project created two methods to collect data:

- Bus listener agent, which takes events generated on the notification bus and transforms them into Ceilometer samples
- Polling agents, poll some API or other tool to collect information at a regular interval; the polling approach is less preferred due to the load it can impose on the API services

The first method is supported by the ceilometer-notification agent, which monitors the message queues for notifications. Polling agents can be configured either to poll the local hypervisor or remote APIs (public REST APIs exposed by services and host-level SNMP/Internet protocol multicast initiative [IPMI] daemons).

Ceilometer monitor measurement samples can be published to multiple collectors. Current capabilities supported by Ceilometer-enabled data to be published include the following:

- Notifier, a notification-based publisher, which pushes samples to a message queue that can be consumed by the collector or an external system
- UDP, which publishes samples using UDP packets
- HTTP, which targets a REST interface
- Kafka, which publishes data to a Kafka message queue to be consumed by any system that supports Kafka

The collected data from polling and notification agents can be accessed via REST API (Figure 16.14).

16.7.2.2 OpenStack Telemetry Service: Ceilometer Examples

Ceilometer can be configured to collect measurements from both the OpenStack services and the hypervisor/virtualization layer. Figure 16.15 shows the configuration used to monitor the performance and capacity compute resources in AIC. A key facet of multitenancy operations is

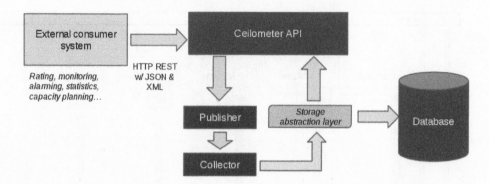

FIGURE 16.14   OpenStack Ceilometer data/event logging architecture.

overbooking of the physical resources such as CPU, memory, memory bus, and I/O. It is critical to monitor the level of overbooking and Ceilometer provides measurements to do that.

- CPU counters
- Memory usage and allocation counters
- Disk I/O counters
- Network I/O (not usually a bottleneck due to VM contention but can be used to check whether network interfaces were successfully established for the VMs). Ceilometer has plugins into SDN local controllers to monitor the east–west traffic between tenants.
- Additional measurements can be added since support is provided for Linux SYSSTAT utilities (also used in Nagios monitoring) and the KVM/Quick EMUlator (QEMU) hypervisor.

A specific example is described below for CPU counters and memory counters (Tables 16.1 and 16.2).

The above counters can be used to monitor the physical resource consumption levels of OpenStack tenants and analytics can reveal when overbooking is causing performance degradation.

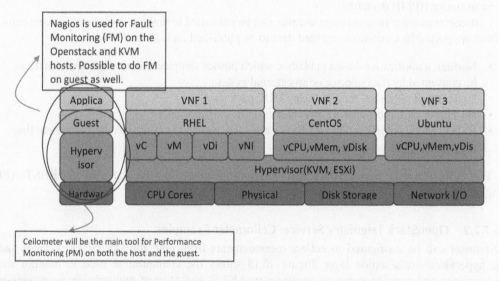

FIGURE 16.15   OpenStack Ceilometer configuration in the NFV stack.

TABLE 16.1
Ceilometer CPU Counter

Measurement Description	Level	OpenStack Ceilometer Parameter	Unit
Total time that VM was ready but could not get scheduled to run on physical CPU[a]	VM level	%CPU steal	ms
Actively used CPU of the host, as a percentage of the total available CPU. Active CPU is approximately equal to the ratio of the used CPU to the available CPU (host view not guest view)[a]	Host level, measured per VM	cpu_util in Libvirt and HyperV	%
Relationship with vCPU usage: virtual CPU usage = (approx)usagemhz / (# of virtual CPUs × core frequency)			

[a] This measurement has been certified available for Xen hypervisors as of July 14, 2016.

TABLE 16.2
Ceilometer Memory Counters

Measurement Description	Level	OpenStack Ceilometer Parameter	Unit
Amount of all physical memory actively used by the powered-on VMs as estimated by VM kernel based on recently accessed memory pages	VM layer	NA	KB
Amount of all physical memory consumed by powered-on VM for guest memory (aka, amount of machine memory used on the host by that VM)	VM level	memory.resident or memory.usage in Libvirt	KB
Current amount of guest physical memory swapped out to the virtual machine's swap file by the VMkernel	VM level	NA	KB
Amount of memory allocated by VM memory control driver (vmmemctl). Amount of guest physical memory currently reclaimed from VM through ballooning	VM level	NA	KB
Sum of all machine memory allocated to powered-on VMs above reserved amount	VM level	NA	KB

16.8 NFV DATA MEASUREMENT FRAMEWORK

In AT&T's vision, virtualized network functions are expected to be instantiated in a significantly dynamic manner that requires the ability to provide real-time responses to actionable events from virtualized resources as well as end-user transactions. NFV lifecycle management (including instantiation, scale-out/in, performance measurements, event correlation, termination, reboots, etc.) requires granular and accurate measurements at the tenant and infrastructure level. Operators need a passive and active measurement framework to gather key performance, usage, telemetry, and events from the dynamic, multivendor virtualized infrastructure in order to compute various analytics and respond with appropriate actions based on any observed anomalies or significant events.

16.8.1 ORGANIC NFV DATA MEASUREMENT MODEL

To realize the full benefits of NFV, it is important for network operators to be able to automate the management of VNF life cycles. ONAP orchestration addresses various VNF lifecycle aspects such as instantiation, elastic scaling, service assurance, automatic recovery from resource failures, resource allocation, etc. To facilitate this, VNFs need to be equipped with capabilities that comply with standards to enable automated measurements. VNFs must provide a compatible interface to ONAP for data collection and event notification. The current options are JSON schema and REST APIs. AT&T excludes the use of SNMP, CLIs, and EMS, so as to ensure vendor independence.

The end-to-end service reliability and availability in a virtualized network will greatly depend on the ability to monitor and manage the behavior of VNFs in near real-time rather than relying on reliability and availability of network elements. The ONAP platform will monitor performance of VNFs using the capabilities provided by these resources to proactively predict potential issues and resolve them automatically by taking appropriate actions such as restart of the resources or providing additional capacity. This means that the VNF developers are required to provide a rich set of monitoring and alerting capabilities to facilitate near real-time monitoring and proactive problem resolution.

VNFs are required to provide the following network data in real time and near real-time streaming:

- Consistent fault alarm and event format in a defined standard
- Events (i.e., host/customer configurations changes, etc.)
- Performance management data (i.e., application KPIs)
- Usage data (i.e., call detail records)
- On-demand access to additional detailed tracing and trouble-shooting data

16.8.2 PASSIVE vPROBE MEASUREMENT DATA MODEL

Passive probe monitoring in legacy networks required an appliance that would tap or port mirror a specific service network interface to capture network packets for service-specific analysis. Passive network monitoring can collect and correlate packets into enriched transactional records. Virtualization has rendered traditional physical probes obsolete. The NFV approach for passive probing transforms the physical probe to a virtual probe (vProbe) network function that would be coresident with the service VNF of interest and use packet mirroring at the shared vSwitch.

To realize the goal of eliminating the need for passive probe-based monitoring at various points in the network, because of inherent inefficiencies and cost penalties, the VNF should natively provide enhanced instrumentation that allows passive measurements (e.g., session analysis) enabling a detailed view into the VF and network. Where a passive vProbe is deployed on a VNF interface, it would enable the following measurements:

- Detailed VNF transaction records (flow records, session records, transaction records, etc.) at configurable intervals
- Control-plane message stream processed by the VF
- User-plane stream (that is, configurable in a flexible manner—e.g., IPs, time interval, etc.)

The above raw measurements can be further processed, aggregated, and analyzed to generate several important indicators. These indicators and relevant underlying data are exposed to the network operators to aid network planning, engineering, and troubleshooting services. Some of the common measurements and capabilities supported by a vProbe include

- Current capacity of the network element and capacity of the underlay network
- KPIs such as throughput, latency, errors
- Statistics describing the volume of transactions and their distribution, by protocol and direction
- Drill-down capability with visual access to troubleshoot root-causes of problems and failures in the network

Traditionally, data from passive probing have been extensively used to help Tier 3 or 4 support teams in troubleshooting and avoiding network problems and bottlenecks.

As is the case with VNFs, orchestrated deployment of vProbes is facilitated through the use of heat orchestration templates. vProbes, once deployed, are controlled by the service orchestrator through the API and all measurements reported are collected via ONAP. In addition, vProbes should generate self-monitoring stats, both for health and performance to enable resilient orchestration.

16.8.3 ACTIVE vPROBE MEASUREMENTS DATA MODEL

Active probe monitoring is a measurement process that injects synthetic transactions into a system or network and monitors the flow of those transactions to detect anomalies, or measures specific performance or availability that can be used for SLA reporting. This is helpful for a simple connectivity testing or performance testing; for example, timing the latency between two hosts operating in the same cloud zone connected via an overlay SDN local controller and leaf-spine underlay, as well as more complex analytical tasks such as collecting measurements to verify QoS and whether availability objectives are being met.

NFV implementations in cloud environments often encounter performance degradation or failures (known as gray failures) where either the problem is not detectable by traditional passive measurements, or the software component does not properly report impairments. These gray failures could silently impair the NFV tenant by causing loss of network or resource connectivity, loss of privileges or lead to operation in a degraded state. Active probes behaving as tenants and generating synthetic heart beats that alarm when the heart beat is impaired, enable operators to get proactive notification of tenant impact.

As is the case with VNFs, orchestrated deployment of active vProbes is facilitated through the use of heat-orchestration templates and the placement of active probes as well as configuration of probe heart beat frequency, which is administered using a centralized management function. Active vProbes once deployed, are controlled by the service orchestrator through the API and all measurements and alarms reported are collected via ONAP.

16.9 VNF REPORTING METRICS

VNF developers have responsibility to report and log key measurements on VNF transactions as well as key event notification of alarms, resource utilization, performance, and availability. The VNF transaction logging can be in the form of traditional session detail records or other standardized form. The event reporting is generally expected to be in streaming event form using a standardized protocol. The alerts can also be in the form of operator programmable threshold crossing alerts.

16.9.1 VNF RESOURCE CONSUMPTION AND OPERATIONAL METRICS

Service providers define and use key capacity indicators (KCIs) to assess the loads and dimensioning profiles of VNFs and to monitor system health as well as unit costs. The metrics are optimized through rigorous capacity planning and management optimization practices that are typically automated using closed-loop mechanisms in the NFV environment.

Examples of KCIs used to optimize both resource consumption and performance, include

- CPU utilization
- Memory utilization
- Storage I/O load
- Network bandwidth utilization

Achieving predictable performance and optimal unit cost is intrinsically complex. The dynamic nature of a cloud environment cannot provide a real-time hard guarantee for performance and resource consumption. There is typically a tradeoff between predictability of resource availability and optimal resource allocation. In a traditional single-tenant architecture and design, trending and profiling can be used for allocating the resources. The problem becomes much harder in multitenant environments when additional dimensions are inserted into predicting both aggregate and individual tenant behavior. The operator's responsibility is to allocate adequate resources and engineer appropriate capacity for all tenants, while also ensuring that the resources are optimized.

Overbooking resources is a critical operational management practice. The overbooking practice relies on the assumed behavior of subscribers and the probability that they will not all operate at their peak SLA at the same time, or at all times. Predictive traffic profiling is an essential part of optimizing resources. Overbooking ratios are adjusted by monitoring KPIs and KCIs.

Another important operational practice is implementing closed-loop automated functions to manage VNFs and the associated resources. Closed-loop operational metrics are used to measure and report automated VNF lifecycle management events. AT&T has implemented closed-loop operations in ONAP and metrics that report effectiveness of these closed-loop events (e.g., detection of VM stall and latency of a VM restart) are important to operational scorecards.

The closed-loop event metrics include measures of frequency, delay, and success rates of the targeted lifecycle management. These can be used as a health and resource consumption indicators (e.g., increase in closed-loop event occurrence frequency such as auto restarts of VNF, can be an indicator of VNF resiliency risk or resource bottleneck).

16.9.2 VNF SLAs and KPI

The SLAs typically specify the targets for the KPIs, which need to be met by the VNF implementation. The SLAs are determined by user expectations and in some cases, by contractual or regulatory constraints. Typical performance KPIs for VNFs include

1. Latency
2. Availability
3. Completion rate
4. Defect rate

Latency bounds are needed to ensure that application transaction flows complete in a timely manner. The application layer latency constraints depend on protocol bounds for expected response times. Protocols that allow for retries of unacknowledged transactions need to be considered as part of the VNF latency bounds.

Availability is a time-based metric that represents the percent of uptime for the VNF. Downtime measures the amount of time the VNF is not available, and is typically expressed in seconds, minutes, or hours.

Completion rate measures the rate of successful transactions through the VNF; transactions are specific to the VNF application.

Defects are unsuccessful VNF transactions; in other words, transactions that do not complete as expected. Defects are measured in absolute counts as well as rates. A standard approach for representing defects for systems that process significant volumes of transactions is to use defects per million (DPM) transactions.

16.9.3 VNF Resiliency Reporting

The ability to measure VNF resiliency and assess the ability of the cloud platform to adapt and recover from any disruption, is a critical function for service operators. Since every VNF may have a different target for availability, its underlying resiliency design may also vary. There are several characteristics of resiliency behavior that can be assessed by measurements:

- *Seamlessness*: Measuring service continuity or seamlessness, requires measuring impact to end users when a VNF is impaired. Service continuity objective is defined as an uninterrupted user experience even when an anomaly event occurs. If the duration of VNF impairment (including failure event, detection, and fail over recovery) is not perceptible to the user, then it can be characterized as seamless service continuity.
- *Robustness*: Measuring robustness can be accomplished by assessing the availability of additional capacity and a wider range of accessible resources in the infrastructure. Topological transparency contributes to higher resiliency; with less topological constraints comes less limitation to finding available resources minimizing the impact from physical failures, or electrical or network outages.
- *Uniformity*: Measuring uniformity requires distributed measurements of performance and comparing VNF performance across different cloud zones and locations. When a service is provided by more than a VNF distributed across multiple VMs and locations, service requests to those VNFs may be imbalanced and drive varying resource constraints. Uniform performance can be achieved by ensuring service requests are distributed and load balanced. This reduces risk of resource scarcity, which enables higher VNF resiliency.

Virtual function resiliency covers both reliability and performance. Typical reliability measurements for VNF resiliency include

1. Availability
2. Defect count
3. DPM
4. Completion rates
5. Latency

Measurements must capture the transactions offered, successfully completed, and unsuccessful or defective transactions. Classification of defective transactions may include transactions that do not complete within a time bound.

For time-based availability measurements, the health of the VNF can be determined with heart beats or light weight active probes that the VNF responds to. A controller mechanism is needed to determine how often probes should be sent, and how many unanswered probes determine whether the VNF is unavailable. In addition, a controller mechanism is needed to determine frequency and number of acknowledged probes required to decide up state of the VNF.

The ONAP data collection and analytics engine provides the mechanism for reporting VF reliability KPIs. In addition, other operations support systems may retrieve data from common data stores provided by ONAP.

16.10 VNF SCALING MEASUREMENTS

One of the key benefits from adopting NFV is its capability to adapt to variable workload by scaling its underlying cloud resources. VNF scaling is accomplished horizontally or vertically [18]. Horizontal scaling, referred to as scaling out/in, adds a new set of resources (virtual servers, storages, and network bandwidth) or removes a set of existing resources, to adapt to the workload changes. On the other hand, vertical scaling, referred to as scaling up/down, increases or decreases the amount of resources assigned to existing virtual instances, for example, allocating more/less physical CPUs or memory, to a virtual machine that is already running. Despite these two mechanisms to accomplish VNF scaling, most cloud providers support only horizontal scaling as the vertical scaling is not supported by most common operating systems without rebooting [19].

16.10.1 NFV Service Auto-Scaling

Auto-scaling aims to dynamically adapt cloud resources to varying workload in an automated fashion. In auto-scaling, a VNF manager monitors the VNF KPIs and triggers VNF scaling when the set of scaling rules/policies defined with the VNF KPIs monitored, are satisfied. There are several types of KPIs used for VNF scaling trigger. Some examples [18–21] are listed below:

1. Workload to VNF (e.g., total number of active subscribers for LTE service)
2. Resource utilization (e.g., CPU, memory, disk I/O and network I/O usage)
3. VNF performance metrics (e.g., throughput, delay)
4. VNF-specific metrics (e.g., queue length)
5. Various events from VNF, virtual infrastructure manager (VIM), EMS

With the scaling-trigger metrics mentioned above, there are different techniques used to build auto-scaling mechanisms. The following is a classification of auto-scaling techniques divided into five groups [18]:

1. Threshold-based rules
2. Reinforcement learning
3. Queuing theory
4. Control theory
5. Time-series analysis

The ultimate goal of an auto-scaling mechanism would be to minimize the resource consumption for the VNF while maintaining the VNF performance above the target level. Therefore, an effective and efficient auto-scaling mechanism should make a judicious choice on the auto-scaling metrics and techniques, being aware of its consequences on both VNF performance and resource consumption. Specifically, over-provisioning resources, an early start of scaling out, and a late start of scaling in may result in resource waste, whereas under-provisioning resources, an early start of scaling in, and a late start of scaling out may result in performance degradation. Also, the auto-scaling mechanism needs to take into account various factors impacting the latency time to complete the execution of VNF scaling, such as the time for auto-scale trigger to be introduced, time for auto-scale trigger to be detected, time for the required resources to be created/deleted, time for the new/deleted virtual machine to start/stop running. Besides, an oscillation effect, that is, scaling actions executed too frequently, should be avoided by carefully employing a cool-down interval—for which no additional auto-scaling is allowed to be executed.

16.10.2 Scaling Triggers

There are multiple types of VNF scaling that differ by the triggers for scaling and the ways of issuing the scaling requests. The following three models of VNF scaling have been identified by ETSI GS NVF SW 001 [23]:

1. Auto-scaling
2. On-demand
3. Manually triggered scaling

In the auto-scaling model, VNF manager and the NFV orchestrator monitor the VNF KPIs and issue a request to trigger scaling; whereas, in the on-demand scaling, the VNF monitors the KPIs of its components (i.e., VNFs) and sends a scaling request to the VNF manager [20]. In manually triggered scaling, the operators from OSS/BSS trigger scaling based on their observation/expectation of an increased/decreased load on the VNF.

16.10.3 SCALE OUT: ILLUSTRATIVE EXAMPLE

The following describes an illustrative example of scaling out. Imagine a virtual machine running a VNF and a VNF manager monitoring the VNF resource consumption, that is, CPU, and network I/O usage, for auto-scaling of the VNF. The virtual machine is currently using its resources at the following levels: CPU usage = 45%, memory usage = 32%, and network usage = 484 Kbps. The load to the VNF keeps increasing, leading to the following virtual machine resource consumption : CPU usage = 75%, memory usage = 51%, and network usage = 806 Kbps. At this point, a VNF manager observes that the CPU usage of the virtual machine reached the CPU utilization threshold (75%) for scaling out. This makes the VNF manager send out a request to the NFV orchestrator for triggering a VNF scaling out action. Then, the NFV orchestrator and cloud infrastructure manager allocate a new virtual machine and a load balancer to the VNF. After the new virtual machine and the load balancer for the VNF are successfully created and are running, the increased load to the VNF is effectively distributed between the two virtual machines by the load balancer.

16.11 VNF EFFICIENCY MEASUREMENTS AND KCI REPORTING

Migration of network functions from physical network elements to a virtualized cloud environment is a complex process and assessing whether the migrated functions are adhering to native cloud principles is a critical task for operators. Simply porting and virtualizing network functions may not optimize resource consumption and performance of the workload.

There are known overheads that impact network functions in a virtualized cloud environment; these need to be assessed and quantified. They include

- Virtualization overhead
- Hypervisor overhead
- Overlay network virtual path overhead

Furthermore, there are VNF design choices that impact VNF resource consumption and performance, including tradeoffs in the following:

- Use of nonuniform memory access (NUMA)
- Context switching between user space and kernel space
- Interrupt driven versus polling

VNF efficiency can be defined as a measure of how efficiently a VNF processes offered load in production as compared to its performance in controlled lab tests. Assuming that utilization of each resource in production varies with the load, one way of computing this is to determine the ratio of offered load to VM resource utilization. If this ratio is the same as what has been benchmarked in the lab, efficiency would be considered 100%. If this ratio is smaller than the lab benchmark, efficiency is lower than 100% and considered degraded. If the ratio is higher than the lab benchmark, efficiency is higher than 100% and performing better than forecasted.

The objective of computing the efficiency metric are multifold:

- Compare network function virtualization implementations across different suppliers
- Formulate recommendations for cost optimization

- Forecast resource consumption under varying workload conditions
- Identify areas in cloud infrastructure and VNF design for performance optimizations

Measuring VNF efficiency in production requires some prerequisites including

- Decomposition of VNF into workloads (data plane, control plane, signal processing, storage).
- Identification of traffic models and performance metrics for each workload.
- Benchmark identified performance metrics for each VNF workload and its defined traffic model in a controlled lab environment.

VNF efficiency can be computed as a function of the efficiency of the VMs that implement it. An average/max/min over the formulated efficiency of the VM would define the VNF efficiency.

16.12   VNF MEASUREMENTS FOR OPTIMAL PLACEMENT AND SIZING

Virtualized networking functions are usually implemented as a single VM or as a disaggregated group of reusable components, each of which is implemented as a VM. In a cloud-computing model, resources are abstracted at the VM level—the abstraction ensures limited control by the VNF regarding which specific host the VM would be located in or what kind of coresident VMs would be along with this VNF.

AT&T's VNF guidelines [22] specify that each VNF must contain at least the following information to support instantiation of the VF in the virtualized environment:

- VF package layout (captures basic network and application connectivity between VF components, clustering/high availability design, scaling capabilities of each component, etc.)
- vCPUs, memory/RAM, disk
- Bandwidth profile
- VF software images
- Multitenant/single tenant
- Placement policy and constraints (including affinity, anti-affinity, latency considerations), redundancy, external/persistent storage, virtual network interface cards (vNICs), and networking
- Vendor affinity or anti-affinity rules must be consistent with AT&T's desire for no single point of failure

The process of selecting which virtual machines should be located and executed on which physical machine or cluster in a cloud environment or zone, is referred to as virtual machine optimal placement [24]. The need for optimal VM placement to maximize resource utilization calls for placing VMs over a minimal set of physical hosts within specified constraints with minimal overhead.

The problem is generally divided into two scenarios—initial VM placement and future VM migration. In simplified form, the initial VM placement can be addressed using heuristic approaches. Machine learning techniques can then be applied after initial placement to minimize future migrations. Some of the primary VNF parameters and measurements driving the VM placement and future VM migrations are as follows:

- Maximize resource utilization by minimizing the number of physical hosts (resources) used.
- Constraints are a combination of any of the following, depending on the VNF:
  - SLA KPIs including throughput, latency, etc.
  - Unit cost KCIs (CPU, memory, storage, and network bandwidth)

- Affinity/anti-affinity rules for scalability and resiliency; for example, it may be required that no two components of a VNF may be on the same host and/or not two instances of the same VNF would be on the same data center or cluster
- Initial VM placement time minimization
- Future VM migration minimization

VNFs may be implemented on any of the AIC flavors. The flavors differ in the resources provided for consumption, for example, number of vCPUs, memory size, disk allocation, etc.

Optimal resource consumption, overbooking, and overhead minimization can guide the determination of choosing the optimal VNF flavor. The main tradeoffs for this determination are as follows:

- Smaller sizes will enable a more granular design, better resource utilization and hence better cost utilization and availability (especially for a larger number of small VFs). It might involve more scaling events that could result in higher overhead, violation of SLAs, and possible loss in revenue.
- Larger sizes could result in poorer resource utilization, but less scaling events and possible lower impact to SLAs. Larger VFs may imply they are less active instances and therefore have a lower reliability in case of an outage.

Workload characteristics can also be used to determine the size of a VNF VM. The main tradeoffs using this criterion are as follows:

- Dynamic workloads with peaked behavior could cause large scaling out (and resulting in events with high overhead) as compared to more predictable static workloads. The scaling overhead depends on the workload type. Generally, a network and storage-intensive workload has a much higher overhead than a compute-intensive workload.
- Small workloads with a high growth rate would imply a larger VM sooner, compared to a VNF with slow growth rate.

16.13   AREAS FOR FURTHER STUDY

In this section, we identify and describe some topics for further exploration:

1. *Optimization algorithm improvements*: There is a continued need for improvement in optimization algorithms that balance between speed of execution and increased efficiency. AT&T collaborates with universities in this regard and organized the AT&T SDN network design challenge in 2015. The task was to come up with a fast and efficient IP/optical network design and routing algorithm based on a fictitious but realistic network topology and traffic matrix, and to use the centralized SDN traffic management capability. Eighteen teams from major US universities participated and the winning submission was a team from Cornell University based on a judging criterion that took into account both the speed of execution and the efficiency of the design.
2. *Measurement scalability*: If there are N endpoints then the measurement complexity increases as N^2. The complexity will increase significantly as we move from core-to-core to edge-to-edge traffic flows and continued research on measurement techniques would be needed. A related capability is the ability to zoom in on a specific aspect of edge-to-edge measurements, which would be faster than sifting through the entire set of end-to-end traffic flows.
3. *SDN controller resiliency*: Typically, the centralized algorithm of the SDN controller will work in conjunction with distributed algorithms used by routers. It is important to make

sure that the decisions made by the centralized and the distributed algorithms are consistent. Also, usually there are multiple SDN controllers with one being active while the others are in a standby mode. It is important to ensure data synchronization among the multiple controllers so that if the active controller fails then one of the standby controllers can become active and start functioning immediately and seamlessly.

ACKNOWLEDGMENTS

The authors would like to acknowledge the significant contributions of Soshant Bali, Gagan Choudhury, Kevin D'Souza, Minh Huynh, Carolyn Johnson, Richard Koch, Bala Krishnamurthy, Zhi Li, Ashima Mangla, Kartik Pandit, Donghoon Shin, Simon Tse, and Kang Xi.

REFERENCES

1. Birk, M., G. Choudhury, B. Cortez, A. Goddard, N. Padi, A. Raghuram, K. Tse, S. Tse, A. Wallace, and K. Xi. Evolving to an SDN-enabled ISP backbone: Key technologies and applications. *IEEE Communications Magazine*, 54(10):129–135, 2016.
2. Jain, S., A. Kumar, S. Mandal, J. Ong, L. Poutievski, A. Singh, S. Venkata et al. B4: Experience with a globally-deployed software defined WAN. *ACM SIGCOMM Computer Communication Review*, 43(4):3–14, 2013, ACM.
3. https://www.opendaylight.org, accessed June 14, 2016.
4. Berde, P., M. Gerola, J. Hart, Y. Higuchi, M. Kobayashi, T. Koide, B. Lantz et al. ONOS: Towards an open, distributed SDN OS. In *Proceedings of the Third Workshop on Hot Topics in Software Defined Networking*, pp. 1–6, Chicago, Illinois, August 22, 2014, ACM.
5. https://www.opennetworking.org/sdn-resources/openflow, accessed on June 14, 2016.
6. Choudhury, G., B. Cortez, A. Goddard, N. Padi, A. Raghuram, S. Tse, A. Wallace, and K. Xi. Centralized optimization of traffic engineering tunnels in a large ISP backbone using an SDN controller. *INFORMS Optimization Society Conference*, March 17–19, 2016, Princeton, NJ.
7. Zhang, Y., M. Roughan, C. Lund, and D. Donoho. An information theoretic approach to traffic matrix estimation. In *SIGCOMM '03, Proceedings of the 2003 Conference on Applications, Technologies, Architectures, and Protocols for Computer Communications*, pp. 301–312, Karlsruhe, Germany, August 25–29, 2003.
8. Feldmann, A., A. Greenberg, C. Lund, N. Reingold, J. Rexford, and F. True. Deriving traffic demands for operational IP networks: Methodology and experience. *IEEE/ACM Transactions on Networking*, 9(3):265–280, June 2001.
9. YANG—A Data Modeling Language: https://tools.ietf.org/html/rfc6020
10. Network Configuration Protocol (NETCONF): https://tools.ietf.org/html/rfc6241
11. Apache AVRO: http://avro.apache.org/
12. Google Remote Procedure Call (gRPC): http://www.grpc.io/docs/
13. TWAMP Function Configuration: https://datatracker.ietf.org/doc/draft-cmzrjp-ippm-twamp-yang/
14. IEEE Connectivity Fault Monitoring (802.1ag) Standards Specification.
15. RESTCONF Protocol: https://tools.ietf.org/html/draft-ietf-netconf-restconf-15
16. Framework for Large-Scale Measurement of Broadband Performance (LMAP): https://datatracker.ietf.org/doc/rfc7594/
17. LMAP Measurement Agent (MA) Collection function: https://datatracker.ietf.org/doc/draft-ietf-lmap-yang/
18. Lorido-Botrán, T., J. Miguel-Alonso, and J. A. Lozano. A review of auto-scaling techniques for elastic applications in cloud environments. *Journal of Grid Computing*, 559–592, 2014.
19. T. Lorido-Botrán, J. Miguel-Alonso, and J. A. Lozano. *Auto-Scaling Techniques for Elastic Applications in Cloud Environments*, University of Basque Country, Technical Report EHU-KAT-IK-09-12, 2012.
20. ETSI, Network Functions Virtualization (NFV); *Pre-Deployment Testing*; Report on Validation of NFV Environments and Services, ETSI GS NFV-TST 001 V1.1.1, available at: https://portal.etsi.org/webapp/workProgram/Report_WorkItem.asp?wki_id=46009
21. Mao, M., J. Li, and M. Humphrey. Cloud auto-scaling with deadline and budget constraints. In *Proceedings of the 11th IEEE/ACM International Conference on Grid Computing*, pp. 41–48, Brussels, Belgium, October 2010.

22. AT&T Vendor VNF Guidelines, 2017, https://wiki.onap.org/pages/viewpage.action?pageId=1015852
23. ETSI, Network Functions Virtualization (NFV); Virtual Network Functions Architecture, ETSI GS NFV-SWA 001 V1.1.1, available at: http://www.etsi.org/deliver/etsi_gs/NFV-SWA/001_099/001/01.01.01_60/gs_NFV-SWA001v010101p.pdf
24. Anand, A., J. Lakshmi, and S. Nandy. Virtual machine placement optimization supporting performance SLAs. *2013 IEEE 5th International Conference on Cloud Computing Technology and Science (CloudCom)*, vol. 1. IEEE, 2013, pp. 298–305. Bristol, United Kingdom, December 2–5, 2013.

17 The Shift to Software

Toby Ford

CONTENTS

17.1 Introduction 389
17.2 The "Shift to Software" 390
17.3 The Viral Nature of UNIX and C 391
17.4 Open Source 392
17.5 Linux and Apache Foundations, OpenStack, ODL, and the OPNFV 395
17.6 Scripting and Concurrent Languages 396
17.7 The Agile Method and DevOps 396
17.8 Web 2.0, RESTful API Exposure, and Interoperability 397
17.9 From the Birth of the "Cloud" to Containers and Microservices 398
References 399

The shift to software is a mindset change that affects all aspects of a telecom operator's operations—from who you hire, to how efficiently you can operate, to what you are able to create, to how quickly you are able to respond to market demands, and so forth. It has the potential to create a major impact—the time to value decreases, decisions are data-driven, there is a marked improvement in quality, operational predictability increases, and costs are reduced. The shift to software can deliver far more than just innovation and speed. As telecom operators see the benefits, they are sharing and collaborating like never before. This chapter explores these themes.

17.1 INTRODUCTION

In his 2011 essay *"Why software is eating the world,"* Marc Andreessen made the case that software was changing every industry, and that a partnership with Apple (implying the importance of smartphones and their apps) was evidence why software was transforming AT&T and Verizon. At that time, it was still a long way to go for AT&T in particular, and the telecom industry in general, before they could claim that they had transformed and were reaping the benefits of "a shift to software." But the general point of Andreessen's essay—software is eating the world—is central to this chapter [1].

Today, across all of telecom, there is a burgeoning notion that "a shift to software" provides generative and accelerative possibilities not available from a slow, bespoke, vertically integrated, siloed, vendor-driven hardware-centric approach. The hardware-centric approach was wrapped in an army of program and project managers and typical "waterfall" methods. The hardware-centric approach was based on standards that took years to agree on, and were not guaranteed to survive. The recent push to a software-centric approach is to apply an "agile" methodology to developing new capabilities constructed with current modern methods, providing a path to change and a time to value that is so frequent it can be thought of as continuous, real-time evolution. As an illustration, many modern e-commerce websites and financial systems are dynamically changing many times an hour, and often very dramatically and significantly, to respond to the market as quickly as is possible. The goal of the transformation to Network Cloud, is to achieve similar benefits for the telecom industry.

Switching from an operations mindset focused on manually configured, vendor-proprietary hardware-based products, to a developer mindset focused on automation and open-source software is the central theme of the transformation to a Network Cloud. Once a commitment to software and a developer mindset is made, then a cadre of useful techniques, technologies, and cultures appear. The UNIX philosophy, agile development methods, Information Technology Infrastructure Library (ITIL) standard for service management, "Design Patterns" for reusable templatization of software, open-source licenses and communities, application programming interface (API) exposure, microservices and DevOps are some examples.

In addition to gaining the benefits of software technologies and methods, developing and cultivating an ecosystem of third-party developers and collaborators, as has been successfully demonstrated with personal computers, Web 1.0 and 2.0, and mobile applications, is a prime objective for the telecom industry in this "shift to software." Open-source software's primary strength is that, in addition to unfettered access to software, it inherently comes with a collaborative ecosystem. In the case of the open-source projects surrounding network functions virtualization (NFV) and software defined networking (SDN), the participation, and involvement of operating entities around the world is unparalleled, both in scope and in achieving results.

For a telecom operator, the "shift to software" can bring significant benefits in the operator's capabilities—rapid response to market demands, flexibility, and extensibility when it comes to augmenting products and services, and streamlined operations, to name a few. The web and cloud providers have already demonstrated how this approach can help to achieve hyper-scale.

While today's telecom network is "software-heavy," it is not "software-centric." Problems that often surface in the "hardware-centric, software-heavy" environment are listed here:

- Premature optimization
- Change is costly—hardware changes are difficult
- Software development regularly fails to deliver, especially when using waterfall development methods
- Development done in silos leads to redundant development—limited horizontal platform and modular thinking
- Unwillingness to take risk and assign accountability
- Lack of modularity and commitment to composable solutions
- Financial models are inadequate and inconsistent
- Thinking which gravitates toward "more is less" not recognizing that simplicity scales [2]

17.2   THE "SHIFT TO SOFTWARE"

The "shift to software" story builds on the notion that one can create something that is perfect from elements that are imperfect. If the software is modular and open (accessible to all), imperfect elements can be independently and iteratively perfected over time, often by the crowd. Software development itself has evolved considerably over the last 60 years. Fundamental to this evolution is the availability of technologies and building blocks, depicted in Figure 17.1.

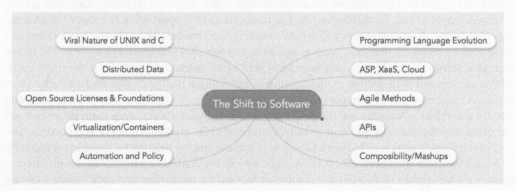

FIGURE 17.1   Shift to software—building blocks.

The technologies that have facilitated (and underlie) the dramatic shift to "modern" software:

- The Viral Nature of UNIX and C
- Open Source Software
- Linux and Apache Foundations, OpenStack, OpenDaylight (ODL), and the Open Platform for NFV (OPNFV)
- Scripting and Concurrent Languages
- The agile method and DevOps
- The "Cloud"
- Web 2.0, API Exposure and Interoperability
- Cloud-Native Workloads: Containers, Microservices, and Composability

17.3   THE VIRAL NATURE OF UNIX AND C

Many key software and standards concepts came from the development of UNIX and C in the early 1970s. Given their prevalence today and the prevalence of their direct descendants, like LINUX and C++, it begs the question—what is the secret behind their success? More than the technology itself, the culture and philosophy around the technology is what helped make UNIX and C so ubiquitous. Their longevity and survivability can be traced to their *viral* characteristics, especially simplicity and community.

Ken Thompson and Dennis Ritchie of AT&T Bell Labs created UNIX and C in 1969. At its core, the combination of UNIX and C was intended to be simple, support more than one user, and serve as a comfortable environment for developers, especially as a community. Today's essential notions of agile refactoring, parallelism, and developer enablement can be seen as being derived from this approach and form a foundational building block for modern software development.

Doug McIllroy, a researcher, programmer, and UNIX contributor at AT&T Bell Labs, further characterizes the Unix philosophy, "This is the Unix philosophy: Write programs that do one thing and do it well. Write programs to work together. Write programs to handle text streams, because that is a universal interface" [3]. This core UNIX philosophy of pipelining tools and ensuring all such tools act as filters of clear text, leads to modern concepts like composability and from that, modularity. These concepts in turn foretell modern methods like microservices and on to simple, clear interfaces between programs such as RESTful APIs and JSON, which act as a universal form of remote procedure calls (RPC). This same "simple and modular with clear, clean interfaces" philosophy has been used as the basis for building even more complex systems [4].

Another UNIX philosophy is expressed by McIllroy. "Design and build software, even operating systems, to be tried early, ideally within weeks. Don't hesitate to throw away the clumsy parts and rebuild them" [5]. This "expect to throw one away" mentality can also be attributed to Fred Brooks [6]. The notion of building software quickly to prototype, to hand to the customer early, to create the shortest possible feedback loop between developer and customer—is an essential tenet of today's agile release planning. The oft repeated mantra "Release early, release often. And listen to your customers" is the result [7].

UNIX "simplicity" was the result of consolidation and "refactoring" to a simpler design from previous operating system efforts like compatible time-sharing system (CTSS) and Multics. In many ways, UNIX inherently manifests the notion best described by the author Antoine de Saint Exupéry "perfection is attained not when there is nothing more to add, but when there is nothing more to remove." The concept of "refactoring" is a central theme of the agile method and will become a key part of the SDN/NFV story later in this chapter. Refactoring or refining code to make it more stable comes from this UNIX philosophy, again from McIlroy who says, "The real hero of programming is the one who writes negative code."

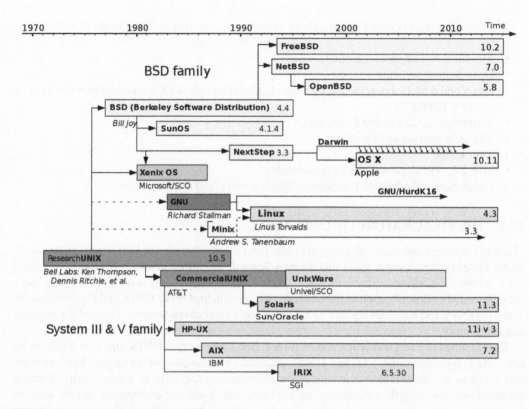

FIGURE 17.2 Timeline for UNIX evolution.

UNIX is also a meaningful first example of diverse community development. To quote Dennis Ritchie from the video *UNIX: Making Computers Easier to Use—AT&T Archives Film from 1982, Bell Laboratories*, "What we wanted to preserve, was not just a good programming environment in which to do programming. But a system around which a community could form fellowship... We knew from experience that the essence of communal computing ... as supplied by remote access time sharing systems... is not just to type programs into a terminal instead of a key punch... but to encourage close communication." Even though UNIX was proprietary licensed software and even though there was inherent internal resistance to distributing the code more widely, the original developers built a community within other telephone companies, with other operating system vendors, and in academic and research community, especially with UC Berkeley Computer Systems Research Group. From this starting point in AT&T Bell Labs, many variations of UNIX were derived as seen is the timeline in Figure 17.2. The Android and IOS operating systems on typical mobile devices, were derived from Linux and OS X, respectively, direct descendants of UNIX.

In addition, the other variations of UNIX and the long history of evolution of C compilers such as gcc and llvm, the communal nature of UNIX and C served as the inspiration for much of the thinking of the original open-source efforts.

17.4 OPEN SOURCE

The benefits of open source are clear—lower long-term costs, less intellectual property (IP) friction, increased collaborative innovation, increased generative innovation, applying coincident need to solve problems less redundantly, understanding what others are up to, and learning from what other more capable developers are doing.

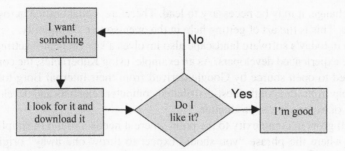

FIGURE 17.3 Simplest form of open-source engagement.

The overall strategy for getting full value from open source can be summarized in three basic operating tenets:

- Consume and learn from it
- Contribute and get help by finding coincident need and collaborating
- Refactor; throw one away and start over

The initiation into using open source typically follows the process as depicted in Figure 17.3.

This simplistic depiction shows how open source inherently allows you to try and inspect multiple software options before choosing either what is available or to create something new. No longer are there sales people or lab owners in your way. You are "free" to experiment. Finding open-source software for a desired function is as simple as doing an Internet search and down loading it from GitHub. Since the code is available to inspect, open source offers the ability to learn from the code—not only how it works but also gives an understanding of how one evolves it. This is a core concept that makes open source differentiated from proprietary code. Open source gives developers an opportunity to learn how others use and develop the software. Given the accessibility, especially through a social version control system like GitHub, of multiple options for any desired software function, it is also possible to differentiate between the various available software sources by further looking at the skillset of the people who developed it.

Linus Torvalds developed the version control software called Git in 2005 to address shortcomings of the then-existing version control tools. It made remote offline operation much more workable with centralized repositories and most importantly, made branching, forking, and merging far less onerous. Git is free software, like the Linux kernel. When GitHub took Git and added the Web 2.0/social aspects to it, the GitHub revolution took off. The combination of the virtues of Git and of social aspects have made GitHub the de facto place for a source code repository. GitHub made forking fashionable—this is another step toward unimpeded progress.

As software gets more and more complex, more help is needed to evolve it. Of course, with software, there is always an initial hump of effort required to make it reliable and functional to an acceptable base level. From there, depending on how extensible the software is and the scope of the software's functionality, the process of evolving it can go on for a long time. In the meantime, the base function should become increasingly more stable—especially, if the software is sufficiently modular and if one adheres to the UNIX dictum "do one thing and do it really well." and the agile tenet of "refactoring."

Why is collaboration so important in open source? Especially given that one can get value out of open-source software without ever making a contribution. That is, until something needs to be changed. If the change is implemented in isolation, and not shared with others, then it also needs to be maintained in isolation forever. It is for this reason it makes sense to collaborate first, sell the other constituents (owners/reviewers/quality assurance team/automation team), the need for change.

To initiate the change, it may be necessary to lead. Therefore, collaboration across the community is an imperative. This is the art of getting help in the open-source community.

Getting help in today's software landscape also involves a key concept—getting help from more capable or more experienced developers. As an example, using Kubernetes, the container-clustering software released to open source by Google derived from their internal Borg tool, means getting software and help from some of the best-distributed computer scientists and developers in the world with 10+ years of experience in production.

Software will grow in complexity to the point where it needs a round of simplification or refactoring. This is where the phrase "you should expect to throw one away" originates. There are always improved methods, programming languages, or new mechanisms to simplify software. Sometimes it takes starting over to take the learnings from the first version and apply it to develop a better second version. An example of this is the recent attempt at rewriting OpenStack in Go. The OpenStack developer community has been using Python—many think of OpenStack as only a Python project. Recently, Intel assembled a group that wrote a second version of the networking and virtual machine management components of OpenStack in an open-source effort called Ciao using the Go programing language and dramatically improved the scaling characteristics of these components [8].

The question of Intellectual Property (IP) is an important one in open source. Whether open source software is patented and subsequently released under a proprietary or open source license hinges upon whether the author or originator believes there is more value to be created in licensing, or from the generative effects of exposure. Once the decision to contribute software into open source is made, it is important to establish which open source license to distribute the software under. In the open source world, in addition to releasing the source code, all open source licenses grant copyright rights to the software being distributed by the originator. Each open source license balances different concerns and attempts to foster a development community that meets the licensor's needs. Certain open source licensing regimes put conditions on downstream users for software "freedom" rights resulting in "copyleft" licenses. For example, strong copyleft licenses prioritize ensuring that all downstream users receive an unrestricted right to the source code and permission to modify the software. By contrast, the permissive licenses guarantee the availability of the license rights only for the first downstream generation; and these licenses are generally understood to permit downstream users to release modified versions under more restrictive terms (including both proprietary and copyleft terms). Some licenses explicitly grant patent rights to avoid patent disputes among users.

Generally, there are four open source licensing regimes:

1. Academic Licenses: These are the simplest licensing regimes where the demands on downstream users of the open source software are minimal, mostly requiring attribution to the originator and the retention of copyright notices (e.g., MIT and BSD licenses).
2. Permissive Licenses: Next in complexity, these licenses provide downstream users substantial rights including, in some cases, explicit patent rights while allowing all downstream uses, including adding the open source software to proprietary products. However, there are certain restrictions put on downstream users of the open source code under permissive licenses, for example, the termination of the permissive license if a downstream user challenges the software with a "submarine" patent (e.g., the Apache 2.0 license).
3. Weak Copyleft Licenses: Here the originator of the open source software makes a conscious choice to copyright its software, and then license it under the so-called "copyleft" terms. Copyleft is a strategy of utilizing copyright law to pursue the policy goal of fostering and encouraging the equal and inalienable right to copy, share, modify, and improve creative works of authorship. One of the conditions the downstream user must satisfy before distributing "copyleft" software is that any changes such user makes to that software be likewise released under the copyleft license. A copyleft license ensures that all modified

versions of the code remain free of restrictions in the same way. (e.g., the Eclipse and the LGPL licenses).

4. Strong Copyleft Licenses: Strong copyleft licenses have significant conditions fostering software sharing. Such licenses are said to keep software "forever free" of use restrictions. In addition to having weak copyright restrictions, strong copyleft licenses have other conditions that may usurp or co-opt adjacent or linked proprietary code, depending on how the copyleft code is combined or linked with the proprietary code. (e.g., the GPLv2 or GPLv3 license, [9]).

17.5 LINUX AND APACHE FOUNDATIONS, OpenStack, ODL, AND THE OPNFV

Initially, open-source projects involved only a small number of devoted people, but the promise of these projects led to success and this brought the involvement of big corporations as consumers and contributors. Big corporation involvement necessitated a need for structure around governance, eventually incorporation of the activity and the reasonable expectation that the project would be marketed, sold, serviced, funded, evolved, and have the requisite legal protections. In almost all cases, these new entities, called Open-Source Foundations, were nonprofit organizations (with notable exception, Mozilla's search engine partnership with Google). The first of these organizations was the Free Software Foundation (FSF) founded in 1985, which was intended to support the GNU project and its enormous library of open-source software. The GNU project software initially was meant to be its own operating system but eventually provided much of the software needed to establish the Linux operating system (officially referred to as GNU/Linux.)

The FSF, the Apache Foundation's growth starting with the Apache web server, and the growth of the Linux Foundation around Linux and the Linux Foundation's Collaborative Projects initiative, all provided motivation to the people originally working on the OpenStack project to create a similar federated foundation. IBM, AT&T, and a few other entities along with Rackspace, actively sought for it to become a separate foundation of its own. OpenStack had an advantage, both in technical terms and in terms of marketing and planning elaborate summits, over other single-vendor or smaller community-backed solutions of 2012, like Nimbula, Eucalyptus, and OpenNebula. These alternative have eventually faded in the shadow of OpenStack. The "federated" plugin model of OpenStack, which leverages the phenomena espoused by Netscape and by UNIX previously, has been a key to OpenStack's success. It provides a way for vendors with proprietary and open solutions to integrate within a standard infrastructure orchestration framework. In the end, different than traditional standards bodies, OpenStack has provided a pathway which creates a de facto "open" standard, implemented in by code that is provisioned, for items such as storage and virtual machines, where the proprietary vendors had no desire to work towards a standard.

ODL project and its foundation provides the same "federated" benefits to the SDN space as OpenStack does in the server and storage virtualization space. ODL project has currently settled into a stable and growing federation. ODL is designed to care for existing environments and new environments. This is one of its key strengths. It allows ODL to be used for a broad set of use-cases, from configuring existing switches and routers, to controlling the configuration of OpenROADM-based devices, to the control over routing and flows. In a similar way, for the hypervisor data plane realm, FD.IO has emerged to fill the "federated" gap, when the incumbents like Open Virtual Switch (OVS) and OpenContrail did not have the scope or the governance model to do so [10].

When the European Telecommunications Standards Institute (ETSI) industry specification group for NFV was finished with its first set of NFV white papers, a group of participants in that effort came up with the idea to create an open-source reference implementation of NFV standards. Based on the lessons learned in creating OpenStack and ODL, an Open-Source Foundation called OPNFV was started within the Linux Foundation Collaborative Projects effort. OPNFV is intended to regularly release a reference platform and reference methodology for implementing NFV and its associated infrastructure. The goal of the OPNFV project is to reduce redundant effort around documenting NFV use cases, creating functional and performance testing standards and tools for NFV, integrating existing software such as OpenStack and ODL, filling in gaps where key functionality is missing, and working to ensure NFV requirements are represented and code accepted in upstream projects. By doing this, consistency and interoperability can be maintained across a diverse set of integrated software.

In the "shift to software," Open Source Foundations, specifically those with federated frameworks, and their collaborative ecosystems, play an integral role in the adoption and evolution of NFV and SDN.

17.6   SCRIPTING AND CONCURRENT LANGUAGES

Long compilation times associated with complex programs led to the origin of scripted languages and with it, the process of continuous build and test. We saw the rise of the interpreted languages— Perl, Python, PHP, and Ruby, to name a few. Though initially dubbed "kiddie scripting" languages, their use in the deployment of hyper-scale systems (Facebook/PHP and OpenStack/Python) saw a rise in their usage.

With multiprocessors and the scale of commodity clusters, the need for languages that support parallel processing led to a revival of functional languages such as Erlang that provide native support for concurrency, as well as Go, which provides native and simpler concurrency.

Go is popular because it provides many of the benefits of the scripting languages, the performance of compiled languages and the concurrency of functional languages. This has led to an enormous growth in Go's usage for cloud infrastructure and container management solutions such as Docker and CoreOS.

17.7   THE AGILE METHOD AND DevOps

Agile methodology is the engine for enabling change that is so frequent that it can be essentially thought of as constant change. The waterfall approach starts out with accumulating as many features in the planning process as possible, assuming the features delivered in a release are concrete, which translates into longer planning time horizons, and with changing market conditions and customer needs, there is constant risk of perpetual slippage. On the other hand, a smaller set of features leads to shorter release cycle, with a greater probability of making release dates, followed by iterations that capture any required market changes. Concrete and frequent releases dates, this is the essence of the agile approach. A modular software architecture makes it easier to continuously improve the quality of the individual elements without disrupting the system-level performance of the software. As such, the agile method promotes a more modular approach.

In addition, testing in the agile methodology, calls for writing tests at the same time as code is written, a concept referred to as *test-driven development*. This ensures that the software is stable for release if all the tests pass. This allows a developer to add more functionality along with new tests. The combination of small iterations and testing led to the concept of *continuous integration/ continuous development (CI/CD)*. These are powerful concepts that have allowed many e-commerce sites and software as a service (SaaS) entities to evolve quickly in response to market and customer needs. There is no reason why the same methodology will not work for telecom infrastructure.

A final characteristic of the agile method is the recognition that after many iteration cycles there needs to be scheduled time to "refactor" or clean up and resolve any built up "technical debt"— short cuts that were used to make releases or to solve problems quickly but that are not consistent with the conceptual integrity of the system on hand (Figure 17.4).

The agile methodology enables constant change. Applying methods such as full test automation, CI, continuous deployment, and peer review to infrastructure automation and orchestration is revolutionizing how typical network services are designed, implemented, and operated. Agile methods

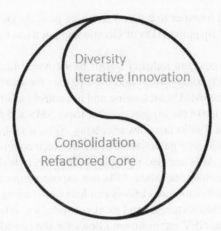

FIGURE 17.4 The yin and yang of agile development.

enable smaller teams to do more, create faster, and to do so much more efficiently. Even the role played by the traditional system administrator or network administrator, is rapidly changing. When you apply the agile methods to the traditional operations role dubbed X administrator [11], where X could be system, network, database, etc., and mix it with the requisite automation needed to scale to support very large-scale deployments, then you get what is referred to as DevOps. The DevOps role can be distilled into three meta-disciplines—policy, predictability, and scalability. Thus, the need to have a policy engine, a consistency checker of some kind, and some mechanism for scaling— both up and down. Many key modern innovations address these areas. For example, OpenStack's Congress for policy (Chapter 7 discusses ONAP's policy engine), or predictability-related innovations (Chef, Puppet, CFEngine, etc.), and the OpenStack Heat project for dynamic scalability.

17.8 Web 2.0, RESTful API EXPOSURE, AND INTEROPERABILITY

One of the integral aspects of modern software is the API. An API is an interface used when

1. One part of a system wants to talk to another part of the same system
2. One system wants to talk to another system over a network
3. "Client" software, like a web browser, wants to talk to a "Server" software, like a web server
4. Multiple "clients" want to subscribe to a service

An interesting use of APIs is when they expose the underlying data of a system to the outside world. As an example, when the SaaS entity, Salesforce.com, began in 2000, they immediately provided not just a web presentation layer but also the entire database via an exposed eXtensible Markup Language (XML) API over Hypertext Transfer Protocol (HTTP). This API "exposure" was one of the first examples of the use of "Web APIs" and what eventually became an integral definition of cloud computing. This API "exposure" allowed customers and third parties to integrate Salesforce data within their applications.

Generally, APIs are technically valuable because APIs drive more modular and scalable designs that allow components to be interchangeable—let the best component win. APIs also allow for multiple presentation layers to be created on top of the same source of data—let the best presentation layer win. As an example, in 2002, when Amazon launched Amazon Web Services (AWSs), it started solely in allowing third parties to integrate Amazon.com content and products behind their presentation layers, into their applications or services. During the era of Web 2.0, the concept of "API mashups" came to the forefront—where an application could be an amalgamation of many different APIs brought together to create something new. As an example, one could take the photo APIs of Flickr and the mapping APIs of Google and put them together to make a visual tour application.

How do APIs impact the telecom industry's shift to software? Late in the 2000s, the telecom industry started exposing APIs to customers and third parties for short message service (SMS) and multimedia messaging service (MMS) messaging and to control calls through applications built by third parties. Facebook was one of the biggest users of these SMS and MMS gateways. An example of this is when an entity like Twilio provides telephony APIs, it makes it possible for application developers of mobile applications or games to incorporate voice communications in unique ways.

In the NFV/SDN context, APIs are also valuable. Within the virtual infrastructure management (VIM) layer of an NFV environment, when APIs are exposed from a VIM (like the OpenStack framework), a developer of automation workflows can lean on standard provisioning APIs for creating virtual machines (VMs), block storage, and local networks for setting up applications or VNFs. This API exposure within the NFV environment allows for the end-to-end automation of network functions through their lifecycle from instantiation, through scaling and descaling, modification, and deletion. The broad adoption of APIs in telecom adds a new type of service whereby the developer is the customer. This is an example of generative innovation, where a broad ecosystem of developers can build upon a variety of exposed services via APIs to create new applications and new mashups.

One of the difficulties of API "exposure" is around how often they get changed, how the changes are communicated, and how others who expose the same or similar API, are changing or not changing the API. When system designers rely on APIs, there is a natural tendency to want an API to stay static and consistent wherever and whenever they are exposed. This is inevitably a place where standards bodies start to appear and attempt to help to create "interoperable" APIs. As APIs have evolved, new techniques have appeared to help in this regard, from "versioned" APIs, to introspection where a developer can rely on a single API to read the definition of an exposed API, and to interfaces that expect change and are ready to adjust accordingly within reason.

17.9   FROM THE BIRTH OF THE "CLOUD" TO CONTAINERS AND MICROSERVICES

Between the invention of the World Wide Web in 1991 and the introduction of the commercial browser in 1995, the need for third parties to "host" a customer website (instead of doing it on a server on premise) emerged. Soon many bulletin board services emerged and the Internet service providers started adding "web hosting" as an option where they would run the customer's website in the service provider's facilities.

In 1998, USinternetworking (USi) became the first application service provider (ASP) by offering to help set up an enterprise application (Customer relationship management [CRM], Enterprise resource planning [ERP]) hosted in its facility with connectivity to it provided over the Internet (or a private connection.) USi also started to offer service level agreements (SLAs) for availability, security, and process guarantees. These SLAs helped enterprises get over the fear of moving their services offsite. Over time, the ASPs realized that they were holding on to underutilized computing assets, which could be shared between multiple tenants. This became a reality with VMware's server virtualization technology. Salesforce.com took the idea of ASPs one step further and created a fully managed CRM offering that ran solely as a multitenant website. This is recognized as the first true SaaS offering and essentially the first modern "Cloud" service.

The idea of Cloud really took off when Amazon started offering web services (AWS) by exposing APIs. In 2006, after a few years of only exposing Amazon.com data, AWS added additional services. When AWS exposed the S3 interface for object storage, and soon after, the EC2 interface for elastic computing, infrastructure as a service (IaaS) was born. A developer could easily spin up servers, new storage allocations, databases, etc. with API calls. Several enhancements (RESTful APIs, active Javascript, etc.) made API use widespread—this was instrumental in launching Web 2.0.

This allowed for a new type of composite design of services, in an elastic manner (spin-up or down) that was not available before. In 2008, Heroku took this concept further and released the first real platform as a service (PaaS) offering, which actually ran on top of AWS.

The concept of a separate processing environment within an operating system that provided process and library separation from other processes, was a UNIX command called "chroot" that was further evolved during the implementation of one of the original firewalls [12]. This concept has today evolved to what we think of as containers. With the introduction of APIs and containers, the rise of microservices was inevitable; it loops back to the UNIX philosophy—do something small and have clean interfaces. It also introduces the next level of modularity and mashup composability which was not previously available. The next step is cloud native applications—building the perfect whole from imperfect parts.

Why is the cloud, containers, and microservices important in this "shift to software"?

Rampant traffic growth, coupled with flat to declining revenues, forces the telecom operators to reassess how networks are designed, deployed, and operated. The shift to software drives a major cultural change in how technology is developed or sourced and integrated into the network. The cloud provides the telecom operator the opportunity to drive to full asset utilization and to composable solutions using resources obtained real time. Containers provide a conceptual simplification and improved resource utilization beyond what server virtualization provides and microservices represents a new application architecture paradigm taken to an optimal extreme.

REFERENCES

1. Andreessen, M. 2011. Why software is eating the world. *The Wall Street Journal.* August 20, 2011.
2. Bias, R. 2012. Clouds are complex, but simplicity scales; a winning strategy for cloud builders. *Cloudscaling Blog.* February 29, 2012.
3. Salus, P. H. 1994. *A Quarter Century of UNIX.* Addison-Wesley, Reading, MA.
4. Voegels, 2007. Dynamo: Amazon's highly available key-value store. *21st ACM Symposium on Operating Systems Principles*, October 14–17, 2007, Stevenson, WA.
5. McIlroy, M. D., E. N. Pinson, and B. A. Tague. 1978. UNIX time-sharing system: Foreword. *Bell System Technical Journal* 57 (6): 1899–1904.
6. Brooks, F. and B. Frederick. 1995. The mythical man-month: After 20 years. *IEEE Software* 12 (5): 57–60.
7. Raymond, E. S. 2001. *The Cathedral and the Bazaar: Musings on Linux and Open Source by an Accidental Revolutionary.* O'Reilly Media, Inc. ISBN 1-56592-724-9.
8. Ciao, https://github.com/01org/ciao
9. Choose a License Website, https://choosealicense.com
10. FD.IO, https://fd.io
11. Burgess, M. 2000. *Principles of Network and System Administration.* Wiley, Hoboken, NJ.
12. Cheswick, B. 1991. An evening with Berferd: In which a cracker is lured, endured, and studied, *USENIX Summer Conference Proceedings*, Vol. 1. USENIX. San Francisco, California: The Association. p. 163.

18 What's Next?

Jenifer Robertson and Chris Parsons

CONTENTS

18.1 Trends 401
18.2 Domain Evolution 402
18.3 Hyper Automation within the Network 403
18.4 Hyper Automation within the Enterprise 403
18.5 Innovation across Entities 404
18.6 Summary 404
References 405

What does a post NFV/SDN world look like? Technology advancements in capturing, collecting, and analyzing data are coming at a rapid clip in an effort to keep up with the exponential growth rate of data generation. The potential value of this data follows a similar exponential trajectory as machine learning (ML) and artificial intelligence (AI) drives hyper automation and augmentation of operations and interactions. The massive data generation of an SDN network when combined with this hyper automation will make network operations invisible. From the vantage of managing and transporting the largest volume of operational and customer data in the world, these technologies will uncover new information that can transform industries and economic models.

18.1 TRENDS

Across the globe, businesses are pursuing the means to benefit from massive amounts of data generation from internal operations as well as customer and business interactions. In the future, data profiles will become so ubiquitous that many more companies will trade in the data economy [1]. The pace at which data is generated and stored is accelerating exponentially, driven in large part by the proliferation of IoT devices. By 2020, there will be an estimated 20–50 billion connected devices and global traffic will exceed 2.3 zettabytes. In the past, the world's data doubled every century, now it doubles every 2 years [2]. This acceleration of data has impacted the ability to derive valuable insights as the amount of data available is often times overwhelming.

As data usage grows, so do concerns about security, identity management, and privacy. The increase in data breaches makes security table stakes for data at end points and in motion. The World Economic Forum has listed cybercrime as the top global risk, estimating that 430 million new types of malware was introduced in 2014, a 40% increase from the prior year [3]. The average cost of a breach has risen to $7 million per incident or $221 per record lost or stolen [4]. Security is critical for both enterprises and consumers. The Department of Justice's Internet Crime Complaint Center recorded 269,422 cybersecurity-related complaints in 2014, an increase of over 1500% since 2000 [5]. Consumer privacy and security concerns have begun to influence internet activity as well. Nearly, half of consumers are more worried about privacy than a year ago and 74% have limited online activity due to privacy concerns [6].

Yet, despite these very real security and privacy concerns, communities are more frequently being used to derive more value from the data. Nearly, every industry is experiencing a trend toward "open innovation" where companies partner with academic institutions, researchers, and other companies, at times, even competitors [7]. The boundaries between a company and its ecosystem are becoming increasingly invisible [8]. For example, Hortonworks along with Arizona

State University, Baylor College of Medicine, and the Mayo Clinic have formed a consortium to define and develop an open source genomics platform to accelerate genomics-based precision medicine [9]. DJ Patil, Chief Data Scientist, White House Office of Science and Technology Policy stated "Unleashing the power of data through open community and collaboration is the right approach to solve a complex problem like precision medicine. Initiatives like this one will break data silos and share data in an open platform across industries to speed genomics-based research and ultimately save lives" [10].

The surge in available data is also one of the key changes that has led to the emergence of practical ML and AI applications. The amount of data available, new training techniques for deep networks, and the use of graphical processing units, used in PCs and video game consoles, to model neural networks [11] have enabled the modern advancement in AI. These AI technologies, when trained on massive amounts of data, are demonstrating significant advancements in the areas such as object detection, speech recognition, and natural language processing. ML and deep learning techniques are designed to be iterative in nature, constantly learning and optimizing outcomes [12]. Instead of a rule or policy-based automation to guide the system on the next best step, AI is able to make a more intelligent recommendation based on data, context, patterns, and outcomes. As the system continues to iterate, more data are collected and the algorithms auto-adjust to minimize errors. Andrew Ng, founder of Coursera reinforces the importance of the data for this technology stating "It's not who has the best algorithm that wins, it's who has the most data."

With these advancements, Ray Kurzweil believes by 2029, we'll reach "singularity" or the point at which AI outpaces human intelligence, a situation that has the potential to launch a host of societal issues [13]. Principles such as designing AI to assist humanity, being transparent, and maximizing efficiencies without destroying the dignity of people are important. There is also an acknowledgment that humans are ultimately accountable for the AI outcomes, implying a need for a measure of control by users of the technology [14].

18.2 DOMAIN EVOLUTION

Over time, significant shifts in capabilities have occurred as a result of the convergence of trends and technological developments. These shifts have had a pervasive impact to the Telecom network—to platforms, products, processes, and people. The evolution from one environment to the next is captured in Figure 18.1, with reference to the structural change in the operating domain (D).

- *Domain 0 (D0)* was driven by single vendor solutions for network and IT environments, and it served the industry well for over 100 years. Functions were contained and optimized within proprietary hardware, usually provided by a single vendor. The solutions were optimized and advanced by adding more complex functions over time; these solutions were designed and implemented for high availability. However, the lack of interoperability across vendors was a constraint that hampered large-scale operations.
- *Domain 1 (D1)* was the evolutionary step that enabled optimization in a multivendor environment. The standardization of solutions across vendors mitigated the complexity, drove greater economies of scale, and achieved interoperability. The scale also created

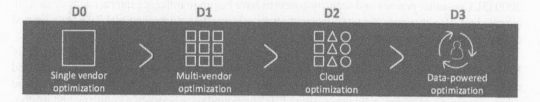

FIGURE 18.1 Historical domain evolution.

significant amounts of data. While only a small portion of the data available was analyzed, it presented meaningful information for use in decision-making.

- *Domain 2 (D2)* was the natural evolution to a cloud centric environment, first showing up in IT, and now in the network. Optimization in this environment is primarily characterized by functional abstraction which enables the complexity of a multivendor environment to be masked by software. The evolution from hardware-based functions to software-based functions running on commodity cloud infrastructure leads to extensibility and agility. It unshackles data from physical architectures, so prevalent in D1, and forms the basis for coherent, closed-loop automation by recognizing events and patterns and taking rule-based actions.
- *Domain 3 (D3)* is a logical next step in the evolution, given four key trends: the accelerating pace of data generation, the need for communities to gain meaningful insights, the requirement for advanced security and identity management, and the utilization of AI to enable hyper automation. In D3, establishing communities of data and capabilities can unlock internal hyper automation opportunities, and can facilitate pertinent innovation across entities. The complexity from the massive data generation of D2 is harvested to power smart optimization and mass personalization, paving the way for new business and operating models across multiple industries.

18.3 HYPER AUTOMATION WITHIN THE NETWORK

In a post NFV/SDN world, communities of internal data and capabilities will form within the enterprise and use massive ML to make network operations invisible. In earlier chapters (Chapters 8, 9, and 16), key applications of ML and AI pertaining to software-defined networking, were described. These include

1. Self-optimizing networks
2. Cybersecurity and threat analytics
3. Fault management
4. Customer experience

Rather than performing reactive analytics, ML helps to predict anomalies or events well in advance based on time-varying signals. ML predicts which machines could potentially go down in the next few days, weeks, or months. When combining those predictions with other data (previous seasonality patterns, weather events, peak demand behaviors), AI can interpret the data to recommend and automate traffic routing actions to mitigate customer impact. This example of hyper automation with ML and AI evolves beyond policies, or rules, that guide the system on what best next action to take. Moreover, as the system continues to iterate and more data are collected, ML and AI are able to learn from success and failures, and adjust their models accordingly to further optimize performance.

18.4 HYPER AUTOMATION WITHIN THE ENTERPRISE

In D3, organizations will transform into learning businesses by forming communities of relevant data and capabilities; in the prevailing ML/AI environment, every community action provides further training resulting in a perpetually optimizing environment. To illustrate this, today call centers open every morning and agents begin the day the same way as they did the previous day. In a hyper-automated world, the 10,000 calls (e.g.) taken the previous day advance the automation and influence outcomes for the current day. Similarly, smarter sales generation tools will help sales personnel better understand their audience's needs and preferences. Field technician dispatch processes will be optimized to send the closest tech with the best track record of success for that specific type of job, and the workflow itself will be augmented with a digital assistant that guides the technician to the most efficient and effective way to complete the job, based on AI recommended actions learned from the best of all technicians. Every business unit (community) can benefit every minute, resulting in smarter operations, with minimal manual touches.

The marriage between humans and AI will drive the need for a different prioritization of skills for knowledge workers. Skills such as the ability to collaborate, empathize, and create, will become vital. These are human skills that can not be replicated by a machine. AI can enrich and augment these skill sets. Cognitive intelligence and a broad understanding of the business can deliver immense value by guiding the use of AI to solve the largest business problems in a structured manner. Judgment will become even more important as humans must ultimately be accountable for the decisions and outcomes of AI [15].

18.5   INNOVATION ACROSS ENTITIES

As previously noted, standing up communities within an enterprise can unlock internal hyper automation opportunities. As those communities spread to include suppliers, partners, customers, universities, and other institutions with an interest in participation, an exponential jump in value creation will be unleashed. While previous evolutionary states presented enough risk to preclude the reward to such communities, D3 needs to provide a safe, trusted environment to move or connect data. In this forthcoming world, success and value is no longer linked to who creates the data, but to who has the right to use the data and for what purpose.

The result is that innovation comes on a much greater scale and at much higher speed by breaking down data silos and enabling the smash up of diverse data sets. D3 will create a way for people and enterprises to share data, access AI, and collaborate in a new and open way. Innovations will easily transfer inward and outward, with the community creator in control over what is shared or held proprietary. This open innovation environment may lead to new business models, redefining the entire logic in which value is created and captured [16]. Enabling this community learning capability could profoundly accelerate advances in solving difficult challenges. This is already happening in smart cities. Though in its early stages, the smart city community shares data which enables smart management of city infrastructure, utilizing AI to optimize utilities, traffic and a variety of other aspects of great interest to the community.

Threat analytics is another area that demonstrates the power of D3. For example, ML in the network could identify an anomaly in traffic patterns between two customer locations and predict that there may be a security threat. That network metadata is shared within the protected information exchange where it can be combined with all available and relevant data. The AI engine identifies that the traffic pattern does not correlate with known explanations based on the collective data and a network anomaly alert is sent to customers notifying of a potential threat. From there, customers could have the ability to combine the shared threat data with their internal data to determine the extent of their risk, or grant permission to automatically adjust the network, using software-defined networking capabilities, to mitigate the risk.

18.6   SUMMARY

A post NFV/SDN world is characterized by data-powered optimization. It harvests the complexity of massive data generation through ML and AI. Communities of data and capabilities drive hyper automation in the network and within the enterprise. Network operations become invisible, augmented employee experiences that deliver a better customer experience. In this future state, a company will not only drive data-powered optimization within its internal communities of applications, employees, and business units, but also within communities of relationships across suppliers, partners, customers, universities, and other institutions. Of the many tantalizing ramifications of such democratized innovation, perhaps chief among them will be the innovative business models that will rise—entirely new theories of value exchange, social engagement, and market conception.

REFERENCES

1. http://adage.com/article/digitalnext/iot-redefine-data-sharing-privacy/300084/
2. Why a Connected Data Strategy is Critical to the Future of Your Data, Hortonworks White Paper. https://hortonworks.com/info/connected-data-platforms/.
3. http://www.nytimes.com/2016/04/07/us/politics/homeland-security-dept-struggles-to-hire-staff-to-combat-cyberattacks.html?_r=0
4. http://fortune.com/2016/06/15/data-breach-cost-study-ibm/
5. http://www.foxbusiness.com/features/2016/04/27/cyber-attacks-on-small-businesses-on-rise.html
6. Meeker Report.
7. http://thomsonreuters.com/en/press-releases/2015/05/tr-analysis-shows-concerning-trend-for-global-innovation.html
8. Zimmermann H.-D. and Pucihar A. Open Innovation, Open Data, and New Business Models. (September 1, 2015). Available at SSRN: https://ssrn.com/abstract=2660692
9. Hortonworks PR release for new consortium on genomics with Arizona State University, Baylor and Mayo Clinic. https://hortonworks.com/press-releases/hortonworks-initiates-precision-medicine-consortium-explore-next-generation-genomics-open-source-platform/
10. ibid.
11. http://www.economist.com/blogs/economist-explains/2016/07/economist-explains-11
12. http://www.forbes.com/sites/louiscolumbus/2016/06/04/machine-learning-is-redefining-the-enterprise-in-2016/#33142a475fc0
13. AT&T SHAPE Conference Speech.
14. http://www.slate.com/articles/technology/future_tense/2016/06/microsoft_ceo_satya_nadella_humans_and_a_i_can_work_together_to_solve_society.html
15. ibid. Hans-Dieter Zimmermann and Andreja Pucihar
16. ibid. Hans-Dieter Zimmermann and Andreja Pucihar

Index

Note: Page numbers followed by "*fn*" indicate footnotes.

A

AAA, *see* Authorization, authentication, and accounting
A&AI function, *see* Active and Available Inventory function
ABAC, *see* Attribute-based access control
Abstraction
    layers, 88–89
    modeling, 94–95
Academic licenses, 394
Access, 17
    scale, 259–260
Access control lists (ACLs), 16, 34
Access network(s), 1, 92, 225, 240
    architecture for, 224, 228
    construction and maintenance of, 229
    control plane functions, 240
    5G, 248
    vOLT hardware into, 237
Access session border controller (A-SBC), 312
Access technologies, 224
    mobile wireless, 241–248
    wireline, 229–241
Accounting request (ACR), 312
Accurate fault detection, 77
    VF error handling, 77
    VF failure detection and alerting, 78
    VF fast recovery, 78
    VF protection from external services, 77
    VF software fault detection, 77–78
    VF software stability, 78
ACLs, *see* Access control lists
ACO, *see* Analog coherent optics
ACR, *see* Accounting request
Action points, 5
Active and Available Inventory function (A&AI function), 93–94, 107, 109, 128
    functionality, 130
    functional view, 128–129
    requirements, 129–130
Active probe monitoring, 379
Active vProbe measurements data model, 379
Adapters, 91, 93
Adaptive poll mode driver, 211
ADC, *see* Analog to digital converters
Adjacency segment, 284–285
ADSL, *see* Asymmetric digital subscriber line
Advanced functions, 5
Advanced multielement antenna structures, 245–246
Advanced persistent threat (APT), 194
AFU testing, *see* Approval for use testing
Agile, 58–59
    method, 396–397
    process, 307
    software development approach, 194
AI, *see* Artificial intelligence
AIC, *see* AT&T Integrated Cloud

"Alien wavelengths," 275
All-flash array, 52
ALL-IP network, 9, 276
    LTE network, 294
    rapid transition to, 10
    transforming ALL-IP network to network cloud, 12–24
Allocate function, 95
Amazon Web Services (AWSs), 397
Ambient Video, 167–168
Analog coherent optics (ACO), 272
Analog telephony, 10
Analog to digital converters (ADC), 272
Analytic(s), 7, 107, 123
    descriptive, 149
    forms, 149
    framework, 132–133, 153, 155–156
    layer, 138–140
    predictive, 149
    reactive, 149
    real-time, 23
    security, 189–193
ANDSL, *see* AT&T Network Domain-Specific Language
Anomaly detection, 190
ANR, *see* Automatic neighbor relations
Anti-affinity, 200
Apache
    CloudStack, 53–54
    foundations, 395–396
APIs, *see* Application programming interfaces
Appliance-based network, 184
Application-specific integration circuits (ASICs), 13, 45, 213, 227
Application controller(s), 240
    information to, 133
    orchestration, 117
Application programming interfaces (APIs), 4, 5, 104, 274, 390
    handler, 93
    mashups, 397–398
Application resiliency, 61–63
Application security, 182–185
Application server (AS), 295
Application service controllers, relationship to, 94
Application service provider (ASP), 398
Approval for use testing (AFU testing), 309
APT, *see* Advanced persistent threat
ARPANET, 9–10
Artificial intelligence (AI), 148, 401, 402
Artificial neural network, 148
AS, *see* Application server
A-SBC, *see* Access session border controller
ASICs, *see* Application-specific integration circuits
ASP, *see* Application service provider
ASTRA, 177, 189, 194
    architecture, 194–195
    defense in depth, 196–198

ASTRA (*Continued*)
   microperimeter, 197
   network perimeter, 198–199
   system, 195–196
   workgroup/tenant-level perimeter, 198
Asymmetric digital subscriber line (ADSL), 228
AT&T, 5–7, 232–233, 250
   approach, 14
   cloud-based architecture, 189
   collaborate, 325
   deploys, 282
   design and implementation, 87
   domestic intercity optical backbone network, 267
   mobile network, 2
   plans, 237, 239
   SDN-Mon framework, 370–372
   SDN ONENET, 5
   VNF guidelines, 384
AT&T Integrated Cloud (AIC), 6, 50, 177–178, 218
   environment, 226–227
   security evolution, 180–182
   site, 50
   zones, 50
AT&T Network Domain-Specific Language (ANDSL),
     95–96
Attribute-based access control (ABAC), 193
Authentication approach, 192
Authorization, authentication, and accounting (AAA), 35
Automated validation techniques, 345
Automatic neighbor relations (ANR), 162
Automation, 16, 51, 157, 183, 347
   for configuration management, 44
   configuring ONAP to enable operations, 343
   control-loop, 157–160
   of infrastructure management, 45
   machine learning for closed-loop, 160–161
   of recovery, 78
Auto-scaling
   model, 382–383
   NFV service, 382
Availability, 69, 266, 380, 381
   to cost tradeoff, 71–72
   to downtime/year conversion, 70
   high, 94
   sensitivity to key parameters, 82
   single-site, 69–71
   tradeoffs, 68
AWSs, *see* Amazon Web Services

B

Backbone, 18
   AT&T's domestic intercity optical backbone
     network, 267
   example IP/optical network with, 356
   IP common, 276–277
   SDN-enabled backbone network, 354
"Bad" automation, 343
"Bad" policy, 344, 345
Bandwidth
   aware routing, 282–283
   "calendared" services, 289–290
   calendaring, 100
Base band unit (BBU), 36, 242

Batch processing, 146
BBF, *see* Broadband Forum
BBU, *see* Base band unit
Beamforming, 245, 247
Behavioral analysis, 190
BGCF, *see* Breakout gateway control function
BGP, *see* Border Gateway Protocol
BGP-LS, *see* Border Gateway Protocol-Link State
Big data, 140
   analytics and ML, 148–151
   control-loop automation, 157–160
   data management, 142–143
   data quality, 141–142
   DCAE, 153–154
   DCAE functional components, 154–156
   dealing with big data's 7Vs, 140–141
   deep learning and SDN, 161–162
   design principles, 151–153
   Hadoop ecosystem, 143–148
   MLs for closed-loop automation, 160–161
   ML toolkit comparisons, 152
   MS design paradigm, 156–157
   and Network Cloud, 151–162
BI systems, *see* Business intelligence systems
"Black box," 56, 255
Blast radius, 42*fn*
Block(s), 52
   function, 95
   storage, 53
BNGs, *see* Broadband Network Gateways
Bootstrapping SDN /NFV deployments leveraging, 348
Border Gateway Protocol (BGP), 6, 14, 93, 286–287
Border Gateway Protocol-Link State (BGP-LS), 355
   protocol, 288
Bottleneck analysis, 71–72
"Bottoms-up" approach, 263
BPMN, *see* Business Process Management Notation
BPON, *see* Broadband passive optical network
Breakout gateway control function (BGCF), 312
Bring your own device (BYOD), 194, 202
Broadband Forum (BBF), 232–233, 241
   G. fast architecture TR-301, 232–233
Broadband Network Gateways (BNGs), 233, 252
Broadband passive optical network (BPON), 231
Broadband Remote Access Server, *see* Broadband
     Network Gateways (BNGs)
Broadcom
   OLT PON MAC SOC, 235
   Trident/Tomahawk, 255
Broad network cloud strategy, 105
BSS, *see* Business support systems
B2B, *see* Business to business
B2C, *see* Business to consumer
Buffer manager, 213
Business intelligence systems (BI systems), 195
Business Process Management Notation (BPMN),
     113, 115
Business support systems (BSS), 15, 30, 61, 99, 104,
     229, 347
   legacy BSSs interactions with ONAP, 133–135
Business to business (B2B), 193
Business to consumer (B2C), 193
Business VoIP (BVoIP), 293, 324–326
BYOD, *see* Bring your own device

C

CaaS, *see* Compute-as-a-Service
Cache-coloring, 200
CALEA, *see* Communications assistance for law
  enforcement act
Call detail record (CDR), 302
Call Home function, 215
Call session control function (CSCF), 81, 311–312
Canonical reference model, 249
Capacity
  design, 323–324
  management process, 289
  planning, 123
"Cap and grow" approach, 346
Capital expenditure (CapEx), 104, 105, 365
Cask Data Application Platform (CDAP), 156
CBB, *see* Common Backbone
CCF, *see* Charging collection function
CCU record, *see* Cell change updates record
CD, *see* Continuous deployment
CDAP, *see* Cask Data Application Platform
CDN, *see* Content distribution network; Content
  distribution network (CDN)
CDR, *see* Call detail record
CE, *see* Coexistence element
Ceilometer, 374–375
  CPU counter, 377
  examples, 375–377
  memory counters, 377
  OpenStack Ceilometer configuration, 376
  OpenStack Ceilometer data/event logging
    architecture, 376
Cell change updates record (CCU record), 139
Centralized controls, 88
Centralized RAN architecture (cRAN architecture), 242–244
Centralized TE, 287–288
  using SDN controller, 358
Central offices, 17
Central processing unit (CPU), 33
CE router, *see* Customer edge router
Certification
  repository, 116
  studio, 116
CFM, *see* Connectivity fault monitoring
C Form-factor Pluggables (CFP), 271
Change management workflows, 342, 343
Charging collection function (CCF), 312
Chromatic dispersion, 272
"Chroot" UNIX command, 399
CI, *see* Continuous integration
Ciao open-source effort, 394
CI/CD, *see* Continuous integration/continuous development
Cinder, 58
Circuit-switched releases (CS releases), 294
Cisco Cloud Server Router (CSR), 219
CLAMP, *see* Control loop automation management platform
Classification, 149, 150
CLI, *see* Command line interface
Client application identity, 193
Clos architecture, 253–254
Closed-loop
  automation, 304
  control, 124

Closed user groups (CUGs), 26
Cloud, 182
  cloud-based applications, 63
  cloud-based paradigms, 299
  cloud-based virtualization technology, 219
  cloud-native, 61
  computing data centers, 227
  data center standardization, 256
  environment, 6
  infrastructure virtual networks, 322
  orchestration mechanisms, 317
  overlay connectivity, 260
Cloud Native Computing Foundation (CNCF), 328–329
Cloud Radio Access Network (C-RAN), 36
Clustering, 149
CM-HA, *see* Continuous monitoring-high available
CNCF, *see* Cloud Native Computing Foundation
CN functions, *see* Core network functions
COBO, *see* Consortium for onboard optics
Coexistence element (CE), 231
Coherent
  optics, 272
  pluggables, 272
Collaborate service complex, 325
Collection framework, 153, 154
Command line interface (CLI), 35, 201, 368
Commercial
  open source, 53–54
  VIM solution, 53
Commercial off-the-shelf (COTS), 28
Commoditization, high volume, 256
Common Backbone (CBB), 276
Common off the shelf (COTS), 12, 218
  hardware, 13, 318
  platform, 208
Common Public Radio Interface (CPRI), 242
Communication(s), 67
  patterns, 20–21
  services, 21
Communications Act (1934), 1
Communications assistance for law enforcement act
    (CALEA), 312
Compatible time-sharing system (CTSS), 391
Compiler function, 91
Completion rate, 380
Component
  component-level measurements, 367–368
  measurements, 365
Compute-as-a-Service (CaaS), 6
Computer networking devices, *see* Networking hardware
Computer processing units (CPUs), 28, 52, 251
  affinity, 200
  inefficiency, 211
Compute virtualization, 28
Concurrent languages, 396
Configure function, 95
Connectivity fault monitoring (CFM), 367
Consortium for onboard optics (COBO), 276
Constrained shortest path first routing (CSPF routing), 282
Consumer services, 22
Container(s), 56–57
  container-based virtualization, 56
  technology, 207
Content distribution network (CDN), 6, 20, 21, 29, 140

Content filtering smart network, customer configurable
　　policy for, 163
Continuous deployment (CD), 188
Continuous integration (CI), 188
Continuous integration/continuous development (CI/CD),
　　13–14, 59, 347, 396
Continuous monitoring-high available (CM-HA), 62
Control
　　framework, 131–132
　　points, 5
　　protocols, 20
Control-loop automation, 157
　　closed-loop modeling and template design, 158–159
　　intelligent service, 157–158
Control loop automation management platform
　　(CLAMP), 158
　　system architecture, 159–160
Control loop systems, 124, 130, 340–342
　　analytic framework, 132–133
　　design framework, 131
　　ONAP closed-loop automation, 131
　　orchestration and control framework, 131–132
　　requirements, 130–131
Control plane, 35
　　functions, 240
Conventional LTE systems, 245
Conventional network architecture, 227
Converged IP messaging server (CPM server), 311
Convergence, 280
Convolutional neural networks, 148
Copper-based technologies, 229
Copyleft, 394–395
Core network functions (CN functions), 310
Core-to-core measurements, 357
Core-to-edge measurements, 357
CoreOS, 396
Core packet network, 287–288
Core router technology evolution, 286
Corrective maintenance, 160, 332
Correlation (CORRELATION), 341
Cost tradeoff, availability to, 71–72
COTS, *see* Commercial off-the-shelf; Common off the shelf
CPE, *see* Customer premises equipment
CPM server, *see* Converged IP messaging server
CPRI, *see* Common Public Radio Interface
CPU, *see* Central processing unit
CPU pinning, *see* Processor affinity
CPUs, *see* Computer processing units
cRAN architecture, *see* Centralized RAN architecture
Critical machine type communication (Critical MTC),
　　244–245
C-RAN, *see* Cloud Radio Access Network
CRM, *see* Customer relationship management
CSCF, *see* Call session control function
CSPF routing, *see* Constrained shortest path first routing
CSR, *see* Cisco Cloud Server Router
CS releases, *see* Circuit-switched releases
CTSS, *see* Compatible time-sharing system
CUGs, *see* Closed user groups
Customer
　　experience improvement, 160
　　premise-based small cells, 253
Customer configurable policy for content filtering smart
　　network, 163

Customer edge router (CE router), 29
Customer premises equipment (CPE), 16, 79–80, 203, 204
Customer relationship management (CRM), 398
Custom silicon, 255
Custom voice/video services, 22
C, viral nature of, 391–392
Cybersecurity, 160, 182

D

Dashboard functions, 117–119
Data
　　distribution bus, 153
　　instrumentation layer, 138
　　migration, 57
　　modeling for network design, 215–216
　　plane, 33–35
　　prefiltering and selection, 190
　　repositories, 115–116
　　router, 155
　　stack, 138
Database as a service (DBaaS), 302–303
Data border element (DBE), 312
Data centers (DCs), 50, 70, 226–227
　　environments, 226
　　merchant silicon, 255
　　networks, 250
　　virtualization of data-center networks, 4–5
Data collection, analytics and events (DCAE), 78–79, 91,
　　106, 119–123, 153–154
　　analytics framework, 155–156
　　collection framework, 154
　　components, 120–121
　　data movement, 155
　　edge, central, and core lakes, 155
　　function, 93–94
　　functional components, 154
　　ONAP portal framework, 118
　　ONAP portal software development toolkit, 120
　　platform approach to DCAE, 121–122
　　platform components, 122–123
　　virtualized network cloud vision, 119
Data leak protection (DLP), 173
Data plane development kit (DPDK), 35, 63, 208, 296
　　DPDK-enabled virtual switch approach, 211
DBaaS, *see* Database as a service
DBE, *see* Data border element
DCAE, *see* Data collection, analytics and events
DCO, *see* Digital coherent optics
DCP, *see* Distributed control plane
DCs, *see* Data centers
DDoS, *see* Distributed Denial of Service
Decomposing network functions, 300, 302
Decomposition of functions, 314
Decoupling
　　control plane, 35
　　data plane, 33–35
　　of long-lived state, 314
　　management plane, 35–37
　　service and network path measurements, 368–370
　　of subscriber data, 314
　　virtual functions, 32
Deep learning, 148, 150–151
　　and SDN, 161–162

Deep packet inspection (DPI), 34, 210, 217
Default gateway IP routing, 260
Defect rate, 380
Defects per millions (DPMs), 69, 380
Defense in depth
    architecture, 173, 174
    ASTRA, 196–198
Denial of service (DOS), 252
Dense metro networks, 267
Dense wavelength division multiplexing (DWDM), 18,
        235, 269, 270
Deployment
    considerations, 305
    models choice, 218
Descriptive analytics, 149
Design
    patterns, 390
    Studio, 114, 115
Device layer, 215
DevOps, 23, 59, 396–397
DG, *see* Directed graph
DGBuilder, graphical tool, 96
DHCP, *see* Dynamic Host Control Protocol
Diameter routing agent (DRA), 312
Differentiated services code point (DSCP), 315
Digital coherent optics (DCO), 272
Digital signal processor (DSP), 227, 272, 296
Digital subscriber line (DSL), 232
Digital subscriber line access multiplexers (DSLAMs), 259
Digital switching systems, 68
Directed acyclic graph, 147
Directed graph (DG), 91
Direct memory access (DMA), 212
Directories, 19, 20, 23
Direct Services Dialing Capability (DSDC), 5
Disaggregated edge platforms, 253
Disaggregation, 238, 254, 338
    fabric, 255
    of hardware and software, 4, 261
Distributed access nodes, 227
Distributed control plane (DCP), 50–51
Distributed controls, 88, 285*fn*
Distributed Denial of Service (DDoS), 6, 100–101, 172
    attack resilience, 175
Distributed local controllers, 88
Distributed nature of access technology, 240
Distributed protocols, 287–288
Distributed RAN architecture (dRAN architecture), 242
Distributed RSVP-TE protocols, 288
Distributed traffic engineering, 282–283
Distributed VNF designs, 41–42
Distribution point unit (DPU), 229, 232
    YANG model, 233
Distribution studio, 116
Diversion, 16
DLP, *see* Data leak protection
DMA, *see* Direct memory access
DMZ, 180
DNS, *see* Domain Name Service
DNS as a service (DNSaaS), 303
Docker, 396
Domain
    domain-specific languages, 264
    evolution, 402–403

Domain Name Service (DNS), 29, 79, 82, 140, 177, 302, 312
Domino language, 263
DOS, *see* Denial of service
DOWNSTREAM automation, 189
DPDK, *see* Data plane development kit
DPI, *see* Deep packet inspection
DPMs, *see* Defects per millions
DPU, *see* Distribution point unit
DRA, *see* Diameter routing agent
DRAM, *see* Dynamic random access memory
dRAN architecture, *see* Distributed RAN architecture
Driver abstraction layer, 239
DSCP, *see* Differentiated services code point
DSDC, *see* Direct Services Dialing Capability
DSL, *see* Digital subscriber line
DSLAMs, *see* Digital subscriber line access multiplexers
DSP, *see* Digital signal processor
DTA, *see* Dynamic time assignment
Dual connectivity, 246–247
DWDM, *see* Dense wavelength division multiplexing
Dynamic
    load management, 168
    programming, 162
    security policy, 194
Dynamic Host Control Protocol (DHCP), 32
Dynamic random access memory (DRAM), 213
Dynamic time assignment (DTA), 232–233

E

E.164 Number to URI mapping (ENUM), 312
EAL, *see* Environment abstraction layer
ECOMP, *see* Enhanced Control, Orchestration,
                Management, and Policy
Economics, NFV
    hardware costs, 45
    operational costs, 45–46
    software costs, 45
Ecosystem identity, 191–192
E-CSCF, *see* Emergency CSCF
Edge
    application, 252–253
    edge-to-core measurements, 357
    edge-to-edge measurements, 357
    vPE VNF, 256–258
Electrical telecommunication systems, 1
    AT&T, 5–6
    cost evolution, 3
    LTE, 7–8
    network functions, 3–4
    ONAP, 6–7
    public networks, 2–3
    virtualization of data-center networks, 4–5
Electro-optical equipment, 18
Element management system (EMS), 14, 105, 296, 336
Element management system layer (EML), 374
Email, 172
eMBB, *see* Enhanced mobile broadband
Embedded measurement functions, 369
Embedded operations channel (EOC), 233
Emergency CSCF (E-CSCF), 312
EML, *see* Element management system layer
EMS, *see* Element management system
End-to-end service path, 342

End user identities, 192–193
Enhanced Control, Orchestration, Management, and
        Policy (ECOMP), 58, 104
Enhanced mobile broadband (eMBB), 244
Enhanced packet core (EPC), 253
Enterprise
    CPE, 16–17
    dedicated Internet service, 252
    Internet, 258
    IT managers, 203
Enterprise networks, 16, 201
    DPI and visibility, 217
    evolution of network complexity, 202–203
    memory and storage resources, 214
    network management and orchestration, 214–217
    NFoD, 217–218
    optimizing virtual environment, 208–212
    optimizing VNF performance, 213
    packet processing capabilities, 207–208
    technology innovations, 203
    uCPE, 218–222
    virtualization of network functions, 204–207
Enterprise resource planning (ERP), 398
ENUM, see E.164 Number to URI mapping
Environment abstraction layer (EAL), 63, 213
EOC, see Embedded operations channel
EPC, see Enhanced packet core; Evolved Packet Core
ERP, see Enterprise resource planning
Error handling, VF, 77
Ethernet, 25–26, 249, 276
    interfaces, 249
    multiplexing, 251
    network traffic, 208
    virtual circuits, 252
    virtual networks using VLANs, 27
Ethernet VPN (EVPN), 26, 258, 260
    BGP signaling, 261, 262
    integrated routing and bridging in, 261
    in nominal operating state, 261
    rerouting psuedowires, 262
"Ethical hackers," 183
ETSI, see European Telecommunications Standards Institute
European Telecommunications Standards Institute (ETSI),
        13, 109–111, 395
    ETSI NFV effort, 15
Evolved Packet Core (EPC), 293
EVPN, see Ethernet VPN
Explicit routing, 284–285
Exposed API, 398
Extended AIC, 248
Extensible Access Control Markup Language
        (XACML), 124
eXtensible Markup Language (XML), 397
Extensible messaging and presence protocol (XMPP), 20
Extension abstraction layer, 239

F

Fabric disaggregation, 255
Facebook, 143, 398
Failure detection and alerting, VF, 78
Faster restoration using segment routing, 285
Fast Fourier Transform (FFT), 36
Fast recovery, VF, 78

Fast reroute (FRR), 19, 280–282
Fault-tolerant building blocks (FTBBs), 61
Fault-tolerant/tolerance, 56, 61–62
    features, 63
    hot VM replication, 76
    passive VM replication, 77
    VM designs, 76
    warm VM replication, 77
Fault configuration, accounting, performance, and security
        (FCAPS), 47, 107, 319, 373
Fault/event correlation, 123
Fault management, see Corrective maintenance
FCAPS, see Fault configuration, accounting, performance,
        and security
FCC, see Federal Communications Commission
FDD, see Frequency-division duplexing
FD-MIMO, see Full dimension MIMO
FEC, see Forward error correction
Federal Communications Commission (FCC), 1, 67, 143
Federal Trade Commission (FTC), 143
"Federated" plugin model of OpenStack, 395
Federation between network controllers, 94
Femtocells, see Customer premise-based small cells
FFA, see First field application
FFT, see Fast Fourier Transform
Fiber optic cables, 267
"Fiber to the home" (FTTH), 17
FIBs, see Forwarding information bases
Field programmable gate array (FPGA), 213, 227
Firewalls (FWs), 172
First field application (FFA), 342
5G networks, 8, 241–242
5G wireless, 244
    advanced multielement antenna structures, 245–246
    dual connectivity, 246–247
    flexible carrier configuration, 247–248
    types of capabilities of 5G, 245
    ultrahigh density and self-backhaul, 247
Fixed network, 1
Flavor Series, 64, 65
Flexibility level, 268
Flexible
    carrier configuration, 247–248
    management models, 218
    software-controlled optical networks, 272–273
Flexible access grooming, 258
    access scale and resiliency, 259–260
    cloud overlay connectivity, 260
    EVPN, 260–261, 262
Flow
    classification, 213
    measurements, 362
    redirection, 100–101
Flume, 144
FMO, see Future mode of operations
Forensics, 16
Forward error correction (FEC), 274
Forwarding information bases (FIBs), 34
4G, see Fourth generation
4G long term evolution (4G-LTE), 223–224, 241
4 Port Hardened OLT, 237
Fourth generation (4G), 17
FPGA, see Field programmable gate array
Frequency band, 232

Frequency-division duplexing (FDD), 232
FRR, *see* Fast reroute
FTBBs, *see* Fault-tolerant building blocks
FTC, *see* Federal Trade Commission
FTTH, *see* "Fiber to the home"
Full dimension MIMO (FD-MIMO), 245
Fundamental network planning, *see* Long-range network
        planning
Future mode of operations (FMO), 90, 309
"Future of Privacy Forum" blog, 143
FWs, *see* Firewalls

G

Gatekeeper function, 198
Gateways, 20
Generic interfaces, 226
Generic routing encapsulation (GRE), 38
Geo-diversity, network controller architecture, 94
Georedundancy, 72–75
Georesiliency, 316–317
G.fast technology, 229, 232–233
Gigabit Passive Optical Networking (GPON), 17, 223, 229
        access technology, 241
GIS, *see* Graphical information systems
Git, 393
GitHub, 393
Global routing table, 27
Global Technology Operations Center (GTOC), 334, 335
GNU project software, 395
Google, 143
Google remote procedure call (gRPC), 366
GPON, *see* Gigabit Passive Optical Networking
GPUs, *see* Graphical processing units
Graphical information systems (GIS), 119
Graphical processing units (GPUs), 52
Graphical user interface (GUI), 115, 117–119, 202
"Graybox," 256
"Gray" optics, *see* Standard wavelength
GRE, *see* Generic routing encapsulation
gRPC, *see* Google remote procedure call
GTOC, *see* Global Technology Operations Center
"Guard rails," 125
GUI, *see* Graphical user interface

H

Hadoop distributed file system (HDFS), 143
Hadoop ecosystem, 143
    Apache software foundation, 143–144
    batch processing, 146
    coordination and workflow management, 147–148
    data ingestion, 144
    real-time and micro-batch processing using STORM
        and SPARK, 146–147
    SOLR, 145–146
    storage, 144, 145
    YARN, 144–145
Hardware (HW), 212, 218–219
    abstraction layers, 256
    costs, 45
    evolving hardware ecosystem, 254
    hardware-centric network design methodology, 4
    layer, *see* Physical layer

HDFS, *see* Hadoop distributed file system
High availability, network controller architecture, 94
High volume message producers, 373
HIVE, 146
Home subscriber server (HSS), 21, 303, 312, 321
Horizontal scaling, 381–382
Host VNF, 40
Hot VM replication, 76
HSS, *see* Home subscriber server
HTTP, *see* Hypertext Transfer Protocol
HTTP-proxy VNF components, 321–322
HW, *see* Hardware
Hybrid architecture, 318
Hyper automation
    within enterprise, 403–404
    within network, 403
Hyper-scale systems, 396
Hyperscan, 214
Hypertext Transfer Protocol (HTTP), 10, 397
Hypervisor, 28, 57, 185–187, 205
    hardening, 199
    technology, 185

I

IaaS layer, *see* Infrastructure as a Service layer
IAM approach, *see* Identity and access management
        approach
I-CSCF, *see* Interrogating CSCF
Identity
    ecosystem, 191–192
    management, 16
Identity and access management approach (IAM
        approach), 189, 191
    client application identity, 193
    end user identities, 192–193
    identity ecosystem, 191–192
    resource identity, 193
IDS, *see* Instruction detection system; Intrusion detection
        system
IETF, *see* Internet engineering task force
IGP, *see* Interior Gateway Protocol
IM, *see* Infrastructure management
IMS, *see* IP multimedia subsystem
IMS web security function (IWSF), 312
Incident response process, 175
Information technology (IT), 13, 45, 51, 105
Information Technology Infrastructure Library standard
        (ITIL standard), 390
Infrastructure as a Service layer (IaaS layer), 303–304, 398
Infrastructure controllers, 240
    orchestration, 116–117
Infrastructure management (IM), 61
Infrastructure orchestration and control (IO&C), 100
Infrastructure resiliency, 61, 62
Innovation across entities, 404
Inorganic measurement functions, 369
Input/output process (I/O process), 52, 257
Instantiation, *see* VNF provisioning
Instruction detection system (IDS), 173
Integrated routing and bridging, *see* Default gateway IP
        routing
Intel DPDK, 213–214
Intellectual property friction (IP friction), 392, 394

Intelligent agents, 167–168
Intelligent Routing Service Control Platform (IRSCP), 6
Intel x8 6-based server technology, 203–204
Interactive Voice Response technology
        (IVR technology), 5–6
Interior Gateway Protocol (IGP), 277
International fiber network, 267
International Telecommunications Union (ITU), 9–10
Internet, 1–2, 7, 172
    access, 22
    broadband service, 252
    cloud data, 140
    companies, 143
    peering, 252
    tunnel gateways, 252–253
Internet engineering task force (IETF), 11, 215, 371
Internet of Things (IoT), 2, 167–168, 244, 326
Internet Protocol (IP), 1, 10, 25–26
    address, 193
    CBB, 276–277
    IP-edge routers, 322–323
    IP-over-WDM architecture, 363
    IP-routing protocols, 26
    IP/optical network with SDN controller, 356–357
    IPv4 internet service, 286–287
    modern IP network, 11–12
    MPLS technology, 252
    network, 293
    rapid transition to ALL-IP, 10
    routing, 26
    transforming ALL-IP network to network cloud, 12–24
    virtual network connections using IP tunnel, 27
Internet protocol multicast initiative (IPMI), 375
Internet Protocol Security (IPSec), 34
Internet security scrubbers, 253
Internet service provider (ISP), 82, 143
Interoffice rerouting, 259
Interoperability, 397–398
"Interoperable" APIs, 398
Interpretation, 191
Interrogating CSCF (I-CSCF), 311–312
Interrupt request (IRQ), 63
Interworking (IWF), 314
Interworking session border controller (I-SBC), 312
Intrusion detection system (IDS), 202
Intrusion prevention system (IPS), 188, 202
IO&C, see Infrastructure orchestration and control
I/O process, see Input/output process
IoT, see Internet of Things
IP, see Internet Protocol
IP friction, see Intellectual property friction
IPMI, see Internet protocol multicast initiative
IP multimedia subsystem (IMS), 10, 81, 293
    access layer, 312
    current implementation, 310
    decomposition of functions, 314
    decoupling of long-lived state, 314
    decoupling of subscriber data, 314
    georesiliency, topology, and scalability, 316–317
    hybrid architecture, 318
    integration with ONAP, 316
    key design principles applied, 314
    layered architecture, 311
    licensing, 315

    maintaining quality of service, 315
    National Subscriber Provisioning, 315
    NPaaS services, 315
    performance, 317–318
    resiliency, 315
    scalability at VNF level, 315
    SDN/NFV implementation, 312–314
    service platform, 310
    smaller failure domains, 315
    VNF catalog, 315–316
IPS, see Intrusion prevention system
IPSec, see Internet Protocol Security
IRQ, see Interrupt request
IRSCP, see Intelligent Routing Service Control Platform
I-SBC, see Interworking session border controller
Isolation, 186–187
ISP, see Internet service provider
IT, see Information technology
ITIL standard, see Information Technology Infrastructure
        Library standard
ITU, see International Telecommunications Union
ITU-T, see Telecommunication Standardization Sector of
        International Telecommunications Union
IVR technology, see Interactive Voice Response
        technology
IWF, see Interworking
IWSF, see IMS web security function

J

Java messaging service (JMS), 77
JFlow, 362
JMS, see Java messaging service
JSON, 145, 378, 391

K

Kafka, 144
    KAFKA-based open source, 155
Kernel-based virtual machine (KVM), 16
    hypervisor, 206
Key capacity indicators (KCIs), 379, 383–384
Key performance indicators (KPIs), 75, 107, 149, 365
    VNF, 380–381
"Kiddie scripting" languages, 396

L

Label distribution protocol (LDP), 19, 278
Label portion, 18
Label switched path (LSP), 277–279, 355
LAG, see Link aggregation group
Lambda architecture and policies, 142–143
LAN, see Local area network
Large-scale measurement of broadband performance
        framework (LMAP framework), 371
"Last mile," 17, 229
Latency bounds, 380
Lawful intercept gateway (LIG), 312
Layer 4 stateful control functions, 81–82
Layered protocols, 2
LBaaS, see Load balancer as a service
LCP, see Local control plane
LDP, see Label distribution protocol

Legacy BSSs interactions with ONAP, 133–135
Libraries for Machine Learning (MLlib), 147
Licensing, 315
LIG, *see* Lawful intercept gateway
Link aggregation group (LAG), 260, 283
Linux, 395–396
    containers, 206, 219
LMAP framework, *see* Large-scale measurement of
        broadband performance framework
Load balancer as a service (LBaaS), 303
Load balancers, 253
Local area network (LAN), 57, 77
Local control plane (LCP), 50–51
Local redundancy with no site failover, 79
Logical dimension, 251
Logical unit number (LUN), 57
Long-haul networks, 267
Long-lived state, decoupling of, 314
Long-range network planning, 332
Long-term evolution (LTE), 7–8, 10, 294
Low software failure rates, 77
    VF error handling, 77
    VF failure detection and alerting, 78
    VF fast recovery, 78
    VF protection from external services, 77
    VF software fault detection, 77–78
    VF software stability, 78
Low volume message producers, 373
LSP, *see* Label switched path
LTE, *see* Long-term evolution
LUN, *see* Logical unit number

M

MaaS, *see* Mobility as a service
MAC, *see* Media access control
Machine learning (ML), 160, 346, 401
    analytics and, 148
    for closed-loop automation, 160–161
    deep learning, 150–151
    descriptive *vs.* reactive *vs.* predictive *vs.* prescriptive
        analytics, 149
    open source distributed processing toolkits for
        ML, 151
    speech recognition, 148–149
    supervised *vs.* unsupervised learning, 149–150
    toolkit comparisons, 152
Maintenance operations model, 334
MAN, *see* Metropolitan area network
Management and orchestration (MANO), 15, 109–111
Management plane, 35–37
MANO, *see* Management and orchestration
MapReduce, 146
Markov decision process, 162
Markov hidden model, 148
Massive IOT traffic, 247–248
Massive machine type communication (Massive MTC),
        244–245
Master reference catalog, 115
Master Service Orchestrator (MSO), 107, 111–112
    comparison of ETSI MANO and ONAP
        architectures, 110
    comparison of MSO and controllers, 112
MCF mechanism, *see* Multicommodity flow mechanism

MDUs, *see* Multi-dwelling units
Mean time between failure (MTBF), 40, 69
Media access control (MAC), 209, 221
Media gateway (MGW), 312
Media gateway control function (MGCF), 312
Media resource function (MRF), 312
Meeting zone demand, 61
Memory
    manager, 213
    resources, 214
Merchant silicon, 234, 255
    competitors, 255
    packet processors, 262
Metadata-driven design time and runtime execution,
        112–115
Metcalfe's Law, 25
Method of procedure (MOP), 342
Metro Ethernet service, 252
Metropolitan area network (MAN), 36, 92–93
MGCF, *see* Media gateway control function
MGW, *see* Media gateway
Micro-batch processing using STORM and SPARK,
        146–147
Micro-loop, 279–280
MicroOLT, 235
Middle box, 202
    model, 219
    VNFs, 40–41
MIMO, *see* Multiple input multiple output
Mitigation strategies, 186
ML, *see* Machine learning
MLlib, *see* Libraries for Machine Learning
MME, *see* Mobility management entities
MMS, *see* Multimedia messaging service
mmWave
    frequencies, 247
    spectrum, 246
Mobile
    cellular services, 252–253
    networks, 1
Mobile wireless access technologies, 241
    5G wireless, 244–248
    LTE RAN configurations, 242–244
Mobility, 50
    3GPP standards organization, 81
    connectivity to mobility VPN, 322–323
    gateway functions, 253
    legacy networks, 322–323
Mobility as a service (MaaS), 328
Mobility management entities (MME), 13, 318,
        319–321
Modern IP network, 11–12
    OSI model, 11
    regulation and standards, 11–12
    services, 20
Modern telecom network transformation, 9
    enterprise CPE, 16–17
    network access, 17
    network core, 18–19
    network data and measurements, 22–23
    network edge, 17–18
    network operations, 23–24
    network security, 15–16
    NFV, 13

Modern telecom network transformation (*Continued*)
  NFVI, 13–14
  ONAP, 15
  rapid transition to ALL-IP, 10
  SDN, 14
  servi ce platforms, 19–22
  transformation to network cloud, 12
  transforming ALL-IP network to network cloud, 12
Modern wireless access networks, 225
Modularity, 270
Monitoring, capacity and scaling design, 323
Moore's law, 227, 228, 254
MOP, *see* Method of procedure
MP-BGP, *see* Multiprotocol Border Gateway Protocol
MPLS, *see* Multiple Protocol Label Switching
MRF, *see* Media resource function
MS design paradigm, 156–157
MSO, *see* Master Service Orchestrator
MTBF, *see* Mean time between failure
M2X, 167–168
Multicommodity flow mechanism (MCF mechanism), 358
Multics, 391
Multi-dwelling units (MDUs), 229
Multilayer control, 288–290
Multimedia, 294
  session border controllers for multimedia services, 253
Multimedia messaging service (MMS), 398
Multipath VNF designs, 41–42
Multiple input multiple output (MIMO), 245
Multiple Protocol Label Switching (MPLS), 14, 32, 252, 265
  core router technology evolution, 286
  distributed traffic engineering, 282–283
  evolution of, 277
  FRR and hitless rearrangements, 280–282
  IP CBB, 276–277
  MPLS-based packet layer, 266
  MPLS-based VPN service, 34
  MPLS transport, 277–280
  packet layer, 276
  route reflection, 286–287
  segment routing, 283–286
  services, 5
  virtual private network service, 22
Multiplexing, 268
Multiprocessing, 28
Multiprotocol Border Gateway Protocol (MP-BGP), 6
Multisite design, stateless network function with, 82–83
Multistep control loop policy, 344
Multitasking, 28
Multi tenancy, 39–40
Multithreading, 28

N

NAS, *see* Network attached storage
NAT, *see* Network address translation
"National Subscriber Provisioning Distribution" model, 315
Native measurement functions, 369
NB FW, *see* Northbound FW
NBI, *see* Northbound interface
NC, *see* Network controller
NCP, *see* Network control point
Near-real time, 47
NEL, *see* Network element layer

NETCONF, *see* Network Configuration
Netflix, 78
NetFlow, 362
Network access, 17, 223–226
  extending NFV and SDN to, 226–229
  mobile wireless access technologies, 241–248
  wireline access technologies, 229–241
Network address translation (NAT), 37, 41, 168, 182, 253
Network attached storage (NAS), 221
Network capacity planning, 360
  benefits from SDN, 361–362
  current network capacity planning process, 360–361
  layer 0 and layer 3 resources, 363–364
  traffic forecast, 362–363
  traffic matrix data, 362
  wavelength circuits, 363
Network cloud, 10–11, 32, 46, 72, 75–76, 138, 149;
        see also Software-defined networking (SDN)
  big data and, 151
  bootstrapping SDN /NFV deployments leveraging, 348
  BSSs, 347
  control-loop automation, 157–160
  DCAE, 153–156
  deep learning and SDN, 161–162
  design principles, 151–153
  disaggregated edge platforms, 253
  edge platforms, 253
  edge vPE VNF, 256–258
  elastic infrastructure, 184–185
  enterprise CPE, 16–17
  environment, 264
  georedundancy, 72–75
  impact of planned downtime, 72–75
  infrastructure, 68
  layers of network cloud, 68
  legacy and SDN /NFV networks coexisting in
        harmony, 348–349
  migrating to, 347
  MLs for closed-loop automation, 160–161
  MS design paradigm, 156–157
  network access, 17
  network cloud infrastructure availability, 69–72
  network core, 18–19
  network data and measurements, 22–23
  network edge, 17–18
  network fabric, 253–256
  network operations, 23–24
  network security, 15–16
  NFV, 13
  NFVI, 13–14
  ONAP, 15
  policies, 115
  rolling out network cloud technologies, 348
  SDN, 14
  service platforms, 19–22
  shared infrastructure, 185
  site, 80
  transformation to network cloud, 12
  transforming ALL-IP network to, 12
  VF classification and examples, 79–83, 84
  VF software design, 75–79
Network cloud infrastructure availability
  to cost tradeoff, 71–72
  single-site availability, 69–71

Network Configuration (NETCONF), 36, 233, 241
Network controller (NC), 88, 99–100, 240
   abstraction modeling, 94–95
   adapters, 93
   ANDSL, 95–96
   API handler, 93
   architecture, 90
   compiler function, 91
   data collection analytic and events function, 93–94
   federation between network controllers, 94
   high availability and geo-diversity, 94
   network controller software components, 91–94
   network resource autonomous control function, 92–93
   orchestration, 117
   policy, 93
   relationship to application service controllers, 94
   SLI function, 92
   software components, 91
   software validation, 94
Network control point (NCP), 5–6, 90–91
Network core, 18–19, 265
   migration to converged network core, 266
   MPLS packet layer, 276–287
   optical layer, 266–276
   SDN control of packet/optical core, 287–291
Network data
   and analytics layer, 138–140
   applications, 162–168
   big data, 140–151
   big data meets Network Cloud, 151–162
   customer configurable policy for content filtering smart
      network, 163
   and measurements, 22–23
   and optimization, 137
   SON, 162–163, 164, 165
   traffic shaping, 163–167
   utilizing SDN to minimizing robocalling, 167–168
Network design, 322
   connectivity to mobility VPN and mobility legacy
      networks, 322–323
   data modeling for, 215–216
   methodology, 4
   overlay network design, 322
Network device abstraction model, 95
Network edge, 17–18, 249; see also Network access
   core paradigm, 249–250
   flexible access grooming and universal cloud overlay,
      258–261, 262
   future evolution, 261
   network cloud edge platforms, 253–258
   open configuration and programing of packet
      processors, 262–263
   open control of packet processors, 263–264
   open packet processors, 262
   traditional edge platforms, 250–253
Network element layer (NEL), 374
Network equipment, see Networking hardware
Network Functions on Demand (NFoD), 217–218
Network functions virtualization (NFV), 7, 13, 25, 46–47,
      50, 68, 105, 158, 171, 224
   benefits, 29–31
   data and measurements, 373
   data measurement framework, 377–379
   data model, 374

   decomposition of VNFs, 32–40
   design enhancements, 173–174
   economics, 44–46
   EML, 374
   extension, 225–226
   impact, 228–229
   and impact on operations, 335–337
   infrastructure telemetry data model, 374–377
   MANO, 109–111
   NFV-related challenges, 337–338
   performance improvements, 175
   real-time capabilities, 175–176
   resiliency and scaling, 40–44
   and SDN, 31–32
   security advantages of, 173–176
   service auto-scaling, 382
   techniques, 241–242
Network functions virtualization infrastructure (NFVI),
      12, 13–14, 32, 50; see also Software-defined
      networking (SDN)
   agile, 58–59
   application resiliency, 62–63
   build, 60
   building NFVI solution, 58
   CI/CD, 59
   components of, 51
   deployment, 61
   design, 59–60
   DevOps, 59
   fault tolerance, 61–62
   infrastructure resiliency, 62
   innovation and integration, 59
   integrating, 60
   leveraging NFVI for VNFs, 63–65
   managing, 60
   meeting zone demand, 61
   operating, 60–61
   operational changes, 58
   physical components, 51–53
   service chain on, 38
   VIM, 53–58
   VNF management, 65
Networking
   functionality, 13
   hardware, 52
Network interface card (NIC), 34, 63
Network interface controllers (NICs), 52
Network management systems (NMS), 105, 214
   Call Home function, 215
   data modeling for network design, 215–216
   VF deployment and management, 216–217
Network measurement reports (NMRs), 139
Network measurements, 365
   AT&T's SDN-Mon framework, 370–372
   network capacity planning, 360–364
   NFV data and measurements, 373–379
   optimization algorithm improvements, 385
   real-time network data with SDN, 356–360
   SDN controller measurement framework, 364–370
   SDN controller resiliency, 385–386
   SDN data and measurements, 354–356
   telemetry measurements, 372–373
   VNF efficiency measurements and KCI reporting,
      383–384

Network measurements (*Continued*)
  VNF measurements for optimal placement and sizing, 384–385
  VNF reporting metrics, 379–381
  VNF scaling measurements, 381–383
Network monitoring, 340
Network on demand solution (NOD solution), 50, 232
Network operating system (NOS), 256
  A&AI system, 128–130
  control loop systems, 130–133
  DCAE, 119–123
  legacy BSSs interactions with ONAP, 133–135
  MSO, 111–112
  ONAP, 105–111
  policy engine, 123–128
  portal, reporting, GUI, and dashboard functions, 117–119
  SDC, 112–116
  software-defined controllers, 116–117
  VNF, 104
Network operations, 23–24, 331
  AT&T, 333, 334
  migrating to network cloud, 347–349
  impact of NFV and SDN on, 335–346
  ONAP-related challenges, 343–346
  operations and engineering teams, 333–335
  operations team transformation, 346–347
  performance requirements, 332
  role of automation, 333
  role ONAP in, 338–343
Network platform as a service layer (NPaaS layer), 302–303, 315
Network(s), 52, 265; *see also* Open network automation platform (ONAP)
  abstraction layer, 215, 238–239
  administration, 332
  administrators, 202
  appliance, 28
  bonding service, 253
  complexity evolution, 202–203
  congestion, 163–167
  connectivity, 2, 94
  control, 88–89
  deployment, 372
  elements, 182
  fabric, 253–256
  feature delivery, 89–90
  function, 3–4, 32, 346, 347
  function virtualization, 28–29
  network-based firewalls, 253
  network-based proxies, 253
  operators, 3
  optimization, 272–273
  perimeter, 198–199
  provisioning, 332
  resource autonomous control function, 92–93
  resource control function, 92
  virtualization, 26–27, 186
Network security, 15–16, 171, 182–185
  advantages of SDN and NFV, 173–176
  architecture, 177–189
  challenges, 176, 177
  components of hardening, 199–200
  future research and development, 199

  perimeter model, 172
  platforms, 189–199
Network service(s), 88, 89
  abstraction model, 94–95
  implementation abstraction model, 95
  order-driven process, 92
Network terminating equipment (NTE), 232–233
Network time protocol (NTP), 253
Neutron, 54–55
Next Generation Mobile Network Alliance, 162
Next Generation x Core (NGxC), 328
"Next Radio," 241
NFoD, *see* Network Functions on Demand
NFV, *see* Network functions virtualization
NFVI, *see* Network functions virtualization infrastructure
NGPON2 technology, 230
NGxC, *see* Next Generation x Core
NIC, *see* Network interface card
NICs, *see* Network interface controllers
NMRs, *see* Network measurement reports
NMS, *see* Network management systems
NOD solution, *see* Network on demand solution
Nonreal time, 47
Nonrelational databases, 145
Nonuniform memory access (NUMA), 317, 383
Northbound FW (NB FW), 180
Northbound interface (NBI), 96
NOS, *see* Network operating system
NoSql cloud databases, 296
Not only Structured Query Language (SQL) (NoSQL), 145
NPaaS layer, *see* Network platform as a service layer
NTE, *see* Network terminating equipment
NTP, *see* Network time protocol
NUMA, *see* Nonuniform memory access

O

OA&M, *see* Operations, administration, and maintenance
OA, *see* Optical amplifier
Object-based storage; *see also* Object storage
Object storage, 53
OCP, *see* Open Compute Project
ODL, *see* Open Daylight
ODN, *see* Optical distribution network
OEMs, *see* Original equipment manufacturers
OEO conversions, *see* Optical–electrical–optical conversions
OFDM, *see* Orthogonal frequency division multiplexing
OLT, *see* Optical line terminal
OMF, *see* Operational Management Framework; Orchestration and management function
ONAP, *see* Open network automation platform
On demand services, 273
1G mobility system, 1–2
ONIE, *see* Open network install environment
ONOS, 101
Open-loop systems, 158
Open-O, 104
Open-Source Foundations, 395
Open Access Language (OpenAL), 228
Open access network software, 240–241
OpenAL, *see* Open Access Language
OpenCL, 228
Open Compute Networking Project, 234

Open Compute Project (OCP), 234, 235, 255
Open configuration, 262–263
OpenContrail, 39, 101
Open control of packet processors, 263–264
Open Daylight (ODL), 14, 101, 391, 395–396
"Open DNS," 140
OpenFlow, 263, 355
OpenGL, 228
Open hardware specifications, 240
Open innovation, 401
Open interfaces, 240
Open network automation platform (ONAP), 6–7, 14, 15,
        46, 104, 105, 106–107, 177; *see also* Network(s)
    component roles, 107–109
    configuring ONAP to enabling operations
            automation, 343
    conflicting control loop policies, 345
    control loops, 340–342
    deploying and deleting VNFs, 339–340
    ETSI, 109–111
    identifying operations policies, 346
    increasing software complexity, 346
    integration with, 316, 324
    legacy BSSs interactions with, 133–135
    network and service monitoring, 340
    network cloud environment, 105–106
    ONAP-related challenges, 343–346
    policy life cycle, 344
    policy testing and automated validation
            techniques, 345
    role, 338–343
    safe policies, 343–345
    security, 187–189
    service design, 306
    VNF change management, 342–343
Open network install environment (ONIE), 256
Open packet processors, 262
Open Platform for NFV (OPNFV), 391, 395–396
OpenROADM-based devices, 395
Open ROADM, *see* Reconfigurable add-drop multiplexer
        (ROADM)
OpenROADM.org specifications, 275
Open shortest path first (OSPF), 35, 277*fn*
Open software
    components, 240–241
    model, 221
Open Source, 392
    GitHub, 393
    IP friction, 392
    licensing regimes, 394–395
    OpenStack, 394
    pure, 54–55
    SDN controllers, 101
    simplest form of open-source engagement, 393
    software, 218, 240
    technologies, 7, 206
Open specifications, 234–235
OpenStack, 50–51, 53–54, 394–396
    environment, 14
    neutron, 38
    neutron plugin, 322
    Object Store project, 58
    software, 177–178
    supported KVM hypervisor, 304

OpenStack resource manager (ORM), 60, 61
OpenStack Telemetry Service
    Ceilometer, 374–377
Open standard model, 251
Open Systems for Interconnection model (OSI model), 11,
        12, 13
Open virtual switch (OVS), 209
    OVS-based approach, 210–211
Open vOLT hardware specifications, 235–237
Open XGS-PON 1RU vOLT, 235
Open XGS-PON 4-port remote vOLT, 237
Operating expenditure (OpEx), 105, 335, 365
Operating system (OS), 25, 72
    operating system-level virtualization, 56
    security, 185–187
Operational costs, 45–46
Operational functions, 19
Operational Management Framework (OMF), 104
Operational security, 183
Operational VNF operational metrics, 379–380
Operations, administration, and maintenance (OA&M),
        35, 109, 182, 249, 314, 372
Operations support systems (OSS), 13, 15, 30, 61, 95, 104,
        195, 229, 249–250, 275, 333
    OSS/BSS orchestration layer, 233
    OSS/BSS systems, 15
Operations team transformation, 346–347
OpEx, *see* Operating expenditure
OPNFV, *see* Open Platform for NFV
Optical amplifier (OA), 18, 268
Optical core, SDN control of, 287–291
Optical distribution network (ODN), 231
Optical–electrical–optical conversions
        (OEO conversions), 269
Optical layer, 266
    AT&T domestic 100G long haul optical backbone, 267
    expressing wavelengths, 271
    fixed optical transport system, 269
    flexible software-controlled optical networks, 272–273
    future work in, 275–276
    100G pluggable form factors, 271
    Open ROADMs, 273–275
    optical fiber types deployed in terrestrial networks, 267
    optical technologies, 268
    ROADMs, 270
Optical line terminal (OLT), 229, 259
    functionality, 235
    hardware designs, 234
Optical transport layer, 266
Optimal placement and sizing, VNF measurements for,
        384–385
Optimization algorithm
    implementation, 291
    improvements, 385
Orchestration, 99–100, 107, 131–132, 214–217
Orchestration and management function (OMF), 99
Orchestrator, 58, 304
Organic measurement functions, 369
Organic NFV data measurement model, 377–378
Original equipment manufacturers (OEMs), 234, 254, 270
ORM, *see* OpenStack resource manager
Orthogonal frequency division multiplexing (OFDM), 232
OS, *see* Operating system
OSI model, *see* Open Systems for Interconnection model

OSP, *see* Outside plant
OSPF, *see* Open shortest path first
OSS, *see* Operations support systems
Outside plant (OSP), 229
    passive OSP, 231
Overbooking resources, 380
Overlay network, 27, 38–39, 322
Overload controls, 16
OVS, *see* Open virtual switch

P

PaaS, *see* Platform as a service
Packet-switched (PS), 294
Packet core, 18
    advantages of virtualizing EPC, 318–319
    capacity and scaling design, 323–324
    evolving, 318
    integration with ONAP, 324
    network design, 322–323
    SDN control of, 287–291
    virtual logical platform architecture and design, 319–322
Packet data network (PDN), 318, 319
Packet delivery network gateway (PGW), *see* Packet
        gateway (P-GW)
Packet edge
    paradigm, 250
    platforms, 249–250
Packet forwarding, 34
    under normal conditions, 283–284
Packet gateway (P-GW), 253, 318, 319
    VNF components, 319–320
Packet networks, 17
Packet processing capabilities, 207–208
Packet processors
    open, 262
    open control of, 263–264
    programing of, 262–263
PAD, *see* Personalized, adaptive, and dynamic
Painstaking process, 263
Partitioning strategies, 186
Party interaction, 20
Passive optical networks (PONs), 229, 252
    PON-specific functions, 235
    technology, 229–232
Passive OSP, 231
Passive VM replication, 77
Passive vProbe measurement data model, 378–379
Path Computation Element Protocol signaling (PCEP
        signaling), 100, 285–286, 288, 355
PBX, *see* Private Branch Exchange
PCI, *see* Peripheral component interconnect
PCIe device, *see* Peripheral component interconnect
        express device
PCM AS, *see* Personal communication manager AS
PCRF, *see* Policy and charging rules function
PDN, *see* Packet data network
PDP, *see* Policy decision point
PDU, *see* Protocol data unit
PE, *see* Provider edge
Peering, 18
Peer-to-peer interactions, 94
PER, *see* Provider-managed edge router
Performability, 69

Performance, 296, 317–318
    metrics, 332
    surveillance, 123
    VNFs performance profiles, 63–64
Perimeter, 184
    network model, 172
Periodic fork-lifting CPU router cards, 251
Peripheral component interconnect (PCI), 34
Peripheral component interconnect express device
        (PCIe device), 63
Peripherals, 226, 227
Permissive Licenses, 394
Persistent management agent (PMA), 233
Personal communication manager AS (PCM AS), 311
Personal communication services, 1
Personalized, adaptive, and dynamic (PAD), 158
PF, *see* Physical function
P4, *see* Programming protocol-independent packet processors
P-GW, *see* Packet gateway
Phone Home
    model, 221
    Service, *see* Call Home service
    uCPE Phone Home Process, 221
Physical components, NFVI, 51
    compute, 52
    network, 52
    storage, 52–53
Physical function (PF), 212
Physical layers (PHY layers), 235, 338
Physical network functions (PNF), 13, 15, 88, 257–258,
        295, 318, 346
Physical network interface cards (pNICs), 208
Physical resource blocks (PRBs), 165
Picocells, *see* Customer premise-based small cells
PIG, 146
Planned downtime, impact of, 72
    design practices for minimizing, 74–75
    example of, 72–74
Platform as a service (PaaS), 399
P leaf, 286
PLMN, *see* Public land mobile network
Plug-and-Play model (PnP model), 221
Pluggable optics, 271–272
PMA, *see* Persistent management agent
PMA Aggregator (PMAA), 233
PMO, *see* Present mode of operation
PNF, *see* Physical network functions
pNICs, *see* Physical network interface cards
PnP model, *see* Plug-and-Play model
Point of local repair, 281
Policy, 91, 93
    control VNF components, 320
    creation, 124, 125
    decision and enforcement, 126–127
    decision distribution, 124
    distribution, 126
    evaluation, 124
    repository, 116
    rules, 14
    technologies, 127–128
    testing, 345
    unification and organization, 127
    validation, 124
Policy and charging rules function (PCRF), 318, 320

Policy decision point (PDP), 115
Policy engine, 123
    policy creation, 125
    policy decision and enforcement, 126–127
    policy distribution, 126
    policy technologies, 127–128
    policy unification and organization, 127
    policy use, 128
PONs, *see* Passive optical networks
Port, 251
Portal functions, 117–119
Post-SDN, 361–362
PRBs, *see* Physical resource blocks
Preboot execution environment (PXE), 256
Predictive analytics, 149
Prescriptive analytics, 149
Pre-SDN environment, 361–363
Present mode of operation (PMO), 90, 295–296
Private Branch Exchange (PBX), 5–6
Privilege escalation detection, 199
Process modeling tools, 115
Processor affinity, 63–64, 213
Process repository, 115
Process specifications, 106
"Product catalogs," 306
Programing of packet processors, 262–263
Programmability, 202
Programming protocol-independent packet processors
    (P4), 263
Protocol abstraction layer, 239
Protocol data unit (PDU), 260
Provider-managed edge router (PER), 29, 80, 252
Provider (P), 276
Provider edge (PE), 17, 249, 276
PS, *see* Packet-switched
PSTN, *see* Public switched telephone network
Psuedowires, *see* Ethernet virtual circuits
Public key
    encryption, 20
    functions, 214
Public land mobile network (PLMN), 294
Public networks, 2–3
Public switched telephone network (PSTN), 294
Public Telecom network, 10
Pure open source, 54–55
PXE, *see* Preboot execution environment
Python project, 394

Q

Quadrature amplitude modulation (QAM), 272
Quality of service (QoS), 18, 34, 226, 294, 362
    maintaining, 315
Quality of Service Class Identifier (QCI), 319
Queue manager, 213
QuickAssist, 214

R

Radio access network (RAN), 139, 241–242, 334
Radio frequency (RF), 242
Radio frequency integrated circuits (RFICs), 245
RAID, *see* Redundant array of independent disks
RAN, *see* Radio access network

RBAC, *see* Risk-based access control
RCS, *see* Rich Communication Services
RDD, *see* Resilient distributed datasets
Reactive analytics, 149
Real-time, 46
    analytics, 23
    centralized TE using SDN controller, 358
    control, 23
    data with SDN, 356
    dynamically managing and reconfiguring
        mapping, 359
    IP/optical network with SDN controller, 356–357
    measuring data, 357–358
    optimization, 23
    packing efficiency of distributed *vs.* centralized
        routing, 358
    processing using STORM and SPARK, 146–147
    use of available spare capacity, 359–360
Real-time media control protocol (RTCP), 20
Real-time transport protocol (RTP), 303
Receive queues (RX queues), 211
Reconfigurable optical add/drop multiplexor (ROADM),
    18, 269–270, 273–275, 338, 354, 355, 359
Redundant array of independent disks (RAID), 62
Refactoring, 391
Regressions, 149
Regulation, 11–12
Relationships, 186–187
Reliability, 69
    analysis, 82
Remote procedure calls (RPC), 36, 96, 366, 391
Reporting functions, 117–119
Request for proposals (RFPs), 3–4
Resiliency, 69, 259–260, 315, 324
    multipath and distributed VNF designs, 41–42
    NFV, 40
    VNF resiliency reporting, 380–381
    vPE example for VNF-specific resiliency design, 42–44
Resilient, 56
Resilient distributed datasets (RDD), 147
Resource
    identity, 193
    images, 116
Resource Reservation Protocol (RSVP), 280
Resource Reservation Protocol for Traffic Engineering
    (RSVP-TE), 19, 285–286
RESTCONF protocol, 371
RESTful APIs, 391, 397–398
Restoration, 272
Reusability, 39
RF, *see* Radio frequency
RFICs, *see* Radio frequency integrated circuits
RFPs, *see* Request for proposals
Rich Communication Services (RCS), 293
Risk-based access control (RBAC), 193
RNC, *see* ROADM Node Controller
ROADM, *see* Reconfigurable optical add/drop multiplexor
ROADM Node Controller (RNC), 289
Robocalling, 167
Role-based-access control function, 185
Rolling out network cloud technologies, 348
Router, 28, 52, 219
Route reflection, 286–287
Route reflector (RR), 19, 32, 265–266, 287

Routing functions, 17–18
RPC, *see* Remote procedure calls
RR, *see* Route reflector
RSVP, *see* Resource Reservation Protocol
RSVP-TE, *see* Resource Reservation Protocol for Traffic
    Engineering
RTCP, *see* Realtime media control protocol
RTP, *see* Real-time transport protocol

S

SaaS, *see* Software as a service
SAE gateway, *see* Service architecture evolution gateway
SAEGW, *see* System Architecture Evolution Gateway
Safe policies, 343–345
SAI, *see* Switch Abstraction Interface
SAN, *see* Storage-area network
SBCs, *see* Session border controllers
SB FW, *see* Southbound FW
SC, *see* Service controller
Scalability, 64–65, 316–317, 323
    at VNF level, 315
Scale out, 383
Scaling
    design, 323
    monitoring, 323
    multipath and distributed VNF designs, 41–42
    NFV, 40
    resiliency, 324
    scalability, 323
    services, 58
    triggers, 382–383
    vPE example for VNF-specific resiliency design,
        42–44
Scaling out/in, *see* Horizontal scaling
Scaling up/down, *see* Vertical scaling
SCC-AS, *see* Service centralization and continuity
Scripting, 396
S-CSCF, *see* Serving CSCF
SDC, *see* Service Design and Creation
SDKs, *see* Software development kits
SDLC, *see* Software development lifecycle
SDN, *see* Software-defined networking
SDNI, *see* Software defined network infrastructure
SDSF, *see* Software-defined service framework
SD-WAN, *see* Software defined-WAN
Searching on Lucene Replication (SOLR), 145–146
Security, 167–168
    ASTRA, 194–199
    by design approach, 187
    IAM, 191–193
    modules, 191
    of platform, 187, 189
    security analytics, 189–191
Security analytics, 189
    components, 190–191
    fundamental functions, 189–190
Security architecture, 177
    AIC security evolution, 180–182
    cloud security, 177–180
    hypervisor and operating system security, 185–187
    network and application security, 182–185
    ONAP security, 187–189
    security controls, 177

Segment routing (SR), 19, 283
    benefits, 285
    explicit routing, 284–285
    faster restoration using segment routing, 285
    packet forwarding under normal conditions, 283–284
    segment routing *vs.* RSVP-TE, 285–286
Self-backhaul, 247
Self-optimizing networks (SONs), 160, 162–165, 349
Semi-supervised techniques, 149
SendMessage function, 95
Server, 52
    hypervisor, 207
Service
    abstraction layer, 215
    assist mechanism, 5–6
    catalog, 306
    chaining, 37–38, 58, 168, 183
    coordination and instantiation, 58
    creation, 306
    decomposition, 63
    design methodology, 307
    images, 116
    logic, 5
    management centers, 334
    measurements, 365
    monitoring, 58, 340
    and network measurements, 368, 369
    reliability, 69
Service-level agreements (SLAs), 34, 93, 317, 364, 398
    VNF, 380–381
Service-oriented architecture (SOA), 33
Service-prover-based IP VPN service, 252
Service-provider-based IP VPN service, 252
Service architecture evolution gateway (SAE gateway), 29
Service centralization and continuity (SCC-AS), 312
Service controller (SC), 99
Service Design and Creation (SDC), 92, 107, 112
    certification studio, 116
    data repositories, 115–116
    distribution studio, 116
    function, 93
    metadata-driven design time and runtime execution,
        112–115
    module, 104
Service level objective (SLO), 368
Service logic interpreter (SLI), 14, 92
Service platforms, 19–22, 293
    architecting for SDN/NFV, 297
    architectural for NGxC, 328
    architecture, 295
    distributed placement of IoT and novel services, 327
    pivot to SDN/NFV, 306–310
    real-time services over IP, 295
    service design solutions with SDN/NFV, 298–305
    3G networks, 294
    virtualization technology, 296
    virtualized service platform use cases, 310–326
Service providers, 206, 379
    networks, 250
    service provider-based NAT, 253
Service quality management (SQM), 7, 339, 341
Serving CSCF (S-CSCF), 311–312
Serving gateway (SGW), 253, 318, 319–320
Session border controllers (SBCs), 310

Session border controllers for multimedia services, 253
Session Initiation Protocol (SIP), 20, 46, 298
Set function, 95
7Vs, *see* Volume, Velocity, Variety, Variability, Validity/
        veracity, Visibility/visualization, Value
SFV, 194
SGW, *see* Serving gateway
Shared network services layer (SNS layer), 302
Shared risk link groups (SRLGs), 281, 363
Shift to software, 389, 390
    agile method, 396–397
    apache foundations, 395–396
    from birth of "cloud" to containers and microservices,
        398–399
    building blocks, 390
    design patterns, 390
    DevOps, 396–397
    interoperability, 397–398
    linux, 395–396
    ODL, 395–396
    open source, 392–395
    OpenStack, 395–396
    OPNFV, 395–396
    RESTful API exposure, 397–398
    scripting and concurrent languages, 396
    vendor-proprietary hardware-based products, 389
    viral nature of Unix and C, 391–392
    Web 2.0, 397–398
    yin and yang of agile development, 397
Short messaging services (SMS), 294
Signaling
    gateways, 76
    protocols, 20
Signal processing workloads, 36
Signature, 344
    signature-based detection, 190
SIM, *see* Subscriber identification module
Simple control loop policy, 344
Simple Network Management Protocol (SNMP), 35, 158,
        217, 355
Simplification, 255
Single-site availability, 69–71
Single network function, 205
Single root I/O virtualization (SRIOV), 34, 63, 208
Single tenancy, 39–40
SIP, *see* Session Initiation Protocol
16 Port OLT Pizza Box, 235
SKU, *see* Stock keeping unit
SLAs, *see* Service-level agreements
SLI, *see* Service logic interpreter
SLO, *see* Service level objective
Small cells, 247
Smaller failure domains, 315
Small form-factor (SFP), 235
SMS, *see* Short messaging services
SNMP, *see* Simple Network Management Protocol
SNS layer, *see* Shared network services layer
SOA, *see* Service-oriented architecture
SOC, *see* System on a chip
Software-centric
    model approach, 203
    network, 4
Software-defined controllers, 116–117
    application controller orchestration, 117

infrastructure controller orchestration, 116–117
    network controller orchestration, 117
Software-defined networking (SDN), 4, 6, 12, 14, 31–32,
        61, 87, 105, 117–119, 171, 180, 182, 223, 354;
        see also Network cloud; Network functions
        virtualization infrastructure (NFVI)
    access controller, 240
    architecture, 201
    AT&T's SDN-Mon framework, 370–372
    centralized TE, 287–288
    component-level measurements, 367–368
    control layer, 266
    controller, 5
    controller measurement framework, 364
    controller resiliency, 385–386
    control of packet/optical core, 287
    data and measurements, 354–356
    decoupling service and network path measurements,
        368–370
    deep learning and, 161–162
    design enhancements, 173–174
    functional overview, 87
    implementing network control, 88–89
    measurement framework, 366–367
    multilayer control, 288–290
    NC and orchestration use-case example, 99–100
    network architecture, 355
    network control, 88
    network controller architecture, 90–96
    network elements, 366
    objectives for measurement framework, 365–366
    open source SDN controllers, 101
    optimization algorithm implementation, 291
    paradigm for network feature delivery, 89–90
    performance improvements, 175
    pivot to, 306
    real-time capabilities, 175–176
    real-time network data with, 356–360
    sample of SDN-Mon deployment scenarios, 372
    SDN-like networks, 6
    SDN/NFV implementation, 312–314
    security advantages, 173
    service and network measurements, 368
    service creation, 306
    service design, 298, 307
    service measurements, 365
    snapshots, 309–310
    techniques, 208, 241–242
    testing cycle time reduction, 308
    testing methodology, 307
    testing vision, 308
    use-case examples of SDN control, 100–101
    utilizing SDN to minimize robocalling, 167
    virtual service framework, 298–305
    YANG network model example, 97–99
    YANG service model example, 96–97
Software, 13
    costs, 45
    fault detection, VF, 77–78
    resiliency engineering, 78–79
    software-based services maturity curve, 75
    in telecom, 5
    validation, 94
    VF software stability, 78

Software as a service (SaaS), 396
Software defined-WAN (SD-WAN), 264
Software defined network infrastructure (SDNI), 237, 240
Software-defined service framework (SDSF), 299
  decomposing network functions, 300
  IaaS, 303–304
  NPaaS, 302–303
  SNS, 302
  VNFaaS, 301–302
Software development kits (SDKs), 106
Software development lifecycle (SDLC), 188
Software-driven functions, 225
Solid-state storage disk system, 52
SOLR, see Searching on Lucene Replication
SONET, see Synchronous optical networking
SONs, see Self-optimizing networks
Southbound FW (SB FW), 180
SPARK, real-time and micro-batch processing using, 146–147
Speech recognition, 148–149
Split RAN architecture (sRAN architecture), 243–244
SQM, see Service quality management
Sqoop, 144
SR-IOV-based approach, 211–212
SR, see Segment routing
sRAN architecture, see Split RAN architecture
SRIOV, see Single root I/O virtualization
SRLGs, see Shared risk link groups
SSMF, see Standard single-mode fiber
Stand-alone appliances, 3–4
Standardizing VNF interfaces, 337
Standard(s), 11–12, 224
  access technologies, 241
  wavelength, 269
Standard single-mode fiber (SSMF), 267
Stateful function with site failover, 79
Stateful network access services, 79–81
Stateless function with site failover, 79
Stateless network function
  with multisite design, 82–83
  resiliency features and impact on resiliency, 84
State management, 63
State persistence service, 303
Stock keeping unit (SKU), 255
Storage-area network (SAN), 53, 57
Storage, 52–53
  data migration, 57
  fewer points of management, 57–58
  improving utilization, 57
  resources, 214
  virtualization, 57
  workloads, 36
STORM, 146–147
Strong Copyleft Licenses, 395
Structured relational databases, 144
Subscriber data, decoupling of, 314
Subscriber identification module (SIM), 20
Supplier independence, 270
Swift, 58
Switch, 28
  function, 95
Switch Abstraction Interface (SAI), 263
Switching, 268
Symmetric cryptography, 214
Synchronous optical networking (SONET), 252, 276

Synthetic measurement functions, see Inorganic measurement functions
System Architecture Evolution Gateway (SAEGW), 319
System on a chip (SOC), 225, 255
  SOC-integrated circuits, 234

T

TCO, see Total cost of ownership
TCP, see Transmission control protocol
TDD, see Time division duplexing
TDM, see Time-Division Multiplexing
TE, see Traffic engineering
Telecom
  network, 1–2, 7, 50
  operator, 2
  software in, 5
Telecommunication(s)
  industry, 67, 294
  providers, 334
Telecommunications management network model (TMN model), 373
Telecommunication Standardization Sector of International Telecommunications Union (ITU-T), 230
Telemetry measurements, 372–373
Terrestrial optical networks, 267
Test
  function, 95
  test-driven development, 396
TF, see Transit function
"Theseus," 148
Third-generation networks (3G networks), 294
Third Generation Partnership Project (3 GPP), 10, 241
Threat analytics, 160
3G networks, see Third-generation networks
3 GPP, see Third Generation Partnership Project
3GPP-specified IMS, 21
Throughput, 251
Tier 1 work tasks, 334
Tier 2 organization, 334
Tier 3 organization, 334
Time-based metrics, 69
Time division duplexing (TDD), 232
Time-Division Multiplexing (TDM), 5–6
Time series data, 332
Time to live (TTL), 41, 279–280
Time-to-market (TTM), 298
TMN model, see Telecommunications management network model
Toll Free Service, 5
TONA, see Tower Outage and Network Analyzer
"Top-down" approach, 263
Top of Rack (ToR), 235
Topology, 316–317
Total cost of ownership (TCO), 44, 256
Tower Outage and Network Analyzer (TONA), 7
Traditional edge platforms
  edge application, 252–253
  vertically integrated edge platforms, 250–251
Traditional IP networks, 22–23
Traditional networking functions, 186
Traffic engineering (TE), 14, 19–20, 282, 354–355
  centralized TE using SDN controller, 358

Traffic forecast, 362–363
Traffic matrix data, 362
Traffic optimization, 160
Traffic shaping, 163
    ANR SON closed-loop function, 166
    DCAE architecture, 165
    feature functional mapping to DCAE, 167
    radio congestion, 163–165
Transaction-based metrics, 69
Transit function (TF), 312
Transmission control protocol (TCP), 10, 282*fn*
Transmission distances, 224
Transponder maps, 269
Transport
    MPLS, 277–280
    technology, 2
TTL, *see* Time to live
TTM, *see* Time-to-market
"Turing Test," 148
Two-Way Active Measurement Protocol
            (TWAMP), 366

U

UCE AS, *see* User capabilities exchange AS
uCPE, *see* Universal CPE
UDC, *see* User data convergence
UDP, *see* User datagram protocol
UDR, *see* User data repository
UE, *see* User endpoint
UI, *see* User interface
UICC, *see* Universal Integrated Circuit Card
Ultrahigh density, 247
Underlay network, 38–39
UNI, *see* User network interface
Uniqueness, 186–187
Universal cloud overlay, 258
    access scale and resiliency, 259–260
    cloud overlay connectivity, 260
    EVPN, 260–261, 262
Universal CPE (uCPE), 218
    CPE and provider network with virtual services, 219
    Phone Home Process, 221
    uCPE call-home process, 222
    uCPE—same hardware with different services, 220
Universal Integrated Circuit Card (UICC), 20
Unix
    timeline for UNIX evolution, 392
    viral nature of, 391–392
UPSTREAM automation, 189
Use-case examples of SDN control, 100
    bandwidth calendaring, 100
    flow redirection, 100–101
User capabilities exchange AS (UCE AS), 311
User data convergence (UDC), 321
User datagram protocol (UDP), 41
User data repository (UDR), 321
UserDefined Node function, 95–96
User endpoint (UE), 311–312, 320
User experience (UX), 109
User interface (UI), 55, 109
User network interface (UNI), 258
USinternetworking (USi), 398
UX, *see* User experience

V

vCSCF, *see* Virtual Call Session Control Function
vDNS-R, *see* Virtual Domain Name System Resolver
vDNSs, *see* Virtual DNSs
VDSL, *see* Very-high-bit-rate digital subscriber line
Vendor neutral interfaces, 217
Vendor-proprietary
    hardware-based products, 389
    load-balancing mechanisms, 300
Vendor-specific load-balancing solutions, 295
Vertically integrated edge platforms, 250–251
Vertical scaling, 381–382
Very-high-bit-rate digital subscriber line (VDSL), 228
    VDSL2 technology, 232
VFs, *see* Virtual functions
vFW, *see* Virtual FW
VIM, *see* Virtual infrastructure manager
VIP addresses, *see* Virtual IP addresses
Virtual Call Session Control Function (vCSCF), 82
Virtual DNSs (vDNSs), 177
Virtual Domain Name System Resolver (vDNS-R), 82
Virtual environment, optimizing, 208
    Linux bridge approach, 209–210
    SR-IOV-based approach, 211–212
    virtual switch based approach, 210–211
Virtual functions (VFs), 32, 47, 105, 205, 212
    classification and examples, 79
    control plane, 35
    data plane, 33–35
    decoupling, 32
    deployment, 216–217
    fault tolerant VM designs, 76–77
    layer 4 stateful control functions, 81–82
    low software failure rates and accurate fault detection, 77–78
    management, 216–217
    management plane, 35–37
    software design, 75
    software resiliency engineering, 78–79
    stateful network access services, 79–81
    stateless network function with multisite design, 82–83, 84
Virtual FW (vFW), 178, 182
Virtual infrastructure manager (VIM), 50–51, 53, 382, 398
    commercial, 53
    commercial open source, 53–54
    components, 55
    containers, 56–57
    hypervisor, 57
    orchestrator, 58
    pure open source, 54–55
    solutions, 53
    storage virtualization, 57–58
    virtualization, 56
    VM, 56
Virtual IP addresses (VIP addresses), 82
Virtualization, 6, 26, 56, 71, 89, 191, 201
    benefits of NFV, 29–31
    compute, 28
    of data-center networks, 4–5
    decomposition of VNFs, 32–40
    hardware virtualization, 205
    linux containers, 206

Virtualization (*Continued*)
  network, 26–29
  service chains across WAN boundaries, 205
  shift to multiple VNFs, 205
  technology, 185, 296
Virtualized functions, 154
Virtualized network architecture, 113
Virtualized network functions (VNF), 13, 14, 15, 28–29,
        50, 79, 88, 104, 105, 117, 123, 138, 204, 260,
        295, 309, 318, 341
  catalog, 315–316
  change management, 342–343
  decomposition, 32
  decoupling virtual functions, 32–37
  deploying and deleting, 339–340
  deployment models, 39
  efficiency measurements, 383–384
  functions, 206–207, 219
  HSS, 321
  HTTP-proxy, 321–322
  Intel DPDK, 213–214
  KCI reporting, 383–384
  leveraging NFVI for, 63
  management, 65
  multi and single tenancy, 39–40
  optimizing VNF performance, 213
  overlay, underlay, and vSs/vRs, 38–39
  P-GW, 319–320
  performance profiles, 63–64
  Policy control, 320
  processor affinity, 213
  provisioning, 339
  reporting metrics, 379–381
  reusability, 39
  scalability, 64–65, 315
  scaling measurements, 381–383
  service chaining, 37–38
  UDR, 321
  VNF-specific resiliency design, vPE example for, 42–44
  VNF measurements for optimal placement and sizing,
        384–385
  zones of advantage, 46
Virtualized service platform use cases, 310
  BVoIP services, 324–326
  evolving packet core, 318–324
  IMS service platform, 310–318
Virtualized solution, 313
Virtualizing EPC, advantages of, 318–319
Virtual LAN (VLAN), 17, 26
  Ethernet virtual networks using, 27
Virtual LCP (vLCP), 60
Virtual logical platform architecture and design, 319
  home subscriber server VNF components, 321
  HTTP-proxy VNF components, 321–322
  mobility management entity VNF components,
        320–321
  policy control VNF components, 320
  S-GW and P-GW VNF components, 319–320
  user data repository VNF components, 321
Virtual machine (VM), 28, 50, 56, 68, 105, 175, 206, 296,
        340
  nonisolation, 199
Virtual network function (VNF), 249
  virtual network-function-centric architecture, 299

Virtual NICs (vNICs), 208
Virtual OLTs (vOLTs), 233
  open vOLT hardware specifications, 235–237
Virtual private local area network service (VPLS),
        26, 260
Virtual private network (VPN), 18, 27, 132
  connectivity to mobility, 322–323
  services, 258
Virtual private wire service (VPWS), 26, 260
Virtual probe (vProbe), 354, 378
  passive vProbe measurement data model, 378–379
Virtual provider edge (vPE), 42, 256
  example for VNF-specific resiliency design, 42–44
Virtual reality (VR), 2, 167–168
Virtual resource identifier and access information (VRID
        information), 117
Virtual route forwarding (VRF), 27
  instances, 34
Virtual router (vR), 28, 38–39, 207, 250
Virtual service
  control loop framework, 304–305
  deployment considerations, 305s
  framework, 298
  SDSF, 299–304
  virtual service control loop framework, 304–305
Virtual switch (vS), 28, 38–39
  virtual switch based approach, 210–211
Virtual zone, 305, 316
Visibility, 217
Visualization, 123
VLAN, *see* Virtual LAN
vLCP, *see* Virtual LCP
VM, *see* Virtual machine
VMware-based cloud, 6
VMWare, 206
VNF-FG, *see* VNF forwarding graph
VNF, *see* Virtual network function; Virtualized network
        functions
VNF forwarding graph (VNF-FG), 37
VNFs as a service (VNFaaS), 301–302
vNICs, *see* Virtual NICs
Voice over Internet Protocol (VoIP), 67
Voice over LTE (VoLTE), 293
  AS, 311
VoIP, *see* Voice over Internet Protocol
VoLTE, *see* Voice over LTE
vOLTs, *see* Virtual OLTs
Volume, Velocity, Variety, Variability, Validity/
        veracity, Visibility/visualization, Value
        (7Vs), 140–141
vPE, *see* Virtual provider edge
VPLS, *see* Virtual private local area network service
VPN, *see* Virtual private network
VPN-SC, *see* VPN service controller
VPN service controller (VPN-SC), 100
vProbe, *see* Virtual probe
VPWS, *see* Virtual private wire service
VR, *see* Virtual reality; Virtual router
VRF, *see* Virtual route forwarding
VRID information, *see* Virtual resource identifier and
        access information
vS, *see* Virtual switch
vSwitch in hypervisor, 296
vTaps as service, 303

W

WAN, *see* Wide area network
WANx, 219, 220
Warm VM replication, 77
Wave division multiplexing (WDM), 354
Wavelength(s), 269, 272
    add/drop, 270
    circuits, 363
    expressing, 271
    fiber wavelength band plan with CE, 231
WDM, *see* Wave division multiplexing
Weak Copyleft Licenses, 394–395
Web
    APIs, 397
    hosting, 398
    Web 2.0, 397–398
"White box," 56, 256
    hardware, 234
    white-box Ethernet switches, 235
Wide area network (WAN), 16–17, 36, 92–93, 255
Wideband radio access, 294
Wireline access
    hardware, 233–234
    network, 229
    software, 237–238
    standards, 234–235
Wireline access technologies, 229, 233
    G.fast technology, 232–233
    merchant silicon, 234
    network abstraction layer, 238–239
    open access network software, 240–241
    open vOLT hardware specifications, 235–237
    PON technology, 229–232
    SDN access controllers, 240
    wireline access hardware, 233–234
    wireline access software, 237–238
Wireline broadband access networks, 225
Workflow, *see* Method of procedure (MOP)
World Economic Forum, 401

X

XACML, *see* Extensible Access Control Markup
        Language
XGS-PON, 231–232, 235, 238
XML, *see* eXtensible Markup Language
XMPP, *see* Extensible messaging and presence protocol

Y

Yet Another Next Generation (YANG), 14, 35, 36, 216,
        233, 241
    model, 368, 371, 373
    modeling language, 91
    network model example, 97–99
    service model example, 96–97
Yet Another Resource Manager (YARN), 144–145
Yin and yang of agile development, 397

Z

Zero Touch Provisioning, *see* Call Home service
Zone of advantage, 257, 258
Zookeeper, 147